Artificial Intelligence-Enabled Digital Twin for Smart Manufacturing

Scrivener Publishing
100 Cummings Center, Suite 541J
Beverly, MA 01915-6106

Publishers at Scrivener
Martin Scrivener (martin@scrivenerpublishing.com)
Phillip Carmical (pcarmical@scrivenerpublishing.com)

Artificial Intelligence-Enabled Digital Twin for Smart Manufacturing

Edited by

Amit Kumar Tyagi

Department of Fashion Technology, National Institute of Fashion Technology, New Delhi, India

Shrikant Tiwari

School of Computing Science and Engineering (SCSE), Galgotias University, Greater Noida, Uttar Pradesh, India

Senthil Kumar Arumugam

Department of Professional Studies, Christ University, Bengaluru, India

and

Avinash Kumar Sharma

Dept. of Computer Science and Engineering, Sharda University, Greater Noida, India

Scrivener Publishing

WILEY

This edition first published 2024 by John Wiley & Sons, Inc., 111 River Street, Hoboken, NJ 07030, USA and Scrivener Publishing LLC, 100 Cummings Center, Suite 541J, Beverly, MA 01915, USA

For more information about Scrivener publications please visit www.scrivenerpublishing.com.

Wiley Global Headquarters
111 River Street, Hoboken, NJ 07030, USA

For details of our global editorial offices, customer services, and more information about Wiley products visit us at www.wiley.com.

Library of Congress Cataloging-in-Publication Data

ISBN 978-1-394-30357-1

Cover image courtesy of Adobe Firefly
Cover design by Russell Richardson

Set in size of 11pt and Minion Pro by Manila Typesetting Company, Makati, Philippines

Printed in the USA

10 9 8 7 6 5 4 3 2 1

Contents

Preface

In the rapidly evolving landscape of modern manufacturing, the integration of cutting-edge technologies has become imperative for businesses to remain competitive and adaptive. Among these technologies, Artificial Intelligence (AI) stands out as a transformative force, revolutionizing traditional manufacturing processes and making the way for the era of smart manufacturing. At the heart of this technological revolution lies the concept of the *Digital Twin*—an innovative approach that bridges the physical and digital realms of manufacturing. By creating a virtual representation of physical assets, processes, and systems, organizations can gain unprecedented insights, optimize operations, and enhance decision-making capabilities.

This timely book explores the convergence of AI and Digital Twin technologies to empower smart manufacturing initiatives. Through a comprehensive examination of principles, methodologies, and practical applications, we explain the transformative potential of AI-enabled Digital Twins across various facets of the manufacturing lifecycle. From design and prototyping to production and maintenance, AI-enabled Digital Twins offer multifaceted advantages that redefine traditional paradigms. By leveraging AI algorithms for data analysis, predictive modeling, and autonomous optimization, manufacturers can achieve unparalleled levels of efficiency, quality, and agility.

This book explains how AI enhances the capabilities of Digital Twins by creating a powerful tool that can optimize production processes, improve product quality, and streamline operations. Note that the Digital Twin in this context is a virtual representation of a physical manufacturing system, including machines, processes, and products. It continuously collects real-time data from sensors and other sources, allowing it to mirror the physical system's behavior and performance.

What sets this Digital Twin apart is the incorporation of AI algorithms and machine learning techniques that enable it to analyze and predict outcomes, recommend improvements, and autonomously make adjustments to enhance manufacturing efficiency. This book outlines essential elements, like real-time monitoring of machines, predictive analytics of machines and data, optimization of the resources, quality control of the product, resource management, decision support (timely or quickly accurate decisions).

Moreover, this book elucidates the symbiotic relationship between AI and Digital Twins, highlighting how AI augments the capabilities of Digital Twins by infusing them with intelligence, adaptability, and autonomy. Hence, this book promises to enhance competitiveness, reduce operational costs, and facilitate innovation in the manufacturing industry. By harnessing AI's capabilities in conjunction with Digital Twins, manufacturers can achieve a more agile and responsive production environment, ultimately driving the evolution of smart factories and Industry 4.0/5.0.

We want to express our deepest appreciation to everyone who dedicated their time and efforts to make this book a success. Furthermore, we wish to gratefully acknowledge the suggestions, help, and support of Martin Scrivener and the team at Scrivener Publishing.

Amit Kumar Tyagi
Shrikant Tiwari
Senthil Kumar Arumugam
Avinash Kumar Sharma

Part 1

FUNDAMENTALS OF AI-BASED SMART MANUFACTURING

1

Machine Learning Fundamentals

Renugadevi A. S.[1]*, R. Jayavadivel[2], Charanya J.[1], Kaviya P.[1] and Guhan R.[1]

[1]*Department of Electronics and Communication Engineering, Kongu Engineering College, Erode, India*
[2]*Department of CSE, Alliance College of Engineering and Design, Alliance University, Karnataka, India*

Abstract

Machine learning (ML) is a topic of study focused on comprehending and developing "learning" methods, or methods that use data to enhance performance on a certain set of tasks. It is considered to be a component of artificial intelligence. The Types of learning in Machine Learning are Supervised Learning: uses labeled data for model training, Unsupervised Learning: uses unlabeled data for model training. When labeled data is not available (there is no result to predict), the learning purpose is to find hidden similarities, groups or clusters among examples, or to determine characteristics in the data structure. Reinforcement Learning: consists of a trained agent that learns on the basis of rewards or penalties. The Model techniques used in machine learning based models are: 1) Classification: prediction task of categorical values in supervised learning. 2) Regression: prediction task of continuous values in supervised learning. 3) Clustering: find groups or similarities in data in unsupervised learning. 4) Dimensionality reduction (DR): reduce the number of variables/features in data in unsupervised learning. Among the types of learning, each machine learning consists of variety of algorithms and performance measures, which is aligned with various model techniques. This chapter focuses on all the types of machine learning algorithms such as Support vector machine, Discriminant Analysis, Naïve Bayes, K nearest neighbor, K Means, Decision tree, principal component analysis, etc.

Keywords: Machine learning, supervised learning, unsupervised learning, reinforcement learning

1.1 Introduction

Computational models known as machine learning algorithms allow computers to learn from data and make judgments or predictions without the need for explicit programming. They may be divided into three primary groups: clustering, regression, and classification. Classification algorithms, like Decision Trees and Support Vector Machines, give labels or categories to incoming data. Regression methods, such as Polynomial and Linear regression, forecast continuous numerical values. Clustering methods such as K-Means and Hierarchical Clustering combine related data points without the need for predetermined labels. These algorithms are essential for a wide range of applications, such as natural

**Corresponding author*: renugame@gmail.com

Amit Kumar Tyagi, Shrikant Tiwari, Senthil Kumar Arumugam and Avinash Kumar Sharma (eds.) Artificial Intelligence-Enabled Digital Twin for Smart Manufacturing, (3–18) © 2024 Scrivener Publishing LLC

language processing and picture identification, since they can adjust to the specific issue and goal at hand. New methods keep coming up as the area develops, which carries on the machine learning progress.

1.2 Classification

Classification algorithms are used to classify test data based on prior learning. Among them, the model learns from previous data and separates the test data into different groups.

Types of Classifiers

Binary classifier: A classifier whose results have exactly two categories is called a binary classifier [1]. Example: Normal or abnormal, yes or no

Multi-class classifier: When the results of the classifier have more than two classes, it is called a binary classifier based on multiple classifiers. Example: Different stages of skin cancer, different types of products

The classification algorithm can be further divided as shown in Figure 1.1.

1.2.1 Linear Model

1.2.1.1 Logistic Regression

The probability of test items is predicted using supervised machine learning techniques like logistic regression. A binary classifier is what logistic regression is. The data is represented as 1 or 0 in the binary output variable. However, the resulting value is between 0 and 1. The S-shaped sigmoid function is used here. Linear regression models theoretically predict values as a function of X. The general linear regression model has a simple equation expressed as shown in below Figure 1.2.

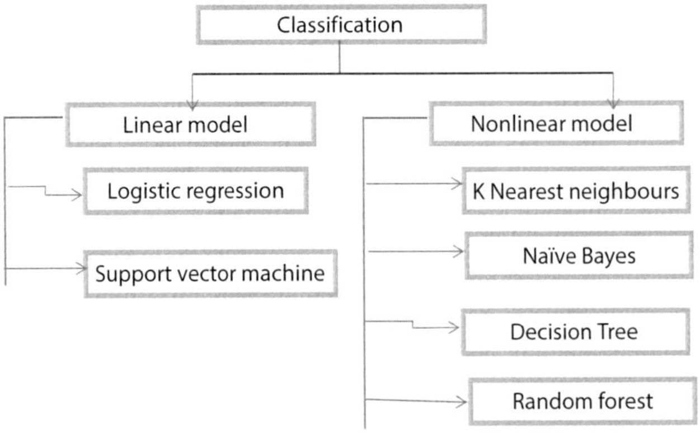

Figure 1.1 Classification of test data using algorithm.

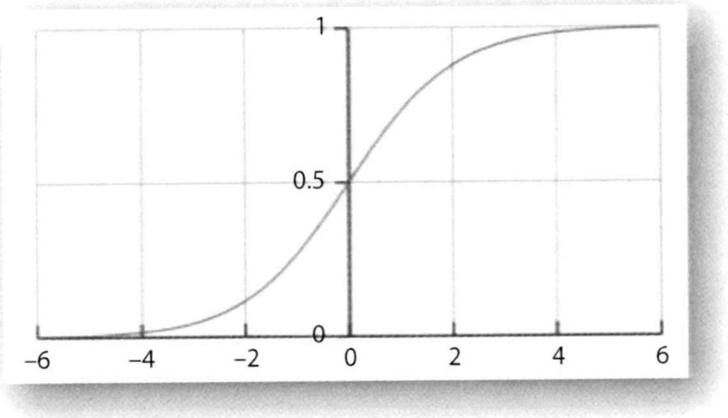

Figure 1.2 Data representation in logistic regression.

1.2.1.2 Support Vector Machine

SVM is a supervised learning system that can transform input data into a higher-order space. Data classification is done using a hyperplane with maximum. SVM can handle large amounts of data and is a widely used tool in machine learning for binary classification problems. Classification methods usually involve a set of training-test datasets.

There are many specifics of the training process and cost estimates for each event. The purpose of SVM is to obtain a model that predicts the objective value of the given data based on training data as shown in Figure 1.3. This function converts the training vectors to higher order. SVM finds the maximum marginal linear separating hyperplane in high-dimensional space. If more than one feature is available, you need to select part of the input before moving to SVM.

1.2.2 Nonlinear Model

1.2.2.1 K-Nearest Neighbor

A supervised machine learning technique called K-Nearest Neighbor is used to categorize an example within a set of kNNs shown clearly in Figure 1.4. At first, the K value corresponds to a low value. Sort the test data into groups based on increased similarity by comparing how similar the test data is to each category. Regression and classification issues can

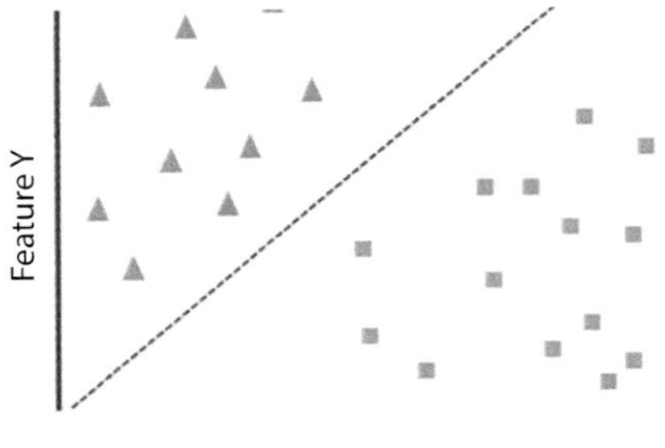

Figure 1.3 SVM system.

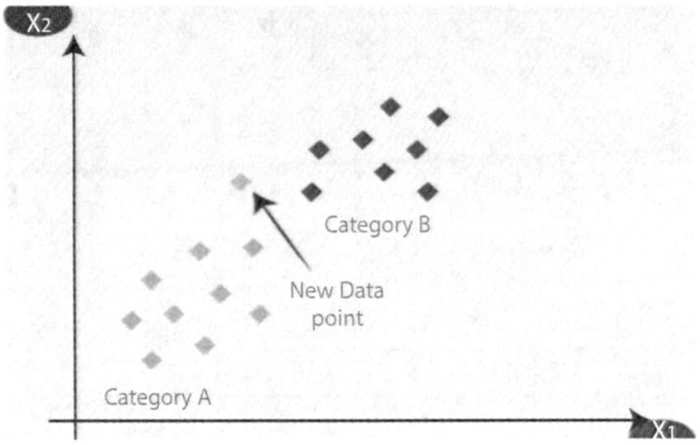

Figure 1.4 Categorization of set using K-nearest neighbor.

benefit from its application. KNN does not learn by training, which is why it is referred to as a lazy learner.

The Euclidean distance between two points is [2]:

$$d(x, y) = \sqrt{\sum_{i=1}^{n} (x_i - y_i)^2}$$

1.2.2.2 Naive Bayes

Naive Bayes is a hedonic method based on the Bayesian manager making predictions and assigning data x to list i with the largest return probability P. It specifically reduces the overhead of including each feature in the prediction class that contains it. Naive Bayes shows a competitive advantage over more traditional and advanced classification methods such as decision trees and neural networks. Due to its short learning time, it is also a good classifier that can easily process high-dimensional data. [3] Bayesian classification is about tracking learning and actual grouping strategies. It takes simple prediction models and allows us to express model uncertainty in a way that calculates the consequences of the event. It can solve many diagnostic and prognostic problems. Bayesian inference is used as a learning method that appears in Naive Bayes text.

A simple statement of Bayes' theorem is as follows:

$$P(A|B) = P(B|A) * P(A)/P(B)$$

1.2.2.3 Decision Tree

The decision tree classifier provides a clear classification model that is valid in many applications.

The nodes of the tree represent tests for a particular item and branches in different directions depending on the value shown in Figure 1.5. A page represents one category of notes. According to the test results, the lines of the educational system are separated by transferring data from the roots to the leaves.

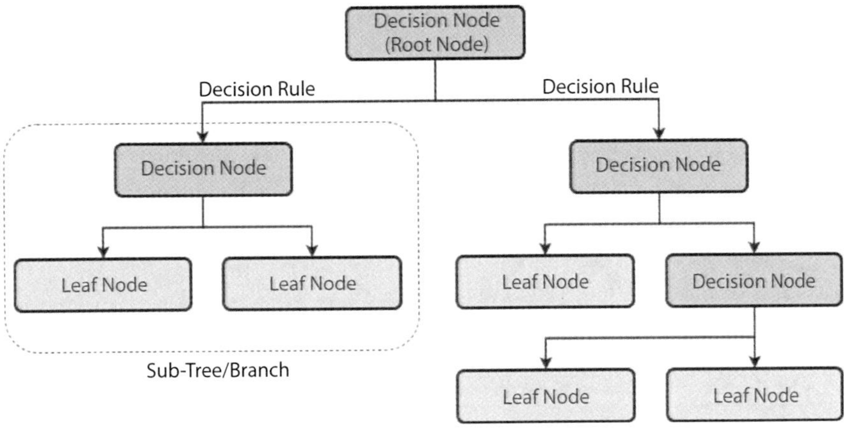

Figure 1.5 Classification of decision rule.

1.2.2.4 Random Forest

A supervised machine learning approach called Random Forest is used to address regression and classification issues. It's an ensemble learning technique that uses many classifiers to tackle challenging issues. To boost accuracy, test data is classified using numerous decision trees in random forests, and then the data is averaged as shown in Figure 1.6. Accuracy will increase with more wood utilized, but overfitting will become an issue.

1.3 Regression

Data scientists can use regression to predict a continuous variable (y) using the mathematical standard for the outcomes of one or more predictions (x). Linear regression is probably the most commonly used form of regression analysis given its ease of estimation and forecasting [4].

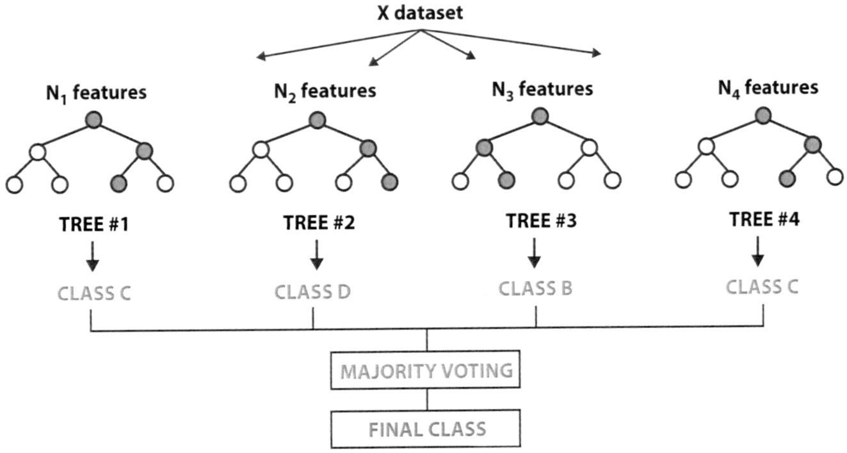

Figure 1.6 Classification of data set.

1.3.1 Linear Regression

One of the most popular and simple machine learning methods used for predictive analytics is linear regression. Predictive regression describes what is predicted, while linear regression estimates constants such as age, salary, and other variables. The variance of the variable (y) varies as a function of the independent variable; this shows the relationship between the variable and the variable (x). A regression line is a line that attempts to fit data between the variables and the independent variables.

The equation of the inverted line is y = a0 + a * x + b.

Here,

- y is a variable.
- x = foreign variable
- a0 is the intersection point of the line.

The following highlights other differences between the two forms of linear regression: Basic Linear Regression An independent variable is utilized in simple linear regression to forecast the value of the dependent variable.

1.3.2 Multiple Linear Regression

Using many independent variables, this technique forecasts a variable's value as shown Figure 1.7.

Several uses for linear regression include:

- Sales forecasting and model analysis
- Salary forecasting
- Real estate forecasting
- ETA-related traffic.

1.3.3 Logistic Regression

Learning is facilitated by regression law. It is employed to ascertain or quantify the likelihood of the equation's result (yes/no). One of the applications of artificial intelligence is

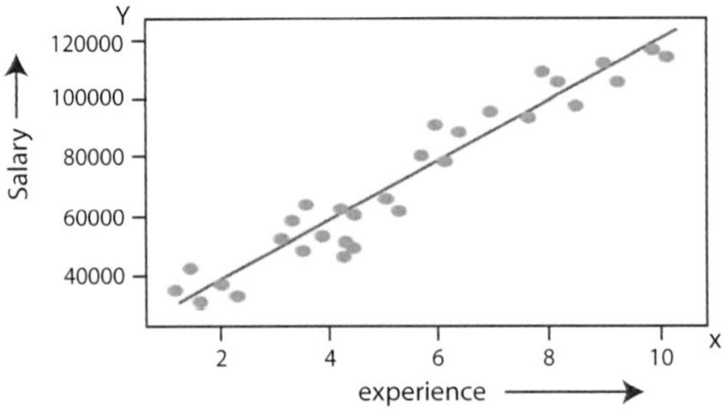

Figure 1.7 Plotting of multiple linear regression.

determining whether a person will catch coronavirus or not, as an example of reverse calculation [5]. Binary classification means that there are only two answers to this question: disease or infection. In this example, a person's ability to contract COVID-19 depends on the virus, symptoms, presence of antibodies, etc. It may depend. These factors (independent variables) may include the disease, symptoms, and immune responses, all of which may have an impact on the predicted outcome (dependent variables).

1.3.3.1 *Three Types of Logistic Regression*

There are three types of logistic regression.

- Binary logistic regression is used when there are only two outcomes; like the old example of whether a 19-year-old is likely to contract COVID-19.
- Multinomial logistic regression is used when there is more than one outcome, such as determining whether a person will get a cold, allergies, flu, or COVID-19.
- Using logistic regression techniques when ranking scores, such as data on classifying COVID-19 infections as minor, medium, and major as shown in Figure 1.8.

1.3.4 Polynomial Regression

Polynomial regression and other probability approaches employ direct models as indirect data. It directly involves a recurring variable even if there is a non-linear curve between the value of x and the relative or positive value of y. [6] Consider a dataset containing data points that are not directly accessible. In this case, direct duplication will not provide the best results for the content of this document. The content of this document requires further elaboration. In polynomial rendering, key points are converted to polynomial highlights to a predetermined degree and then a direct model is used to display them. This indicates that the linear polynomial provides the best fit to the data of interest.

The polynomial iteration formula is added directly with equations; where the polynomial iteration Y = b0 + b1x + b2x2 + b3x3 + ... + bnxn is derived directly from the conditions repeating Y = b0 + b1x.

Therefore Y is the normal or desired result and b0, b1, ..., bn are the iteration coefficients. Our independent variable or input variable is x.

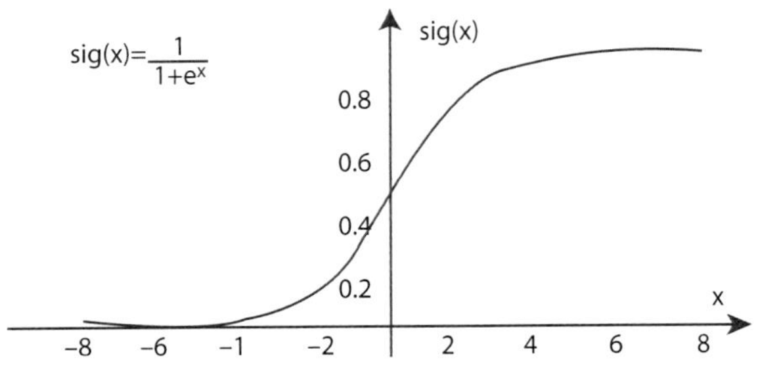

Figure 1.8 Logistic regression representation in graph form.

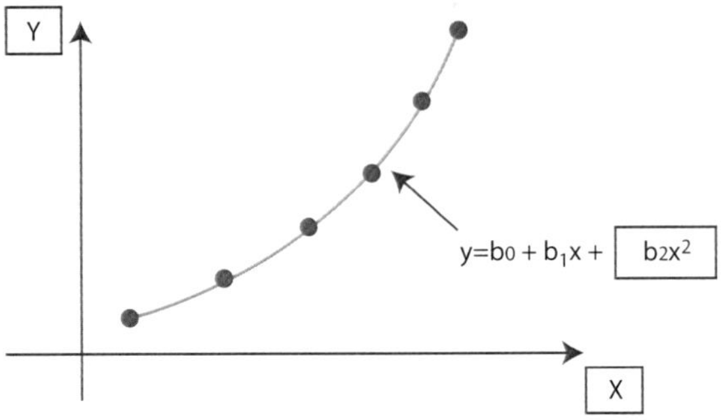

Figure 1.9 Plotting of polynomial regression using equation.

Since the coefficients are linear with the quadratic term, the model is still linear as shown in Figure 1.9.

1.3.5 Support Vector Regression

Support any supervised learning technique called Vector Machine is appropriate for issues that are descriptive and recurrent. If we use it to create related problems in this way, it is called help vector generation. The optimization method for continuous data is support vector reconstruction as shown in Figure 1.10. Next, a few words about propagating the help vector: use the fragment function to index lower-level data to higher levels. In SVM, the hyperplane is usually a line separating two groups; however, one row in the SVR supports the prediction of change and covers most of the data of interest.

1.3.6 Decision Tree Regression

Train with AutoML tools to use decision trees and artificial intelligence technology to group data using true or false answers for ambiguous or repetitive questions. Considering its structure, it looks like a tree with roots, inside and certain parts of the leaves as shown in Figure 1.11. The selected tree starts from the central hub and branches towards the inner hubs and leaf hubs. The final group or actual quality is tracked in the middle of the page. Tree selection is easy to understand and easy to understand. Start by defining the material that will be the center of the selected tree [7].

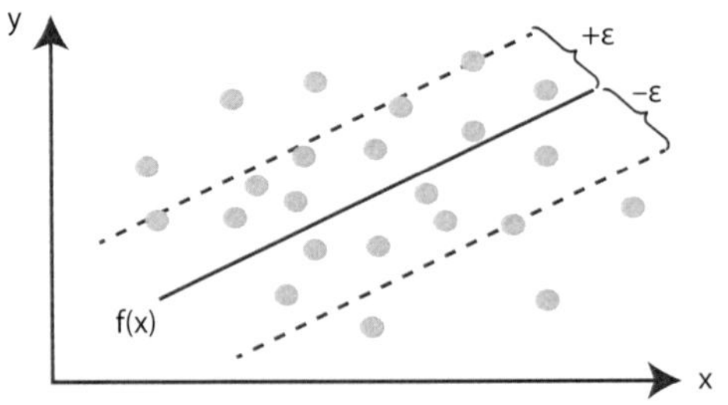

Figure 1.10 Vector generation using vector machine.

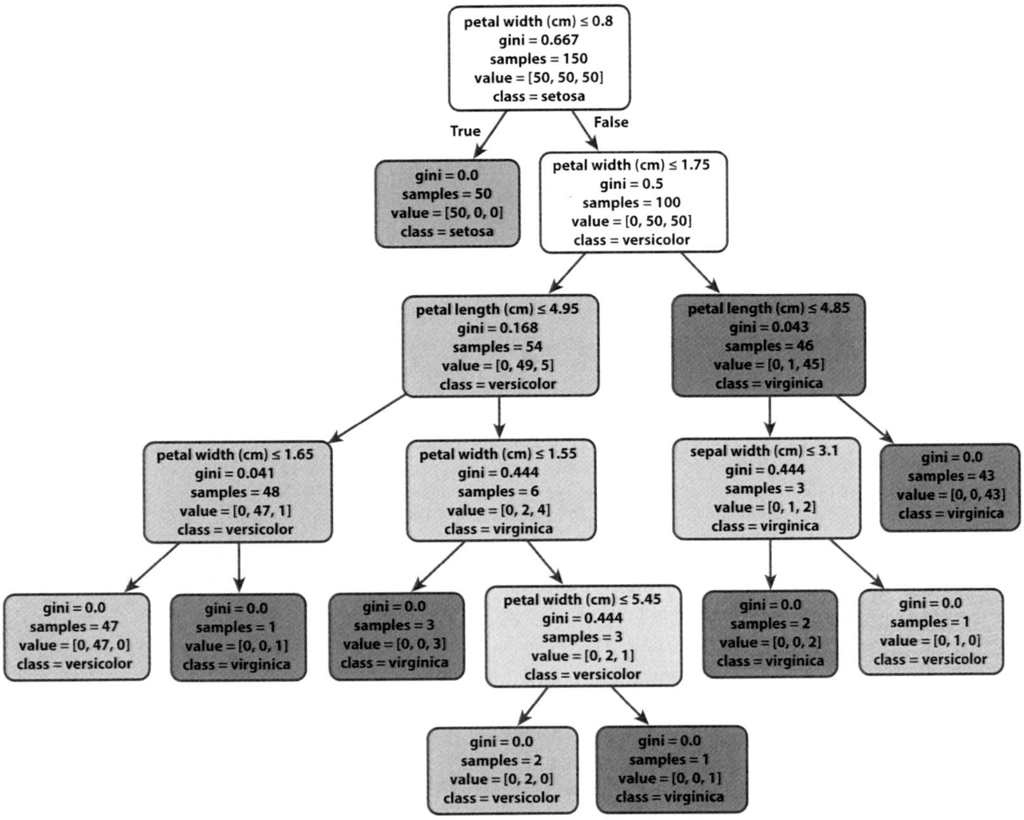

Figure 1.11 Tree selection representation.

In general, no one can predict the final course with complete certainty; This is called transmission. This spread is estimated using methods such as the Gini coefficient, entropy, and information gain, which indicate how well items relate to the given data. At each stage, the core is selected as the least polluting component. For equipment with mathematics, the Gini coefficient is determined by analyzing the data on the request ramp before calculating the average of the continuous equipment. To find the Gini coefficient, use the following formulas: K is the total number of classes, and p is the level of classes that are now in existence.

The weighted normally distributed Gini coefficient is not the same for all values of the leaf. Select the price with the lowest price for this trademark. The tool has been restructured according to some important values so that points and values similar to the base can be selected. This cycle is repeated at each depth and each location until all records are identified. After creating the tree, you can predict the values or properties of important data by pulling the tree from each location. When tree selection is used for replication, transmission is estimated using transformed or squared residuals rather than the Gini coefficient. The remaining steps of the interaction can be compared.

1.3.7 Random Forest Regression

Random forest is a well-liked technique for regression and classification. Given that regression and classification are the two most crucial components of machine learning, the random forest method is among the most significant algorithms in the field. Determining whether a user would purchase a product or how to grant credit are just two examples of the numerous commercial applications where the capacity to do precise analysis comes in extremely handy. Naive Bayes classifiers, logistic regression, decision trees, and support vector machines are examples of data science classifiers. Conversely, the Random Forest classifier occupies a position close to the summit of the classifier hierarchy.

1.3.7.1 Random Forest Algorithm in Practice

The Random Forest algorithm functions rather sensibly. [8] The method is used in two steps: first, a random forest is generated and N decision trees are combined; second, a prediction is produced for each tree generated in the first step as shown in Figure 1.12.

The steps listed below can be used to illustrate how this approach functions:

Step 1: Choose M spots at random from the training.
Step 2: For the data points (subset) you choose in Step 1, create a decision tree.
Step 3: Every decision tree will provide results. Look at this for sure.
Step 4: A majority vote or average, whichever is suitable, determines the categorization and regression's ultimate outcome.

1.3.7.2 Random Forest Algorithm Application Examples

For example, the following four industries frequently use the Random Forest Algorithm: credit risk of the banking industry.

Hospital: Identify dangerous diseases and their consequences.

Land use: To find regions with similar land use patterns, a random forest distribution technique was also employed.

Business trends: Business trends may be found using this method.

The random forest algorithm has the following advantages:

- Overfitting is eliminated because the results are based on the mean or majority vote.
- The parallel feature is seen with the independence of each decision tree.
- It preserves diversity by not taking into account all advantages, although this is not always the case when creating every decision tree.

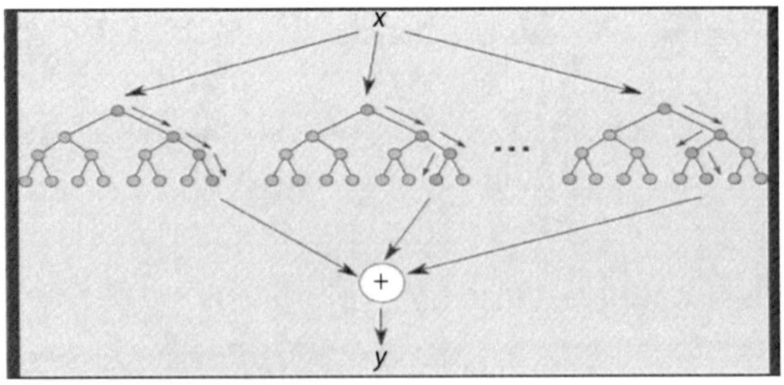

Figure 1.12 Flow diagram of random forest algorithm.

- It preserves diversity by not taking into account all advantages, although this is not always the case when creating every decision tree.
- Unaffected by Curse of Dimension.
- It is very stable since the average response of large trees is used. The unique position is reduced because not all attributes are determined for each tree.
- Despite being among the greatest algorithms for handling regression and classification issues, Random Forest has a few drawbacks that you should be aware of before utilizing it.
- Random forests are more complex than decision trees chosen by walking through the tree.
- Due to its complexity, this model takes longer to train than other models. When a decision tree needs to make a prediction, it needs to produce input data.

1.3.8 Lasso Regression

The normal method is Rope Regression. It is preferred over recursive for more accurate numbers. In this model, shrinkage is used. [9] When data values jump towards the mean, this is called shrinkage. Simple, redundant models can be improved by coupling methods (i.e. models with fewer boundaries). This particular type of iteration works best when the sample shows a large difference or you want to take several steps on the selected sample, such as deciding and escaping the boundary. L1 normalization for string recovery. Since it expresses choice, it is used in situations where there are many things.

1.4 Clustering

Clustering, commonly known as cluster analysis, is a problem in unsupervised learning. It is often used as a data analysis tool to analyze interesting data, such as grouping customers based on their behavior. There are dozens of algorithms to choose from, and more than one algorithm works best in every situation. Instead, it's good to work on different integration algorithms and different settings for each method.

1.4.1 Clustering Algorithms

There are many types of clustering algorithms. It uses many similar algorithms to find a dense observation area or measure the distance between samples at a particular location. That's why it's usually a good idea to scale data before using the integration.

1.4.2 K-Means Clustering Algorithm

Unmonitored Learning System Unsupervised data is divided into many groups using K-Means clustering. Here, K is the number of presets that should be generated during the process; for instance, if K = 2, there will be two groups, and if K = 3, there will be three groups.

It enables us to automatically recognize groups in unlabeled data without the need for training and to divide data into distinct groups. Every cluster is given a centroid since the method is based on them. This algorithm's primary objective is to shorten the distance between each data point and its matching cluster. The method takes as input unlabeled data, splits it into k groups, and keeps going until it runs out of groups. The k value in this algorithm has to be known ahead of time.

The K-means clustering technique serves two primary purposes, which are:

To determine the ideal value for the K-means point or centroid, an iterative procedure is employed. The closest k-center is allocated to each data point. Data points that are near particular places k are used to construct clusters as shown in Figure 1.13.

So each group is different from the other groups and has some different data points.

Here are the steps that show how the K-word algorithm works:

Step 1: First, select K to determine the number of groups.
Step 2: Select random centroids or K locations. It may not be informative.
Step 3: Match each data point to the nearest center point; This will create K pre-set groups.
Step 4: Determine the number of changes and transfer them to the center of each group.
Step 5: Repeat step three to reassign all data points to the new center of each cluster.
Step 6: If the address exists, go to Step 4; otherwise go to Done.
Step 7: Complete the model.

1.4.3 Mean Shift Clustering Algorithm

Meanshift is a clustering algorithm that, unlike unsupervised learning, regroups data points by moving the points to mode (in the context of Meanshift, mode is the fastest data points in the region). For this reason it is sometimes called the pattern search algorithm. Context-changing algorithms can be used in computer vision and image processing. [10] Meanshift

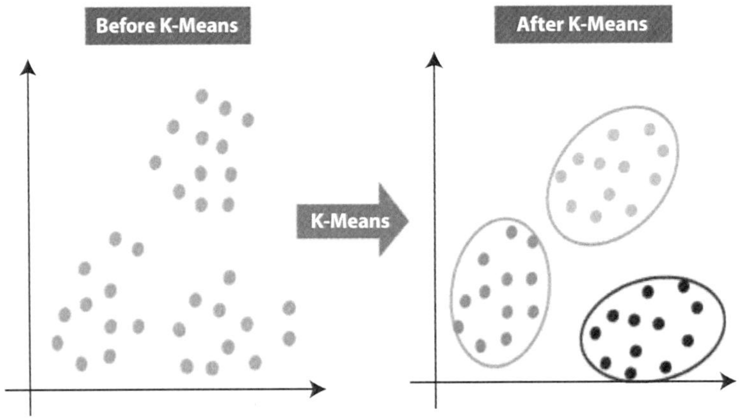

Figure 1.13 Data before and after of K-means.

clustering technique does not need a predefined number of clusters, in contrast to the popular K-Means clustering algorithm. This program counts the number of groups by analyzing the input.

The process of the center transformation algorithm can be done as follows:

- Make the data points cluster centers the first priority.
- Repeat the following steps until you converge or reach a predetermined decision:
- For each data center, select the average of all points within a specific radius (e.g., "core") in the data center.

In the subject guide.

- The point that does not change after convergence is the group center of gravity.
- For the final cluster center and cluster function of the data points.

One significant benefit of average function clustering is that the number of clusters need not be predetermined. Furthermore, it is capable of managing groups of any size or form and does not assume any particular arrangement of items. It could, however, be dependent on the core's radius and choice of core.

Mean-Shift clustering can be applied to many types of data, including object tracking, image and video processing, and bioinformatics.

1.4.4 DBSCAN Clustering

Convex or spherical clusters can be found using hierarchical clustering or partitioning techniques (K-means, PAM clustering). [11] Stated differently, they are only appropriate for small, evenly distributed groups as shown in Figure 1.14.

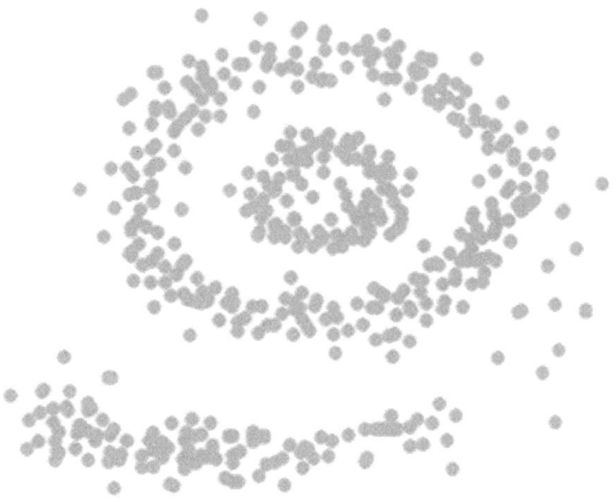

Figure 1.14 Example for DBSCAN clustering.

The data's noise and outliers will also have a significant effect on this.

- Real world data can contain errors, for example:
- Clusters can have shapes like the example below.
- The file may contain noise.

1.4.4.1 Two Elements Required by DBSCAN

The region surrounding a data point is represented by eps. Two points are said to be neighbors if their distance is less than or equal to eps. A substantial percentage of the data will be regarded as outliers if the Eps value is very low. The categories will be combined if it is determined to be too big, resulting in the majority of the data points falling into one category. an approach to eps value determination using k-distance graphs [10].

MinPts are the closest neighbors (data points) inside the Eps radius. The MinPts number should increase with the size of the file. Generally speaking, MinPts>= D + 1 may be used to determine the minimal MinPts based on the length D of the data collection. MinPts should be set to a minimum value of 3 [12].

1.4.4.2 Steps Followed to do DBSCAN Algorithm

- Indicate landmarks or points that visit more than MinPts neighbors in eps by finding all neighbors.
- If necessary, create a new category for each priority if not already done.
- Recursively find all related items by velocity and add them to the same group.
- Points a and b are said to as having a rapid connection if there is a point c, sufficient points are nearby, and both points are eps away. There are chains in this system. Consequently, b must be a neighbor of someone if b is a neighbor of c, c is a neighbor of d, d is a neighbor of e, and e is a neighbor of a.
- Now, skip over sections of the file in search of unexplored material. All items that are not part of a group are considered noise.

1.5 Conclusion

This study concludes with a thorough examination of many machine learning methods, including approaches for grouping, regression, and classification. The significance of learning techniques and tactics in the context of artificial intelligence is highlighted in the abstract, which establishes the scene. The classification section explores both linear and non-linear models, including Decision Tree, Random Forest, K-Nearest Neighbor, Support Vector Machine, Naive Bayes, and Logistic Regression, and makes a distinction between binary and multi-class classifiers. The section on regression provides information on both linear and non-linear regression techniques, such as Random Forest, Polynomial, Decision Tree, Support Vector, and Logistic regression. In the section on clustering, the concepts of K-Means clustering, Mean Shift clustering, and DBSCAN are introduced along with clustering as an unsupervised learning issue.

References

1. Tan, H., Machine Learning Algorithm for Classification. *J. Phys. Conf. Ser.*, 1994, 012016, 2021, 10.1088/1742-6596/1994/1/012016.
2. Osisanwo, F.Y. *et al.*, Supervised Machine Learning Algorithms: Classification and Comparison. *Int. J. Comput. Trends Technol.*, 48, 2017, 128–138, 2017.
3. Charbuty, B. and Abdulazeez, A., Classification Based on Decision Tree Algorithm for Machine Learning. *J. Appl. Sci. Technol. Trends*, 2, 01, 20–28, Mar. 2021, doi: 10.38094/jastt20165.
4. Mahesh, B., Machine Learning Algorithms -A Review, 2019, 10.21275/ART20203995.
5. Sander, J., Ester, M., Kriegel, H.P. *et al.*, Density-Based Clustering in Spatial Databases: The Algorithm GDBSCAN and Its Applications. *Data Min. Knowl. Discovery*, 2, 169–194, 1998, https://doi.org/10.1023/A:1009745219419.
6. Tran, T., Drab, K., Daszykowski, M., Revised DBSCAN algorithm to cluster data with dense adjacent clusters, in: *Chemometrics and Intelligent Laboratory Systems*, vol. 120, pp. 92–96, 2013, 10.1016/j.chemolab.2012.11.006.
7. Laurent, S., Machine Learning Algorithms: A Review. *Inf. Syst. J.*, 2022, ISJ-RA-3392. 6.
8. Sarker, I.H., Machine Learning: Algorithms, Real□World Applications and Research Directions. *SN Comput. Sci.*, 2, 160, 2021.
9. Ahmed, N. and Abdul, R.T., IJARCCE An Overview of Various Improvements of DBSCAN Algorithm in Clustering Spatial Databases. *Int. J. Adv. Res. Comput. Commun. Eng.*, 5, 2016, 10.17148/IJARCCE.2016.5277.
10. Kumar, K.M. and Reddy, A.R.M., A fast DBSCAN clustering algorithm by accelerating neighbor searching using Groups method. *Pattern Recognit.*, 58, 39–48ISSN 0031-3203, 2016, https://doi.org/10.1016/j.patcog.2016.03.008.
11. Kinaneva, D., Hristov, G., Kyuchukov, P., Georgiev, G., Zahariev, P., Daskalov, R., Machine Learning Algorithms for Regression Analysis and Predictions of Numerical Data. *3rd International Congress on Human-Computer Interaction, Optimization and Robotic Applications (HORA)*, Ankara, Turkey, pp. 1–6, 2021, doi: 10.1109/HORA52670.2021.9461298.
12. Shi, R., Leng, X., Wu, Y. *et al.*, Machine learning regression algorithms to predict short-term efficacy after anti-VEGF treatment in diabetic macular edema based on real-world data. *Sci. Rep.*, 13, 18746, 2023, https://doi.org/10.1038/s41598-023-46021-2.

2

Industry 4.0 in Manufacturing, Communication, Transportation, Healthcare

P. Nancy[1], S. Gnanavel[1]*, V. Sudha[2], G. Deepika[2] and Mahmoud Elsisi[3]

[1]*Department of Computing Technologies, Faculty of Engineering and Technology, SRM Institute of Science and Technology, Kattankulathur, Tamil Nadu, India*
[2]*Department of Computer Engineering, Government Polytechnic College, Coimbatore, Tamil Nadu, India*
[3]*Department of Electrical Engineering, National Kaohsiung University of Science and Technology, Kaohsiung City, Taiwan*

Abstract

The Fourth Industrial Revolution, often known as Industry 4.0, refers to the incorporation of new technology into industrial processes, resulting in more efficient and automated systems. This transformation is having a significant impact across various sectors, including manufacturing, communication, transportation, and healthcare. In manufacturing, Industry 4.0 technologies such as the Internet of Things (IoT), Artificial Intelligence (AI), and 3D printing are being used to create smarter and more connected factories. These technologies are enabling manufacturers to optimize production processes, reduce costs, and improve quality, leading to increased competitiveness in the global marketplace. In the communication sector, Industry 4.0 is revolutionizing the way we communicate with each other and with machines. With the advent of 5G networks, ultra-fast data transfer speeds are enabling real-time communication, which is critical for many applications such as autonomous vehicles and remote surgery. In transportation, Industry 4.0 technologies are transforming the way we move goods and people. Connected and autonomous vehicles are becoming increasingly common, with many benefits such as reduced traffic congestion, improved safety, and lower emissions. In healthcare, Industry 4.0 is bringing about significant advancements in medical treatment and patient care. Technologies such as AI and the IoT are being used to develop personalized medicine, improve diagnostic accuracy, and streamline hospital operations, leading to better patient outcomes. Overall, Industry 4.0 is driving a major transformation across various sectors, resulting in significant improvements in efficiency, productivity, and quality. As these technologies continue to develop and mature, they are expected to bring about even more profound changes in the years to come.

Keywords: Artificial intelligence, internet of things, virtual reality, augmented reality

**Corresponding author*: gnanaves1@srmist.edu.in

Amit Kumar Tyagi, Shrikant Tiwari, Senthil Kumar Arumugam and Avinash Kumar Sharma (eds.) Artificial Intelligence-Enabled Digital Twin for Smart Manufacturing, (19–38) © 2024 Scrivener Publishing LLC

2.1 Introduction

The term "Industry 4.0," or "Fourth Industrial Revolution," describes the ongoing transformation of the manufacturing and industrial sectors. To build an ecosystem that is more intelligent and networked, digital technology, automation, and data sharing are all integrated. Industry 4.0, which was first proposed by the German government in 2011, expands on earlier industrial revolutions to usher in a new era powered by cyber-physical systems. It makes use of cloud computing, robotics, 3D printing, artificial intelligence (AI), big data analytics, Internet of Things (IoT), and virtual/augmented reality technology. Improving industrial processes' productivity, sustainability, adaptability, and efficiency are the main goals of Industry 4.0. It envisions smart factories where machines and systems can optimize performance, diagnose issues, and self-correct. The concept of digital twins, virtual replicas of physical assets, is also central to Industry 4.0, enabling simulation, monitoring, and predictive maintenance in various sectors.

2.1.1 The Significance of Industry 4.0 Across Multiple Domains

The advent of Industry 4.0 is of crucial importance in various sectors due to its transformative potential and the wide range of benefits it offers. The significance of Industry 4.0 in different industries are discussed below:

Manufacturing: By enabling smart factories and cutting-edge automation, Industry 4.0 revolutionizes the manufacturing sector. As a result, production, efficiency, and flexibility all increase. Manufacturers can increase product quality, decrease downtime, and optimize production processes, which lowers costs and boosts competitiveness.

Logistics and Supply Chain: By offering data-driven insights and real-time visibility, Industry 4.0 technologies improve supply chain management. This makes it possible to estimate demand, manage inventories effectively, and improve logistics processes. Businesses can more efficiently meet consumer expectations, cut expenses, and expedite delivery.

Healthcare: Industry 4.0 has the potential to revolutionize healthcare delivery. Through IoT devices, wearable technology, and data analytics, healthcare providers can offer remote monitoring, personalized treatment plans, and preventive care. AI-powered diagnostics and predictive analytics improve disease detection, drug development, and patient outcomes.

Energy and Utilities: In order to maximize energy production, distribution, and consumption, Industry 4.0 is essential. Real-time monitoring and intelligent energy resource management are made possible by smart grids, Internet of Things sensors, and AI algorithms. This contributes to a more sustainable energy ecosystem by increasing integration of renewable energy sources, decreasing prices, and improving efficiency.

Transportation and Automotive: Industry 4.0 technologies are transforming the transportation sector. Connected vehicles, autonomous systems, and data-driven analytics enhance safety, traffic management, and fuel efficiency. Fleet management, supply chain operations, and logistics may all be optimized by organizations to save money, increase sustainability, and improve customer satisfaction.

Agriculture: Industry 4.0 uses drones, IoT sensors, and AI analytics to make precision farming possible in the agriculture sector. Real-time crop health, soil, and weather monitoring by farmers allows for focused interventions and efficient use of available resources. As a result, yields rise, resource waste is reduced, and sustainable farming methods are used.

Retail: Industry 4.0 innovations are reshaping the retail industry. With AI-powered personalization, data analytics, and IoT devices, retailers can deliver personalized customer experiences and optimize supply chains. Technologies such as augmented reality (AR) and virtual reality (VR) improve the online and in-store shopping experience, boosting sales and consumer engagement. The importance of Industry 4.0 extends to other sectors like finance, construction, education, and more. Organizations may obtain a competitive edge, increase operational effectiveness, and satisfy changing consumer and market demands by implementing Industry 4.0 technology.

2.2 Industry 4.0 in Manufacturing: Overview

Industry 4.0 is characterized by the integration of digital technologies into manufacturing processes to create smart, automated, and interconnected systems. It revolutionizes the manufacturing sector by leveraging various technologies to enhance efficiency, productivity, and flexibility. Internet of Things (IoT) enables the connection of sensors and devices which enable to monitor and control equipment, track inventory, optimize maintenance schedules, and enable real-time data analysis for process optimization. The vast amount of data generated in manufacturing can be analyzed to identify patterns, anomalies, and correlations, enabling better decision-making, predictive maintenance, quality control, and supply chain optimization. AI technologies are used to automate processes, perform complex data analysis, and enable cognitive capabilities. Demand forecasting, quality assurance, predictive maintenance, production scheduling, and even autonomous decision-making can all benefit from the use of AI [1]. Robotics and Automation also play a vital role in manufacturing performing complex tasks with precision and speed improving efficiency, reduces errors, and enables flexible manufacturing processes. Additive Manufacturing (3D Printing) allows the creation of objects by adding successive layers of materials based on a digital design which in turn enables rapid prototyping, customization, on-demand production, and reduced waste. It can be used for both small-scale production and large-scale manufacturing.

Technologies like augmented reality (AR) and virtual reality (VR) provide immersive experiences that can be integrated into production processes. Assembling, maintaining, and training are made easier with the help of augmented reality. For teaching, design reviews, and simulation, VR produces virtual environments. With increased connectivity and data exchange, robust cyber security measures are essential. Protecting critical manufacturing infrastructure, sensitive data, and intellectual property from cyber threats is vital to make certain the reliability and consistency of systems. Cloud platforms provide scalable and secure storage and processing capabilities for manufacturing data. They enable real-time access to information, facilitate collaboration, and support advanced analytics and AI algorithms. Digital twin allows instantaneous monitoring, investigation, and replication of physical assets throughout their lifecycle. Digital twins help optimize performance, predict maintenance needs, and improve product development and quality. Blockchain technology can enhance transparency, traceability, and security in supply chains. It enables decentralized and immutable records of transactions, improving trust, and reducing fraud in areas like inventory management, product provenance, and supplier verification. All these technologies are interrelated and often implemented in combination to create intelligent, interconnected, and data-driven manufacturing systems that drive the Industry 4.0 transformation. Their adoption can lead to increased efficiency, productivity, and agility, enabling manufacturers to stay competitive in the rapidly evolving global market.

2.2.1 Benefits of Industry 4.0 in Manufacturing

The manufacturing sector can profit greatly from Industry 4.0 technologies, which include automation, real-time data monitoring, and predictive analytics, resulting in improved production efficiency. Processes can be optimized, bottlenecks can be identified and addressed promptly, and resource allocation can be optimized, leading to higher productivity and reduced costs. The utilization of superior technologies such as AI, IoT and data analytics allows for instantaneous quality monitoring along with predictive maintenance. Manufacturers can detect defects or deviations in the production process early on, leading to higher product quality and reduced waste. Predictive maintenance helps prevent equipment failures and ensures optimal performance. Industry 4.0 enables adaptable and nimble production procedures. With smart automation and digital technologies, manufacturers can swiftly become accustomed to varying requirements of customer, efficiently handle small batch production, and offer customization options. This agility allows for faster time-to-market and a competitive edge. It also generates immense data from a variety of sources. Through the use of data analytics and AI algorithms, manufacturers can gain actionable insights in real-time. This enables informed decision-making, process optimization, and proactive problem-solving [17]. It can facilitate cost reduction in multiple ways. Automation and robotics eliminate the need for manual labor in repetitive or hazardous tasks, reducing labor costs and improving worker safety. Predictive maintenance lowers maintenance expenses and unscheduled downtime. Inventory holding costs and waste are reduced through supply chain optimization and real-time data-driven inventory management. It encourages innovation and accelerates product development. Digital twin simulations, augmented reality, and virtual reality technologies aid in product design, prototyping, and testing. These tools enable rapid iterations, reducing time-to-market and facilitating the introduction of new and improved products.

Industry 4.0 technologies enhance the supply chain's traceability, transparency, and coordination. Real-time visibility into inventory levels, demand trends, and logistics is made possible by IoT sensors and data analytics. This makes it possible to estimate demand accurately, manage inventories effectively, and optimize logistics, which lowers costs and delays while raising customer satisfaction [18]. It can contribute to sustainable manufacturing practices. Optimization of energy usage, waste reduction through predictive maintenance and process optimization, and the ability to monitor and minimize environmental impacts are facilitated by these technologies. This supports sustainability goals and compliance with regulations. Automation and robotics can take over dangerous or repetitive tasks, reducing the risk of workplace accidents. Collaborative robots (cobots) can labor along with workers, improving ergonomics and security. IoT-enabled safety systems and real-time monitoring also enhance worker safety by detecting and addressing potential hazards. Embracing Industry 4.0 technologies can provide a competitive edge in the global market. Manufacturers that leverage these technologies are better equipped to meet customer expectations, deliver high-quality products efficiently, and adapt to changing market conditions swiftly. This positions them as industry leaders and enhances their market share and profitability.

2.2.2 Examples of Industry 4.0 in Manufacturing

Smart factories utilize IoT, data analytics, and automation to create interconnected and intelligent manufacturing systems. Machines and devices are connected, thus promoting real-time exchange of data along with monitoring which in turn allows predictive maintenance, optimized production planning, and efficient resource allocation [10]. Industry 4.0 enables the implementation of predictive maintenance strategies. Sensors and IoT devices continuously monitor equipment performance, predicting potential failures. Maintenance can then be scheduled proactively, reducing downtime and optimizing maintenance costs. Robotics and automation play a significant role in Industry 4.0 manufacturing. These robots can work together with human or autonomously, enhancing efficiency, reducing errors, and improving workplace safety. Additive Manufacturing enables the creation of complex and lightweight designs, reduces material waste, and shortens production lead times [11]. Digital twins help optimize processes, improve product quality, and enable predictive maintenance. For example, a digital twin of a machine can identify potential bottlenecks and suggest process modifications to improve efficiency. AR and VR technologies are employed in manufacturing for various purposes. AR can overlay digital information onto the real-world environment, aiding in assembly, maintenance, and training. VR creates immersive virtual environments for simulation, design reviews, and training, enabling workers to visualize and interact with products and processes before physical production. Supply Chain Optimization enables efficient demand forecasting, inventory management, and streamlined logistics, reducing costs and improving customer satisfaction. Machines and equipment can be controlled and adjusted to minimize energy waste, contributing to sustainability goals and cost savings.

Cloud platforms are utilized to store and process large volumes of manufacturing data. Cloud-based solutions enable real-time access to information, remote monitoring, and collaboration among stakeholders. Blockchain technology is used in supply chain management to improve traceability and transparency. It ensures the integrity of supply chain data by enabling safe, unchangeable transaction records. Blockchain can be used to track and confirm raw material origins, enhance supplier oversight, and stop counterfeiting. These examples show how Industry 4.0 technologies are boosting productivity, efficiency, and innovation in manufacturing across several industries. The application of these technologies is customized to meet the unique requirements and objectives of every sector and company.

2.2.3 Challenges in Implementation of Industry 4.0 in Manufacturing

While Industry 4.0 offers tremendous opportunities, its implementation in manufacturing also presents certain challenges. It includes high Implementation Cost. Many manufacturing facilities yet operate on legacy systems and equipment which lack the necessary connectivity and compatibility with Industry 4.0 technologies. Integrating these systems and retrofitting existing infrastructure can be complex and time-consuming. Protecting sensitive data from cyber threats and ensuring compliance with regulations can be challenging. Manufacturers must employ vigorous cyber security procedures and create data authority frameworks [2]. There is often a skills space in the existing workforce, by means of a lack of expertise in areas like data analytics, AI, and automation. Upskilling and reskilling programs are necessary to prepare the workforce for the digital transformation. Organizational culture, workflows, and procedures all need to be significantly altered in order to implement Industry 4.0 technologies. Successful adoption of new technologies can be hampered by traditional thinking, ignorance, and resistance to change. Organizations must concentrate on change management techniques, offer assistance and training, and promote an innovative and ever-learning culture. Industry 4.0 calls for the integration of several technologies and systems from various manufacturers. It might be difficult to ensure that various components are compatible and interoperable with one another since data formats and communication protocols lack a uniform standard. Enabling smooth integration and data interchange requires the establishment of industry-wide standards and protocols.

Scaling up Industry 4.0 implementations from pilot projects to full-scale operations can be complex. The interconnectedness of systems, the amount of data generated, and the difficulty of algorithms and analytics pose scalability challenges. Manufacturers need to carefully plan and design their systems to ensure scalability and manage the complexity effectively. The rapid advancements often outpace the advancement of regulatory and legal frameworks. Manufacturers need to navigate evolving regulations related to data privacy, intellectual property, liability, and safety standards. Adapting to these frameworks while staying innovative can be a challenge. Industry 4.0 relies on collaboration and integration across the entire value chain. Establishing effective partnerships and collaboration with suppliers, customers, and ecosystem partners is crucial. However, aligning goals, sharing data, and coordinating processes can be challenging, especially when dealing with multiple stakeholders. The embracing of Industry 4.0 technologies is not evenly distributed, and there is a digital divide between companies and regions. Limited internet connectivity, inadequate infrastructure, and varying levels of technological readiness can hinder the accomplishment in less urbanized areas. It will take a comprehensive strategy, stakeholder participation, infrastructure and training investments, and a change management-focused mindset to overcome these obstacles. To fully reap the benefits of Industry 4.0 and maintain their competitiveness in the changing manufacturing scene, organizations must solve these issues.

2.3 Industry 4.0 in Communication: Overview

A significant impact on communication has been made using Industry 4.0 technologies enabling faster, more efficient, and more interconnected communication channels in various industries. IoT plays a vital role in communication by linking devices, apparatus, and sensors to the internet. It promotes synchronized data exchange, distant monitoring, and control of assets. IoT devices gather and broadcast data, providing valuable insights for decision-making and enabling automated actions. Cloud computing is instrumental in communication. It provides a measurable and flexible structure which enables keeping and dealing of large volumes of data generated by IoT devices. Cloud platforms facilitate real-time access to information, data analysis, and collaboration across geographically dispersed teams. AI technologies enhance communication capabilities in Industry 4.0. Machines can now comprehend and analyze human language thanks to natural language processing, or NLP, which improves communication between humans and machines. Artificial intelligence (AI)-driven chatbots and virtual assistants can improve customer service and assistance by answering questions, offering information, and helping with troubleshooting. Edge computing, which moves data processing and analysis closer to the data source, is a complement to cloud computing. It lowers latency, boosts communication effectiveness, and facilitates better real-time decision-making. By processing and filtering data locally, edge devices can reduce the amount of bandwidth needed by only transmitting pertinent information to the cloud.

Unified Communication and Collaboration (UC&C) systems combine many communication channels into one cohesive platform, including text, video, phone, and collaboration capabilities. Regardless of where they are physically located, these platforms allow for seamless communication and cooperation between departments, individuals, and organizations and facilitates real-time information sharing, project collaboration, and remote team collaboration.

AR and VR technologies enhance communication and collaboration in Industry 4.0. AR overlays digital information onto the real-world environment, allowing for assistance in remote, training virtually and in visualizing real-time data. VR creates virtual settings that are realistic, enabling remote collaboration, design reviews, and simulations. Wearable devices like smart glasses and smart watches, enhance communication and enable hands-free access to information. Wearables can display real-time data, provide notifications, and facilitate communication in real-time, enabling workers to access critical information without interruption. Video conferencing tools have become essential in Industry 4.0 communication. They enable face-to-face meetings and remote collaboration among geographically dispersed teams. High-quality video and audio capabilities facilitate real-time discussions, presentations, and administrations, dipping the necessity for physical travel. Mobile devices and connectivity play a crucial role in Industry 4.0 communication. They enable real-time access to information, communication on the go, and remote monitoring and control of processes. Mobile apps offer workers with immediate contact to data, alerts, communication tools, improving productivity and responsiveness. Blockchain technology has the prospect to transform communication. It provides secure and transparent communication channels for data exchange. These Industry 4.0 technologies have transformed communication in manufacturing and other industries, enabling real-time data exchange, seamless collaboration, remote monitoring, and enhanced customer interactions. They facilitate efficient and effective communication across different stakeholders, leading to improved productivity, faster decision-making, and enhanced customer experiences.

2.3.1 Benefits of Industry 4.0 in Communication

Industry 4.0 brings several benefits to communication in various industries. It enable seamless connectivity between devices, machines, and systems. This improves communication channels, allowing real-time data exchange and collaboration among stakeholders. Improved connectivity enables faster decision-making, streamlined processes, and better coordination among teams. Improved Collaboration among employees, teams, and organizations are fostered which enables efficient collaboration, document sharing, and project management, regardless of physical location.

Industry 4.0 enables remote communication and monitoring of processes and assets. Remote communication reduces the need for physical presence, improving efficiency and flexibility. Personalized and Interactive Customer Engagement is made possible by Industry 4.0 technologies AI-powered chatbots can interrelate with clients in an informal manner, addressing their inquiries, providing information, and offering personalized recommendations. This enhances customer satisfaction and improves the overall customer experience. With Industry 4.0, troubleshooting and support become more efficient. Remote experts can assist field technicians using augmented reality (AR) or video conferencing tools. They can guide technicians through step-by-step instructions, overlaying visual cues or sharing real-time video feeds. This reduces downtime, speeds up issue resolution, and improves customer support. Industry 4.0 enables the rapid communication of alerts and notifications. IoT devices and sensors can detect anomalies or critical events and trigger automated alerts. This ensures that relevant stakeholders are immediately notified, allowing them to take prompt action and mitigate risks.

Streamlined communication processes, reduced travel requirements, and efficient collaboration lead to cost reductions. Real-time data and improved decision-making shorten response times, optimize operations, and reduce downtime, resulting in time savings and increased productivity.

They can accommodate changing business needs, evolving technologies, and growing data volumes. As organizations expand or introduce new processes, the communication infrastructure can easily scale up to support increased demand. By leveraging Industry 4.0 in communication, businesses can enhance collaboration, improve responsiveness, streamline processes, and provide better customer experiences. These benefits lead to increased efficiency, productivity, and competitiveness in the evolving digital landscape.

2.3.2 Examples of Industry 4.0 in Communication

The technologies utilized in Industry 4.0 have revolutionized communication in various industries. Voice, video, chat, and document sharing are just a few of the communication modalities that UC&C platforms combine into one cohesive interface. Regardless of where they are physically located, these platforms allow for seamless communication and collaboration among staff. Examples include Microsoft Teams, Slack, and Cisco Webex. IoT devices are used to monitor and communicate real-time information in the supply chain. For instance, RFID tags and sensors can track inventory levels, location, and condition of goods, enabling efficient communication between suppliers, manufacturers, and distributors. This ensures transparency, reduces errors, and optimizes logistics operations. AR technology enables remote assistance and support. Field technicians can wear AR glasses that overlay digital information onto their field of view.

Chatbots powered by AI and natural language processing are used for customer support and communication. They answer queries, provide product information, and offer personalized recommendations. Examples include chatbots used in customer service for e-commerce platforms or banking institutions. Video conferencing tools facilitate real-time communication and collaboration among geographically dispersed teams. High-quality video and audio capabilities make remote communication feel more immersive and enhance team collaboration. Industry 4.0 has transformed communication in marketing, advertising, sharing product information, and gathering feedback. Digital marketing strategies leverage data analytics and AI algorithms to deliver personalized advertisements and targeted messaging based on customer preferences and behavior. Mobile apps enable real-time communication and information sharing.

Employees can access critical data, receive notifications, and collaborate with colleagues using their smartphones or tablets. Mobile apps also facilitate field communication, allowing remote workers to stay connected with the central team and access important information on the go. Remote Monitoring and Control allows for proactive maintenance and reduces downtime. Remote control capabilities enable operators to adjust settings and parameters without physically being present at the location. Blockchain-based platforms enable trusted communication and transactions between parties without the need for intermediaries. Wearable devices, such as smart watches or smart glasses, enhance communication in industries like healthcare and logistics. They provide hands-free access to information, receive notifications, and facilitate communication through voice commands or gesture-based interfaces. Wearables enable real-time communication and data exchange without disrupting tasks. These examples demonstrate how Industry 4.0 technologies are transforming communication across industries, enabling efficient collaboration, personalized customer interactions, remote assistance, and real-time data exchange. These advancements enhance productivity, decision-making, and customer experiences in the digital era.

2.3.3 Challenges of Implementing Industry 4.0 in Communication

The main obstacles to using Industry 4.0 to communication are that it relies on robust and reliable connectivity, including high-speed internet and network infrastructure. However, inadequate connectivity in certain regions or outdated infrastructure can pose challenges in implementing Industry 4.0 communication solutions. It may require investments in infrastructure upgrades and ensuring connectivity across all operational areas. It involves the exchange of sensitive data and information. Ensuring data protection and isolation is critical, as cyber threats and data breaches can lead to significant consequences. Implementing robust security measures, encryption protocols, and data protection policies becomes essential to safeguard communications and maintain trust. Interoperability among different Industry 4.0 communication technologies is vital for seamless integration and communication across systems and platforms. However, diverse communication protocols, standards, and interfaces pose challenges in achieving interoperability. The lack of universal standards can lead to compatibility issues, requiring efforts to establish common protocols and facilitate integration. Integrating these legacy systems with modern communication solutions can be complex and require additional investments. Organizations may need to carefully plan and implement integration strategies to ensure smooth transitions without disrupting existing operations. There may be a scarcity of skilled professionals with expertise in areas like IoT, AI, cyber security, and data analytics. Upskilling and reskilling programs are necessary to equip the workforce with the required competencies.

Industry 4.0 communication technologies must abide by a number of laws and rules pertaining to privacy, data security, and communication standards. It can be difficult to stay on top of changing legislation and make sure you're in compliance, particularly if you operate in numerous jurisdictions. While putting communication solutions into place, organizations must manage these regulatory complexities. Evaluating vendors, their track record, support services, and alignment with organizational goals can be a complex process. Organizations must carefully assess and choose partners who can provide reliable and scalable solutions for long-term success.

Industry 4.0 communication technologies can disrupt traditional communication dynamics within organizations. The shift towards digital communication and increased automation may require employees to adapt to new ways of working and communicating. Ensuring proper training and support during this transition is crucial to minimize resistance and maximize the benefits of Industry 4.0 communication. Overcoming these challenges requires a strategic approach, careful planning, strong partnerships, and a commitment to address the technological, operational, and cultural aspects of implementing Industry 4.0.

2.4 Industry 4.0 in Transportation: Overview

Industry 4.0 has had a profound impact on transportation across different sectors, including rail, air, water, and road transportation. The manufacturing and transportation sectors have seen a radical transformation thanks to the incorporation of cutting-edge technologies including the Industrial Internet of Things (IIoT), Artificial Intelligence (AI), robotics, drones, blockchain, autonomous cars, and machine learning [3]. These technologies have not only transformed the way goods and passengers are moved but have also revolutionized the interaction between humans and machines. With the increasing demands of customers, the timely and transparent delivery of goods has become essential. The implementation of Industry 4.0 technologies has made it easier to achieve successful transportation of passengers, cargo, and mail by leveraging emerging technologies. Cyber-Physical Systems (CPS) and the inclusion of technologies like IIoT devices, AI, drones, advanced robotics, blockchain, and autonomous vehicles have enhanced the flexibility and improved customer service in transportation and logistics. For example, automated driver-less trucks and vans powered by machine learning algorithms enable efficient delivery of goods. Real-time tracking of goods has become effortless through the digitalization of data, facilitated by technologies like RFID chips and IoT devices in vehicles. As a result, businesses have moved closer to achieving end-to-end visibility in their supply chains [10–12]. The impact of Industry 4.0 on transportation can be categorized according to different transportation modes:

Rail Transportation: Track conditions, train performance, and maintenance requirements can all be monitored in real time by intelligent infrastructure systems that are fitted with sensors and Internet of Things devices. Train schedules are optimized by AI algorithms, which also increase capacity and decrease delays. Passengers also benefit from improved comfort and real-time information through connected onboard systems.

Air Transportation: AI and machine learning algorithms optimize flight routes, reducing fuel consumption and improving operational efficiency. Drones and autonomous aircraft are being explored for last-mile deliveries and inspections. Advanced robotics streamline baggage handling processes at airports, and predictive maintenance techniques enhance flight safety [13].

Water Transportation: Real-time monitoring of environmental conditions, fuel consumption, and vessel performance is made possible by IoT devices and sensors. AI algorithms optimize ship routing and cargo loading, reducing fuel consumption and improving operational efficiency. Automated systems streamline cargo handling and container tracking in ports.

Road Transportation: Autonomous vehicles equipped with AI and advanced sensors enhance road safety, reduce congestion, and optimize fuel consumption. IoT devices and RFID technology enable real-time tracking of goods, facilitating efficient supply chain management. Machine learning algorithms analyze data to optimize route planning and improve delivery efficiency. Additionally, shared mobility services and ride-hailing platforms have transformed the passenger experience [14].

2.4.1 Industry 4.0's Advantages for Transportation

There are numerous advantages to using industrial 4.0 technologies in the transportation sector, which have a positive impact on a number of industrial elements. Enhanced Efficiency is one of Industry 4.0's main advantages for transportation. Transport processes are optimized by Industry 4.0 technology, increasing overall operational efficiency. AI algorithms and data analytics help optimize route planning, resource allocation, and scheduling, reducing travel time, fuel consumption, and costs. Automated systems and robotics streamline tasks such as cargo handling, warehousing, and maintenance, improving overall efficiency [15]. Industry 4.0 brings advancements in safety measures for transportation. Technologies like autonomous vehicles, intelligent transportation systems, and real-time monitoring enable safer operations. AI algorithms and sensors help detect potential hazards, monitor driver behavior, and alert operators in real-time, reducing accidents and improving overall safety for passengers, drivers, and pedestrians. Industry 4.0 technologies enable a more personalized and seamless customer experience in transportation. Real-time information on schedules, routes, and seat availability improves passenger convenience and satisfaction. Mobile apps and digital platforms facilitate easy booking, ticketing, and payment processes, simplifying the customer journey. Additionally, end-to-end visibility in supply chains ensures timely and transparent delivery of goods, meeting customer expectations [16]. Implementing Industry 4.0 technologies can lead to significant cost savings in transportation. Optimized route planning, reduced fuel consumption, and improved resource utilization result in lower operational costs. Predictive maintenance and real-time monitoring help identify maintenance needs proactively, minimizing downtime and costly repairs. Effective supply chain optimization and inventory management lower the expenses associated with keeping inventory on hand and raise overall cost effectiveness. Industry 4.0 contributes to a more sustainable transportation industry. Optimized routing and resource allocation reduce fuel consumption and greenhouse gas emissions. Autonomous vehicles can be programmed for eco-friendly driving patterns, further reducing environmental impact. Additionally, digital platforms and shared mobility services promote efficient resource utilization and reduce the need for private vehicle ownership, supporting sustainable urban mobility. Industry 4.0 technologies produce enormous volumes of data that are useful for making well-informed choices. Data analytics and AI algorithms help transportation companies gain insights into operations, demand patterns, and customer preferences. This data-driven approach enables better strategic planning, resource allocation, and service improvements, resulting in more effective decision-making and improved business outcomes.

End-to-end supply chain visibility and optimization are made possible by Industry 4.0 technology. Improved inventory control, real-time tracking of items, and logistics optimization result in less stockouts, increased supply chain efficiency, and enhanced customer service. Transparent and collaborative platforms facilitate seamless coordination among stakeholders, reducing bottlenecks and delays in the supply chain. All things considered, the transportation sector gains a great deal from Industry 4.0, including better efficiency, safer operations, better customer service, cost savings, sustainability, data-driven decision-making, and supply chain optimization. These benefits drive positive transformations in transportation systems, making them more reliable, sustainable, and customer-centric.

2.4.2 Application of Industry 4.0 in Transportation

Industry 4.0 has brought numerous advancements to the transportation sector, revolutionizing various aspects of operations and services. One of the most notable applications of Industry 4.0 in transportation is the autonomous vehicle, which includes self-driving cars, trucks, and drones. These autonomous vehicles navigate and function with the use of AI algorithms, connection, and sensors. They might minimize traffic jams, increase fuel efficiency, and enhance road safety. Intelligent Transportation Systems incorporates various Industry 4.0 technologies to enhance transportation efficiency and safety. It includes systems for real-time traffic monitoring, congestion management, dynamic routing, and smart traffic signal controls. ITS leverages data from sensors, cameras, and connected vehicles to make informed decisions and optimize the flow of traffic. Industry 4.0 makes predictive maintenance possible in the transportation sector, which lowers downtime and boosts asset dependability. Transportation businesses can keep an eye on the condition of its cars, tracks, and equipment in real time by utilizing IoT devices, sensors, and data analytics. Proactive maintenance measures are made possible by predictive maintenance algorithms, which examine data patterns to anticipate possible breakdowns. Industry 4.0 technologies, such as blockchain, IoT, and AI, have completely changed supply chain management in the transportation sector. End-to-end visibility is made possible by real-time tracking of items utilizing RFID and Internet of Things devices, which lowers inventory levels and improves logistics processes. Route planning, warehouse operations, and inventory management can all be optimized using AI algorithms, which can save costs and increase productivity. Ports are integrating Industry 4.0 technologies to enhance efficiency and throughput. Automated container handling systems, robotic cranes, and autonomous guided vehicles streamline cargo operations. Real-time tracking and control of containers is made possible by IoT-enabled sensors and RFID tags, which increase transparency and cut down on delays. Advanced analytics and AI help optimize port operations, vessel scheduling, and cargo handling. Industry 4.0 has facilitated the emergence of digital freight marketplaces, connecting shippers and carriers through online platforms. These platforms utilize AI algorithms to match cargo with available capacity, optimize route planning, and provide real-time visibility. Digital freight marketplaces improve efficiency, reduce empty miles, and enable more transparent and collaborative freight management. Industry 4.0 technologies are transforming public transportation systems by introducing smart ticketing, mobile apps, and connected services. Passengers can access real-time information on schedules, routes, and seat availability through mobile applications. Smart ticketing systems integrate contactless payment options, making travel more convenient and efficient.

These examples demonstrate the transformative impact of Industry 4.0 in transportation, enhancing efficiency, safety, and customer experience. By leveraging advanced technologies, transportation systems are becoming more connected, automated, and data-driven, ushering in a new era of intelligent transportation.

2.4.3 Challenges of Implementing Industry 4.0 in Transportation

Industry 4.0 presents the transport industry with a number of potential and benefits, but its adoption will face a number of difficulties. For Industry 4.0 technologies to be successfully implemented in the transportation sector, reliable infrastructure and seamless connectivity are essential. However, many regions still lack the necessary infrastructure, such as high-speed internet and communication networks, to support the integration of advanced technologies. Addressing these infrastructure gaps is crucial for widespread adoption. Industry 4.0 calls for the gathering, analyzing, and exchange of enormous volumes of data across various transportation ecosystem participants. One of the biggest challenges is making sure that this data is secure and private. It is imperative that important concerns including safeguarding private data from online dangers, preventing unauthorized access, and making sure data protection laws are followed be taken into account. Industry 4.0 technologies introduce automation, AI, and robotics into transportation processes, which can potentially disrupt traditional job roles. The workforce needs to be adequately trained and upskilled to adapt to the changing technological landscape. This calls for funding training initiatives and encouraging a culture of continuous learning in order to provide staff members with the know-how needed to collaborate with cutting-edge technologies. Major investments in software, hardware, infrastructure, and training are frequently needed to use Industry 4.0 technology. To justify adoption, organizations must thoroughly assess the costs and anticipated returns on investment. Balancing the upfront costs with long-term benefits is a challenge that needs to be carefully managed. The integration of various Industry 4.0 technologies in transportation, such as IoT devices, AI systems, and robotics, often involves multiple vendors and platforms. For smooth data interchange and collaboration, it is essential to ensure standardization and interoperability across various systems and technologies. In order to facilitate efficient integration and communication throughout the transportation ecosystem, common standards and protocols must be defined. Regulatory and legal issues are frequently brought up when Industry 4.0 technologies are implemented in the transportation sector. Issues such as liability, safety regulations, data governance, and privacy laws need to be addressed to ensure compliance and build trust among stakeholders. Governments and regulatory bodies play a crucial role in establishing frameworks that support the safe and responsible adoption of these technologies. Collaboration between legislators, technology providers, and industry stakeholders is necessary to overcome these obstacles. To fully realize Industry 4.0's promise in the transportation sector, comprehensive strategies that cover workforce development, interoperability, data security, infrastructure demands, and regulatory frameworks must be developed.

2.5 Industry 4.0 in Healthcare: Overview

"Healthcare Industry 4.0" describes the state-of-the-art operational techniques applied in the medical field. Large-scale group decision-making (GDM), which allows several

Decision-Makers (DMs) from various locations or departments to participate in decision-making, has quickly evolved and been utilized in the healthcare business due to the rapid development of information technology. You can make judgments swiftly and wisely with the help of 4. In order to properly apply GDM strategies to promote the expansion of the medical industry, it is essential to thoroughly examine the prior pertinent achievements. GTM method implementations in the Healthcare Industry 4.0 are assessed using the general GTM process paradigm, which comprises information representation, dimension reduction, consensus establishment, and elicitation [4]. The term "Healthcare Industry 4.0" describes how advanced technologies like big data and machine learning are being used in the medical industry. Both in the medical sector and in people's lives, making judgments is crucial. Decision science principles and approaches have been used in various healthcare business settings [5]. One essential element of decision science is group decision making, or GDM. GDM uses consensus measurement and result expression to produce a final response based on the assessment data of numerous decision makers. In a complex set with significant details, GDM techniques can assist in making optimal or compromise decisions. Moreover, GDM approaches can be used to satisfy the demand for democracy made by contemporary decision makers. Because of the advantages mentioned above, GDM techniques have drawn the interest of many academics and been used in a variety of sectors. With the advancement of information technology, large-scale GDM (LSGDM), which permits a greater number of DMs to engage in decision-making, has grown in popularity. LSGDM techniques can considerably increase output quality and efficiency. The use of LSGDM techniques fosters the innovation of the Healthcare Industry 4.0. Since decisions in the medical area directly impact life safety, they are more significant than in other professions [6]. Everyone should be f of medical decision-making. GDM helps to reduce errors brought on by limited knowledge and emotional responses because it takes into account the viewpoints of numerous experts. To properly utilize GDM approaches to advance the clinical sector, a review of pertinent papers is required.

2.5.1 Benefits of Industry 4.0 in Healthcare

Health 4.0 applications serving patients differ in their objectives; however, they overlap and complement each other [7].

- Patient Administration.
- Public Healthcare
- Inpatient service.
- Home health care.
- Personalized healthcare.

2.5.2 Healthcare Professional Applications

- Management.
- Scheduling.
- Reducing "burnout".
- Access to Resources
- Collaboration.
- Remote Access

2.5.3 Challenges of Implementing Industry 4.0 in Healthcare

Everyone should be informed of the best ways to increase the accuracy of medical judgments. Because it considers the opinions of various experts, GDM helps to eliminate errors caused by ignorance and emotional reactions. A review of relevant publications is necessary to effectively use GDM methodologies to progress the clinical sector. It was combined in the next year's poll. Cyber attacks and fraud/data theft rank two of the top five threats, according to a 2018 World Economic Forum report. A completely networked healthcare system might be feasible, according to Health 4.0. For example, the Internet of Things (IoT) can be used to network and provide data through a hospital bed. It seems improbable that a sizable portion of the global population will soon have access to these revolutionary advancements, as they have not yet been made available to everyone. Compared to 81% of people in wealthy countries, only 17.5% of people in the least developed countries have access to the Internet.

2.6 Future of Industry 4.0 in Terms of Emerging Trends and Technologies

Emerging trends and technologies in Industry 4.0 are continually shaping the future of industrial processes and systems. Edge computing lowers latency and facilitates real-time decision-making by processing data closer to the source or edge of the network. Faster data analysis and response times make it perfect for Industry 4.0 time-sensitive applications like smart grids, remote monitoring, and autonomous cars. 5G and Wireless Connectivity: Industry 4.0's connectivity will be completely transformed with the introduction of 5G networks. With ultra-low latency, high bandwidth, and the ability to connect a massive number of devices, 5G will enable real-time communication and enhance the capabilities of IoT devices. This technology will support advanced applications like remote control of machinery, augmented reality (AR), and collaborative robotics. Digital twins are digital copies of real-world resources, operations, or setups. They construct a digital picture of the physical world by combining real-time data from sensors, Internet of Things devices, and simulations. Digital twins make it possible to simulate, optimize, and do predictive maintenance, which lowers downtime and boosts productivity. Technologies like augmented reality (AR) and virtual reality (VR) have the potential to fundamentally alter several aspects of Industry 4.0. Workers can receive real-time information overlays from AR, which enables them to visualize instructions, carry out difficult activities, and get remote support. Immersive training, simulation, and virtual prototyping are all possible with VR, allowing for efficient design and optimization [8]. Industry 4.0 is seeing a rise in additive manufacturing, also known as 3D printing. Decentralized manufacturing is enabled, waste is reduced, and complicated and customized parts can be produced on demand. Rapid prototyping, mass personalization, and the capacity to produce replacement components on-site are all made possible by 3D printing. Through the combination of sensors, actuators, and computer systems, CPS unifies the real and virtual worlds while monitoring and managing physical processes. CPS facilitate automation, optimization, and adaptive processes in Industry 4.0 by enabling real-time data capture, analysis, and decision-making. Blockchain technology offers secure and transparent decentralized record-keeping and transactional capabilities. It can enhance supply chain traceability, improve product authentication, and enable secure and trusted data exchange among multiple parties in Industry 4.0. Blockchain can ensure data integrity, reduce fraud, and streamline transactions.

Complex problems could potentially be solved by quantum computing and perform computations at an unprecedented speed. In Industry 4.0, quantum computing can be utilized for optimization, machine learning, and solving complex mathematical models, enabling breakthroughs in areas like logistics optimization, material design, and drug discovery. Improvements in human-machine interaction are shaping Industry 4.0. Natural language processing, gesture recognition, and haptic interfaces allow for intuitive and seamless communication between humans and machines. This facilitates collaborative workflows, enhances user experience, and improves overall productivity. Sustainability is a growing concern in Industry 4.0. Technologies focusing on renewable energy, energy-efficient systems, waste reduction, and circular economy principles are gaining importance. These include renewable energy integration, energy harvesting from IoT devices, closed-loop manufacturing, and eco-friendly materials. These emerging trends and technologies in Industry 4.0 demonstrate the continuous evolution and innovation in the field. Implementing and integrating these technologies effectively will unlock new possibilities for enhanced productivity, efficiency, and sustainability in industries.

2.7 Implications of Industry 4.0 on Various Sectors

The economy as a whole will be significantly impacted by Industry 4.0, which will change how industries function and open up new avenues for growth and innovation. Industry 4.0 is revolutionizing manufacturing by introducing advanced automation, robotics, and digital technologies. It makes it possible for connected machinery to operate in smart factories, analyze data in real time, and optimize production procedures. Increased productivity, fewer downtime due to preventive maintenance, better quality assurance, and adaptable and agile production systems that can satisfy specific needs are all implications. Logistics and transportation are changing as a result of Industry 4.0 technologies, which make supply chains more intelligent and linked. Real-time tracking, optimization of routes and delivery schedules, autonomous vehicles, and drones enhance efficiency, reduce costs, and improve overall logistics management. Additionally, technologies like Blockchain provide transparency and security in supply chain transactions. In the healthcare sector, Industry 4.0 is transforming patient care, diagnostics, and drug development. IoT-enabled medical devices, remote patient monitoring, and wearable technologies enhance healthcare delivery and enable personalized medicine. Advanced analytics and AI-powered systems facilitate disease detection, treatment planning, and drug discovery, leading to improved patient outcomes and cost-effective healthcare. Industry 4.0 technologies are causing a revolution in the energy sector. IoT devices and smart grids make it possible to manage energy efficiently, respond to demand, and monitor in real time. Enhancing sustainability and optimizing energy generation and distribution are possible through the integration of renewable energy sources, energy storage technologies, and predictive maintenance of energy infrastructure. Industry 4.0 is revolutionizing the agricultural sector with smart farming practices. IoT sensors, drones, and AI-based analytics enable precision farming, optimizing resource usage, crop monitoring, and automated irrigation. It improves crop yields, reduces wastage, and enhances food traceability and safety. Additionally, smart packaging and supply chain management ensure quality and reduce food loss. Industry 4.0 technologies are reshaping the retail sector through personalized customer experiences and efficient supply chains. Artificial intelligence (AI)-driven chatbots and virtual assistants improve customer experience, and real-time demand forecasting and inventory tracking maximize stock levels. Technologies like augmented reality enable virtual shopping experiences and customization, enhancing customer engagement. Industry 4.0 impacts the financial sector by improving customer experience, risk assessment, and fraud detection. Advanced analytics and AI algorithms enable personalized financial advice and faster credit evaluations. Blockchain technology provides secure and transparent transactions, simplifies cross-border payments, and streamlines compliance processes. Industry 4.0 technologies are transforming the construction sector with digital planning, monitoring, and automation. Building information modeling (BIM), drones, and IoT devices improve project management, enhance safety, and optimize construction processes. Robotics and 3D printing enable efficient and cost-effective construction methods. The workforce's skill requirements have to change as a result of Industry 4.0. The education and skill development sectors must adjust to provide training in cutting-edge technologies like cyber security, AI, robotics, and data analytics. Upskilling and lifetime learning become essential to meet Industry 4.0 demands and close the skills gap. Overall, Industry 4.0 has a profound impact on various sectors, driving innovation, efficiency, and sustainability. Organizations will need to embrace new technologies and modify their business plans to fully utilize their advantages if they want to prosper in the quickly changing digital landscape and remain competitive.

2.8 Opportunities for Businesses and Industries

Enhanced Productivity and Efficiency: Industry 4.0 technology, such robots, automation, and artificial intelligence, can greatly increase the productivity and efficiency of corporate operations. Businesses can achieve greater productivity levels and cost reductions by optimizing resource utilization, cutting down on manual labor, and streamlining procedures.

Industry 4.0 enables businesses to offer customized and personalized products and services. Advanced data analytics and AI algorithms allow businesses to analyze customer preferences and behavior, leading to tailored offerings that meet individual needs. This can enhance customer satisfaction and loyalty. Businesses may make data-driven decisions because to Industry 4.0's wealth of real-time data and sophisticated analytics. Businesses may respond quickly and make well-informed decisions by using predictive analytics and machine learning algorithms to get insights into market trends, client needs, and operational performance. Industry 4.0 technologies offer opportunities to improve the customer experience. From chat bots and virtual assistants for personalized customer support to immersive technologies like AR and VR for interactive product experiences, businesses can engage customers in innovative ways and create memorable interactions. Industry 4.0 creates opportunities for new revenue streams and company models. For example, subscription-based services, outcome-based pricing models, and product-as-a-service approaches allow businesses to monetize their offerings in different ways. Value-added services can also be developed as a result of data-driven insights.

Supply Chain Optimization: Supply chain optimization is made easier by Industry 4.0's automation, real-time tracking, and predictive analytics. Businesses can achieve greater visibility, traceability, and efficiency in their supply chains, resulting in reduced costs, improved inventory management, and faster response times to market demands.

2.8.1 Challenges for Businesses and Industries

Cyber security Risks: Industry 4.0's enhanced connection and data interchange raise serious concerns about the possibility of cyber attacks and data breaches. To guard against unauthorized access and cyber attacks, businesses must invest in strong cyber security measures for their systems, data, and customer information. Industry 4.0 technology adoption necessitates a skilled labor force that can use, maintain, and operate these technologies efficiently. It may be difficult for businesses to identify and nurture personnel with the requisite digital capabilities. Reskilling and up skilling programs become crucial to bridge the skills gap. Industry 4.0's massive data generation and usage creates issues with compliance and privacy. To keep customers' trust and adhere to legal obligations, businesses must implement transparent data governance policies and make sure they are in line with data protection legislation [9]. Industry 4.0 technology implementation sometimes necessitates a large initial investment as well as integration work. Businesses may face challenges in financing these initiatives and integrating new technologies with existing systems and processes. It requires careful planning and strategic decision-making. Embracing Industry 4.0 technologies require changes in processes, workflows, and organizational culture. Businesses may face resistance to change, requiring effective change management strategies and a culture that fosters innovation, collaboration, and continuous learning. As Industry 4.0 advances, businesses need to address ethical considerations related to AI, automation, and data usage. Ethical concerns, such as algorithmic bias, job displacement, and responsible AI deployment, need to be addressed to ensure that technology is used ethically and responsibly.

Navigating these challenges and capitalizing on the opportunities offered by Industry 4.0 requires strategic planning, adaptability, and a proactive approach to innovation and technology adoption. Companies that effectively tackle these obstacles might attain a competitive advantage and prosper in the digital.

2.9 Conclusion

Industry 4.0 is having a profound and revolutionary impact on how organizations and industries function. It brings forth a new era of digitization, connectivity, and automation, offering immense opportunities and challenges. On the positive side, Industry 4.0 presents businesses with the opportunity to enhance efficiency, productivity, and competitiveness. By using cutting-edge technology like artificial intelligence, the Internet of Things, robotics, and data analytics, companies may enhance customer satisfaction, cut expenses, and streamline processes. Businesses are able to stay ahead of the curve in a market that is changing quickly because to the increasing prevalence of customization, personalization, and data-driven decision-making. Additionally, Industry 4.0 creates new revenue sources and business models. Subscription-based services, outcome-based pricing models, and product-as-a-service approaches enable businesses to innovate and create value in new ways. The supply chain becomes more efficient and transparent, with real-time tracking, predictive analytics, and automation improving logistics and inventory management.

However, the impact of Industry 4.0 also comes with challenges. Cyber security risks and data privacy concerns require businesses to prioritize robust security measures and ethical data practices. The skills gap and workforce adaptation requirements make it difficult to identify and nurture talent with the requisite digital skills. Effective adoption and utilization of Industry 4.0 technologies necessitate change management and organizational culture changes. However, Industry 4.0 has wider ramifications for society, including the necessity for fair access to digital innovations and the possibility of employment displacement. To address these societal issues and make sure that everyone benefits equally from Industry 4.0, businesses and legislators must collaborate.

To summarize, Industry 4.0 is having a significant impact on business models, industries, and innovation. Embracing the opportunities and proactively addressing the challenges can empower businesses to thrive in the digital era, while also considering the ethical, societal, and environmental implications. Through a responsible adoption of this technological revolution, we can fully utilize Industry 4.0 to build a future that is more productive, sustainable, and interconnected.

References

1. Zadjali, S.A. and Ullah, A., Impact of Industry 4.0 in Manufacturing Sector. *Int. J. Manage. Sci. Bus. Adm.*, 7, 6, 25–33, 2021.
2. Yang, F. and Gu, S., Industry 4.0, a revolution that requires technology and national strategies. *Complex Intell. Syst.*, 7, 1311–1325, 2021.
3. Holubčík, M., Koman, G., Soviar, J., Industry 4.0 in Logistics Operations. *Transp. Res. Proc.*, 53, 2021.
4. Lu, K. and Liao, H., A survey of group decision making methods in Healthcare Industry 4.0: bibliometrics, applications, and directions. *Appl. Intell.*, 52, 13689–13713, 2022.
5. Larrucea, X., Moffie, M., Asaf, S., Santamaria, I., Towards a GDPR compliant way to secure European cross border Healthcare Industry 4.0. *Comput. Stand. Interfaces*, 69, 103408, 2020.
6. Rehman, M.U., Andargoli, A.E., Pousti, H., *Healthcare 4.0: trends, challenges and benefits*, 2019.
7. Al-Jaroodi, J., Mohamed, N., Abukhousa, E., Health 4.0: On the Way to Realizing the Healthcare of the Future. *IEEE Access*, 8, 211189–211210, 2020.
8. Bhuiyan, A.B., Ali, M.J., Zulkifli, N., Kumarasamy, M.M., Industry 4.0: Challenges, Opportunities, and Strategic Solutions for Bangladesh. *Int. J. Bus. Manage. Future*, 4, 2, 41–56, 2020.
9. Okwu, M., Tartibu, L., Maware, C., Enarevba, D.R., Emerging Technologies of Industry 4.0: Challenges and Opportunitie. *International Conference on Artificial Intelligence, Big Data, Computing and Data Communication Systems*, 2022.
10. Paul, S., Riffat, M., Yasir, A., Mahim, M.N., Sharnali, B.Y., Naheen, I.T., Rahman, A., Kulkarni, A., Industry 4.0 Applications for Medical/Healthcare Services. *J. Sens. Actuator Netw.*, 10, 3, 43, 2021. https://doi.org/10.3390/jsan10030043
11. Neborachko, M., Pkhakadze, A., Vlasenko, I., Current trends ofdigital solutions for diabetes management. *Diabetes Metab. Syndr.*, 13, 5, 2019.
12. Akram, S.V., Implementation of Digitalized Technologies for Fashion Industry 4.0: Opportunities and Challenges. *Sci. Program.*, 2022.
13. Jeddi, Z. and Bohr, A., Remote patient monitoring using artificial intelligenc e, in: *Artificial Intelligence in Healthcare*, Academic Press, 2020.

14. Mian, S.H., Adapting universities for sustainability education in industry 4.0: Channel of challenges and opportunities. 12, 15, 2020.
15. Elmunsyah, H., Adaptive learning system in open educational resource digital sharing community as a media for learning autonomous students. *IOP Conference Series: Materials Science and Engineering*, vol. 732, IOP Publishing, 2020.
16. Peraković, D., Periša, M., Cvitić, I., Analysis of the possible application of assistive technology in the concept of industry 4.0. *Proceedings the Thirty-Sixth Symposium on Novel Technologies in Postal and Telecommunication Traffic—PosTel*, 2018.
17. Amer-Yahia, S., Towards AI-powered data-driven education. *Proceedings of the VLDB Endowment*, vol. 15, 2022.
18. Schumacher, A., Erol, S., Sihn, W., A maturity model for assessing Industry 4.0 readiness and maturity of manufacturing enterprises. *Proc. Cirp.*, 52, 2016.

3

Data Analytics and Big Data Analytics

Neeta Nathani* **and Jagdeesh Kumar Ahirwar**

Gyan Ganga Institute of Technology and Sciences, Jabalpur (M.P.), India

Abstract
Manufacturers have been under constant pressure to reduce expenses and boost efficiencies for many years. The increasing cost of raw materials, the desire for reduced pricing from consumers, and the complexity of supply chains have forced firms to look for innovative solutions to comply with these requirements. The need to improve efficiency and agility has become more pressing because to the COVID-19 pandemic. Employing data analytics may benefit manufacturing companies in many ways, such as improved inventory management, more accurate demand projections, improved quality control, and complete end-to-end visibility. After decades of exponential technological progress, commonly known as Industry 4.0, contemporary manufacturers are capable of collecting enormous volumes of data from each and every link in the supply chain, from sourcing raw materials to making the final delivery to the customer. Data platforms can receive real-time data feeds from IoT sensors and edge devices installed on machinery for processing. Sensor-equipped warehouses can collect data from forklift operations that can be utilized to improve operational efficiency. Future product design improvements can be informed by the analysis of data gathered from online consumer reviews, which can help detect quality control issues. Manufacturers have a huge opportunity as a result of the recent tremendous growth in the amount and variety of data available.

Keywords: Data analytics, manufacturing industry, artificial intelligence

3.1 Introduction to Data Analytics

Data analytics is the process of generating actionable insights from raw data. It includes a range of tools, techniques, and protocols for using data to find trends and fix problems. Data analytics improves decision-making, shapes organizational procedures, and fosters business growth. Data analytics may help businesses view and understand their operations and offerings more clearly. It gives them a thorough grasp of the problems and experiences of their clients. Businesses that move beyond data and link insights with action can customize customer experiences, create relevant digital goods, optimize processes, and increase employee productivity.

**Corresponding author*: neetanathani@ggits.org

Amit Kumar Tyagi, Shrikant Tiwari, Senthil Kumar Arumugam and Avinash Kumar Sharma (eds.) Artificial Intelligence-Enabled Digital Twin for Smart Manufacturing, (39–54) © 2024 Scrivener Publishing LLC

3.1.1 Types of Data Analytics Techniques

Many computational methods are used in data analytics. The following are a few of the most well-liked ones:

a) Natural language processing: Natural language processing is the technology that allows computers to understand and respond to spoken and written human language. Data analysts manage voice commands, chat messages, and dictated notes using this way.
b) Text mining: Data analysts use text mining as a method to look for patterns in textual data, such as emails, tweets, blog posts, and research publications. It can be used for sorting news content, customer feedback, and client emails.
c) Sensor data analysis: Examining the data generated by different sensors is part of the process of analyzing sensor data. It is used in business operations when machinery generates data, such as freight tracking and preventative maintenance.
d) Outlier analysis: Anomaly identification, also called outlier analysis, is the process of identifying data points and occurrences that deviate from the rest of the data.

3.1.2 Distinction between Big Data, Data Analytics, and Data Science

Now that we have a basic understanding of data science, data analytics, and big data, let's try to compare and identify the key differences between the three; which is also shown in Table 3.1. To put it simply:

a) Big data: The field of "big data" involves gathering and analyzing enormous volumes of structured, semi-structured, and unstructured data. Working with massive data requires strong knowledge of computer languages and tools like Python, Java, R, Scala, Apache Spark, Apache Hadoop, Zoho Analytics, Samza, and RapidMiner in addition to mathematical and statistical skills. Large databases will need to be created and kept up to date in order to exploit big data effectively.
b) Data science: Data science is a generic term that refers to the application of data analytics, machine learning, artificial intelligence, and other algorithms to process both structured and unstructured data, extract insightful information, and predict future trends. To develop machine learning models, data mining tools, and unstructured data processing techniques, data scientists must possess advanced levels of skill in mathematics, statistics, analysis, and programming. The most widely used technologies include Tableau, SAS, R, Pear, Scala, Julia, Python, and Data Ladder in addition to Java. It also calls for familiarity with database management systems and Hadoop platforms.
c) Data analytics: To find patterns and turn them into insights, it mostly uses organized historical data. Data analytics requires knowledge in mathematics and statistics as well as skills in data mining, data modeling, data analysis, database management, and reporting. Commonly utilized programming languages include SQL, Python, SAS, Matlab, Excel, and Apache Spark.

Table 3.1 Difference between data science, data analytics and big data.

Parameters	Data Analytics	Data Science	Big Data
Goal	Obtaining pertinent information from a limited, organized data set	Validate or refute a particular theory, providing ultimately useful information	Utilize massive data sets, derive insightful conclusions, and spot fresh chances
Skills	Reports, database administration, data mining, and data modeling and analysis	ML model creation, unstructured data approaches, data modeling, reporting, and visualization	Constructing dispersed data gathering systems, large-scale data processing systems, and unstructured data methods
Popular tools	Matlab, Excel, SAS, SQL, and Apache Spark	Python, Scala, Tableau, Java, SAS, R, Pear, Julia, and data ladder	Rapidminer, Zoho, Samza, Analytics, Python, Scala, R, Pear, Java, and Apache Spark
Working with data	Get inquiries, spot trends, and use past data to determine the answers to these queries	Create inquiries, decide which sources to use for information, and forecast future events	Gather a lot of data, analyze it, and share it to inform decisions in the future
Use Cases	gaming, energy management, and e-commerce buying experience optimization	systems for recommendations, diagnosis, finance, and fraud control, route optimization, and client data analysis	Retail, communication, financial services (compliance, risk management, fraud detection)

3.2 Literature Survey

In order to create manufacturing intelligence, modern production facilities are data-rich settings that facilitate the transmission, distribution, and analysis of information across ubiquitous networks [1–3]. Among the possible advantages of manufacturing intelligence are enhancements in environmental impact, process innovation, and operational efficiency [4, 5]. The present information systems supporting business and manufacturing intelligence are, nevertheless, facing comparable challenges to other sectors and domains: the need to store ever-larger data sets, or "Big Data," and to facilitate the real-time processing of this "Big Data" through the use of advanced analytics [5–10]. The number of instruments that capture measurements from physical surroundings and processes, as well as the frequency at which these devices record and persist measurements, will increase, leading to the anticipated exponential expansion in data generation. The legacy automation and sensor networks, as well as new and developing paradigms like the Internet of Things (IoT) and Cyber Physical Systems (CPS), will be among the technologies that communicate this raw data [1, 11, 12]. These technologies capture low-level granular data, which may be used by analytics and modeling tools to help manufacturers gain a better knowledge of their operations and processes and generate insights that can enhance current ones.

An emerging field of multidisciplinary research that combines automation, engineering, information technology, and data analytics is the focus on big data technologies in manufacturing settings. In order to support the next-generation infrastructure and manufacturing technologies, it is critical to comprehend the current state of research on big data technologies in manufacturing and to pinpoint areas on which future efforts should be concentrated.

3.3 An Overview of Big Data Analytics

Big data analytics is the process of gleaning important insights from massive data sets, such as obscure linkages, hidden patterns, market trends, and customer preferences. Big data analytics has various advantages, such as better decision-making and the capacity to thwart fraud. Massively diverse, continuously generated, high-speed, high-volume volumes of structured, unstructured, and semi-structured data are referred to as "big data." Big data is often measured in terabytes or petabytes. One million gigabytes is equal to one petabyte. To put this in perspective, bear in mind that 4 gigabytes of data are usually needed for an HD movie. A terabyte is equivalent to 2.550,000 video files. Extensive datasets can span several hundred to millions of petabytes in size. Big data analytics is the process of finding connections, patterns, and trends in enormous databases. These complex analytics require a lot of computing power, specialized tools and technology, and vast amounts of data storage.

3.3.1 Difference between Big Data and Data Analytics

The following details illustrate the distinctions between data analytics and big data:

1. Technologies utilized in Big Data vs. Data Analytics: Big data is a more involved and extensive process, which calls for the usage of sophisticated and complex technologies. Useful data sets can be created from unstructured data by employing techniques like parallel computing and automation. Data analytics, in the meantime, makes use of fundamental techniques like statistical and predictive modeling. The analysis and interpretation of data also make extensive use of numerous mathematical and statistical formulas.
2. Types of businesses employing Big Data and Analytics: Many companies are implementing state-of-the-art scientific methods, such as data science, big data, and analytics, to promote growth and innovation. The service industries—tourism, healthcare, and information technology (IT)—use data analytics to enhance and transform their administration. Analytics are utilized by hospitals, for instance, to enhance the effectiveness of their administrative duties, which include patient management, patient treatment, and equipment management. Big data is utilized, nevertheless, in sectors that interact

directly with data, such as retail and finance. Big data is used by customer-focused sectors primarily because contemporary technology allows for the systematic tracking of consumer requirements and demand and aids in the identification of occasionally missed patterns.

3. Type of Data in Big Data vs. Data Analytics: Big data contains unstructured and raw data. The main objective of big data is to convert unstructured data into useful data sets that may be used to deliver insightful findings or tackle difficult business problems. Structured data makes up the bulk of the data utilized in data analytics. It examines the structured data to answer complex business questions, fix business issues, etc.

3.3.2 The Three Vs of Big Data

a) Volume: The amount of data generated by users interacting with any online program is indicated by this term.
b) Velocity: This represents the rate of data generation. The velocity increases with the number of users.
c) Variety: This encompasses all forms of online data, both structured and unstructured. A few instances include voicemail, images, videos, and tweets.

Normal software is unable to process the three Vs of big data. Instead, difficulties are addressed and important outcomes are obtained from them using a new field known as big data analytics. Big data fusion, when combined with analytics, enables them to bring together information from various sources to create a more complete and cohesive model, which improves their comprehension of the data. Businesses also spend money on machine learning (ML) and artificial intelligence (AI) to sort through data from several sources and produce accurate and coherent conclusions.

3.3.3 Big Data Analytics' Advantages and Benefits

1. Hazard Assessment: It can aid in anticipating potential hazards.
   Use Case: To identify irregularities and fraudulent behavior, the Philippine bank Banco de Oro uses big data analytics. The organization employs it to lower the quantity of possible suspects or the root causes of problems.
2. New Product Development and Innovations: Companies that seek to support consumer-driven innovation in research and development (R&D) can leverage big data analytics. Companies can develop products that are more appealing to customers by better understanding the needs of their customers.
   Use Case: One of the biggest producers of jet engines for military applications and airlines worldwide, Rolls-Royce, use Big Data analytics to assess the effectiveness of its engine designs and identify areas for improvement.
3. Faster and More Effective Decision-Making in Organizations: By concentrating more on their profitable business areas, big data analytics can assist firms in optimizing their operations. In contrast to conventional analysis techniques, it aids firms in enhancing their decision-making models.

Use Case: Starbucks uses big data analytics to make strategic decisions. For example, the company utilizes it to assess if a particular neighborhood would be a suitable location for a new outlet. They will look at a lot of factors, such as population, demography, and site accessibility.

4. Boost Client Experience: It gives organizations information into the purchasing habits of their customers and aids in the prediction of future trends.
   Case: Big Data analysis is used by Delta Air Lines to enhance passenger experiences. They keep an eye on tweets to learn about the experiences of their clients with regard to travel, delays, and other issues. When an airline sees bad tweets, it takes the appropriate action to address the issue. The airline is able to foster positive customer connections by openly addressing these problems and providing remedies.

3.3.4 Steps for Big Data Analytics

To analyze any large dataset, big data analytics uses the following five steps:

I. Data collection: This includes locating and obtaining data from data sources. Data collection follows ETL or ELT processes.
   a) Extract Transform Load, or ETL: Before being loaded into storage, the generated data from ETL is first transformed into a standard format.
   b) ELT, or Extract Load Transform: This method involves loading the data into storage and formatting it in accordance with specifications.

II. Data storage: Depending on how complex it is, data can be moved to cloud data lakes or warehouses. Business intelligence tools can access it when needed. A data warehouse is a type of database created especially for analyzing relational data from transactional and business application systems. The data structure and schema are designed for expedient reporting and searching. Data is improved, altered, and purified in order to become the trustworthy "single source of truth." Client profiles and product details are two examples of data.
   A data lake differs in that it can store both structured and unstructured data without the need for extra processing. This implies that you can preserve all of your data without having a suitable design, which is particularly useful in situations where you are not sure how the data will be used. When data is recorded, its schema, or organization, is not yet established. Examples of data include nonrelational data from mobile apps, data from IoT devices, and content from social media platforms.
   For data analytics, organizations usually need both data lakes and data warehouses. Your data needs can be met via Amazon Redshift and AWS Lake Formation.

III. Data processing: Once the data is collected, it must be transformed and arranged in order for analytical queries to yield precise findings. To do this, there are various choices for data processing. The analytical and computational resources available for data processing determine which approach is best.

a) Distributed processing: Data is dispersed among several servers for storage.
b) Centralized processing: All of the data is stored on a customized central server where all operations occur.
c) Batch processing: As data fragments accumulate over time, they are handled in batches.
d) Real Real-time processing: Data is continuously processed while computational tasks are finished in a couple of seconds.

IV. Data cleansing: Data cleansing entails removing any mistakes, including duplicates, inconsistencies, redundant information, and incorrect format. Additionally, it is employed to remove any unnecessary data for analytics.

V. Data analysis: This stage involves turning raw data into insights that can be put to use. Four categories of data analytics are as follows:
a) Descriptive analytics: Data scientists examine data to understand what has happened or is happening in the data environment. It is defined by data visualizations such as tables, line graphs, pie charts, bar charts, and generated narratives.
b) Diagnostic analytics: This refers to a comprehensive or in-depth data analytics process that identifies the cause of an event. Among its distinguishing features are methods like drill-down, data mining, correlations, and discoveries. Raw data is analyzed using a variety of data operations and transformations in each of these methods.
c) Predictive analytics: With the aid of historical data, predictive analytics is able to accurately predict future patterns. It is characterized by methods such as machine learning, forecasting, pattern matching, and predictive modeling. In each of these approaches, computers are trained to reverse engineer causality correlations in the data.
d) Prescriptive analytics: This technique expands on the application of predictive data. It predicts likely results and suggests the best course of action to address them. It may evaluate the potential outcomes of different choices and recommend the best course of action. Neural networks, recommendation engines, complicated event processing, graph analysis, and simulation are its defining features.

3.4 Process of Generation of Big Data Analytics for Manufacturing

The sector uses a variety of applications to gather data in addition to standard sources such as social media monitoring, loyalty programs, and online marketing analysis. The manufacturing industry generates big data through the integration of several software sets, such as CMMS, MES, and CRP, with its machines. These software and machine-generated data sets can then be utilized to create patterns, pinpoint issue areas, and develop remedies that are supported by evidence. In order for the industry to produce this kind of enormous amounts of data, a strong and user-friendly technology stack is needed.

3.4.1 Big Data Analysis's Function in Manufacturing

In the manufacturing sector, big data has several benefits for prediction and preventive decision-making. Let's look at the several strategies that show how important data analytics is to the manufacturing industry.

1. Enhanced competitive advantage: The manufacturing industry has been the center of technological innovation. The data generated by the various channels—industrial IoT, mobile, or next-generation technologies, for example—helps raise the bar for competition. The data enables better understanding of customer needs, more insights into market trends, and future trend projections. When all is said and done, it gives manufacturing houses a big competitive edge.
2. Minimal downtime: In the manufacturing sector, hardware downtime can pose a serious threat to productivity. In addition to taking up employees' time, It also requires extensive maintenance and troubleshooting. The industry's current response to the issue is to use industrial data analysis to do predictive and preventative maintenance on their equipment. It helps the manufacturers keep an eye on the appraisal of hardware quality by testing them and using them every day.
3. Greater CX: These days, manufacturing organizations use data-driven reports that offer exact recommendations for improving customer service, sophisticated sensors to transmit big-data-powered warnings to field technicians about necessary maintenance, and RFID tags to track the state of their equipment.
4. Chain of supply: Using big data analytics, manufacturers may locate products in the manufacturing process. Products that vanish from the market or become difficult to discover can be found thanks to technologies like barcode scanners and radio frequency transmission devices. Customers gain from this since businesses can now give them a delivery schedule that is more precise.
5. Manufacturing oversight: One of the main measures of productivity in a manufacturing facility is the ability to identify the needs of the market and the quantity of goods needed to meet those needs. In the days before big data in production, companies relied on human estimations, which resulted in either an excess or a shortfall of things being produced. Big data helps firms make better decisions by giving them important predictive insights.
6. Adaptable reaction to changes in customer needs: By specifically incorporating real-time manufacturing analytics into their CRM systems, manufacturing organizations are able to predict future trends in real time. CRM data analysis can highlight variations in order and consumption trends that can be utilized to motivate production adjustments. Furthermore, by knowing what the clients want and scheduling the production cycle so that response times are as short as possible, the big data-driven insight acquired from the CRM may assist.
7. Increasing assembly speed: Thanks to big data analytics in the manufacturing sector, businesses can now divide their production and identify the items that are manufactured more swiftly. This helps businesses in the industrial sector

decide where to focus their efforts to get the best results. They might assess where they are strongest and where they still need to make improvements.

8. Hidden risk detection in the process: By analyzing historical equipment failure data, manufacturers may anticipate the equipment's lifecycle and establish usage- or time-based predictive maintenance programs that are accurate. As a result, all of this facilitates gap identification, lowers waste and downtime, and helps businesses create a recovery plan in case of an unplanned malfunction. Furthermore, the manufacturers can automate operations with big data and AI to enable self-optimization without the need for human interaction.
9. Product customization made possible: Traditionally, manufacturing facilities have concentrated on producing large quantities of goods while leaving the modification to businesses catering to niche markets. By forecasting demand and providing producers with advance notice to build customized goods on a large scale, data analysis for manufacturing enables customization at the manufacturing stage. By forecasting demand and removing waste, firms may optimize their manufacturing process with the help of big data. This simplification saves them the time needed to mass customize the products.
10. Improvement increased yield and throughput: Manufacturers can more confidently pursue their programs for continuous improvement when they use big data technologies to uncover hidden trends in their processes. Increases in yield and throughput are the outcome of this.
11. Pricing optimization: Big data can be used to determine a product's pricing point. In order to find the ideal price point that works for both customers and businesses, the system may gather and analyze data from a variety of stakeholders, including suppliers and customers.
12. Picture recognition: A manufacturing company can use big data to uncover a variety of use cases that are particular to picture recognition. Give us an example. Let's say you need a certain spare part, but you have no idea what it's called or how much it will cost. Businesses can capture the image and provide the details to the makers with the aid of big data-driven image recognition software.

3.4.2 Several of the Most Well-Known Actual Applications of Big Data in Manufacturing

The manufacturing sector has demonstrated the numerous advantages that big data brings to the field. But how are those benefits truly being leveraged in the real world? Let's investigate using a few actual business cases as mentioned in Table 3.2.

3.4.3 Utilizing Big Data in the Manufacturing Industry

Even though each project is unique, there are a few basic procedures that are involved in all projects that need the use of big data in manufacturing.

1. Determine the business KPIs: Knowing what is anticipated of big data inclusion should be the first step in every big data initiative. Once you understand

Table 3.2 Examples of actual applications of big data in manufacturing.

Company	Result of applying big data
Colfax	✓ Identification of irregularities and trends in applications, ✓ Enhanced use of assets
National Engineering Industries Limited (NEI)	✓ Enhancement of shop floor, line, plant, and business performance visibility; ✓ Preventive measures to avoid unscheduled breakdowns
Kia Motors	✓ Predicting maintenance expenses and failure rates, cutting down on manufacturing time, ✓ classifying and extracting complaints from customer surveys to identify quality problems
Siemens Healthineers	✓ Prognosis of product failure, ✓ 36% reduction in system downtime
Deutsche Bahn	✓ A 25% decrease in maintenance expenses, ✓ a decrease in malfunctions that cause delays

the key performance metrics to compare the technology's profitability and viability against, you can only proceed with validating its application in your manufacturing business.

2. Examine the manufacturing problems: The next stage would be to gather information about your present demands and requirements for production. You won't be able to identify opportunities for big data inclusion until you have an understanding of how your manufacturing unit is doing right now. Developing a robust method for quality improvement will also benefit from a study of your present situation.
3. Conduct a project cost-benefit analysis: Finding out the project's cost comes next, after you have established the technology's KPIs and examined the business problems. Consider all project development, integration, and maintenance when estimating this budget. When you're done, compare this expense against any potential benefits the manufacturing unit may receive.
4. Integrate big data into the manufacturing process: The next step would be to collaborate with a reputable big data business after you have determined which processes you will be using big data in and have examined the cost-benefit analysis. They will support you in integrating technology in production houses in a seamless manner.

3.5 Utilizing Big Data Analytics for Manufacturing Market Analysis

Based on projections and mentioned in Figure 3.1, the manufacturing big data analytics market is expected to reach USD 23.72 billion by 2029, growing at a compound annual growth rate (CAGR) of 21.60% from USD 8.93 billion in 2024.

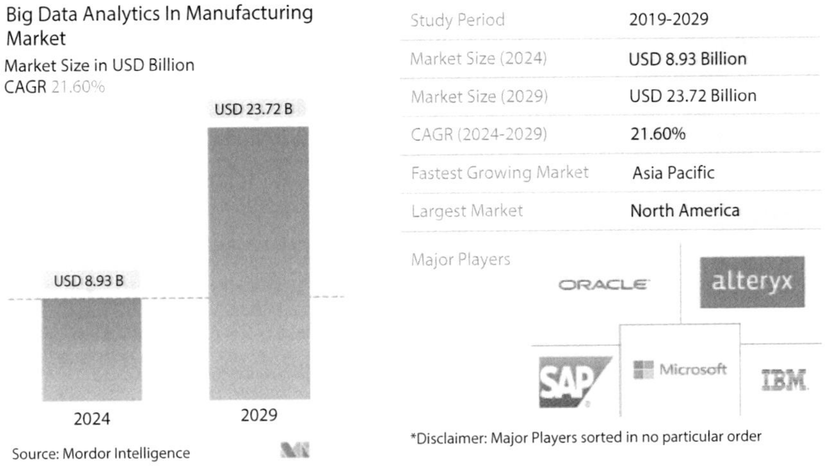

Figure 3.1 Utilizing Big Data Analytics for manufacturing market analysis [Source: Morder Intelligence, Big Data Analytics in Manufacturing Industry Size & Share Analysis – Growth Trends & Forecasts (2024–2029)].

Artificial intelligence (AI) technologies are being used more and more by the manufacturing sector to manage complicated workflows effectively. Making well-informed decisions is crucial for manufacturers, as it affects the entire product life cycle, ensures timely delivery, and guards against unplanned outages, faults, or safety issues.

The global IoT manufacturing industry is expected to reach USD 1.5 trillion by 2030. By instantly converting unstructured and semi-structured data into useable information, artificial intelligence (AI) systems like intelligent document processing (IDP) assist manufacturers in reducing the amount of time spent processing documents.

a) According to a McKinsey report, data-driven firms have a 19 times higher chance of profitability and a 23 times higher chance of attracting clients. By gathering and displaying data in real-time from multiple operational areas, manufacturing organizations can benefit from data analytics. For instance, Schneider Electric moved to Oracle Analytics Cloud, which assisted in gathering, analyzing, and processing 90% of its sell-out data automatically. This system minimized errors in data collecting, eliminated manual data aggregation, and enhanced accuracy and timeliness.
b) Manufacturing companies need to anticipate staffing, scheduling, training, and productivity issues in order to address workforce management roadblocks. Predictive analytics solutions from big data businesses assist industrial companies plan maintenance and repairs, cut downtime, and boost output. For example, throughout the car-making process, BMW uses a data analytics tool to track the quality of its vehicles. Data from the sensors fitted at BMW's manufacturing facility can be used to identify quality problems before they become serious ones.
c) The manufacturing sector was rocked by the COVID-19 outbreak, making inventory management and supply chain management difficult. These

challenges in the manufacturing sector forced companies to work with big data analytic providers for technologies like RFID and barcode scanning, which sit on the manufacturer's property and help track goods in dispatch and warehouse placement.

d) The availability of a huge volume of data and the requirement for high-quality filtering processes present a significant barrier to the adoption of big data analytics in the industrial sector. Integrating data from several departments should allow them to function as a single unit to quickly detect and address issues. The industrial sector may streamline processes, boost output, and maintain its competitiveness in the global market by implementing AI and data analytics technologies.

3.6 Global Big Data Insights for the Manufacturing Sector

Big data analytics is a framework for obtaining vast amounts of data for data mining and trend analysis. Industrialization has quickened over time, and manufacturing output has been steadily rising. The global big data in manufacturing industry market for big data analytics is therefore being driven by the significant shift in data generation in this area.

Businesses are adopting the idea of a "smart industry", one in which real-time data creation and visualization occurs. This plethora of data has potential benefits that the industry is now aware of, thanks to analytics' development from descriptive to predictive. In the industrial sector, a metrics-based economy is developing, which could enhance data-driven decision-making.

Robust processing power is needed at the edge of networks due to the rise in connected devices, industrial digitalization, and the need to satisfy the expanding demands of workplaces for simple and efficient operations. Flexible, reliable, and quick delivery of on-premises and cloud-based applications is another requirement for IT staff. Systems in the industrial sector that use big data can achieve these objectives by automatically identifying devices on the network, automating and monitoring network configuration, and troubleshooting network problems. More advanced big data in the manufacturing industry configures and manages networks using big data, artificial intelligence, and machine learning. Moreover, several open-source efforts are centered on developing standards for virtualization-based big data in the manufacturing industry.

The industrial sector has not yet fully embraced Big Data analytics, despite the technology's encouraging outcomes. The manufacturing industry now has a lot of room to develop when it comes to the application of big data analytics. According to estimates, there are several variables at work in the manufacturing sector, and similar verticals have different levels of efficiency. For example, in order to obtain the numerous parts required to assemble an automobile, a carmaker may collaborate with a huge number of vendors and suppliers.

The size of the big data in manufacturing industry market was estimated at USD 576.47 billion in 2021 and is expected to increase at a compound annual growth rate (CAGR) of 26.7% from USD 731.13 billion in 2022 to USD 4617.78 billion by 2030, as mentioned in Figure 3.2.

Asia Pacific accounted for the greatest global market share of big data in the manufacturing sector in 2021 as shown in Figure 3.3. As smart manufacturing becomes more popular in

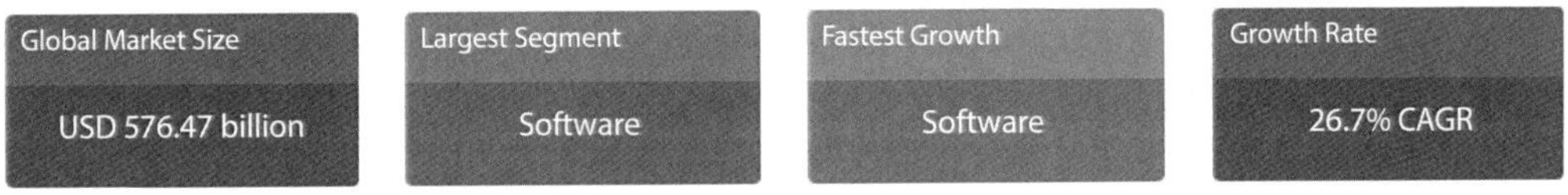

Figure 3.2 Big Data in the Manufacturing Sector: The market size was estimated at USD 576.47 billion in 2019 and is expected to expand at a compound annual growth rate (CAGR) of 26.7% from USD 731.13 billion in 2023 to USD 4617.78 billion by 2031. (2024–2031) [Source: As of April 2023, the Global Big Data in Manufacturing Industry Forecast 2023–2030 provides size, share, growth analysis by type (software, services)].

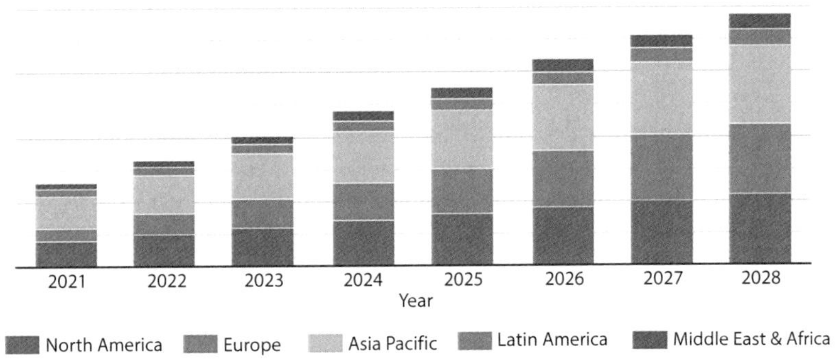

Figure 3.3 Global Big Data Insights for the Manufacturing Sector [Source: As of April 2023, the Global Big Data in Manufacturing Industry Forecast 2023–2030 provides size, share, growth analysis by type (software, services)].

regional company operations, the manufacturing sector in the Asia-Pacific region is changing. This is particularly true for the main manufacturing hubs in the region—China, India, South Korea, and Japan, which are also strong users of big data analytics. The increasing Asia-Pacific region's industrial facilities continue to differ widely in terms of infrastructure, technical skills, government regulations, and research and development (R&D) spending. As a result, there is still an uneven distribution of technological innovation.

3.7 Conclusion

Manufacturing has greatly benefited from the arrival of big data. Clearly, it controls the future of manufacturing, as evidenced by the findings of the 'Manufacturing and the Data Conundrum' Economist Intelligence Study, which Wipro commissioned. Of the survey participants, 86% reported significant increases in data collection, and 90% said their organizations had sophisticated data analysis capabilities for a wide range of manufacturing processes. Having said that, there are still some unanswered questions. How are producers making use of these massive amounts of data to get insightful knowledge that will boost profitability? Is gathering data simple or fraught with issues pertaining to transparency and quality? Is combining data from several sources easy?

It's time to stop using data to just predict issues and start using data to answer the questions outlined above. Despite the gaps between shop-floor data theory and practice, companies surveyed have found a range of comfort zones where the benefits of real-time machine-generated information are available. As was already mentioned, a large number of them possess advanced data analysis abilities. In terms of production efficiency and quality expenses, two-thirds of enterprises report yearly savings of 10% or more through the use of knowledge gained from production-data analysis. According to a McKinsey research, almost 1.8 billion people will become part of the global consuming class over the next 15 years, and global expenditure will nearly triple to $64 trillion. Data analytics present manufacturers with a fantastic opportunity to foresee, create, and execute in a scenario such as this one. Right now, the challenge is to compile information from various sources and draw wise conclusions.

Even if state-of-the-art digital systems can predict problems and provide solutions before they happen, more than half of manufacturers believe their analytical skills are insufficient for the task. Professor Daniel Apley of Northwestern University, who specializes in teaching industrial engineering and management sciences, says it is very difficult to find intelligent young people who want to work in manufacturing. They would prefer to work for banks, Google, and Facebook. Merely 22% of the organizations examined possess analytical capabilities that can anticipate production throughput, and only 16% possess advanced analytical ability that can produce viable solutions.

References

1. Davis, J., Edgar, T., Porter, J., Bernaden, J., Sarli, M., Smart manufacturing, manufacturing intelligence and demand-dynamic performance. *Comput. Chem. Eng.*, 47, 145–156, 2012.
2. Chand, S. and Davis, J., What is smart manufacturing? Time magazine, 2010.
3. Lee, J., Kao, H.A., Yang, S., Service innovation and smart analytics for Industry 4.0 and big data environment. *Proc. CIRP*, 16, 3–8, 2014.
4. Hazen, B.T., Boone, C.A., Ezell, J.D., Jones-Farmer, L.A., Data quality for data science, predictive analytics, and big data in supply chain management: An introduction to the problem and suggestions for research and applications. *Int. J. Prod. Econ.*, 154, 72–80, 2014.
5. Fosso Wamba, S., Akter, S., Edwards, A., Chopin, G., Gnanzou, D., How 'big data' can make big impact: Findings from a systematic review and a longitudinal case study. *Int. J. Prod. Econ.*, 165, 1–13, 2015.
6. Lee, J., Lapira, E., Bagheri, B., Kao, H., Recent advances and trends in predictive manufacturing systems in big data environment. *Manuf. Lett.*, 1, 1, 38–41, 2013.
7. Kumar, P., Dhruv, B., Rawat, S., Rathore, V.S., Present and future access methodologies of big data. *Int. J. Adv. Res. Sci. Eng.*, 8354, 3, 541–547, 2014.
8. McKinsey, Big data: The next frontier for innovation, competition, and productivity, 2011.
9. Philip Chen, C.L. and Zhang, C.-Y., Data-intensive applications, challenges, techniques and technologies: A survey on Big Data. *Inf. Sci. (NY)*, 275, 314–347, 2014.
10. Vera-baquero, A., Colomo-palacios, R., Molloy, O., Towards a process to guide Big data based decision support systems for business processes, in: *Conference on ENTERprise information systems towards*, vol. 00, p. 2212, 2014.

11. Lee, J., Bagheri, B., Kao, H., A cyber-physical systems architecture for industry 4. 0-based manufacturing systems. *Manuf. Lett.*, 3, 18–23, 2015.
12. Wright, P., Cyber-physical product manufacturing. *Manuf. Lett.*, 2, 2, 49–53, 2014.

4

Artificial Intelligence Empowered Smart Manufacturing for Modern Society: A Review

Amit Kumar Tyagi[1]*, Pooja Bhatt[2], Chidambaram N.[3] and Shabnam Kumari[4]

[1]*Department of Fashion Technology, National Institute of Fashion Technology, New Delhi, Delhi India*
[2]*CSE Department, PIET, Parul University, Gujarat, India*
[3]*School of Business and Management, CHRIST (Deemed to be University), Pune, Maharashtra, India*
[4]*Department of Computer Science, Faculty of Science and Humanities, SRM Institute of Science and Technology, Kattankulathur, Chennai, Tamil Nadu, India*

Abstract
Artificial Intelligence (AI) has emerged as a transformative force in the realm of smart manufacturing, shaping the landscape of modern society. This paper delves into the application of AI in smart manufacturing and its profound impact on various aspects of society, from industrial processes to daily life. We discuss how AI-driven technologies optimize efficiency, sustainability, and quality in manufacturing, enabling Society 5.0's vision of a harmonious convergence between technology and humanity. From intelligent automation to predictive analytics and personalized experiences, we uncover the multifaceted role of AI in shaping the future of smart manufacturing and its broader implications for a modern, interconnected society.

Keywords: Artificial intelligence, smart manufacturing, Society 5.0, smart era

4.1 Introduction to AI, Smart Manufacturing

Artificial Intelligence (AI) is revolutionizing the landscape of smart manufacturing, also known as Industry 4.0 or the Fourth Industrial Revolution [1]. It represents the convergence of advanced technologies to optimize industrial processes, enhance productivity, and drive innovation in manufacturing industries. AI, as a core component of smart manufacturing, plays a pivotal role in transforming traditional factories into intelligent and data-driven production systems. Now few key aspects of AI in Smart Manufacturing (refer to Figure 4.1) are included as:

- Data Analytics: AI algorithms analyze large amounts of data generated by sensors, machines, and production lines in real-time. This data-driven

**Corresponding author*: amitkrtyagi025@gmail.com

Amit Kumar Tyagi, Shrikant Tiwari, Senthil Kumar Arumugam and Avinash Kumar Sharma (eds.) Artificial Intelligence-Enabled Digital Twin for Smart Manufacturing, (55–84) © 2024 Scrivener Publishing LLC

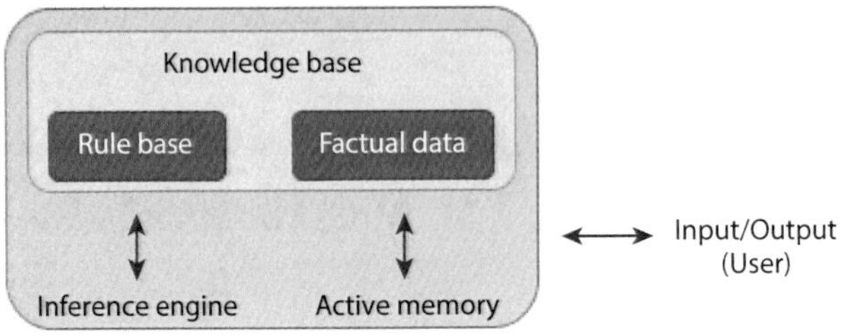

Figure 4.1 AI in manufacturing.

approach helps manufacturers gain insights, detect patterns, and make data-driven decisions to improve efficiency.

- Predictive Maintenance: AI can predict equipment failures and maintenance needs by analyzing data from sensors and historical maintenance records. This proactive approach reduces downtime and extends the lifespan of machinery.
- Process Optimization: AI optimizes manufacturing processes by adjusting parameters in real-time to improve efficiency, reduce energy consumption, and enhance product quality.
- Quality Control: AI-powered systems inspect products for defects and anomalies, ensuring that only high-quality items reach consumers. Machine vision and AI can identify even subtle flaws.
- Supply Chain Management: AI optimizes supply chain operations, predicting demand, optimizing inventory levels, and streamlining logistics for cost savings and timely deliveries.
- Robotic Automation: AI-driven robots and cobots (collaborative robots) perform tasks ranging from repetitive assembly to complex machining, enhancing precision and flexibility in manufacturing.
- Human-Machine Collaboration: AI facilitates human-machine collaboration, where workers can interact with AI systems to enhance decision-making, safety, and productivity.
- Customization: AI enables mass customization by allowing manufacturers to efficiently produce customized products at scale, meeting individual customer preferences.
- Energy Efficiency: AI manages energy consumption by monitoring and optimizing usage in real-time, reducing costs and environmental impact.
- Cyber security: AI helps secure smart manufacturing systems by identifying and mitigating cyber security threats, ensuring the integrity and confidentiality of data.

4.1.1 Challenges and Opportunities

While AI provides numerous benefits in smart manufacturing, it also presents challenges related to data privacy, cyber security, workforce readiness, and ethical issues. Striking a balance between automation and human involvement is important for successful implementation. In summary, AI is at the forefront of smart manufacturing, driving innovation, efficiency, and competitiveness in modern manufacturing industries [2]. As AI continues to evolve, it holds the promise of transforming traditional factories into highly adaptive, data-driven, and intelligent manufacturing ecosystems.

4.2 AI Applications in Manufacturing

Artificial Intelligence (AI) applications in manufacturing are diverse and are transforming the industry by optimizing processes, improving efficiency, and enhancing product quality. These applications leverage AI's capabilities in data analysis, machine learning, computer vision, and automation. Here are some key AI applications in manufacturing:

- Predictive Maintenance: AI analyzes data from sensors and equipment to predict when machines are likely to fail. This allows for proactive maintenance, reducing unplanned downtime and extending the lifespan of machinery.
- Quality Control: AI-powered systems use machine vision and deep learning to inspect products for defects. They can identify even subtle imperfections, ensuring high-quality production and minimizing waste.
- Process Optimization: AI algorithms continuously analyze data from production processes to optimize parameters, reduce energy consumption, and improve product quality.
- Supply Chain Management: AI optimizes supply chain operations by predicting demand, managing inventory levels, optimizing logistics, and improving supplier relationships.
- Robotic Automation: AI-driven robots and cobots (collaborative robots) perform tasks with precision and efficiency, from repetitive assembly to complex machining [3].
- Customization: AI enables mass customization by efficiently producing customized products at scale, catering to individual customer preferences.
- Human-Machine Collaboration: AI facilitates collaboration between humans and machines, enhancing decision-making, safety, and productivity on the factory floor.
- Energy Efficiency: AI manages energy consumption by monitoring and optimizing usage in real-time, reducing costs and environmental impact.
- Inventory Management: AI helps manufacturers optimize inventory levels, reducing carrying costs and minimizing stockouts.
- Demand Forecasting: AI analyzes historical data and market trends to provide accurate demand forecasts, allowing manufacturers to align production with customer needs.
- Maintenance Scheduling: AI-driven maintenance scheduling optimizes downtime, ensuring that maintenance activities are performed at the most opportune times.
- Root Cause Analysis: AI identifies the root causes of production issues by analyzing data from various sources, helping manufacturers address underlying problems.

- Product Design and Development: AI aids in product design through generative design, simulation, and virtual testing, accelerating the development process.
- Quality Assurance: AI can predict potential quality issues in real-time, allowing for immediate corrective actions.
- Worker Safety: AI-enhanced safety systems use sensors and computer vision to detect unsafe conditions and protect workers from accidents.
- Compliance and Reporting: AI ensures that manufacturing processes adhere to regulatory standards, automating compliance checks and reporting.
- Supply Chain Visibility: AI provides real-time visibility into the supply chain, helping manufacturers respond to disruptions quickly and efficiently.
- Cost Reduction: By optimizing various aspects of production and supply chain management, AI contributes to cost reduction, increasing competitiveness.
- Innovation: AI fosters innovation by enabling the development of new products, services, and business models.
- Cybersecurity: AI helps protect manufacturing systems from cyber threats by detecting and mitigating vulnerabilities.

These AI applications not only enhance efficiency and productivity but also pave the way for more sustainable and responsive manufacturing processes, aligning with the demands of the modern industry.

4.3 Benefits and Challenges with AI

AI (Artificial Intelligence) provides a wide range of benefits and opportunities across various industries and applications. However, it also presents several challenges and issues that need to be addressed. Here's a summary of the benefits and challenges associated with AI (refer to Figure 4.2):

Benefits of AI:

- Automation: AI can automate repetitive and mundane tasks, freeing up human workers to focus on more creative and strategic activities.
- Efficiency: AI-driven systems can process and analyze large amounts of data quickly and accurately, leading to increased efficiency and productivity.

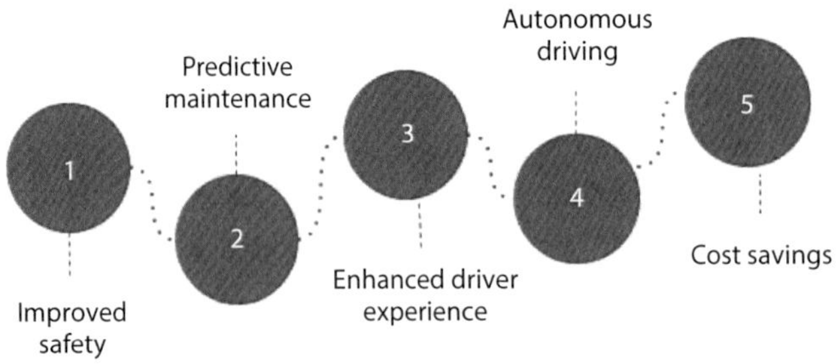

Figure 4.2 AI benefits in manufacturing.

- Cost Reduction: By automating tasks and improving efficiency, AI can help organizations reduce operational costs.
- Improved Decision-Making: AI provides data-driven insights and predictive analytics, helping organizations make more informed and timely decisions.
- Personalization: AI algorithms can tailor products, services, and content to individual preferences, enhancing customer experiences.
- Quality Control: In manufacturing and other industries, AI-powered quality control systems can identify defects and anomalies with high accuracy.
- Healthcare Advancements: AI aids in medical diagnosis, drug discovery, and personalized treatment plans, improving healthcare outcomes.
- Safety and Security: AI enhances security through threat detection, fraud prevention, and cyber security applications.
- Natural Language Processing: AI-powered chatbots and virtual assistants enable more natural and efficient interactions with computers and systems.
- Autonomous Systems: AI enables the development of autonomous vehicles, drones, and robots, with potential applications in transportation, delivery, and more.

Challenges with AI:

- Data Privacy: AI systems rely on large amounts of data, raising issues about data privacy, security, and the ethical use of personal information [4].
- Bias and Fairness: AI algorithms can perpetuate biases present in training data, leading to unfair or discriminatory outcomes.
- Transparency and explainability: AI models, especially deep learning models, can be complex and difficult to interpret, making it challenging to explain their decisions.
- Lack of Skilled Workforce: There is a shortage of professionals with the necessary AI skills, leading to talent gaps in the job market.
- Regulatory and Legal Issues: The rapid development of AI has outpaced the development of regulations, creating legal and ethical dilemmas.
- Security Risks: AI systems can be vulnerable to attacks, and malicious AI can be used for cyber attacks and misinformation.
- Job Displacement: Automation through AI can lead to issues about job displacement and the need for reskilling and upskilling the workforce.
- Data Quality: AI relies on high-quality data, and errors or biases in training data can lead to inaccurate or biased AI models.
- High Costs: Developing and implementing AI systems can be costly, particularly for small businesses and organizations.

In summary, AI provides significant benefits in terms of efficiency, decision-making, and innovation, but it also comes with challenges related to ethics, privacy, security, and workforce readiness. Addressing these challenges and ensuring responsible AI development is important to realizing the full potential of this transformative technology.

4.4 Emerging Technologies Enabling Smart Manufacturing

4.4.1 Internet of Things (IoT) in Smart Manufacturing

Internet of Things (IoT) is revolutionizing the manufacturing industry by enabling smart manufacturing processes. IoT in smart manufacturing involves the use of interconnected devices and sensors to collect, analyze, and share data across the manufacturing ecosystem [5–8]. This connectivity allows for real-time monitoring, automation, and optimization of various manufacturing operations. Here are key aspects of IoT in smart manufacturing:

- Sensor Integration: IoT devices and sensors are embedded in manufacturing equipment, machinery, and products. These sensors collect data on temperature, humidity, pressure, machine performance, and more.
- Data Collection: Sensors transmit data to a centralized platform or cloud-based system, where the data is collected, processed, and stored. This data can include information about equipment status, production quality, and energy consumption.
- Real-time Monitoring: Manufacturers can monitor their production processes in real-time, allowing them to identify issues, track progress, and make informed decisions promptly.
- Predictive Maintenance: IoT-enabled sensors provide data that can be analyzed with machine learning algorithms to predict equipment failures. This enables proactive maintenance, reducing downtime and repair costs.
- Quality Control: IoT sensors monitor product quality throughout the production process, identifying defects and ensuring that only high-quality products reach consumers.
- Supply Chain Visibility: IoT provides visibility into the supply chain, enabling manufacturers to track raw materials, components, and finished products as they move through the production process and distribution network.
- Energy Efficiency: IoT sensors monitor energy consumption in real-time, helping manufacturers identify opportunities for energy-saving measures and sustainability initiatives.
- Inventory Management: IoT systems optimize inventory levels by providing real-time data on stock levels and demand, reducing excess inventory and associated costs.
- Process Automation: IoT can trigger automated responses based on predefined conditions, such as adjusting machine settings or reordering supplies when inventory is low.
- Product Customization: IoT facilitates mass customization by enabling manufacturers to produce customized products efficiently, meeting individual customer preferences.
- Worker Safety: IoT enhances worker safety by monitoring environmental conditions, equipment status, and providing real-time alerts for potential hazards.
- Compliance and Reporting: IoT systems can automate compliance checks and generate reports to ensure that manufacturing processes adhere to regulatory standards.

- Remote Monitoring: Manufacturers can remotely monitor and control manufacturing operations, allowing for greater flexibility and responsiveness, including remote maintenance and troubleshooting.
- Data Analytics: Advanced analytics tools analyze the massive volumes of data generated by IoT devices, providing valuable insights for process optimization and decision-making.
- Enhanced Collaboration: IoT fosters collaboration among stakeholders in the manufacturing ecosystem, including suppliers, partners, and customers, by providing real-time data sharing and communication.

IoT in smart manufacturing not only improves operational efficiency but also enhances competitiveness, sustainability, and innovation in the manufacturing sector. As IoT technologies continue to evolve, they will play an important role in shaping the future of manufacturing by creating more adaptive, data-driven, and intelligent production systems.

4.4.2 Big Data and Analytics in Smart Manufacturing

Big data and analytics play an important role in smart manufacturing by harnessing the power of data to optimize processes, enhance decision-making, and drive innovation [4, 9–11]. Here's how big data and analytics are transforming smart manufacturing:

- Data Collection and Integration: Big data analytics in smart manufacturing involves collecting data from various sources, including IoT sensors, equipment, production lines, and supply chains. Data integration platforms aggregate and process diverse data types, including structured and unstructured data, to create a unified view of manufacturing operations.
- Real-time Monitoring: Big data analytics enable real-time monitoring of manufacturing processes. Data from sensors and machines are analyzed in real-time to detect anomalies, deviations, or inefficiencies. Manufacturers can respond swiftly to issues, reducing downtime and optimizing production.
- Predictive Maintenance: Advanced analytics and machine learning models predict when equipment is likely to fail based on historical data and current conditions. Predictive maintenance schedules maintenance activities, reducing unplanned downtime and optimizing maintenance costs.
- Quality Control: Big data analytics and machine vision systems inspect products throughout the manufacturing process, identifying defects or variations in real-time. Manufacturers can take immediate corrective actions to ensure high-quality production.
- Process Optimization: Data analytics continuously optimize manufacturing processes by adjusting parameters, improving energy efficiency, and enhancing product quality. Algorithms identify opportunities for process improvement and automation.
- Supply Chain Visibility: Big data analytics provide end-to-end visibility into the supply chain, tracking the movement of raw materials, components, and finished products. Manufacturers can make data-driven decisions regarding inventory, demand forecasting, and logistics.

- Energy Efficiency: Data analytics monitor energy consumption patterns and identify energy-saving opportunities. Manufacturers can implement energy-efficient practices and sustainability initiatives.
- Inventory Management: Analytics optimize inventory levels by considering demand patterns, lead times, and production schedules, reducing carrying costs and stockouts.
- Demand Forecasting: Advanced analytics models analyze historical data and market trends to provide accurate demand forecasts, allowing manufacturers to align production with customer needs.
- Root Cause Analysis: When issues arise, data analytics tools help identify the root causes by analyzing data from various sources, allowing for effective problem-solving.
- Worker Safety: Data analytics can monitor worker safety by analyzing sensor data, ensuring compliance with safety protocols, and alerting to potential hazards.
- Data-Driven Decision-Making: Analytics tools provide actionable insights and visualizations that aid decision-makers in optimizing manufacturing operations, resource allocation, and strategy development.
- Continuous Improvement: Big data analytics support a culture of continuous improvement by providing data-driven feedback on processes, products, and performance.
- Innovation and Product Development: Analytics help identify trends and customer preferences, aiding in the development of new products and services.
- Scalability and Flexibility: Cloud-based analytics platforms enable scalability, allowing manufacturers to adapt to changing business needs and scale their analytics capabilities.

In summary, big data and analytics are fundamental to smart manufacturing, driving efficiency, quality, and innovation. By leveraging data-driven insights, manufacturers can optimize operations, enhance competitiveness, and adapt to the demands of a rapidly changing manufacturing landscape.

4.4.3 Robotics and Automation in Smart Manufacturing

Robotics and automation are integral components of smart manufacturing, reshaping the way products are produced, and factories operate [12]. These technologies enhance efficiency, flexibility, and competitiveness in manufacturing processes. Here's an overview of the role of robotics and automation in smart manufacturing:

- Industrial Robots: Industrial robots are used for various tasks, including material handling, welding, assembly, and inspection. Collaborative robots (cobots) work alongside human operators, enhancing safety and efficiency.
- Automated Guided Vehicles (AGVs): AGVs are autonomous vehicles that transport materials and goods within the factory. They improve material flow, reduce manual handling, and increase productivity.

- Automated Manufacturing Cells: Automated cells combine various robotic and automation systems to perform specific manufacturing tasks or processes. These cells can be reconfigured for different products, increasing flexibility.
- 3D Printing and Additive Manufacturing: 3D printing and additive manufacturing technologies enable the production of complex parts and prototypes with high precision. They reduce waste and lead times while allowing for design flexibility.
- Computer Numerical Control (CNC) Machining: CNC machines are automated machine tools controlled by computer programs. They provide high precision and repeatability in machining processes.
- Automated Inspection Systems: Vision systems and sensors are used for quality control and inspection tasks. These systems identify defects and anomalies in real-time, ensuring product quality.
- Automated Assembly Systems: Automated assembly lines use robotics to assemble products efficiently and consistently. They reduce labor costs and improve product quality.
- Pick-and-Place Robots: These robots are used to pick items from one location and place them in another. They are common in packaging and material handling applications.
- Automated Material Handling Systems: Conveyor systems, sorting machines, and robotic palletizers automate material handling and logistics. They optimize material flow and reduce manual labor.
- Flexible Manufacturing Systems (FMS): FMS integrate multiple automated machines and robots, allowing for versatile and adaptive production. They can quickly switch between different product configurations.
- Internet of Things (IoT) Integration: Robotics and automation systems are often connected to the IoT, allowing for real-time monitoring and control. IoT sensors provide data for predictive maintenance and process optimization.
- Autonomous Vehicles and Drones: Autonomous vehicles and drones are used for tasks such as inventory management, inspections, and material transport in large facilities.
- Robotics Process Automation (RPA): RPA involves software robots that automate repetitive, rule-based tasks in administrative processes, improving efficiency in non-manufacturing areas.
- Human-Robot Collaboration: Smart manufacturing emphasizes collaboration between humans and robots, enhancing productivity and safety.
- Scalability and Adaptability: Robotics and automation systems are designed to be scalable and adaptable to changing production requirements.

Hence, by integrating robotics and automation into smart manufacturing, companies can achieve higher productivity, quality, and agility while reducing operational costs and enhancing their competitive edge in a rapidly evolving industry landscape.

4.4.4 AI-IoT-Cloud in Smart Manufacturing

The integration of AI, IoT, and Cloud computing in smart manufacturing is a transformative approach that leverages the collective power of these technologies to enhance efficiency, flexibility, and innovation in manufacturing processes [13–17]. Here's how AI-IoT-Cloud is shaping smart manufacturing:

- Data Collection and Connectivity: IoT sensors and devices collect data from manufacturing equipment, products, and supply chain operations. This data is transmitted to the cloud for storage and processing.
- Cloud-Based Data Processing: Cloud computing provides scalable and robust infrastructure for storing and processing large volumes of IoT data. AI algorithms analyze this data to gain insights, detect patterns, and make predictions.
- Real-time Monitoring and Control: AI-driven analytics in the cloud enable real-time monitoring of manufacturing processes. Manufacturers can remotely control and optimize operations, making data-driven decisions.
- Predictive Maintenance: AI models in the cloud predict equipment failures by analyzing IoT sensor data and historical maintenance records. Manufacturers can schedule maintenance proactively, reducing downtime.
- Quality Control and Assurance: IoT sensors collect data on product quality throughout the production process. AI algorithms in the cloud analyze this data to ensure that only high-quality products are shipped.
- Supply Chain Optimization: Cloud-based AI analyzes supply chain data to optimize inventory management, demand forecasting, and logistics. Manufacturers can respond quickly to changes in demand and supply chain disruptions.
- Process Optimization: AI algorithms in the cloud continuously optimize manufacturing processes by adjusting parameters for efficiency and quality. Manufacturers can identify opportunities for process improvement.
- Energy Efficiency and Sustainability: IoT sensors monitor energy consumption in real-time. AI in the cloud identifies opportunities for energy savings and sustainability initiatives.
- Scalability and Flexibility: Cloud-based solutions are scalable, allowing manufacturers to adapt to changing production requirements. Manufacturers can easily scale up or down as needed.
- Remote Collaboration: Cloud-based AI facilitates collaboration among global teams and stakeholders. Engineers and experts can access real-time data and provide remote support.
- Data Security and Compliance: Cloud providers provide robust security measures and compliance frameworks to protect sensitive manufacturing data. AI can identify and respond to cyber security threats.
- Data Analytics and Machine Learning: AI in the cloud employs data analytics and machine learning to gain insights from IoT data. Manufacturers can uncover hidden patterns and make data-driven decisions.

- Cost Efficiency: Cloud computing eliminates the need for extensive on-premises infrastructure, reducing capital expenses. Manufacturers pay for the resources they use, promoting cost efficiency.
- Innovation and Product Development: AI-driven analytics in the cloud help identify trends and customer preferences, aiding in the development of new products and services.
- Disaster Recovery and Redundancy: Cloud-based solutions provide built-in disaster recovery and data redundancy, ensuring data availability and continuity in case of failures.

The integration of AI, IoT, and Cloud technologies in smart manufacturing enhances competitiveness, agility, and sustainability. By harnessing the power of data and intelligent analytics, manufacturers can make informed decisions, optimize operations, and drive innovation in an increasingly data-driven and interconnected manufacturing landscape.

4.4.5 AI Blockchain-Based Smart Manufacturing

The integration of AI (Artificial Intelligence) and Blockchain technology in smart manufacturing is a promising approach that combines the power of data analytics and decentralized, secure ledger systems [15, 16]. This combination provides several advantages for enhancing efficiency, transparency, and security in manufacturing processes. Here's how AI-Blockchain is shaping smart manufacturing:

- Supply Chain Traceability: Blockchain provides an immutable and transparent ledger for tracking the movement of raw materials, components, and finished products throughout the supply chain. AI analytics can be applied to this data to gain insights into supply chain performance and identify areas for optimization.
- Quality Control and Assurance: IoT sensors collect data on product quality and process conditions, which is recorded on the blockchain. AI algorithms analyze this data to ensure product quality, and the immutable nature of the blockchain ensures data integrity.
- Provenance and Authentication: Blockchain records the origin and history of products, making it possible to verify the authenticity and origin of goods. AI can be used to detect counterfeit products or anomalies in product histories.
- Smart Contracts: Smart contracts, powered by blockchain, automate and enforce agreements between parties in manufacturing processes. AI can be used to optimize and execute these contracts efficiently.
- Intellectual Property Protection: Blockchain can be used to timestamp and protect intellectual property, such as designs and patents. AI can help identify potential IP infringements through data analysis.
- Predictive Maintenance: IoT sensors collect data on machine performance, which is recorded on the blockchain. AI can analyze this data to predict equipment failures and optimize maintenance schedules.

- Data Sharing and Collaboration: Blockchain facilitates secure and permissioned data sharing among stakeholders in the manufacturing ecosystem. AI-driven analytics can be applied to shared data for collaborative decision-making.
- Quality Auditing and Compliance: Blockchain records all changes and transactions, providing an audit trail for regulatory compliance. AI analytics can assist in automating compliance checks and reporting.
- Data Security and Privacy: Blockchain provides enhanced data security through encryption and decentralized data storage. AI can be used to detect and respond to cyber security threats.
- Transparent Production Records: Blockchain records production data, including machine settings, process parameters, and environmental conditions. AI analytics can uncover patterns and trends in production records.
- Efficient Data Marketplaces: Blockchain-based marketplaces for manufacturing data can enable secure and efficient data exchange between organizations. AI algorithms can analyze data from these marketplaces for insights and predictions.
- Environmental and Sustainability Tracking: Blockchain can track environmental data, such as emissions and resource consumption, promoting sustainability. AI can provide insights into sustainability performance and suggest improvements.

The combination of AI and Blockchain in smart manufacturing provides enhanced data security, transparency, and automation. This integration promotes trust among stakeholders, reduces fraud, and enables efficient data-driven decision-making throughout the manufacturing ecosystem. It also addresses key challenges in supply chain management, quality control, and intellectual property protection.

4.5 AI-Driven Smart Manufacturing

4.5.1 Predictive Maintenance and Quality Control in Smart Manufacturing

Predictive maintenance and quality control are two important aspects of smart manufacturing that leverage advanced technologies such as IoT, AI, and data analytics [18–21]. These approaches enable manufacturers to optimize operations, reduce downtime, enhance product quality, and improve overall efficiency. Here's how predictive maintenance and quality control work in smart manufacturing:

Predictive Maintenance:

- Data Collection: IoT sensors and devices are deployed on manufacturing equipment to collect real-time data on various parameters, such as temperature, vibration, pressure, and performance.
- Data Transmission: Collected data is transmitted to a centralized system, often in the cloud, where it is stored and processed.

- Data Analysis: AI and machine learning algorithms analyze historical and real-time data to identify patterns, anomalies, and early signs of equipment degradation or failure.
- Predictive Models: Predictive maintenance models are developed based on the analyzed data. These models predict when equipment is likely to fail and the remaining useful life (RUL) of important components.
- Alerts and Notifications: When the predictive model detects potential issues or maintenance needs, it triggers alerts and notifications to maintenance teams.
- Proactive Maintenance: Maintenance teams can proactively schedule maintenance activities, such as servicing or replacing components, during planned downtime rather than reacting to unexpected failures.
- Downtime Reduction: Predictive maintenance reduces unplanned downtime, extends equipment lifespan, and optimizes maintenance costs.

Quality Control:

- Data Collection: IoT sensors and cameras are used to collect data at various stages of the manufacturing process, capturing information on product dimensions, characteristics, and defects.
- Data Transmission: Collected data is transmitted to a centralized system or the cloud for real-time or post-process analysis.
- Real-time Inspection: AI-powered machine vision systems can inspect products in real-time, identifying defects, deviations, or variations from quality standards.
- Data Analysis: AI algorithms analyze collected data to ensure product quality. Machine learning models learn from historical data to improve accuracy.
- Automated Rejection: When a product does not meet quality standards, automated systems can trigger the rejection of the defective item, ensuring that only high-quality products continue downstream.
- Process Adjustment: Real-time quality data can be used to adjust manufacturing processes, making immediate corrections to prevent further defects.
- Quality Reporting: Quality control data is used to generate reports and documentation for compliance and traceability purposes.
- Continuous Improvement: Insights from quality data contribute to continuous process improvement and product innovation.

Benefits of Predictive Maintenance and Quality Control:

- Cost Reduction: Predictive maintenance reduces maintenance costs and avoids costly unplanned downtime. Quality control minimizes the production of defective products, reducing waste.
- Enhanced Product Quality: Quality control ensures that products meet high-quality standards, improving customer satisfaction and reducing recalls.
- Efficiency Improvement: Predictive maintenance and quality control enhance overall manufacturing efficiency and productivity.

- Downtime Reduction: Predictive maintenance minimizes unplanned downtime, ensuring that manufacturing processes run smoothly.
- Resource Optimization: Both approaches optimize the allocation of maintenance resources and reduce the use of raw materials in defective products.

In summary, predictive maintenance and quality control are important components of smart manufacturing, enabling manufacturers to achieve higher efficiency, lower costs, and improved product quality. Leveraging IoT and AI technologies, manufacturers can proactively manage equipment and ensure the consistent production of high-quality products.

4.5.2 Human–Machine Collaboration

Human-machine collaboration is at the core of Industry 5.0, which represents the next evolution of smart manufacturing. Unlike its predecessors, Industry 5.0 emphasizes the symbiotic relationship between humans and machines, recognizing that each brings unique strengths to the manufacturing process. Here's how human-machine collaboration contributes to smart manufacturing in Industry 5.0:

- Augmented Workforce: In Industry 5.0, machines and robots augment the capabilities of human workers. Machines handle repetitive and physically demanding tasks, allowing humans to focus on more complex, creative, and value-added activities.
- Collaborative Robots (Cobots): Cobots are designed to work alongside human operators. They are equipped with sensors and safety features that enable safe interaction with humans. Human operators and cobots collaborate in tasks such as assembly, quality control, and material handling.
- Skill Enhancement: Human workers acquire new skills to operate, program, and maintain advanced manufacturing machines and robots. Training programs and educational initiatives prepare the workforce for roles that require human-machine collaboration.
- Decision Support: AI-powered analytics provide real-time insights to human operators, helping them make informed decisions on the factory floor. Humans use their experience and judgment to interpret AI-generated insights and adjust processes as needed.
- Problem Solving: When issues arise, human operators and engineers collaborate with machines to diagnose problems and develop solutions. AI and machine learning assist in root cause analysis and suggest corrective actions.
- Flexibility and Adaptability: Human workers are more adaptable and flexible in handling unforeseen situations and changes in production requirements. Machines and robots are programmed to be versatile but may require reconfiguration for major changes.
- Continuous Improvement: Humans and machines collaborate in identifying opportunities for process improvement and innovation. Data analytics and AI-driven insights inform decisions regarding process optimization and product innovation.

- Quality Control: Humans oversee quality control processes and leverage machine vision and AI systems for automated inspection. Collaboration ensures that high-quality standards are maintained throughout production.
- Safety: Collaborative robots are designed with safety features such as sensors, vision systems, and collision avoidance technology. Human safety is a top priority in human-machine collaboration.
- Maintenance and Repair: Human technicians work alongside machines to perform preventive maintenance and repair tasks. Predictive maintenance systems powered by AI help plan maintenance activities.
- Creativity and Innovation: Humans contribute creativity, problem-solving skills, and innovation to product design and process improvement. Machines assist in design simulations, rapid prototyping, and data-driven decision-making.
- Cross-Functional Teams: Cross-functional teams consisting of engineers, data scientists, technicians, and operators collaborate to optimize manufacturing processes. Data sharing and communication are facilitated by digital technologies.

In Industry 5.0, human-machine collaboration enhances manufacturing agility, quality, and efficiency. It leverages the strengths of both humans and machines, promoting a more adaptive and responsive approach to production. By nurturing this symbiotic relationship, smart manufacturing in Industry 5.0 aims to achieve greater levels of productivity, innovation, and sustainability while maintaining a focus on the well-being and skills development of the workforce.

4.6 Popular Challenges and Issues Towards AI-Based Smart Manufacturing Systems

AI-based smart manufacturing systems provide numerous benefits, but they also come with several challenges and issues that need to be addressed. Here are some of the popular challenges and issues faced in the implementation of AI in smart manufacturing, as mentioned in Table 4.1.

In summary, these challenges and issues is important for the successful implementation of AI-based smart manufacturing systems. Manufacturers must carefully plan and strategize their AI adoption, taking into account data management, security, workforce development, etc.

4.7 AI-Based Smart Manufacturing Systems for the Future

4.7.1 AI-Enhanced Supply Chain Management for the Future

AI-enhanced supply chain management represents a promising future for businesses looking to optimize their operations, improve efficiency, reduce costs, and enhance customer satisfaction. Here's how AI is transforming and will continue to impact supply chain management in the future:

- Demand Forecasting: AI algorithms analyze historical sales data, market trends, and external factors to provide accurate demand forecasts. In the

Table 4.1 Challenges and issues towards AI-based smart manufacturing systems.

Sr. no	Types	Challenges	Issues
1	Data Quality and Availability	AI systems require high-quality and consistent data. In manufacturing, data may come from various sources and may not always be accurate or complete.	Poor data quality can lead to inaccurate AI predictions and decisions.
2	Data Security and Privacy	Smart manufacturing systems generate and collect a large amount of sensitive data. Ensuring data security and protecting intellectual property is a significant challenge.	Data breaches and cyber security threats can lead to data loss, production disruptions, and legal issues.
3	Integration with Legacy Systems	Many manufacturing facilities have existing legacy systems that may not be compatible with AI technologies.	Integrating AI with legacy systems can be complex and costly, requiring careful planning and execution.
4	Lack of Skilled Workforce	There is a shortage of professionals with expertise in both manufacturing and AI.	The lack of skilled workers can hinder the adoption and effective use of AI-based systems.
5	Interoperability	Different AI solutions and platforms may not always be interoperable, leading to integration difficulties.	Lack of interoperability can limit the flexibility and scalability of AI implementations.
6	Cost of Implementation	Implementing AI-based systems in manufacturing can require significant investments in technology, training, and infrastructure.	High implementation costs may deter smaller manufacturers from adopting AI solutions.
7	Ethical Issues	AI-based systems may raise ethical issues, especially when it comes to the use of AI in decision-making processes.	Ensuring ethical AI practices and transparency is essential to address issues related to bias, fairness, and accountability.

(Continued)

Table 4.1 Challenges and issues towards AI-based smart manufacturing systems. (*Continued*)

Sr. no	Types	Challenges	Issues
8	Regulatory Compliance	Manufacturing operations are subject to various industry-specific regulations and standards.	Ensuring that AI-based systems comply with regulatory requirements can be complex and time-consuming.
9	Scalability	As manufacturing operations grow, AI systems must scale accordingly to handle increased data and complexity.	Inadequate scalability can lead to system inefficiencies and performance bottlenecks.
10	Change Management	Introducing AI-based systems may require changes in processes, workflows, and organizational culture.	Resistance to change from employees can hinder the successful adoption of AI solutions.
11	Maintenance and Support	AI systems require ongoing maintenance, updates, and support.	Ensuring the reliability and availability of AI systems over time can be challenging.
12	Data Ownership and Sharing:	Collaborative manufacturing environments may involve sharing data among partners and suppliers.	Defining data ownership and sharing agreements can be complex and require trust among stakeholders.

future, AI will incorporate real-time data from IoT devices, social media, and other sources for even more accurate predictions.

- Inventory Optimization: AI optimizes inventory levels by considering demand variability, lead times, and holding costs. Future AI systems will enable dynamic and automated inventory management that responds to real-time changes in demand and supply.
- Supplier Management: AI assesses supplier performance, identifies potential risks, and recommends supplier diversification strategies. Future AI systems will assess supplier sustainability and ethical practices to align with corporate social responsibility goals.
- Predictive Maintenance: AI predicts equipment failures in the supply chain, reducing downtime and maintenance costs. In the future, AI-driven predictive maintenance will become more proactive and precise, further improving asset reliability.
- Route Optimization: AI optimizes transportation routes, reducing fuel consumption and transportation costs. Future AI systems will consider environmental factors and traffic conditions to make real-time route adjustments.

- Warehouse Automation: AI-powered robots and autonomous vehicles automate warehouse tasks like picking, packing, and sorting. In the future, warehouses will become fully autonomous, utilizing AI for efficient storage, retrieval, and order fulfillment.
- Quality Control: AI-enhanced quality control systems identify defects and deviations in real-time Future AI systems will leverage advanced computer vision and sensors for even more precise quality checks.
- Supply Chain Visibility: AI provides end-to-end supply chain visibility by aggregating data from various sources. Future AI systems will provide real-time tracking of shipments, allowing for more agile decision-making.
- Risk Management: AI assesses and mitigates risks by analyzing data on supply chain disruptions and market fluctuations. In the future, AI will incorporate machine learning models for more accurate risk assessments and rapid response strategies.
- Customer Experience: AI personalizes the customer experience by optimizing order fulfillment and delivery schedules. Future AI systems will enable hyper-personalization, providing tailored products and services based on individual preferences and behaviors.
- Sustainability: AI supports sustainability efforts by optimizing transportation routes, reducing waste, and enhancing energy efficiency. In the future, AI will play a central role in achieving environmentally sustainable supply chains.
- Compliance and Regulation: AI ensures compliance with regulatory requirements and ethical standards. Future AI systems will adapt to evolving regulations and automate compliance reporting.

AI-enhanced supply chain management is poised to create more agile, resilient, and sustainable supply chains. As AI technologies continue to evolve, they will provide businesses with the tools needed to navigate an increasingly complex and competitive global marketplace.

4.7.2 Smart Factory Concepts for Next Generation

Smart factory concepts for the next generation represent the evolution of manufacturing facilities, leveraging advanced technologies to optimize operations, improve flexibility, and enhance competitiveness. Here are key concepts and components of smart factories for the future:

- IoT Integration: IoT plays a central role, connecting machines, equipment, and sensors throughout the factory. IoT sensors collect real-time data on machine performance, production processes, and environmental conditions.
- Big Data Analytics: Advanced analytics tools process and analyze the large amount of data generated by IoT devices. Predictive and prescriptive analytics provide insights for decision-making, predictive maintenance, and process optimization.
- Artificial Intelligence (AI) and Machine Learning: AI-powered algorithms and machine learning models enhance automation, quality control, and

predictive maintenance. AI systems continuously learn from data to make informed decisions and optimize processes.

- Digital Twin Technology: Digital twin models replicate physical factory assets in the digital realm, allowing for real-time monitoring and simulation. Changes or optimizations can be tested in the digital twin environment before implementation.
- Augmented Reality (AR) and Virtual Reality (VR): AR and VR technologies provide immersive training, maintenance, and troubleshooting experiences. Workers can receive real-time guidance and visualize complex processes.
- 5G Connectivity: High-speed and low-latency 5G networks enable seamless communication between devices and systems. It supports real-time data transmission, remote monitoring, and edge computing.
- Edge Computing: Edge devices and edge computing capabilities bring data processing closer to the source, reducing latency. It enables real-time decision-making on the factory floor.
- Autonomous Robots and Cobots: Autonomous robots and collaborative robots (cobots) work alongside human operators. They handle repetitive tasks, material handling, and assembly, improving efficiency and worker safety.
- Additive Manufacturing (3D Printing): 3D printing technologies provide rapid prototyping and on-demand production of complex components. It reduces lead times and supports customization.
- Blockchain for Supply Chain Transparency: Blockchain technology ensures transparent and secure supply chain management. It provides end-to-end visibility and traceability of products and materials.
- Energy Efficiency and Sustainability: Smart factories prioritize energy efficiency and sustainable practices. Renewable energy sources and energy management systems are integrated to reduce environmental impact.
- Human-Machine Collaboration: Smart factories foster collaboration between human workers and machines. Humans focus on tasks that require creativity, problem-solving, and decision-making, while machines handle repetitive work.
- Cybersecurity and Data Privacy: Robust cyber security measures protect sensitive data and factory operations. Data privacy is a top priority to comply with regulations and ensure trust.
- Continuous Improvement Culture: Smart factories embrace a culture of continuous improvement, leveraging data and feedback for ongoing optimization. Employees are encouraged to contribute ideas and innovations.
- Agile Manufacturing: Agile manufacturing principles are adopted, allowing rapid adaptation to changing market demands and customer preferences. Smart factories are designed for flexibility and scalability.

Smart factory concepts for the next generation aim to create more efficient, sustainable, and adaptable manufacturing environments. By harnessing cutting-edge technologies and fostering collaboration between humans and machines, these factories will drive innovation, enhance product quality, and remain competitive in a rapidly evolving industrial landscape.

4.7.3 Quality Assurance and Inspection in Real Time

Quality assurance and inspection in real-time are important components of smart manufacturing, ensuring that products meet the highest quality standards while minimizing defects and waste. Here's how quality assurance and inspection work in the context of smart manufacturing:

- Automated Quality Control: IoT sensors and cameras are strategically placed throughout the production process to monitor and capture data in real-time. Sensors measure parameters such as temperature, pressure, dimensions, and material properties. Cameras perform visual inspections to detect defects, anomalies, and deviations from quality standards.
- Data Collection and Analysis: Collected data is transmitted to a centralized system, often in the cloud or at the edge, for real-time processing and analysis. AI and machine learning algorithms analyze data to identify quality issues, trends, and patterns.
- Real-time Alerts and Notifications: When quality deviations or defects are detected, the system triggers real-time alerts and notifications. Alerts are sent to relevant personnel or control systems for immediate action.
- Automated Rejection and Sorting: In case of quality issues, automated systems can trigger the rejection of defective products. Products can be sorted based on quality criteria, ensuring that only high-quality items continue through the production process.
- Root Cause Analysis: Real-time data and historical records are used for root cause analysis to identify the source of quality problems. AI-driven analysis helps determine whether the issue is due to equipment, materials, or process variations.
- Process Adjustments: Real-time quality data can be used to make immediate process adjustments Machines and robots can adapt their operations to correct deviations and ensure product quality.
- Continuous Improvement: Insights from real-time quality data contribute to continuous process improvement and product innovation. Manufacturers can make data-driven decisions to optimize production parameters.
- Remote Monitoring and Control: Quality assurance can be managed remotely, allowing experts to monitor and control quality from anywhere. Remote experts can provide guidance to on-site personnel in real-time.
- Compliance and Documentation: Real-time quality data is used to ensure compliance with regulatory standards and customer requirements. Automated documentation and reporting processes simplify compliance efforts.
- Predictive Quality: AI models can predict quality issues before they occur based on historical data and real-time sensor information. Preventive actions can be taken to avoid defects.
- Supplier Quality Control: Real-time quality data can be shared with suppliers to ensure that incoming materials meet quality standards. Supplier performance can be continuously monitored.

- Customer Feedback Integration: Customer feedback and quality data from the field can be integrated into the quality control process in real-time. This ensures that product improvements are driven by customer needs.
- Real-time quality assurance and inspection in smart manufacturing result in higher product quality, reduced waste, and improved customer satisfaction. By leveraging IoT, AI, and automation, manufacturers can detect and address quality issues in real-time, leading to more efficient and competitive production processes.

4.8 Improving Operational Efficiency and Environmental Sustainability in AI Based Smart Manufacturing

Improving operational efficiency and environmental sustainability in AI-based smart manufacturing is essential for both economic competitiveness and environmental responsibility. By optimizing operations and adopting sustainable practices, manufacturers can reduce waste, energy consumption, and environmental impact. Here are strategies to achieve these goals:

A. Energy Efficiency:
   - Smart Energy Management: Implement AI-driven energy management systems to monitor, analyze, and optimize energy consumption in real-time. This includes adjusting machine operation schedules, lighting, and HVAC systems for maximum efficiency.
   - Renewable Energy Integration: Invest in renewable energy sources, such as solar panels and wind turbines, to reduce reliance on fossil fuels. AI can optimize the use of renewable energy sources based on availability and demand.

B. Waste Reduction:
   - Predictive Maintenance: AI-driven predictive maintenance reduces unexpected equipment failures and material waste. Machines are maintained only when necessary, minimizing downtime and waste.
   - Inventory Optimization: AI helps optimize inventory levels, reducing overstock and obsolete goods, which can contribute to waste reduction.

C. Supply Chain Optimization:
   - Real-time Visibility: Leverage AI to gain real-time visibility into the supply chain, allowing for better inventory management, demand forecasting, and streamlined logistics.
   - Sustainable Sourcing: Use AI analytics to identify sustainable suppliers, materials, and transportation options that align with environmental goals.

D. Process Optimization:
   - Process Efficiency: AI-driven process optimization identifies bottlenecks and inefficiencies, reducing resource consumption and waste generation.
   - Quality Control: Implement AI-powered quality control systems to minimize product defects and waste.

E. Circular Economy Practices:
- Product Lifecycle Management (PLM): Use AI to extend product lifecycles by designing products for disassembly, repair, and recycling. Implement reverse logistics to recover and reuse materials from end-of-life products.
- Remanufacturing and Upcycling: Discuss opportunities to remanufacture or upcycle components and materials to reduce waste.

F. Environmental Monitoring:
- IoT Environmental Sensors: Deploy IoT sensors to monitor environmental parameters, such as air quality, emissions, and water usage. AI analytics can identify opportunities for improvement.

G. Green Technology Adoption:
- Efficient Equipment: Invest in energy-efficient machinery and equipment. AI can optimize the operation of these assets for maximum efficiency.

H. Sustainability Reporting:
- AI-powered Reporting: Use AI for sustainability reporting and compliance with environmental regulations. AI can automate data collection and reporting processes.

I. Employee Engagement:
- Training and Awareness: Educate employees about sustainability goals and encourage their participation in energy-saving and waste-reduction initiatives.

J. Regulatory Compliance:
- AI for Compliance: Utilize AI to monitor and ensure compliance with environmental regulations and standards.

K. Collaborative Innovation:
- Cross-Industry Collaboration: Collaborate with other industries, research institutions, and startups to exchange knowledge and develop innovative solutions for sustainability challenges.

Improving operational efficiency and environmental sustainability in AI-based smart manufacturing is a continuous journey. By embracing AI-driven technologies and integrating sustainability practices into manufacturing processes, companies can reduce their environmental footprint, enhance their reputation, and achieve long-term economic and environmental benefits. Additionally, adopting sustainable practices can lead to increased customer loyalty and a competitive edge in the marketplace.

4.8.1 Cost Reduction and Resource Optimization in AI-Based Smart Manufacturing

Cost reduction and resource optimization in AI-based smart manufacturing are important objectives for manufacturers looking to improve competitiveness and sustainability. Here are strategies to achieve these goals:

- Predictive Maintenance: Use AI-driven predictive maintenance to schedule equipment servicing when needed, reducing downtime and avoiding unnecessary maintenance costs.

- Energy Efficiency: Implement AI-based energy management systems to optimize energy consumption by machines, HVAC, and lighting. Utilize AI to analyze energy data and identify areas for improvement.
- Waste Reduction: Employ AI to monitor and optimize production processes, minimizing material waste and scrap. Implement AI-driven quality control to reduce defects and rework.
- Inventory Optimization: Utilize AI for demand forecasting and inventory management to reduce carrying costs and minimize excess stock. Implement just-in-time (JIT) manufacturing principles supported by AI.
- Supply Chain Optimization: Use AI for real-time supply chain visibility to optimize logistics, reduce transportation costs, and improve delivery times. Implement AI-driven supplier performance monitoring for cost-effective sourcing.
- Labor Optimization: Apply AI for workforce scheduling and task allocation to optimize labor resources. Use AI-driven HR analytics to match skills to tasks and reduce labor costs.
- Process Efficiency: Implement process optimization using AI to identify bottlenecks and inefficiencies. Use AI to automate routine tasks, reducing manual labor costs.
- Equipment Utilization: Maximize machine and equipment utilization with AI-driven production scheduling and real-time monitoring. Implement AI for condition-based equipment maintenance to extend the lifespan of machines.
- Simulation and Digital Twins: Use AI-powered digital twins to simulate and optimize manufacturing processes, reducing the need for physical prototyping. Optimize resource allocation based on digital twin simulations.
- Resource Allocation: Implement AI for resource allocation, including materials, labor, and equipment, to ensure efficient use. Use AI-driven analytics to allocate resources based on real-time demand.
- Scalability and Flexibility: Design manufacturing systems with scalability and flexibility in mind to adapt to changing production volumes. Utilize AI to dynamically adjust resource allocation as demand fluctuates.
- Green Manufacturing: Incorporate sustainability principles into manufacturing processes, such as recycling and using eco-friendly materials. Use AI to optimize processes for minimal environmental impact.
- Data-Driven Decision-Making: Promote a culture of data-driven decision-making, with AI analytics supporting cost reduction initiatives. Use AI for real-time cost tracking and budget optimization.
- Supplier Collaboration: Collaborate with suppliers to identify cost-saving opportunities and negotiate favorable terms. Share demand forecasts and production plans using AI-driven supply chain platforms.
- Continuous Improvement: Foster a culture of continuous improvement, with AI providing insights for ongoing cost reduction efforts. Use AI to track Key Performance Indicators (KPIs) related to cost efficiency.

Hence, Effective cost reduction and resource optimization in AI-based smart manufacturing require a complete approach that leverages data, analytics, and AI technologies to make informed decisions and continuously improve processes. Manufacturers should embrace AI as a powerful tool to enhance efficiency, reduce waste, and enhance overall profitability while contributing to sustainability goals.

4.9 Future Research Opportunities and Research Gaps Towards AI-Empowered Smart Manufacturing

Future research opportunities and addressing research gaps in AI-empowered smart manufacturing are important to advance the field and enable manufacturers to fully harness the potential of artificial intelligence. Here are some key areas for future research, mentioned in Table 4.2.

Future research in these areas will play an important role in advancing the field of AI-empowered smart manufacturing, enabling the industry to address emerging challenges, improve efficiency, and achieve sustainability goals. Collaboration between academia, industry, and government organizations will be essential to drive innovation and bridge existing research gaps.

4.10 The Evolution of Industry 5.0 and Industry 6.0

Industry 5.0 and Industry 6.0 represent the next phases in the ongoing evolution of industrial and manufacturing practices. Here's an overview of each stage:

Industry 4.0:

- Industry 4.0, also known as the Fourth Industrial Revolution, marked the integration of digital technologies into manufacturing.
- Key technologies included IoT, AI, big data analytics, and automation.
- The focus was on creating smart factories with interconnected machines, data-driven decision-making, and predictive maintenance.
- Industry 4.0 aimed to improve operational efficiency, reduce costs, and enhance product quality.
- Human-machine collaboration and cyber-physical systems were central to this stage.

Industry 5.0:

- Industry 5.0 represents a shift from complete automation back to emphasizing the role of humans in manufacturing.
- This stage recognizes that humans bring unique skills in creativity, problem-solving, and decision-making.
- Industry 5.0 envisions a collaborative environment where humans and machines work together symbiotically.
- AI, robotics, and automation continue to play a significant role, but their purpose is to assist and augment human workers rather than replace them.

Table 4.2 Research opportunities and research gaps towards AI-empowered smart manufacturing.

Types	Research gap	Opportunity
AI-Driven Sustainability	More research is needed to develop AI algorithms that optimize manufacturing processes for sustainability, including energy efficiency, waste reduction, and emissions reduction.	Investigate AI's role in achieving circular economy principles within smart manufacturing, such as product remanufacturing and materials recycling.
Human-AI Collaboration	Discuss the most effective ways to facilitate collaboration between humans and AI systems in smart manufacturing environments.	Study the impact of AI on the workforce, including the development of AI-augmented skills and training programs for manufacturing personnel.
Explainable AI (XAI)	Develop and enhance explainable AI techniques specific to manufacturing processes to ensure transparency and trustworthiness.	Investigate methods for providing interpretable explanations of AI-driven decisions to human operators and decision-makers.
Edge AI and Fog Computing	Discuss the potential of edge AI and fog computing for real-time decision-making and data processing in resource-constrained manufacturing environments.	Develop algorithms and architectures for AI inference at the edge to reduce latency and enhance responsiveness.
AI for Small and Medium-sized Enterprises (SMEs)	Investigate ways to make AI technologies more accessible and affordable for small and medium-sized manufacturing enterprises.	Develop AI solutions tailored to the specific needs and constraints of SMEs in the manufacturing sector.
AI-Enabled Product Customization	Discuss how AI can enable mass customization of products while maintaining cost-effectiveness.	Develop AI-driven design and production systems that adapt to individual customer preferences without compromising efficiency.
AI-Enhanced Quality Assurance	Investigate AI-based quality control and inspection methods that can ensure high-quality products in complex manufacturing processes.	Develop AI algorithms for real-time defect detection and quality monitoring with high accuracy.

(*Continued*)

Table 4.2 Research opportunities and research gaps towards AI-empowered smart manufacturing. (*Continued*)

Types	Research gap	Opportunity
Supply Chain Resilience	Discuss how AI can enhance supply chain resilience, especially in the face of disruptions like pandemics and natural disasters.	Develop AI-driven supply chain models that can quickly adapt to unexpected challenges and optimize resource allocation.
Ethical AI in Manufacturing	Address ethical issues and potential biases in AI applications within smart manufacturing.	Investigate the development of ethical AI frameworks and guidelines tailored to the manufacturing industry.
AI-Driven Regulatory Compliance:	Discuss how AI can help manufacturers ensure compliance with evolving environmental and safety regulations.	Develop AI-powered systems for monitoring and automating compliance-related tasks.
Standardization and Interoperability	Investigate standardization efforts to ensure interoperability and compatibility among AI systems and equipment from different vendors.	Contribute to the development of industry-wide standards for AI in manufacturing.
AI for Sustainable Materials	Investigate how AI can assist in the identification and development of sustainable materials and supply chain sources.	Discuss AI-driven approaches to assess the environmental impact of materials throughout their lifecycle.

- Customization and personalization of products for individual customers are emphasized.
- The goal is to achieve a balance between automation and human craftsmanship.

Industry 6.0 (Emerging):

- Industry 6.0 is an emerging concept that builds upon the foundations of Industry 4.0 and Industry 5.0.
- It is characterized by even greater integration of advanced technologies, including quantum computing, nanotechnology, and advanced materials.
- Industry 6.0 aims to create highly adaptive and self-optimizing manufacturing systems.
- It emphasizes sustainability, with a focus on environmentally friendly practices, circular economy principles, and reduced resource consumption.

- Supply chains become more resilient and responsive, leveraging AI for real-time risk management and demand forecasting.
- The concept of "lights-out" manufacturing, where factories operate with minimal human intervention, may become more prevalent but not at the expense of human employment.
- Ethical issues, data privacy, and cyber security continue to be important aspects.

Note that the boundaries between these industrial stages are not rigid, and the transition from one to another is gradual. Companies may adopt elements of Industry 5.0 and Industry 6.0 while still benefiting from Industry 4.0 technologies. The evolution of industry is driven by the rapid advancement of technology, changing market dynamics, and the need to address emerging challenges and opportunities in manufacturing and beyond.

4.11 Conclusion

The integration of Artificial Intelligence (AI) into smart manufacturing stands as a remarkable testament to the transformative potential of technology in modern society. AI has not only revolutionized industrial processes but has also extended its influence into various facets of daily life. Through intelligent automation, predictive analytics, and data-driven decision-making, AI has paved the way for more efficient, sustainable, and higher-quality manufacturing practices. As we embrace the principles of Industry 5.0 and Society 5.0, the synergy between AI and smart manufacturing promises a future where human ingenuity and machine capabilities harmoniously coexist. The benefits extend far beyond the factory floor, touching areas such as healthcare, transportation, energy management, and education, enhancing the overall quality of life in modern society. However, it is essential to recognize that this journey towards an AI-empowered smart manufacturing society also comes with ethical, security, and regulatory challenges. Striking the right balance between technological advancement and societal well-being remains an important imperative. Responsible AI development, robust cyber security measures, and thoughtful regulations will be essential to navigate these challenges. In summary, AI-powered smart manufacturing is not just a technological evolution; it is a societal transformation that holds the promise of a more sustainable, efficient, and interconnected modern world. By harnessing the full potential of AI, we can aspire to create a future where machines and humans collaborate seamlessly to achieve unprecedented levels of innovation and prosperity.

References

1. Hsieh., W.-H., Garza-Reyes., J.A., Wong., P.K., Editorial for the special issue on manufacturing technology for modern factories. *Proc. Inst. Mech. Eng., Part B: J. Eng. Manuf.*, 2022, doi: 10.1177/09544054221127816.
2. Chen., M.Y., Lughofer, E., Egrioglu, E., Deep learning and intelligent system towards smart manufacturing. *Enterp. Inf. Syst.*, 2022, doi: 10.1080/17517575.2021.1898050.

3. Kalarani, P., Empowering Artificial Intelligence and Cyber Security Challenges in Smart Manufacturing, 2021, doi: 10.17762/TURCOMAT.V12I6.1280.
4. Ding., H., Gao, R.X.,., Isaksson., A.J., Landers, R.G.,., Parisini., T., Yuan, Y., State of AI-Based Monitoring in Smart Manufacturing and Introduction to Focused Section. *IEEE/ASME Trans. Mechatron.*, 2020, doi: 10.1109/TMECH.2020.3022983.
5. Oztemel, E., *Intelligent Manufacturıng Systems, Smart Factories And Industry 4.0: A General Overview*, 2019, doi: 10.1201/9780429464768-1.
6. Dubey., G.P. and Tahiliani, K., A LSTM based Deep Learning Model for Smart Manufacturing, 2023, doi: 10.1109/ISCON57294.2023.10111964.
7. Petrillo, A., Caiazzo, B., Piccirillo, G., Santini, S., Murino., T., An IoT-based and cloud-assisted AI-driven monitoring platform for smart manufacturing: design architecture and experimental validation. *J. Manuf. Technol. Manage.*, 2022, doi: 10.1108/jmtm-02-2022-0092.
8. Zeba., G., Dabić., M., Čičak., M., Daim, T.U., *Artificial Intelligence in Manufacturing: Bibliometric and Content Analysis*, 2020, doi: 10.1109/AI4G50087.2020.9311087.
9. Wan., J., Li., X., Dai., H.-N., Kusiak., A., Martinez-Garcia., M., Li., D., Artificial Intelligence-Driven Customized Manufacturing Factory: Key Technologies, Applications, and Challenges. *arXiv Artif. Intell.*, 2021, doi: 10.1109/JPROC.2020.3034808.
10. Kumar., M., Pandey., S., Upadhyay., K., Taunk., P., Tamrakar., R.K., Recent technical breakthroughs enable smart manufacturing: a review. *Al-Bahir J. Eng. Pure Sci.*, 2022, doi: 10.55810/2312-5721.1010.
11. Karim, H., Azeddine, K., Yassine, I., Ayoub, B., Hassan, O., Recent technological progress to empower Smart Manufacturing: Review and Potential Guidelines. *IEEE Access*, 2023, doi: 10.1109/access.2023.3246029.
12. Jiafu, W., Xiaomin, L., Hong-Ning, D., Kusiak, A., Martinez-Garcia, M., Li, D., Artificial-Intelligence-Driven Customized Manufacturing Factory: Key Technologies, Applications, and Challenges, 2021, doi: 10.1109/JPROC.2020.3034808.
13. Asunción, F. and Morales, Implementation of Artificial Intelligence (AI) in Smart Manufacturing: A Status Review, 2022, doi: 10.1007/978-3-031-22915-2_7.
14. Feiyu, J., Ali, J., Yongsheng, M., Rafiq, A., An Intelligent Manufacturing Approach Based on a Novel Deep Learning Method for Automatic Machine and Working Status Recognition. *Appl. Sci.*, 2022, doi: 10.3390/app12115697.
15. Vahid, M. and Saeid, M., Artificial Intelligence in the Production Process, 2019, doi: 10.1016/B978-0-12-815258-4.00002-0.
16. Singh, R., Tyagi, A.K., Arumugam, S.K., Imagining the Sustainable Future With Industry 6.0: A Smarter Pathway for Modern Society and Manufacturing Industries, in: *Machine Learning Algorithms Using Scikit and TensorFlow Environments*, IGI Global, 2024, doi: 10.4018/978-1-6684-8531-6.ch016.
17. Tyagi, A.K. and Richa, Digital Twin Technology: Opportunities and Challenges for Smart Era's Applications, in: *Proceedings of the 2023 Fifteenth International Conference on Contemporary Computing (IC3-2023). Association for Computing Machinery*, New York, NY, USA, pp. 328–336, 2023, https://doi.org/10.1145/3607947.3608015.
18. Tyagi, A.K. and Tiwari, S., The Future of Artificial Intelligence in Blockchain Applications, in: *Machine Learning Algorithms Using Scikit and TensorFlow Environments*, IGI Global, 2024, doi: 10.4018/978-1-6684-8531-6.ch018.
19. Tyagi, A.K., Richa, Smart Manufacturing Using Internet of Things, Artificial Intelligence, and Digital Twin Technology, in: *Global Perspectives on Robotics and Autonomous Systems: Development and Applications*, 2023.

20. Tyagi, A.K., Mishra, A.K., Vedavathi, N., Kakulapati, V., Sajidha, S.A., Chapter 21 Futuristic Technologies for Smart, in: *Manufacturing, Research Statement and Vision for the Future*, Tyagi, A.K. (Ed.), 2023, https://doi.org/10.1002/9781394213948.ch21.
21. Nair, M.M., Tyagi, A.K., Press, C.R.C., 6G: Technology, Advancement, Barriers, and the Future, in: *6G-Enabled IoT and AI for Smart Healthcare*, 2023.

5

Use Cases of Digital Twin in Smart Manufacturing

Vijayakumar Ponnusamy[1]*, Dilliraj Ekambaram[1] and Nemanja Zdravkovic[2]

[1]Department of Electronics and Communication Engineering, SRM Institute of Science and Technology, Kattankulathur Campus, Changalpattu District, Tamilnadu, India
[2]Faculty of Information Technology, Metropolitan University, Belgrade, Serbia

Abstract
The Digital Twin (DT) is a revolutionary advent in manufacturing industries to achieve Industry 4.0 by interacting with real-world types of machinery that can be controlled in an immersive environment. The idea is appealing, but many people, particularly those working for small and medium-sized enterprises, still need to figure out how to put it into practice in actual manufacturing processes. This chapter explores the several digital twin applications in smart manufacturing that significantly influence a complete solution for improving effectiveness, efficiency, and decision-making processes on product design and prototyping to predictive maintenance, quality control, and more concerning the standard framework defined by the International Organization for Standardization (ISO)/Draft International Standard (DIS) 23247. Here, it also summarizes the DT vision for production processes and its guiding principles, concepts, and technical solutions, a concise overview of the current research on DT concepts, practical implemented operations in industries, and potential avenues for further enhancement in the DT.

Keywords: Smart manufacturing, prototyping, product design, predictive maintenance, industry IoT (IIoT), quality assurance

5.1 Introduction

The most recent innovations in digital technology have made possible a shift in production methodology. To oversee, enhance, and manage operations, digital systems can generate an electronic replica of the real world and make decisions independently of any entity. According to this model, every production resource used in manufacturing must have an electronic copy, or "Digital Twin," created [1]. However, many people, particularly in smaller and medium-sized businesses [2], are still unsure about the idea and how to apply it to production systems. Digital Twins, in conjunction with smart algorithms [3], allow organizations to accomplish data-driven operation optimization and monitoring [4], and create new business models [5]. At the same time, smart manufacturing services have the potential to optimize all production-related business processes and operational procedures,

**Corresponding author*: vijayakp@srmist.edu.in

Amit Kumar Tyagi, Shrikant Tiwari, Senthil Kumar Arumugam and Avinash Kumar Sharma (eds.) Artificial Intelligence-Enabled Digital Twin for Smart Manufacturing, (85–102) © 2024 Scrivener Publishing LLC

leading to an unprecedented increase in productivity [6]. Product design, production, use, maintenance, repair, and optimization (MRO) would all undergo profound transformations if digital twins and smart manufacturing services were combined [7].

A federated learning system that can increase learning performance under resource restrictions by adaptively adjusting the aggregation frequency [8]. In DT, physical systems, objects, or assets are digitally reflected, replicated, or identified, process optimization and tracking, industrial IoT, prototyping, industrial health monitoring, bio-engineering, production environment, and so on [9–16]. Figure 5.1 shows the concept of digital twin development for production.

Due to the ever-increasing capabilities of information and communication technology and the pervasive Internet, recent developments such as Industry 4.0, cyber-physical production systems (CPPS), and the Internet of Things (IoT) have centered on automating and exchanging data in the manufacturing sector. A digital twin, or virtual copy of a physical one, is an important component of the advancements that CPPSs claim to bring about [17–21]. The most recent industrial data, including asset status and pertinent historical data, is represented by many DTs that are in operation. Deployed DTs can work together smartly to anticipate dispersed manufacturing system risks by analyzing DT-driven operational data. Specifically, the DTs work together to analyze previous operating data based on DTs, and then forecast future dangers in real-time using predictive data analytics [22]. New standards for digital twins are opening the door to interoperability by connecting separate systems.

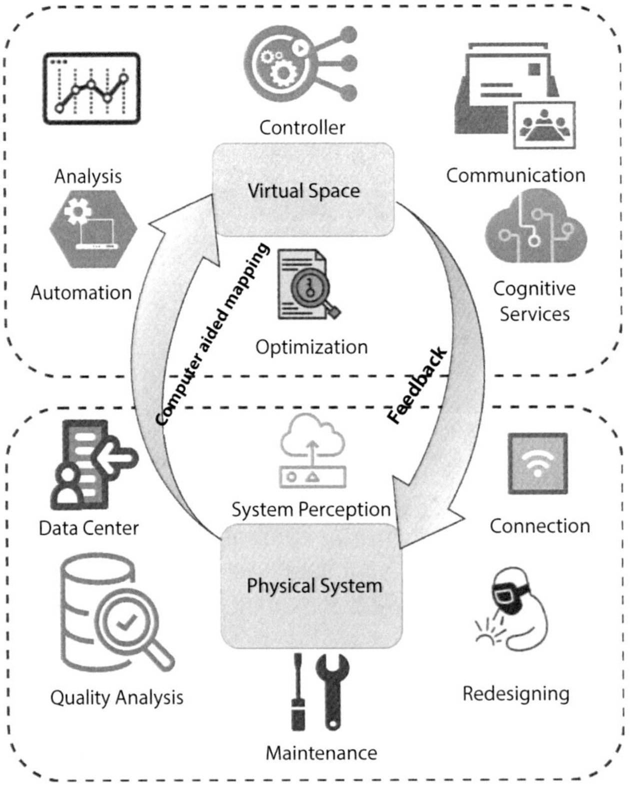

Figure 5.1 A production-oriented digital twin framework.

A "universal translator" capable of interpreting data from the physical and virtual realms and fusing them into a single cyber-physical reality is the end goal [23]. Systematic and thorough integration of domain-specific knowledge is even more critical for the grounding of DT and AI in industrial sectors [24].

Consumers around the world are increasingly concerned about the growing expenses and negative effects of energy consumption on the environment. There is an immediate need to reduce energy prices for industries that use a lot of energy, such as manufacturing, because of the increased impact on these sectors [25]. Researchers are looking into the Digital Twin, a new development in industry 4.0 technologies that provides a smart tool and technology for analyzing and optimizing energy consumption, as a means to lower energy expenses on the shop floor [26, 27]. Smart manufacturing systems are slowly incorporating digital twin technology. Having said that, studies in this field are still lacking [28]. IIoT necessitates smart connectivity, processing of data in real-time, collaborative monitoring, and automated information processing to enhance product dependability and quality. Despite its growing popularity, heterogeneous IIoT frequently necessitates multi-party collaborative information processing. A major obstacle in information security is an efficiency, security of all parties utilized in interactions with other partners [29, 30]. Product development leaders expect digital twins to speed up procedures, improve outcomes, and decrease costs. Quick investments are in the works because the digital-twin technologies industry is anticipated to experience a compound annual growth rate (CAGR) of 60% between 2019 and 2027, reaching $73.5 billion [31].

5.1.1 Smart Manufacturing Streams Utilizing Digital Twin Technology

Digital twins simplify operations, decrease downtime, and increase overall efficiency in smart manufacturing across the board, from design and simulation to operation and maintenance. Figure 5.2. Represents the various smart manufacturing streams utilizing DT technology.

A DT incorporates dynamic data to build an accurate model that evolves, going beyond basic static representations. The scope of this digital representation can go well beyond physical things; it might encompass whole production lines, product lifetimes, or even intricate systems such as smart cities. Studies with a strong thematic match delve deeper into the real world in DT, such as 3D printing [36], airplanes [37], or point machines [38]. In contrast, the virtual representation provides useful insights on DT in smart manufacturing [39–42]. The research addresses DT descriptions and definitions [32–34] and centers on real-time data presentation [43], prototypes [44], or replication features [45].

A focus area is human-robot collaboration (HRC), and study fields include things like smart wearable devices [46, 47], robot motion predictability [48], and automobile assembly lines [49]. Sustainable development [50], hydraulic support virtual monitoring [51], and WEEE recovery [52] are some of the other areas where DT has been addressed. Improving efficiency and productivity is just the beginning; this comprehensive strategy also paves the way for industrial ecosystems that are more robust, sustainable, and adaptable. Matchmaking and relationship extraction within a social manufacturing setting, and the paradigm shift from centralization to socialization inside manufacturing. In this area, specialists examine ways to improve production process efficiency through the transformation of unstructured data. Discrete manufacturing systems (DMS) [53], parts machining outsourcing (PMO) [54], service planning, and collaboration are the main areas of concentration.

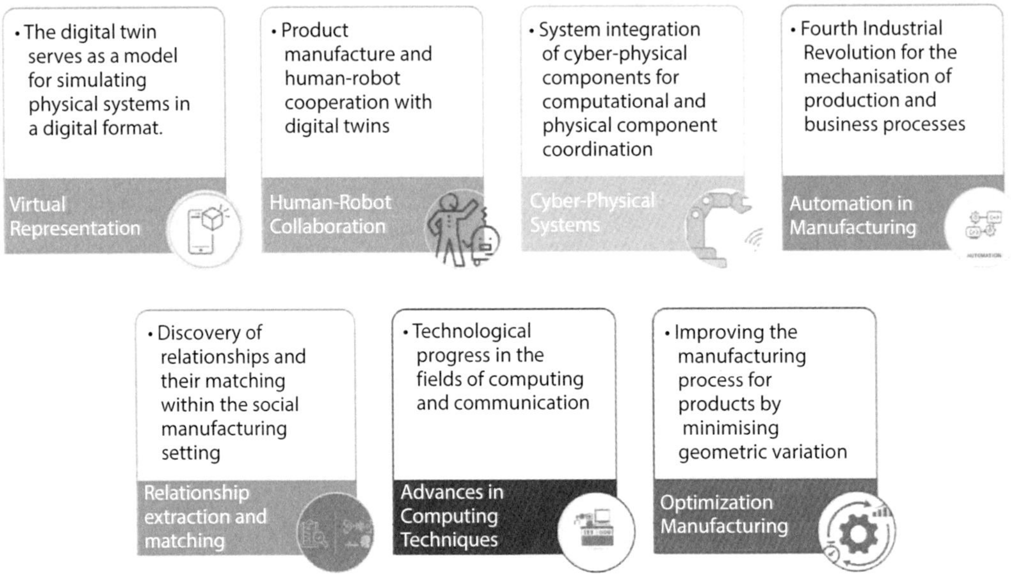

Figure 5.2 Smart manufacturing streams that utilize DT.

Recent developments in computer and communication technologies, such as the IoT [8], CPS [29, 33], Industry 4.0 [34, 52], and machine-to-machine (M2M) communication protocols [9], are reviewed. The possibilities of sustainable manufacturing, techniques, and uses of CPS, edge computing, picture categorization using neural networks, and the Internet of Things are some of the broad topics covered in the studies. Reducing geometric variance in the manufacturing process is an important step towards better products, more efficient production, and happier customers. The number of very good reasons why reducing these differences is an important objective in production. First, the uniformity of the product is guaranteed by lowering geometric variation. Products must be consistent to meet design standards and functional needs, which in turn leads to improved performance and dependability. The automotive and aerospace industries, where accuracy is of the utmost importance, place a premium on this consistency. Second, reducing waste is heavily dependent on minimizing geometric variance. The possibility of making components that are flawed or not up to standard can be reduced by increasing manufacturing precision and obtaining tighter tolerances. A more sustainable and eco-friendly manufacturing method is the result of less material wastage and reduced production costs.

5.1.2 Technological Advancement That Makes Digital Twins Ideal for Smart Manufacturing System Design

To successfully use digital twins in smart manufacturing, the following technologies are essential:

Connectivity in Industry: With a physical process's digital twin, it's possible to run simulations of the process in a controlled setting [55]. The IIoT is a network of interconnected sensors and actuators that allows for better modeling and simulations by providing data for digital twins and enabling real-time physical system awareness [56]. Figure 5.3 shows the facilitating technologies for digital-twin based smart manufacturing.

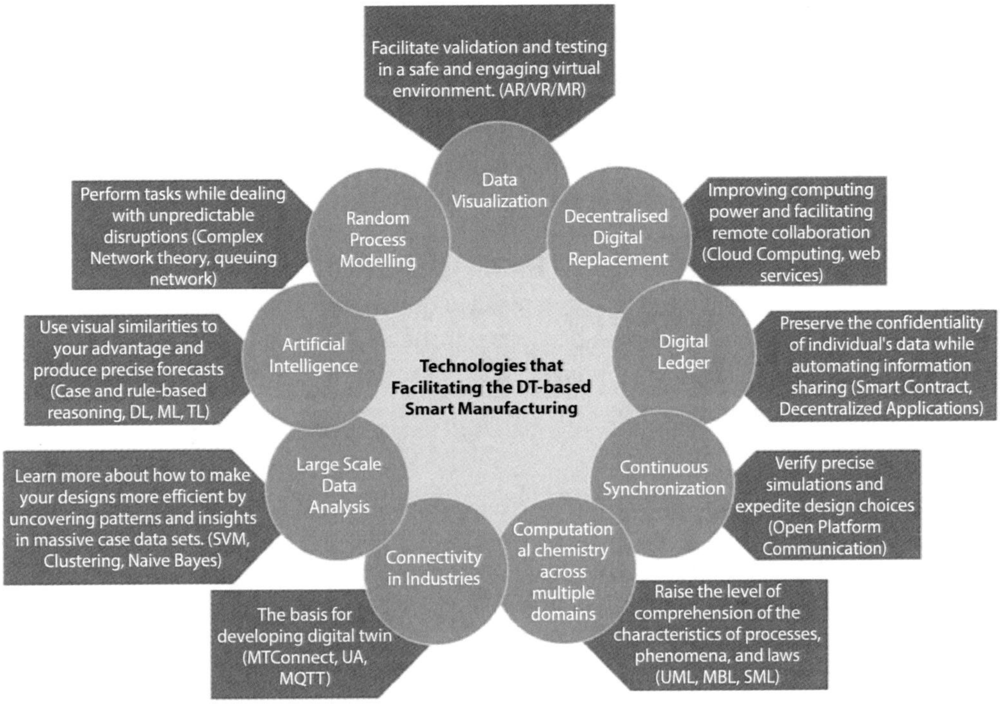

Figure 5.3 Facilitating technologies for digital twin-based smart manufacturing.

Computational Process on Multiple Domains: Model-Based Systems Engineering (MBSE) serves as a central location for models from different fields of study. As an example, one software developer created a tool modeling language called SysML (System Modelling Language) to complement preexisting professional analysis programs like CAD and Finite Element Analysis [7]. Thus, there is a need for a fresh wave of digital design methodologies that include information from other disciplines through simulation and multi-domain modeling.

Continuous Synchronization: It is well-built and given the relevant environmental and contextual data, a model can faithfully represent the properties and parameters of the real world. One key distinction between digital twins and more traditional forms of simulation is the ability to synchronize cyber and physical systems in real time. One important measure of a digital twin system's synchronization capacity is network latency.

Data Visualization: A human-computer interaction process, SMSD is complex and error-prone. It is rarely possible for designers to have a fully immersive experience with the final SMS using traditional 3D simulation methods. There have been suggestions for intelligent systems incorporating VR, AR, or MR to assist engineers with design and operation.

Artificial Intelligence: Specifically, SMSD aims to meet each client's needs in terms of manufacturing capacity, product quality, production efficiency, site area, and other related measures. We incrementally transfer the product domain's customization needs to the SMS domain.

Digital Ledger: Current digital twin systems have limited data auditability and traceability due to their centralized nature. Under cyber-attack, data security in this networked runtime becomes a major issue. Secure collaboration SMSD has a new solution in blockchain technology. By securing network transmission and access using cryptographic technology and distributed consensus procedures, blockchains increase cyber-credit among different participants.

Random Process Modelling:
For a wide range of industrial issues, both academic institutions and businesses have turned to discrete event simulation methods. Due to the dramatic shifts occurring in the world's manufacturing sectors in the late 1990s, discrete event simulation had hit a slump. Another sector that underwent concentration was the simulation software industry. New difficulties, possibilities, and threats have emerged as a result of the advances in discrete event simulation [57]. The global industrial industries have been through drastic transformations, and with them, discrete event simulation technology. Rather than relying on a single model, manufacturers are increasingly using supply-chain-wide distributed modeling and simulation.

Large Scale Data Analysis:
The term "smart manufacturing" (SM) refers to a way of improving production processes through the integration of systems, the connection of physical and cyber capabilities, and the utilization of information, particularly the growth of big data. Due to the uneven adoption of SM throughout industries, sectors like biochemistry and biology can learn from other sectors' problems and potential solutions [58]. To better comprehend their capabilities, constraints, and any gaps in technology, analytics approaches can be characterized along certain parameters.

Decentralized Digital Replacement:
Cooperatively sharing design tasks across several similar units located in remote places is made possible by cloud computing and web service platforms. Another reason to introduce cloud computing is the need for HPC for SMSD simulation work. This might lead to flexible expansion, dynamic allocation of computing resources, and low-cost sharing of those resources [59]. With the help of cloud computing and web services, designers may distribute and work remotely on SMSD projects.

This chapter is structured as follows: an introduction to digital twins (DTs), a review of various DT designs based on artificial intelligence (AI), an examination of the integration of DTs across different domains in the manufacturing industry, a discussion of the technologies that enable smart manufacturing through DTs, and a global DT development scenario. Section 5.2 provides a brief overview of the literature on DT for smart manufacturing. Section 5.3 focuses on the different ways DT is used in smart manufacturing. Section 5.4 explains how DT can be used to create a sustainable smart manufacturing information management system. Section 5.5 lists the different ways DTs are used for energy management and smart manufacturing. Section 5.6 discusses the difficulties and potential future directions of DT for smart manufacturing. Lastly, we end the chapter by outlining the benefits and drawbacks of DTs for smart manufacturing.

5.2 Review of Relevant Literature

Researchers are aggressively tackling obstacles to lay the groundwork for more general adoption, while also identifying many prospects and accomplishments. Continued work on standards, human-machine collaboration, and real-time visualization highlights the direction of future research in this game-changing field.

Židek *et al.*, 2020 [4] address the development of an electronic replica of an automated manufacturing line used for quality control in an experimental assembly system. The focal point is a 3D-printed Bowden holder assembly with a stepper motor, plastic parts, and a few fastening pieces. Dimension, part presence, and error checking were all part of the inspection system's industrial vision system, which was used both before and after the assembly activity. A PLC system received data from the assembly system, which was then synchronized with a digital twin model and a cloud platform (CP) through an OPC server.

A federated learning framework is suggested by Q. Song *et al.* (2021) [8] to enhance learning performance when resources are limited by adaptively adjusting the aggregation frequency. The suggested framework achieves better learning accuracy, convergence, and energy savings than the benchmark, according to the numerical results. Vatankhah Barenji *et al.* (2020) [10] developed a method for robotic cellular navigation that relies on digital twins and takes into account agents that make decisions about motion planning in real time. On top of that, a generic method for creating a virtual twin was suggested based on the previously mentioned strategy.

The suggested DT establishment framework by Zheng and Sivabalan, 2020 [11] can be used as a blueprint for future DT implementations, taking into account the essentials. In comparison to traditional DAQs, it offers a more attractive paradigm for data gathering and use. The data-driven methodology, as suggested by Friederich *et al.*, 2022 [12], makes use of recent developments in process mining and machine learning, and it also provides a foundation for ongoing model validation and refinement. The framework's objective is to make the extraction of related simulation models as easy as possible while completely defining or eliminating the necessity for expert knowledge. Two case studies were conducted in Southern and Northern China by Ma *et al.*, 2022 [16] to prove that the method worked. The results showed that after applying the strategy, Companies A and B were able to save energy and money. A 3% reduction in both Company A's energy cost of production and unit energy usage was achieved through the implementation of an energy management system. Company B also saw a considerable drop in environmental protection expenditures. Lastly, this article examines and discusses the suggested strategy's efficacy as well as some management insights for energy-intensive industries (EIIs) in China.

As an example of a digital twin, the design proposed by Redelinghuys, Basson, and Kruger, 2019 [17] uses an Internet of Things (IoT) gateway layer, databases in the cloud, a layer housing simulations and emulations, and local data are all components of the design. Present installations can be minimally disrupted throughout the architecture's implementation. Implementing the architecture for a tiny but typical physical production system component allowed us to assess it. Manufacturers, especially SMEs, can benefit from Jwo, Lee, and Lin's (2022) [20] data-centric approach to process modeling and Digital Twin simplification. To conduct simulations and predictions, the CPF architecture provides a versatile implementation. Nonetheless, there are a few outstanding issues that will be further explored in subsequent studies. To evaluate the proposed framework's resilience and efficacy, additional testing and implementation on various use cases are required. Secondly, it is important to think about the security risks associated with data storage and exchange. The foundational elements for resolving this issue are data formats and communication methods. Third, it is critical to investigate the DTS's sophisticated machine-learning approaches because our architecture's performance is highly dependent on the capability of machine-learning models. Following an analysis of the literature cited above, we will now delve into the numerous applications of digital twins in smart manufacturing.

5.3 Various Use Cases of Digital Twin in Smart Manufacturing

Digital twins are now widely used in the engineering and manufacturing industries for creating precise 3D models of physical items and running simulations of business processes. Related publications explore the relevance of digital twins in operations and supply chain management, specifically in terms of operations traceability, transport maintenance, remote support, asset visualization, and design customization [18]. Product performance can be studied and altered with the help of digital twin applications in the manufacture of parts. It is possible to mimic the whole production cycle, from sourcing raw materials to shipping completed goods, using a digital twin. The system's behavior can be predicted under various settings by simulating different scenarios [32]. Because of this, makers can find problems before they happen and improve the system's performance. In this section, we deliberated some of the use cases of DTs in smart manufacturing.

5.3.1 Data-Driven Smart Factory

As a result, the CPF design presented by Jwo, Lee, and Lin (2022) [20] offers an efficient approach and transparent communication, enabling the adaptation of appropriate technology, the adaptive integration of diverse production resources, and the regular execution of interaction operations. Figure 5.4 depicts the general structure for a data-driven smart factory.

Creation, queries, updates, and deletions are all within the scope of supported operations. The configuration of the DTSs is the responsibility of the CPF Composer component.

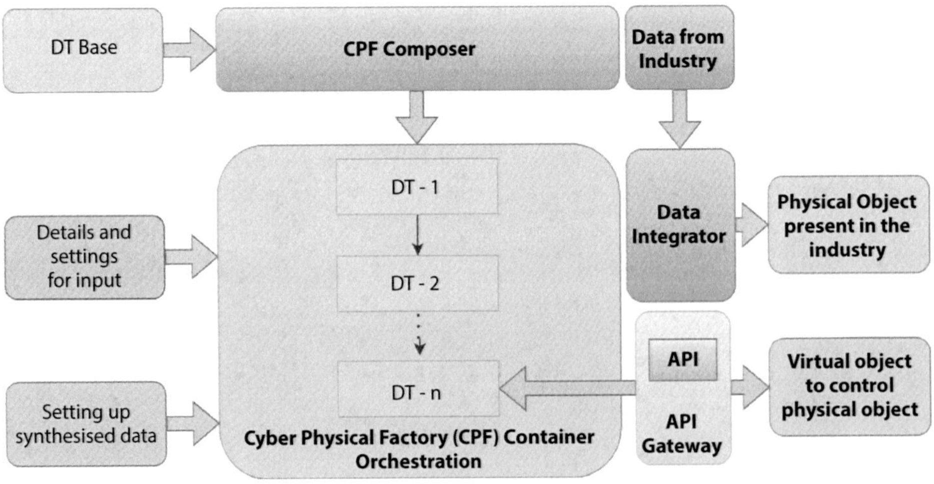

Figure 5.4 General structure of data-driven smart factory.

After choosing a DTS, users can configure its five parts to make them work together. The input data component of DTS explains how data sources, such as real-time, historical, and synthesized data, are assigned. Data and information are constantly being pushed to CPF Orchestration from the physical units, which serves as the foundation of CPF. With the benefits of container technology, such as consistent deployment and ease of use, each DTS configuration may achieve high processing throughput and flexibility [20].

5.3.2 Cyber-Physical System (CPS) Integration

The digitization of production systems and the integration of various systems for collaborative tasks are both facilitated by cyber-physical systems or CPS. A DT is a digital replica of a physical thing or system that includes all of its attributes and its connection to its surroundings. The flow process of CPS is represented in the Figure 5.5.

In the physical layer, many machines may use distinct communication protocols and interfaces for data transport. The data extraction and consolidation layer transmits data packets received from physical systems to the cyber layer after processing and making them machine-readable. Consolidating the acquired data into a common machine-readable format is necessary since processing data from each physical system item separately is exceedingly wasteful. This module not only houses the processing device—a microcontroller or computer—but also provides the framework for the physical layer's numerous communication ports.

5.3.3 Human-Robot Collaboration

An HRC system's virtual or digital twin should be created as early as feasible throughout the ideation phase to enhance the design process. It all starts with a basic animation and then evolves into an event-based simulation for system-level performance analysis. The digital twin framework for the human-robot collaboration is shown in the Figure 5.6.

As the physical system undergoes evolutionary changes, DT must be regularly updated as well. The ideation process has resulted in a simulation that is more refined and more integrated with the real world. A whole run-time cycle is executed swiftly by the digital space during operation for each production changeover, allowing it to make optimal judgments for production control. Because the DT framework is flexible enough to accommodate changes in design, development, and operation, the DT's behavior can change at any stage of the product lifecycle.

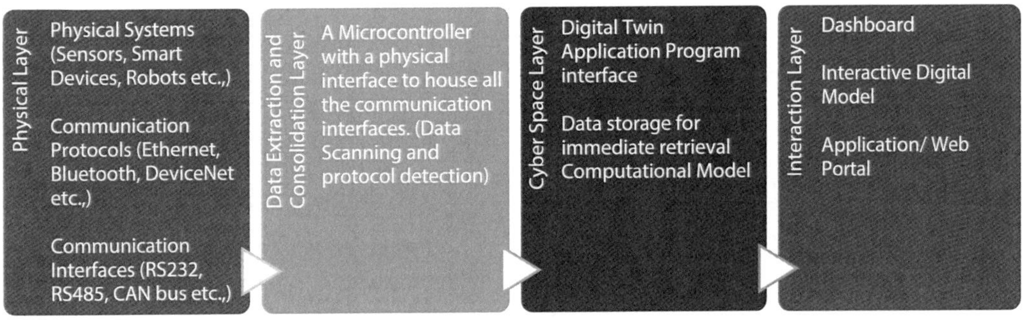

Figure 5.5 Flow process of cyber physical systems (CPS).

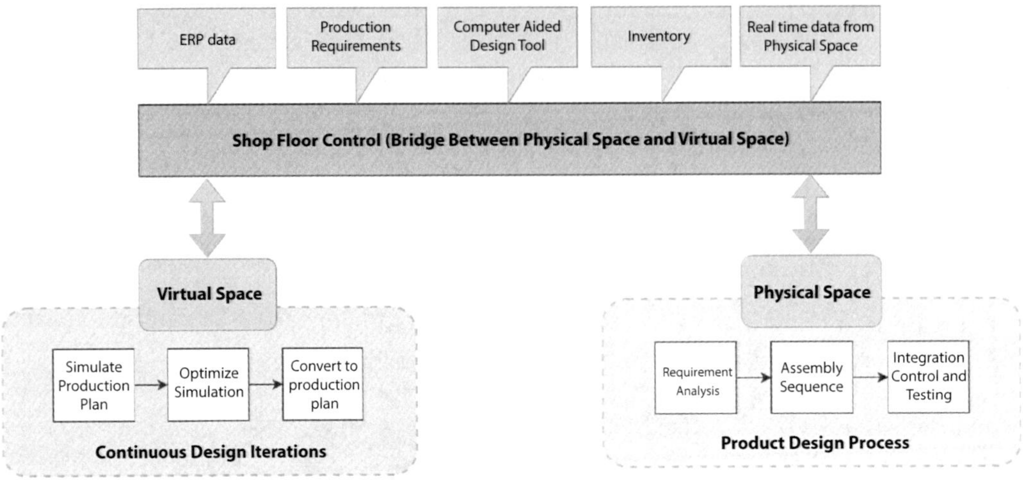

Figure 5.6 An architecture for digital twins in human–robot interaction.

To improve design, an HRC system's virtual or digital twin should be created early in ideation [62]. Start with a simple animation, then add an event-based simulation for system performance analysis. To keep up with physical system evolution, DT must be upgraded. The simulation built during ideation evolves in information and physical system connection. At every production changeover, the digital space swiftly completes the run-time cycle and makes appropriate production control decisions. Because the DT framework supports design, development, and operation, its behavior might vary during the lifecycle.

5.3.4 Adaptive Federated Learning for Industrial IoT

With federated learning, a distributed machine learning system may be built with the help of multiple parties while maintaining the privacy of their data. Federated learning can now guarantee privacy while lowering communication costs because to these features. Consequently, federated learning is an effective method for dealing with problems like data islands, costly or unreliable connections, privacy concerns, and regulatory mandates. The general architecture of adaptive federated learning is shown in the Figure 5.7.

The Industrial Internet of Things (IoT) is enabled by data discovery technologies, including DTs of industrial devices, curator servers, and different kinds of industrial devices [8]. Through wireless communication networks, devices with limited processing and communication capacity can establish communication with servers. Distributed sensors update the physical condition of heterogeneous devices in real-time by mapping them into digital space.

In Industrial IoT federated learning, parameters sent by clients with a good reputation should be aggregated more heavily to protect against harmful assaults and boost learning performance. This model takes into account all potential influences on learning, including DT deviation, learning effect, and dataset quality, in contrast to the conventional reputation model which focuses just on security issues.

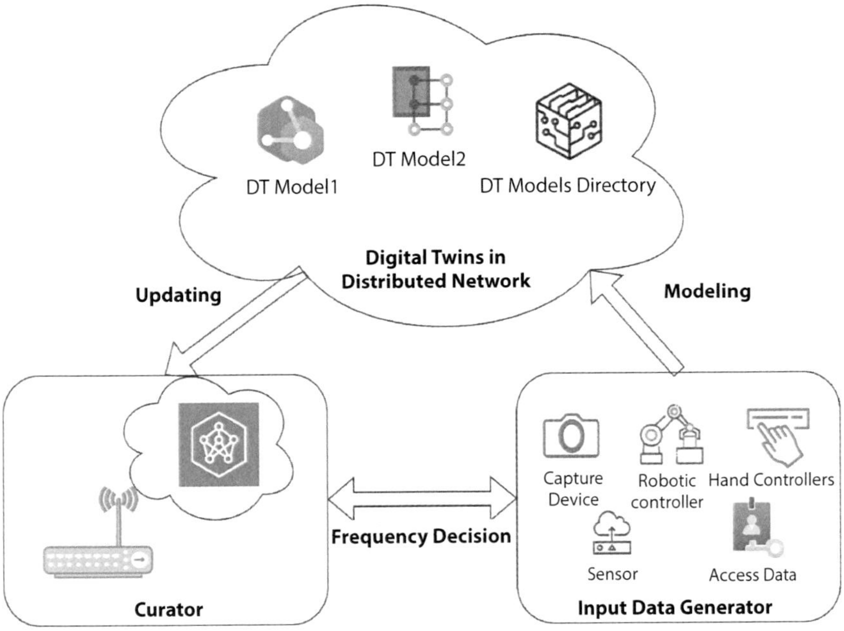

Figure 5.7 General architecture of adaptive federated learning technique with DTs.

5.4 Information Management System-Based Digital Twins and Big Data for Sustainable Smart Manufacturing

The smarter, more efficient, and more environmentally friendly production processes made possible by Internet of Things (IoT) technologies are attracting more and more attention from both businesses and universities. Sustainable smart manufacturing is a major Internet of Things (IoT) use case that optimizes product lifecycle management in terms of cost, productivity, flexibility, quality, and sustainability. Smart manufacturing and sustainable manufacturing have been front-and-center initiatives for many businesses lately. The employment of a "digital twin" to accomplish smart manufacturing is common in the Industry 4.0 setting, yet this method frequently disregards sustainability. Digital twins and big data, when combined, offer essential tools for data collecting in energy-intensive manufacturing settings, for mining and prediction in uncertain contexts, and real-time control in complicated work environments.

The information management systems (IMSs), together with the virtual and real workplaces, make up the digital twin technology, as shown in Figure 5.8. Power Control System (PCS), Production Management Software (PMS), Planning for Materials Needed (PMN), Business Resource Planning (BRP), Data Management for Products (DMP), and Product Life-cycle Management (PLM) are all products of the IMSs, which are responsible for business resource planning. Information management systems (IMSs) regulate and mimic the virtual workshop by collecting data in real-time from the physical workshop [60].

Upon receiving an assignment, the physical workshop immediately begins producing the associated products. Data on people, machines, goods, and energy usage is subsequently gathered and stored by the IMSs. To mimic the physical workshop's product manufacturing process and give production feedback, the virtual workshop has been developed. In the

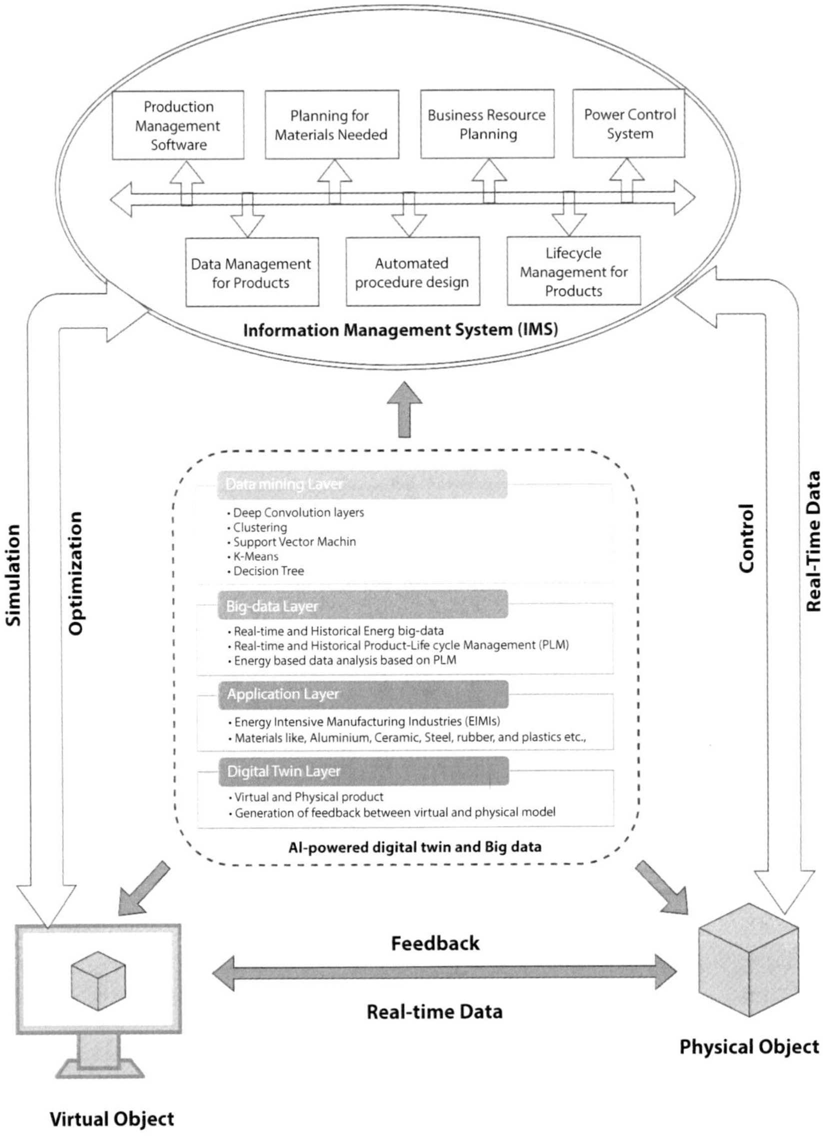

Figure 5.8 EIMI operational mechanism driven by digital twins and big data.

future, smart factories will be able to replicate the workshop's capacity for self-organization, self-adaptation, and self-study [61].

5.5 Challenges and Future Avenues

While there have been some early adopters who have shown how digital twins can be used in manufacturing, there are still a lot of obstacles to overcome. For one, there is a lack of reference models for digital twins, which means that people only have a basic understanding of the concept. Secondly, most research on digital twins has focused on product operation and maintenance, so people only have a superficial understanding of the research questions and challenges. Lastly, most of the applications of digital twins are still in their early stages, so there is a lack of comprehensive guidelines for their implementation. Based on the smart manufacturing development trend, this paper's research seeks to critically analyse the aforementioned elements in order to ensure the sustainable development of digital twin-driven smart manufacturing. This section deliberates the various current research challenges and future avenues where, researchers can indulge themselves to explore the solutions for these.

Challenges:

- Data access for production resources is a primary concern while implementing the Digital Twin vision. The real and virtual worlds need to be kept in sync and consistent with each other [1].
- In most cases, one of two approaches will be used: either constantly monitoring the asset for changes or taking periodic "pictures" of its current state [2]. Although there have been numerous suggested methods and approaches in the literature—for instance, making use of EtherNet/IP's multicast functionality—none of them fully automate the process [33].
- It is necessary to standardise a broad variety of technologies that are used for data collecting, insight development, and decision-making. But widespread use of digital twins is sluggish since standards development is typically sluggish as well [25].
- Mindsphere, Thingworxs Kepware, Vuforia Spatial Toolbox, and others are among the platforms currently available, each with its own set of features and degrees of integration. A solid foundation for integrating data from many domains may be found in these solutions, which also embrace the IoT concept. However, integrating diverse and varied legacy systems into a single, cohesive platform remains the primary obstacle [34].
- It is important to evaluate the cost-benefit analysis to ensure their economic and financial interests are protected. This isn't a "technical" issue per se, but it bears mentioning: as is the case with any technology, it can't be widely used in production and manufacturing settings unless it significantly reduces or eliminates operating expenses (OpEx) or improves capital expenditures (CapEx) for the end process [35].
- Building a digital twin probably involves assembling a mixture of technology and solutions from several suppliers. Given the present state of technology maturity, which affects the time and money needed to develop new technologies, development efforts should prioritise those technologies that provide the most substantial gains in efficiency and effectiveness.

Future Avenues:

- Connected physical objects with built-in sensors, software, and network connectivity form what is known as the Internet of Things (IoT). The Internet of

Things (IoT) and digital twin technology can work together to let us monitor, analyze, and control production processes in real time [8, 30].

- The manufacturing process can be better understood with the help of multi-scale digital twins, which incorporate many levels of information ranging from the microscopic to the macroscopic. This method can improve process optimization by assisting in the detection and resolution of issues of various sizes.
- Data on production procedures and machinery performance can be found in abundance with the use of digital twin technology. Machine learning, artificial intelligence, and digital twin technology all work together to help manufacturers better understand their processes and make decisions.
- The importance of protecting these digital twins from cyber assaults is growing as their use in manufacturing continues to rise. Cybersecurity protocols and technology that safeguard digital twins from malicious assaults can be the subject of future research.
- Manufacturing processes can have a substantial influence on the environment, and digital twin technology can help find ways to reduce that impact. Many sectors are making use of digital twin technology. Transportation, medicine, and aerospace are just a few examples. The development of digital twin systems that are both scalable and adaptable to various production settings should be the goal of future studies.
- It will be critical to think about the function of human workers in fully automated production settings. Ways to enhance collaboration and decision-making through the integration of digital twin technologies with human-machine interfaces could be the subject of future research.

5.6 Conclusion

In conclusion, the Digital Twin's (DT) arrival heralds a paradigm shift in production and a groundbreaking step towards Industry 4.0. The idea is appealing, but really putting DT into practice in manufacturing processes is difficult, especially for workers working for small and medium-sized businesses. The chapter explores several uses of digital twins in smart manufacturing, resulting in a comprehensive solution that influences product life-cycle decision-making, efficacy, and efficiency. Following the guidelines laid out by the International Organisation for Standardisation (ISO)/Draft International Standard (DIS) 23247, this chapter delves into a standardised framework that covers all aspects of product development, from initial concept to final quality control and predictive maintenance. In it, the DT vision for manufacturing processes is laid out in detail, with guiding principles, important concepts, and technical solutions for integration laid out in detail.

Further, the chapter is a collection of recent studies on DT ideas, with an emphasis on real-world applications in different sectors. These practical examples demonstrate the usefulness and adaptability of DT in our everyday lives. The chapter provides a road map for future research and development by identifying possible ways to improve DT in light of the changing manufacturing scenario. Overall, this chapter summarises where digital twins are at in smart manufacturing and provides a road map for practitioners to follow as they deal with the challenges of implementation. It also encourages further study and development to ensure that digital twins continue to advance in the business world.

References

1. Lattanzi, L., Raffaeli, R., Peruzzini, M., Pellicciari, M., Digital twin for smart manufacturing: a review of concepts towards a practical industrial implementation. *Int. J. Comput. Integr. Manuf.*, 1–31, 2021, doi: https://doi.org/10.1080/0951192x.2021.1911003.
2. Shao, G. and Helu, M., Framework for a digital twin in manufacturing: Scope and requirements. *Manuf. Lett.*, 24, 105–107, 2020, doi: https://doi.org/10.1016/j.mfglet.2020.04.004.
3. Lu, Y., Liu, C., Wang, K.I.-K., Huang, H., Xu, X., Digital Twin-driven smart manufacturing: Connotation, reference model, applications and research issues. *Rob. Comput. Integr. Manuf.*, 61, 101837, 2020, doi: https://doi.org/10.1016/j.rcim.2019.101837.
4. Židek, K., Piteľ, J., Adámek, M., Lazorík, P., Hošovský, A., Digital Twin of Experimental Smart Manufacturing Assembly System for Industry 4.0 Concept. *Sustainability*, 12, 9, p.3658, 2020, doi: https://doi.org/10.3390/su12093658.
5. Lu, Y. and Xu, X., Cloud-based manufacturing equipment and big data analytics to enable on-demand manufacturing services, Robot. *Comput. Integr. Manuf*, 57, 92–102, 2019, https://doi.org/10.1016/j.rcim.2018.11.006.
6. Qi, Q., Tao, F., Zuo, Y., Zhao, D., Digital Twin Service towards Smart Manufacturing. *Proc. CIRP*, 72, 237–242, 2018, doi: https://doi.org/10.1016/j.procir.2018.03.103.
7. Leng, J., Wang, D., Shen, W., Li, X., Liu, Q., Chen, X., Digital twins-based smart manufacturing system design in Industry 4.0: A review. *J. Manuf. Syst.*, 60, 119–137, 2021, doi: https://doi.org/10.1016/j.jmsy.2021.05.011.
8. Song, Q., Lei, S., Sun, W., Zhang, Y., Adaptive Federated Learning for Digital Twin Driven Industrial Internet of Things. *2021 IEEE Wireless Communications and Networking Conference (WCNC)*, 2021, doi: https://doi.org/10.1109/wcnc49053.2021.9417370.
9. Ante, L., Digital twin technology for smart manufacturing and industry 4.0: A bibliometric analysis of the intellectual structure of the research discourse. *Manuf. Lett.*, 27, 96–102, 2021, doi: https://doi.org/10.1016/j.mfglet.2021.01.003.
10. Vatankhah Barenji, A., Liu, X., Guo, H., Li, Z., A digital twin-driven approach towards smart manufacturing: reduced energy consumption for a robotic cellular. *Int. J. Comput. Integr. Manuf.*, 1–16, 2020, doi: https://doi.org/10.1080/0951192x.2020.1775297.
11. Zheng, P. and Sivabalan, A.S., A generic tri-model-based approach for product-level digital twin development in a smart manufacturing environment. *Rob. Comput. Integr. Manuf.*, 64, 101958, 2020, doi: https://doi.org/10.1016/j.rcim.2020.101958.
12. Friederich, J., Francis, D.P., Lazarova-Molnar, S., Mohamed, N., A framework for data-driven digitial twins of smart manufacturing systems. *Comput. Industry*, 136, p.103586, 2022, doi: https://doi.org/10.1016/j.compind.2021.103586.
13. Son, Y.H., Kim, G.-Y., Kim, H.C., Jun, C., Noh, S.D., Past, present, and future research of digital twin for smart manufacturing. *J. Comput. Des. Eng.*, 9, 1, 1–23, 2021, doi: https://doi.org/10.1093/jcde/qwab067.
14. Wang, Y., Kang, X., Chen, Z., A Survey of Digital Twin Techniques in Smart Manufacturing and Management of Energy Applications. *Green Energy Intell. Transp.*, 100014, 2022, doi: https://doi.org/10.1016/j.geits.2022.100014.
15. Nguyen, H.X., Trestian, R., To, D., Tatipamula, M., Digital Twin for 5G and Beyond. *IEEE Commun. Mag.*, 59, 2, 10–15, 2021, doi: https://doi.org/10.1109/mcom.001.2000343.

16. Ma, S., Ding, W., Liu, Y., Ren, S., Yang, H., Digital twin and big data-driven sustainable smart manufacturing based on information management systems for energy-intensive industries. *Appl. Energy*, 326, 119986, 2022, doi: https://doi.org/10.1016/j.apenergy.2022.119986.
17. Redelinghuys, A.J.H., Basson, A.H., Kruger, K., A six-layer architecture for the digital twin: a manufacturing case study implementation. *J. Intell. Manuf.*, 31, 1383–1402, 2020, https://doi.org/10.1007/s10845-019-01516-6.
18. Attaran, M. and Celik, B.G., Digital Twin: Benefits, use cases, challenges, and opportunities. *Decis. Anal. J.*, 6, 100165, 2023, doi: https://doi.org/10.1016/j.dajour.2023.100165.
19. Onaji, I., Tiwari, D., Soulatiantork, P., Song, B., Tiwari, A., Digital twin in manufacturing: conceptual framework and case studies. *Int. J. Comput. Integr. Manuf.*, 1–28, 2022, doi: https://doi.org/10.1080/0951192x.2022.2027014.
20. Jwo, J.-S., Lee, C.-H., Lin, C.-S., Data Twin-Driven Cyber-Physical Factory for Smart Manufacturing. *Sensors*, 22, 8, 2821, 2022, doi: https://doi.org/10.3390/s22082821.
21. Novák, P., Vyskočil, J., Wally, B., The Digital Twin as a Core Component for Industry 4.0 Smart Production Planning. *IFAC-PapersOnLine*, 53, 2, 10803–10809, 2020, doi: https://doi.org/10.1016/j.ifacol.2020.12.2865.
22. Sahal, R., Alsamhi, S.H., Brown, K.N., O'Shea, D., McCarthy, C., Guizani, M., Blockchain-Empowered Digital Twins Collaboration: Smart Transportation Use Case. *Machines*, 9, 9, 193, 2021, doi: https://doi.org/10.3390/machines9090193.
23. O'Connell, E., O'Brien, W., Bhattacharya, M., Moore, D., Penica, M., Digital Twins: Enabling Interoperability in Smart Manufacturing Networks. *Telecom*, 4, 2, 265–278, 2023, doi: https://doi.org/10.3390/telecom4020016.
24. Huang, Z., Shen, Y., Li, J., Fey, M., Brecher, C., A Survey on AI-Driven Digital Twins in Industry 4.0: Smart Manufacturing and Advanced Robotics. *Sensors*, 21, 19, 6340, 2021, doi: https://doi.org/10.3390/s21196340.
25. VanDerHorn, E. and Mahadevan, S., Digital Twin: Generalization, characterization and implementation. *Decis. Support Syst.*, 145, 113524, 2021, doi: https://doi.org/10.1016/j.dss.2021.113524.
26. Billey, A. and Wuest, T., Energy digital Twins in smart manufacturing Systems: A literature review. *Manuf. Lett.*, 35, 1318–1325, 2023, doi: https://doi.org/10.1016/j.mfglet.2023.08.047.
27. Park, K.T., Son, Y.H., Ko, S.W., Noh, S.D., Digital Twin and Reinforcement Learning-Based Resilient Production Control for Micro Smart Factory. *Appl. Sci.*, 11, 7, 2977, 2021, doi: https://doi.org/10.3390/app11072977.
28. Feng, Q., Zhang, Y., Sun, B., Guo, X., Fan, D., Ren, Y., Song, Y., Wang, Z., Multi-level predictive maintenance of smart manufacturing systems driven by digital twin: A matheuristics approach. 68, 443–454, 2023, doi: https://doi.org/10.1016/j.jmsy.2023.05.004.
29. Qian, C., Liu, X., Ripley, C., Qian, M., Liang, F., Yu, W., Digital Twin—Cyber Replica of Physical Things: Architecture, Applications and Future Research Directions. *Future Internet*, 14, 2, 64, 2022, doi: https://doi.org/10.3390/fi14020064.
30. Chen, H., Jeremiah, S.R., Lee, C., Park, J.H., A Digital Twin-Based Heuristic Multi-Cooperation Scheduling Framework for Smart Manufacturing in IIoT Environment. *Appl. Sci.*, 13, 3, 1440, 2023, doi: https://doi.org/10.3390/app13031440.
31. Digital Twin market by enterprise, application (predictive maintenance, business optimization), industry (aerospace, automotive & transportation, healthcare, infrastructure, energy & utilities) and geography - Global forecast to 2027, in: *Research and Markets*, July 2022.
32. Mylonas, G., Kalogeras, A., Kalogeras, G., Anagnostopoulos, C., Alexakos, C., Muñoz, L., Digital Twins From Smart Manufacturing to Smart Cities: A Survey. *IEEE Access*, 9, 143222–143249, 2021, doi: 10.1109/ACCESS.2021.3120843.

33. Astarhi, T.B., Jung, T., Lindemann, B., Sahlab, N., Jazdi, N., Schloegl, W., Weyrich, M., An architecture of an Intelligent Digital Twin in a Cyber-Physical Production System. *at – Autom.*, 67, 9, 762–782, 2019. https://doi.org/10.1515/auto-2019-0039.
34. Qi, Q. and Tao, F., Digital Twin and Big Data Towards Smart Manufacturing and Industry 4.0: 360 Degree Comparison. *IEEE Access*, 6, 3585–3593, 2018, doi: 10.1109/ACCESS.2018.2793265.
35. Boehm, B. and Valerdi, R., The ROI of Systems Engineering: Some Quantitative Results. *2007 IEEE International Conference on Exploring Quantifiable IT Yields*, Amsterdam, Netherlands, pp. 79–86, 2007, doi: 10.1109/EQUITY.2007.14.
36. Debroy, T., Zhang, W., Turner, J., Babu, S.S., Building digital twins of 3D printing machines. *Scr. Mater.*, 135, 119–24, 2016, https://doi.org/10.1016/j.scriptamat.2016.12.005.
37. Seshadri, B.R. and Krishnamurthy, T., Structural Health Management of Damaged Aircraft Structures Using Digital Twin Concept, 2017, doi: https://doi.org/10.2514/6.2017-1675.
38. Rosen, R., Fischer, J., Boschert, S., Next Generation Digital Twin: an Ecosystem for Mechatronic Systems? *IFAC-PapersOnLine*, 52, 15, 265–270, 2019, doi: https://doi.org/10.1016/j.ifacol.2019.11.685.
39. Tao, F., Cheng, J., Qi, Q., Zhang, M., He, Zhang, Sui, F., Digital twin-driven product design, manufacturing and service with big data. *Int. J. Adv. Manuf. Technol.*, 94, 9-12, 3563–76, 2018, https://doi.org/10.1007/s00170-017-0233-1.
40. Yildiz, E., Møller, C., Bilberg, A., Virtual Factory: Digital Twin Based Integrated Factory Simulations. *Proc. CIRP*, 93, 216–221, 2020, doi: https://doi.org/10.1016/j.procir.2020.04.043.
41. Glaessgen, E.H. and Stargel, D.S., The digital twin paradigm for future NASA and U.S. Air force vehicles. *Collect. Tech. Pap. - AIAA/ASME/ASCE/AHS/ASC Struct. Struct. Dyn. Mater. Conf. 2012*, https://doi.org/10.2514/6.2012-1818.
42. Rosen, R., von Wichert, G., Lo, G., Bettenhausen, K.D., About the importance of autonomy and digital twins for the future of manufacturing. *IFAC PapersOnLine*, 48, 3, 567–72, 2015, https://doi.org/10.1016/j.ifacol.2015.06.141.
43. Schroeder, G., Steinmetz, C., Pereira, C.E., Muller, I., Garcia, N., Espindola, D. *et al.*, Visualising the Digital Twin using Web Services and Augmented Reality. 2016. *IEEE 14th Int. Conf. Ind. Informatics*, pp. 522–7, 2016, https://doi.org/10.1109/INDIN.2016.7819217.
44. Brenner, B. and Hummel, V., Digital twin as enabler for an innovative digital shopfloor management system in the ESB Logistics Learning Factory at Reutlingen - University. *Proc. Manuf.*, 9, 198–205, 2017, https://doi.org/10.1016/j.promfg.2017.04.039.
45. Boschert, S. and Rosen, R., Digital Twin — The Simulation Aspect, in: *Mechatron. Futur*, pp. 59–74, Springer, 2016, https://doi.org/10.1007/978-3-319-32156-1.
46. Zheng, P., Lin, T., Chen, C., Xu, X., A systematic design approach for service innovation of smart product-service systems. *J. Clean Prod.*, 201, 657–67, 2018, https://doi.org/10.1016/j.jclepro.2018.08.101.
47. Zheng, P., Xu, X., Chen, C.-H., A data-driven cyber-physical approach for personalised smart, connected product co-development in a cloud-based environment. *J. Intell. Manuf.*, 31, 1, 3–18, 2020, https://doi.org/10.1007/s10845-018-1430-y.
48. Oyekan, J.O., Hutabarat, W., Tiwari, A., Grech, R., Aung, M.H., Mariani, M.P. *et al.*, The effectiveness of virtual environments in developing collaborative strategies between industrial robots and humans. *Robot. Comput. Integr. Manuf.*, 55, 41–54, 2019, https://doi.org/10.1016/j.rcim.2018.07.006.
49. Caputo, F., Greco, A., Fera, M., Macchiaroli, R., Digital twins to enhance the integration of ergonomics in the workplace design. *Int. J. Ind. Ergon.*, 71, 20–31, 2019, https://doi.org/10.1016/j.ergon.2019.02.001.
50. Park, K.T., Im, S.J., Kang, Y.-S., Noh, S.D., Kang, Y.T., Yang, S.G., Service-oriented platform for smart operation of dyeing and finishing industry. *Int. J. Comput. Integr. Manuf.*, 32, 3, 307–326, 2019, doi: https://doi.org/10.1080/0951192x.2019.1572225.

51. Xie, J., Wang, X., Yang, Z., Hao, S., Virtual monitoring method for hydraulic supports based on digital twin theory. *Min. Technol.*, 128, 2, 77–87, 2019, doi: https://doi.org/10.1080/25726668.2019.1569367.
52. Wang, X.V. and Wang, L., Digital twin-based WEEE recycling, recovery and remanufacturing in the background of Industry 4.0. *Int. J. Prod. Res.*, 57, 12, 3892–3902, 2018, doi: https://doi.org/10.1080/00207543.2018.1497819.
53. Leng, J. and Jiang, P., Dynamic scheduling in RFID-driven discrete manufacturing system by using multi-layer network metrics as heuristic information. 30, 3, 979–994, 2017, doi: https://doi.org/10.1007/s10845-017-1301-y.AU: Please provide journal title.
54. Leng, J., Jiang, P., Ding, K., Implementing of a three-phase integrated decision support model for parts machining outsourcing. *Int. J. Prod. Res.*, 52, 12, 3614–3636, 2014, doi: https://doi.org/10.1080/00207543.2013.879344.
55. Souza, V., Cruz, R., Silva, W., Lins, S., Lucena, V., A digital twin architecture based on the industrial internet of things technologies, in: *2019 IEEE International Conference on Consumer Electronics (ICCE)*, 2019, January, IEEE, pp. 1–2.
56. Cheng, J., Zhang, H., Tao, F., Juang, C.F., DT-II: Digital twin enhanced Industrial Internet reference framework towards smart manufacturing. *Rob. Comput. Integr. Manuf.*, *62*, 101881, 2020.
57. Babulak, E. and Wang, M., Discrete Event Simulation: State of the Art, A. Goti (Ed.), August 18, 2010. Eduard Babulak and Ming Wang (2010) Discrete Event Simulation: State of the Art, Discrete Event Simulations ISBN: 978-953-307-115-2, InTech, Available from: http://www.intechopen.com/books/discrete-event-simulations/discrete-event-simulation-state-of-the-art, Available at SSRN: https://ssrn.com/abstract=3904191.
58. Moyne, J. and Iskandar, J., Big Data Analytics for Smart Manufacturing: Case Studies in Semiconductor Manufacturing. *Processes*, 5, 4, 39, 2017, doi: https://doi.org/10.3390/pr5030039.
59. Cohen, Y., Faccio, M., Pilati, F. *et al.*, Design and management of digital manufacturing and assembly systems in the Industry 4.0 era. *Int. J. Adv. Manuf. Technol.*, 105, 3565–3577, 2019. https://doi.org/10.1007/s00170-019-04595-0.
60. Liu, B., Chen, C.-H., Zheng, P., Zhang, G., An Adaptive Parallel Feature Learning and Hybrid Feature Fusion-Based Deep Learning Approach for Machining Condition Monitoring. *IEEE Trans. Cybern.*, 53, 12, 7584–7595, Dec. 2023, doi: 10.1109/TCYB.2022.3178116.
61. Li, X., Zheng, P., Bao, J., Gao, L., Xu, X., Achieving cognitive mass personalization via the self-X cognitive manufacturing network: An industrial-knowledge-graph- and graph-embedding-enabled pathway. *Engineering*, 2021, doi: https://doi.org/10.1016/j.eng.2021.08.018.
62. Malik, A.A. and Bilberg, A., Digital twins of human robot collaboration in a production setting. *Proc. Manuf.*, 17, 278–285, 2018, doi: https://doi.org/10.1016/j.promfg.2018.10.047.

Part 2

METHODS AND APPLICATIONS

6

Distributed Systems and Distributed Ledger Technology - An Introduction

Amit Kumar Tyagi[1]*, Shrikant Tiwari[2] and Kanchan Naithani[2]

[1]Department of Fashion Technology, National Institute of Fashion Technology, New Delhi, Delhi, India
[2]School of Computing Science and Engineering (SCSE), Galgotias University, Greater Noida, Uttar Pradesh, India

Abstract
The landscape of modern computing has witnessed a paradigm shift with the proliferation of Distributed Systems and the revolutionary emergence of Distributed Ledger Technology (DLT). This chapter moves towards the synergistic relationship between these two technological frontiers, unraveling the identified implications they bear on the fabric of contemporary information systems. Note that Distributed Systems, characterized by the decentralized distribution of processing and storage across interconnected nodes, have evolved to address the challenges faced by scalability, fault tolerance, and resilience in the face of diverse and dynamic workloads. As our reliance on interconnected devices and cloud computing intensifies, understanding the principles governing Distributed Systems becomes imperative. Complementing this narrative is the rise of Distributed Ledger Technology (DLT), epitomized by blockchain, which introduces a novel approach to secure, transparent, and tamper-resistant record-keeping. DLT not only transforms traditional notions of trust but also fosters a new era of decentralized applications and smart contracts. Its impact extends beyond the financial realm into sectors such as supply chain, healthcare, and governance. Hence, this chapter navigates through the conceptual intersections of Distributed Systems and DLT, exploring how the amalgamation of these technologies catalyzes innovation. We explain several challenges and opportunities presented by their convergence, touching upon issues of consensus mechanisms, scalability, and interoperability. Additionally, we analyze real-world use cases where the symbiosis of Distributed Systems and DLT has led to transformative solutions, ushering in a new era of distributed, transparent, and resilient information architectures.

***Keywords*:** Distributed systems, distributed ledger technology (DLT), blockchain, decentralization, consensus mechanisms, interoperability

**Corresponding author*: amitkrtyagi025@gmail.com

Amit Kumar Tyagi, Shrikant Tiwari, Senthil Kumar Arumugam and Avinash Kumar Sharma (eds.) Artificial Intelligence-Enabled Digital Twin for Smart Manufacturing, (105–124) © 2024 Scrivener Publishing LLC

6.1 An Introduction

In recent years, the confluence of Distributed Systems and Distributed Ledger Technology (DLT) has reshaped the landscape of information technology. Distributed Systems, characterized by decentralized processing and storage, have become instrumental in addressing scalability and fault tolerance challenges. Simultaneously, Distributed Ledger Technology, notably blockchain, introduces novel paradigms for secure, transparent, and tamper-resistant record-keeping [1, 2]. This introduction discusses the interplay between these two domains, unraveling the symbiotic relationship that underpins their convergence. As we move towards the intricate web of decentralized architectures and distributed ledgers, the transformative potential and challenges of this dynamic intersection become apparent, ushering in a new era of resilient, transparent, and collaborative computing [4, 5].

An introduction to Distributed System: Distributed systems refer to a network of interconnected computers or devices that work together to achieve a common goal. Instead of relying on a single centralized system, distributed systems distribute data, processing power, and tasks across multiple nodes in the network. This decentralization provides benefits such as improved performance, fault tolerance, scalability, and resource sharing. However, it also introduces challenges related to coordination, consistency, and communication between the nodes. By effectively managing these challenges, distributed systems enable various applications and services, ranging from cloud computing and big data processing to online gaming and social networks.

An introduction to Decentralized Systems (DS): Decentralized systems are a type of network where control and decision-making authority are distributed among multiple nodes or participants, rather than being concentrated in a central authority. These systems operate without a single point of failure or control, allowing for increased resilience, transparency, and trust. In decentralized systems, each node in the network maintains its own copy of the system's data and participates in the decision-making process through consensus mechanisms. This distributed approach provides several advantages, such as enhanced security, censorship resistance, and the ability to operate in a trustless environment. Decentralized systems are often associated with blockchain technology, which enables the creation of decentralized applications (DApps), crypto currencies, and smart contracts. In general, decentralized systems have applications in various domains, including finance, supply chain management, voting systems, and decentralized internet protocols. They provide opportunities for individuals and organizations to interact directly, removing the need for intermediaries and fostering peer-to-peer interactions. While decentralized systems provide several benefits, they also present challenges such as scalability, efficiency, and governance that need to be addressed for widespread adoption [3, 29, 30].

6.2 Related Work

Several studies and research efforts have been conducted in the field of distributed systems to address the challenges and discuss new possibilities [5, 6]. Here are some notable areas of related work in distributed systems shown in the Figure 6.1:

Consensus Protocols: Consensus protocols are fundamental to distributed systems, ensuring agreement among nodes on the state of the system. Classic protocols like Paxos

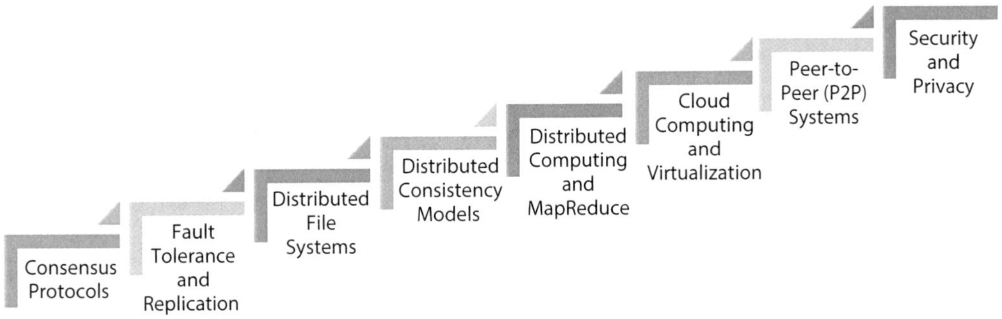

Figure 6.1 Areas of related work in distributed systems.

and Raft, as well as newer ones like Practical Byzantine Fault Tolerance (PBFT) and Proof of Stake (PoS), have been extensively studied and optimized.

Fault Tolerance and Replication: Ensuring system availability and reliability in the presence of failures is an important aspect. Research focuses on fault tolerance techniques, such as replication, where data is stored and processed across multiple nodes, enabling fault recovery and load balancing.

Distributed File Systems: Distributed file systems like Google File System (GFS) and Hadoop Distributed File System (HDFS) tackle the challenge of storing and accessing large amounts of data across multiple nodes. They provide fault tolerance, scalability, and efficient data processing capabilities.

Distributed Consistency Models: Research discusses different consistency models to maintain data consistency across distributed systems, such as strong consistency (e.g., linearizability), eventual consistency, and causal consistency. Techniques like distributed transactions, conflict resolution, and quorum systems have been investigated in this context.

Distributed Computing and MapReduce: Distributed computing frameworks like MapReduce, popularized by Hadoop, enable large-scale data processing across clusters of machines. Research in this area focuses on optimizing resource allocation, load balancing, and data partitioning techniques.

Cloud Computing and Virtualization: Distributed systems form the backbone of cloud computing platforms. Research efforts concentrate on resource management, scheduling, virtualization techniques, and elasticity to efficiently utilize the underlying distributed infrastructure.

Peer-to-Peer (P2P) Systems: P2P systems facilitate direct communication and resource sharing between nodes without relying on centralized servers. Research discusses routing algorithms, distributed hash tables (DHTs), and incentive mechanisms for efficient data dissemination and discovery.

Security and Privacy: Protecting data, communications, and user privacy in distributed systems is an important area of research. Encryption techniques, access control mechanisms, Byzantine fault tolerance, and anonymity protocols are studied to enhance security and privacy in distributed environments.

Note that these are examples of the longest range of research in distributed systems. The field is continuously evolving, with ongoing efforts to address scalability, performance, fault tolerance, security, and other challenges associated with distributed computing. Now we will discuss few essential terms:

Decentralized systems: Decentralized systems have gained huge attention in recent years, leading to a diverse range of research and development efforts. Here are some notable areas of related work in decentralized systems:

Blockchain Technology: Blockchain is a basic technology behind many decentralized systems. Research focuses on consensus algorithms, such as Proof of Work (PoW), Proof of Stake (PoS), and Byzantine fault tolerance (BFT), to ensure agreement and security in decentralized networks. Scalability solutions, privacy enhancements, and smart contract frameworks are also active research areas.

Distributed Ledger Technologies (DLTs): Besides blockchain, other DLTs like Directed Acyclic Graphs (DAGs) and Hashgraph have emerged as alternatives for decentralized systems. Research discusses their scalability, consensus mechanisms, and performance characteristics to address the limitations of traditional blockchain architectures.

Decentralized Applications (DApps): DApps are applications built on top of decentralized systems, using the benefits of transparency, immutability, and peer-to-peer interactions. Research in this area focuses on designing and optimizing DApp architectures, user experience, decentralized storage solutions, and integration with existing systems.

Cryptocurrencies and Tokenomics: Cryptocurrencies, such as Bitcoin and Ethereum, have driven extensive research on their economic aspects, including tokenomics, incentive mechanisms, game theory, and decentralized finance (DeFi) protocols. Topics like token design, stablecoins, decentralized exchanges, and lending platforms are actively studied.

Governance and Decentralized Autonomous Organizations (DAOs): With decentralized systems, new models of governance are being discussed, where decision-making processes are distributed among network participants. Research investigates decentralized governance mechanisms, voting protocols, dispute resolution, and the creation of DAOs to enable decentralized decision-making.

Interoperability and Cross-Chain Communication: As the number of decentralized systems and blockchains grows, achieving interoperability and enabling seamless communication between different networks becomes important. Research efforts are dedicated to developing protocols and frameworks that allow the transfer of assets and information across diverse decentralized systems.

Privacy and Security in Decentralized Systems: Preserving user privacy and ensuring secure transactions are key issue in decentralized systems. Research focuses on techniques such as zero-knowledge proofs, secure multi-party computation, privacy-preserving smart contracts, and decentralized identity solutions to address privacy and security challenges.

Decentralized Internet Protocols: Decentralized systems extend beyond blockchain and encompass protocols like InterPlanetary File System (IPFS) and BitTorrent. Research discusses peer-to-peer content distribution, decentralized naming systems, decentralized messaging, and Web3.0 concepts to build a more decentralized and censorship-resistant internet.

These are examples of the active research areas in decentralized systems. The field is rapidly evolving, with continuous efforts to improve scalability, efficiency, usability, governance models, and privacy-enhancing features to enable the widespread adoption of decentralized technologies.

6.3 Blockchain - In General

Blockchain technology is a decentralized and distributed ledger that records transactions across multiple computers or nodes. It enables secure and transparent transactions without the need for intermediaries [7, 8]. Some key aspects related to blockchain are shown in Figure 6.2:

Consensus: Blockchain achieves consensus among network participants by using various consensus algorithms. The most well-known consensus algorithm is Proof of Work (PoW), used by Bitcoin, which requires participants to solve complex mathematical puzzles to validate transactions. Other consensus algorithms, such as Proof of Stake (PoS) and Practical Byzantine Fault Tolerance (PBFT), have also been developed to address scalability and energy efficiency issues.

Security: Blockchain ensures the security of data through cryptographic techniques. Each transaction is digitally signed and linked to the previous transaction, forming a chain of blocks. This immutability makes it extremely difficult to alter past transactions, providing a high level of security and tamper resistance.

Smart Contracts: Blockchain platforms, like Ethereum, support smart contracts, which are self-executing agreements with predefined rules and conditions. Smart contracts enable automation, transparency, and enforceability of agreements, eliminating the need for intermediaries in various fields, including finance, supply chain management, and decentralized applications.

Decentralization: Blockchain operates in a decentralized manner, with no central authority controlling the network. Transactions are validated and recorded by multiple nodes, ensuring that no single point of failure exists. Decentralization enhances system resilience, censorship resistance, and trust among participants.

Use Cases: Blockchain technology has a wide range of applications beyond crypto currencies. It is utilized in industries such as finance, healthcare, supply chain management, voting systems, intellectual property, and more. Blockchain enables transparency, traceability, and efficiency in these sectors by providing an immutable and auditable record of transactions.

Challenges: Blockchain faces challenges such as scalability, energy consumption, interoperability, and regulatory frameworks. Research and development efforts are focused on improving scalability through techniques like sharding and layer 2 solutions, as well as addressing environmental issues related to energy consumption. Interoperability protocols and regulatory frameworks are being discussed to ensure seamless integration and compliance with existing systems and regulations.

In summary, blockchain technology provides the potential to revolutionize various industries by providing trust, security, and transparency in a decentralized manner. Ongoing research and innovation aim to address its limitations and unlock the full potential of this transformative technology.

Figure 6.2 Key aspects related to blockchain.

6.4 Evolution of Blockchain

The evolution of blockchain has witnessed a transformative journey from its inception as the underlying technology for crypto currencies to becoming a foundational pillar of decentralized applications and smart contracts. From Bitcoin's pioneering blockchain implementation, subsequent iterations have focused on scalability, interoperability, and sustainability, shaping blockchain into a versatile and disruptive force across industries [9, 10]. Today, the evolution continues with ongoing advancements in consensus mechanisms, privacy solutions, and the integration of emerging technologies, paving the way for a more mature and diversified blockchain ecosystem which is shown in Figure 6.3.

Blockchain evolution was traced as:

a) *Bitcoin:* The first blockchain, Bitcoin, was introduced in 2009 by an anonymous person or group known as Satoshi Nakamoto. It was primarily designed as a decentralized digital currency system.
b) *Blockchain Explorations:* As the concept of blockchain gained attention, researchers and developers started exploring its potential beyond crypto currencies. They began experimenting with different blockchain use cases and exploring the possibilities of applying the technology to various industries.
c) *Ethereum and Smart Contracts:* Ethereum, launched in 2015, introduced a programmable blockchain platform that enabled the execution of smart contracts. This development opened new opportunities for decentralized applications (DApps) and expanded the use cases of blockchain technology.
d) *Alternative Blockchain Platforms:* Following Ethereum's success, alternative blockchain platforms emerged with their own unique features and capabilities. Examples include platforms like EOS, NEO, Cardano, and many others.

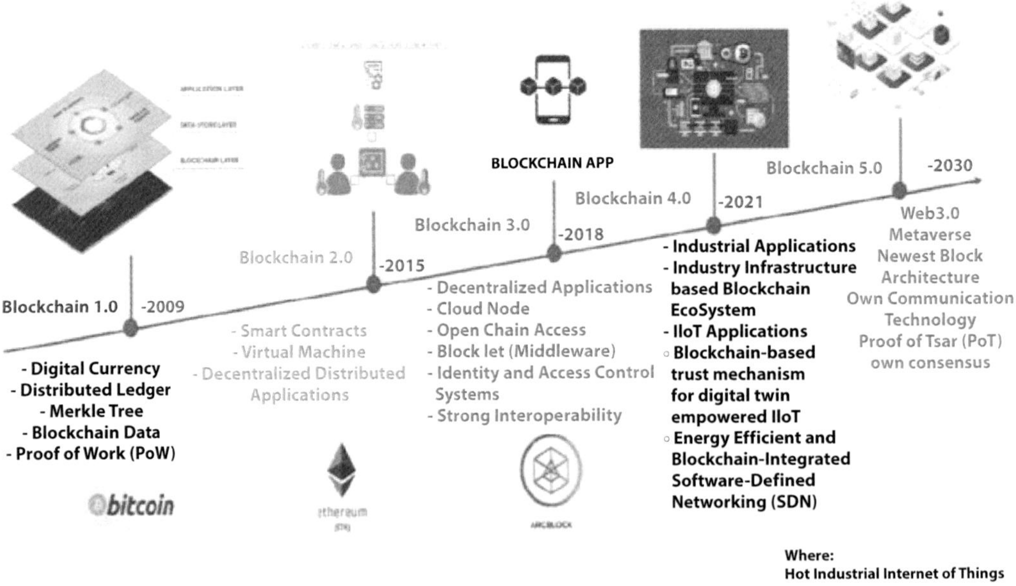

Figure 6.3 Evolution of blockchain [11].

These platforms aimed to address limitations such as scalability, interoperability, and governance.

e) *Interoperability Protocols:* With the growth of multiple blockchain platforms, the need for interoperability between different networks became apparent. Interoperability protocols like Cosmos, Polkadot, and ICON were developed to enable communication and data transfer between disparate blockchain ecosystems.
f) *Permissioned and Consortium Blockchains:* While public blockchains like Bitcoin and Ethereum gained popularity, there was also a rise in the development of permissioned and consortium blockchains. These blockchains were designed for specific use cases, often within industries or organizations, and provided more control over access and governance.
g) *Scalability Solutions:* As blockchain adoption increased, scalability became an important challenge. To address this, various scaling solutions were proposed and implemented, such as layer-two protocols (e.g., Lightning Network for Bitcoin) and sharding (e.g., Ethereum 2.0).
h) *Integration with Traditional Systems:* To facilitate the adoption of blockchain technology, efforts were made to integrate it with existing traditional systems and infrastructure. This integration involved bridging the gap between blockchain and legacy systems, enabling seamless data flow and interoperability.
i) *Privacy and Confidentiality Enhancements:* Privacy and confidentiality of transactions and data on the blockchain became a focus area. Technologies like zero-knowledge proofs (ZKPs) and private or confidential blockchains were developed to address privacy issues and ensure selective data disclosure.
j) *Regulatory Frameworks:* Governments and regulatory bodies worldwide started acknowledging and addressing the regulatory challenges posed by blockchain technology. Some jurisdictions established frameworks and guidelines to govern blockchain-based activities and crypto currencies.

6.4.1 Progress of Blockchain 1.0 to Blockchain 4.0

The concept of blockchain 1.0 to 4.0 refers to different stages or generations of blockchain technology [12, 13]. Here is an overview of each stage:

Blockchain 1.0

a) *Bitcoin:* Blockchain 1.0 represents the initial stage of blockchain technology, primarily associated with the creation of Bitcoin and its underlying blockchain. Bitcoin introduced the concept of a decentralized digital currency and demonstrated the potential of using blockchain as a distributed ledger for financial transactions.
b) *Blockchain 2.0: Programmable Blockchains:* Blockchain 2.0 emerged with the introduction of Ethereum in 2015. It introduced the concept of programmable smart contracts, which expanded the capabilities of blockchain beyond simple financial transactions. Ethereum allowed developers to create and deploy decentralized applications (DApps) by writing code on the

blockchain, enabling various use cases such as decentralized finance (DeFi), decentralized governance, and more.

Blockchain 3.0

Scalability and Interoperability: Blockchain 3.0 aims to address the scalability and interoperability challenges faced by earlier generations. This stage focuses on enhancing blockchain networks to handle a higher transaction volume, improve consensus mechanisms, and enable seamless communication and data exchange between different blockchains. Solutions like sharding, layer-two protocols, and interoperability frameworks are being developed to achieve these goals.

Blockchain 4.0

Advanced Features and Integration: Blockchain 4.0 represents an envisioned stage where blockchain technology incorporates advanced features and integrates with other emerging technologies. This stage focuses on integrating blockchain with artificial intelligence (AI), Internet of Things (IoT), big data, and other technologies to create more robust and versatile decentralized systems. Blockchain 4.0 aims to facilitate the development of decentralized applications with enhanced scalability, privacy, security, and intelligence.

Hence, the classifications of blockchain generations, such as 1.0 to 4.0, are not universally standardized and may vary depending on different perspectives and interpretations within the blockchain community.

6.5 Generic Elements of a Blockchain

A blockchain typically consists of several generic elements that work together to form a decentralized and secure system [14]. The key elements of a blockchain are shown in the Figure 6.4:

- *Distributed Ledger:* The blockchain is a distributed ledger that records a chronological sequence of transactions or data entries. It is maintained across multiple nodes or computers in a network, with each node holding a copy of the entire ledger.
- *Blocks:* Transactions or data entries are grouped into blocks, which are added to the blockchain in a linear and chronological order. Each block contains a set of transactions, a timestamp, a reference to the previous block (forming a chain), and a unique identifier called a hash.
- *Cryptographic Hash:* A cryptographic hash function is used to create a unique digital fingerprint, or hash, for each block. The hash is derived from the data

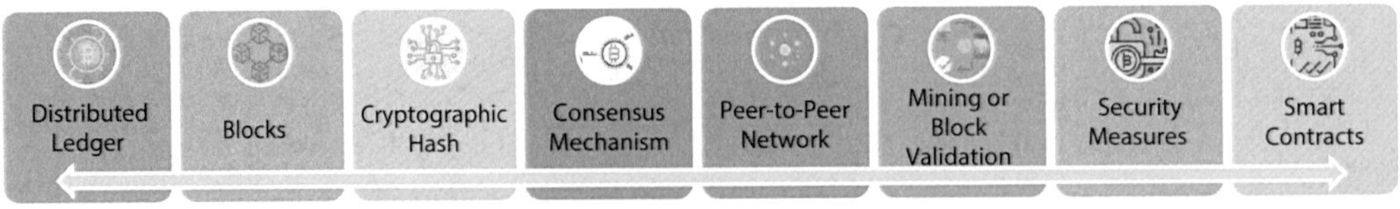

Figure 6.4 Key aspects related to blockchain.

within the block and is important for maintaining the integrity and security of the blockchain.

- *Consensus Mechanism:* To reach an agreement on the validity and order of transactions, a consensus mechanism is employed. Consensus algorithms, such as Proof of Work (PoW), Proof of Stake (PoS), or Practical Byzantine Fault Tolerance (PBFT), ensure that a majority of participants agree on the state of the blockchain.
- *Peer-to-Peer Network:* The blockchain operates on a peer-to-peer network, where nodes communicate and share information. Each node maintains a copy of the blockchain and participates in transaction validation, block creation, and consensus formation.
- *Mining or Block Validation:* In some blockchain systems, such as Bitcoin, a process called mining is used to validate and add new blocks to the blockchain. Miners compete to solve complex mathematical puzzles, and the first one to find the solution is rewarded and gets to add the next block to the chain. This process ensures security, prevents double-spending, and maintains the blockchain's integrity.
- *Security Measures:* Blockchain employs various cryptographic techniques to ensure security. Transactions are digitally signed to verify authenticity and prevent tampering. Consensus mechanisms and the decentralized nature of the blockchain make it resistant to attacks and manipulation.
- *Smart Contracts:* Some blockchains, like Ethereum, support smart contracts. Smart contracts are self-executing agreements with predefined rules and conditions. They enable automation and enforceability of contractual obligations, reducing the need for intermediaries.

Hence, these elements collectively provide the base for a decentralized and secure blockchain system, enabling transparent and tamper-resistant record-keeping and transaction processing. Note that different blockchain implementations may have variations and additional features based on their specific use cases and requirements.

6.6 Benefits and Limitations of Blockchain

Blockchain technology provides several benefits and advantages, but it also has some limitations [15, 16]. Here are the key benefits and limitations of blockchain:

Benefits of Blockchain: Few of Benefits of Blockchain are shown in Figure 6.5.

a) *Security:* Blockchain provides a high level of security through cryptographic techniques. Transactions are digitally signed, and the decentralized nature of the blockchain makes it resistant to tampering and fraud.
b) *Transparency and Immutability:* Once a transaction is recorded on the blockchain, it is difficult to alter or delete. This immutability ensures transparency and auditability, making blockchain suitable for applications where trust and transparency are important.

Figure 6.5 Benefits of blockchain.

c) *Decentralization:* Blockchain operates on a decentralized network, eliminating the need for intermediaries or central authorities. This decentralization enhances resilience, censorship resistance, and reduces the risk of a single point of failure.
d) *Efficiency and Cost Reduction:* Blockchain can streamline and automate processes, reducing the need for manual reconciliation and paperwork. It eliminates intermediaries, reduces transaction costs, and enables faster settlement times.
e) *Trust and Identity Management:* Blockchain can provide a trust layer by establishing the identity and reputation of participants through cryptographic signatures. It enables secure and self-sovereign identity management, reducing reliance on centralized identity systems.
f) *Disintermediation and Peer-to-Peer Transactions:* Blockchain enables direct peer-to-peer transactions without the need for intermediaries, reducing costs, and increasing efficiency. It empowers individuals and promotes financial inclusion by bypassing traditional gatekeepers.

Limitations of Blockchain: Few of limitation of Blockchain are shown in the Figure 6.6:

a) *Scalability:* Blockchain faces scalability challenges in terms of transaction throughput and network capacity. Public blockchains, in particular, can experience slower transaction speeds and higher costs as the network grows.
b) *Energy Consumption:* Some blockchain consensus algorithms, like Proof of Work (PoW), require huge computational power, resulting in high energy consumption. This environmental impact has raised issues about sustainability.
c) *Governance and Regulatory Challenges:* Blockchain technology often operates outside traditional regulatory frameworks, raising challenges related to legal compliance, jurisdiction, and governance of decentralized networks.
d) *Data Privacy:* While blockchain provides transparency, it can also raise privacy issues. Public blockchains store transaction data publicly, potentially

Figure 6.6 Limitation of blockchain.

exposing sensitive information. Private and permissioned blockchains address this to some extent, but privacy challenges remain.

e) *Interoperability:* Interoperability between different blockchain networks and legacy systems is still an ongoing challenge. Efforts are underway to develop standards and protocols to enable seamless data exchange and interoperability.

f) *Adoption and User Experience:* Blockchain technology is still relatively new and complex for the average user. User-friendly interfaces and improved user experience are needed to drive wider adoption and make blockchain more accessible.

Note that these benefits and limitations while assessing the suitability of blockchain for specific use cases and to be aware of ongoing research and development efforts to address its limitations.

6.7 Tiers of Blockchain Technology

Blockchain technology can be categorized into different tiers based on various factors such as functionality, accessibility, and consensus mechanisms [17, 18]. Here are three commonly used tiers:

- *First-tier (Public Blockchains):* Public blockchains, like Bitcoin and Ethereum, are open and permissionless networks that anyone can join and participate in. They operate on a decentralized model, allowing anyone to validate transactions, create new blocks, and participate in the consensus mechanism. These blockchains are transparent, secure, and resistant to censorship. However, they often face scalability and energy consumption challenges.
- *Second-tier (Private/Permissioned Blockchains):* Private or permissioned blockchains restrict participation to a selected group of known and trusted entities. These blockchains are typically used within organizations or consortiums where participants require more control over access, data privacy, and governance. Permissioned blockchains provide faster transaction speeds, improved scalability, and greater privacy compared to public blockchains. Examples include Hyperledger Fabric and R3 Corda.
- *Third-tier (Hybrid/Interoperable Blockchains):* Hybrid or interoperable blockchains aim to combine the strengths of both public and private blockchains. They facilitate interoperability between different blockchain networks, enabling the seamless exchange of assets and data across multiple chains. These blockchains use interoperability protocols and standards to bridge different networks and allow for the transfer of value or information. Examples include Cosmos Network and Polkadot.

Note that these tiers are not strictly defined and can vary based on different interpretations and evolving technologies. Additionally, new variations and combinations of blockchain tiers continue to emerge as the technology advances and evolves.

6.8 Features of a Blockchain

Blockchain technology encompasses several key features that contribute to its unique characteristics and functionalities [19, 20]. Here are the fundamental features of a blockchain are shown in Figure 6.7:

- *Decentralization:* Blockchain operates on a decentralized network of computers or nodes. There is no central authority controlling the system, and transactions are validated and recorded by multiple participants. Decentralization enhances resilience, censorship resistance, and removes the need for intermediaries.
- *Distributed Ledger:* A blockchain is a distributed ledger that maintains a record of all transactions or data entries across multiple nodes in the network. Each node has a copy of the entire ledger, ensuring transparency and redundancy.
- *Transparency and Immutability:* Once a transaction is recorded on the blockchain, it becomes part of an immutable and transparent record. Transactions are typically visible to all participants, creating a high level of transparency and accountability. The immutability of the blockchain prevents tampering and ensures data integrity.
- *Security through Cryptography:* Blockchain employs cryptographic techniques to secure transactions and data. Transactions are digitally signed, ensuring

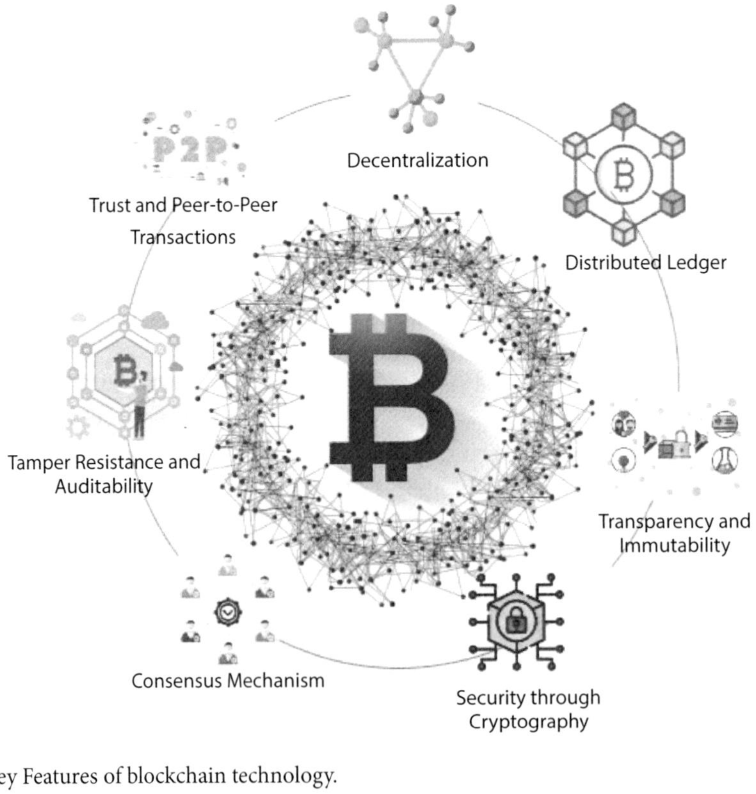

Figure 6.7 Key Features of blockchain technology.

authenticity and preventing unauthorized modifications. Hash functions and cryptographic puzzles are used for data integrity and to secure the consensus process.

- *Consensus Mechanism:* Consensus algorithms are employed to reach agreement among participants on the validity and order of transactions. Various consensus mechanisms, such as Proof of Work (PoW), Proof of Stake (PoS), and Practical Byzantine Fault Tolerance (PBFT), are used to achieve consensus and maintain the integrity of the blockchain. Smart Contracts: Smart contracts are self-executing agreements with predefined rules and conditions. They are stored and executed on the blockchain, enabling automation, transparency, and enforceability of contractual obligations. Smart contracts eliminate the need for intermediaries in various applications and enable trustless interactions.
- *Tamper Resistance and Auditability:* Once a transaction is added to the blockchain, it becomes nearly impossible to alter or delete it. The combination of cryptographic techniques, consensus mechanisms, and decentralization ensures that the blockchain is resistant to tampering and provides an auditable trail of transactions.
- *Trust and Peer-to-Peer Transactions:* Blockchain enables trust between participants without relying on centralized authorities. Through cryptographic signatures, participants can verify the authenticity and integrity of transactions. Blockchain facilitates direct peer-to-peer transactions, reducing the need for intermediaries and enabling efficient value transfer.

These features collectively make blockchain a unique technology with applications in various industries, including finance, supply chain management, healthcare, and more. Note that different blockchain implementations and variations may emphasize certain features more than others, depending on their specific use cases and requirements.

6.9 Types of Blockchain

The types of blockchain can be categorized based on the level of access and control granted to participants [21]. Here are the four main types:

- *Public Blockchain:* Public blockchains are open and permissionless networks where anyone can participate and validate transactions. Examples include Bitcoin and Ethereum. Public blockchains are decentralized and provide high transparency, security, and censorship resistance. They are typically used for crypto currencies, decentralized applications, and global peer-to-peer transactions.
- *Private Blockchain:* Private blockchains are restricted networks that limit participation to selected entities or individuals. Permission to join and validate transactions is controlled by a centralized authority. Private blockchains are often used within organizations or consortiums to enhance privacy, scalability, and control over the network. They are suitable for applications that

require a higher degree of trust among participants, such as supply chain management or interbank transactions.

- *Consortium/Shared Blockchain:* Consortium or shared blockchains are a hybrid approach that involves multiple organizations or entities coming together to form a shared network. In this model, participants collectively validate transactions and maintain the blockchain. Consortium blockchains provide a balance between privacy and decentralization, as they involve a predefined set of trusted nodes. They are commonly used for industry-specific applications and collaborations, such as supply chain consortiums or healthcare networks.
- *Proprietary/Permissioned Blockchain:* Proprietary or permissioned blockchains are controlled by a single organization or entity. Access to the blockchain and validation of transactions are limited to authorized participants. These blockchains provide higher efficiency, privacy, and control compared to public blockchains. Proprietary blockchains are suitable for internal use within an organization or for specific business applications that require a controlled environment, such as enterprise resource planning or data management.

Note that hybrid models and variations of these types also exist, as organizations and developers tailor blockchain solutions to meet their specific requirements. These types of blockchains provide flexibility in choosing the appropriate level of transparency, control, and access for different use cases.

6.10 Open Issues in Blockchain Technology

Blockchain technology, despite its potential, still faces several open issues that need to be addressed for its widespread adoption and effectiveness [22]. Here are some of the key open issues (refer Figure 6.8) in blockchain technology:

- *Scalability:* Scalability remains a major challenge for blockchain networks. As the number of transactions and participants increases, blockchain networks

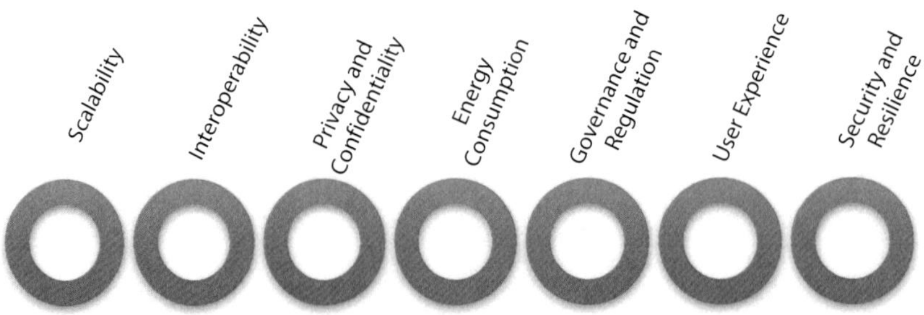

Figure 6.8 Key issues in blockchain technology.

struggle to handle the load, resulting in slow transaction speeds and increased costs. Finding scalable solutions that can handle a high volume of transactions without compromising decentralization is an important issue.

- *Interoperability:* Interoperability between different blockchain networks is an important challenge. Blockchain platforms often operate in isolation, making it difficult for them to communicate and share data seamlessly. Developing standards, protocols, and frameworks that enable interoperability across various blockchain networks is essential for unlocking the full potential of blockchain technology.
- *Privacy and Confidentiality:* While blockchain provides transparency and immutability, privacy and confidentiality of sensitive data are still open issues. Finding ways to protect user privacy and sensitive information while ensuring data integrity and transparency poses an important challenge. Advances in privacy-enhancing technologies, such as zero-knowledge (zk) proofs, are being discussed to address this issue.
- *Energy Consumption:* The energy consumption associated with blockchain consensus mechanisms, particularly in proof-of-work (PoW) based networks, is an important issue. As blockchain adoption grows, the environmental impact of energy-intensive mining operations becomes more important. Exploring alternative consensus mechanisms, such as proof-of-stake (PoS), and improving energy efficiency are important areas of research.
- *Governance and Regulation:* Blockchain's decentralized nature raises questions around governance and regulatory frameworks. Determining the right balance between decentralization and regulatory compliance is a challenge. Establishing clear governance models and regulatory frameworks that promote innovation, protect users, and address such issues such as anti-money laundering (AML) and know your customer (KYC) requirements is important for blockchain technology to thrive.
- *User Experience:* Blockchain technology still suffers from complex user experiences, making it less accessible to non-technical users. Improving the user interface, simplifying key management, and enhancing overall user experience are important factors in driving mainstream adoption of blockchain applications.
- *Security and Resilience:* Blockchain networks are not immune to security vulnerabilities. Smart contract bugs, 51% attacks, and other forms of malicious activities face risks to the integrity and security of blockchain networks. Continual research, robust security measures, and best practices are necessary to address these challenges and ensure the resilience of blockchain systems.

Hence, addressing above mentioned open issues requires collaborative efforts from researchers, developers, businesses, governments, and regulatory bodies. Ongoing innovation, research, and development in these areas will make the way for a more mature and effective blockchain technology that can revolutionize various industries and drive broader adoption.

6.11 Important Challenges with Blockchain Technology

Blockchain technology faces important challenges that need to be overcome for its widespread adoption and successful implementation [23–28]. Some of the key challenges is shown in the Figure 6.9:

- *Scalability:* Blockchain scalability is an important challenge. As transaction volumes increase, blockchain networks struggle to handle the load, resulting in slow transaction speeds and higher costs. Scaling solutions, such as sharding, sidechains, or layer 2 protocols, need to be developed to enable higher throughput and improved performance.
- *Interoperability:* Interoperability between different blockchain networks is important for realizing the full potential of blockchain technology. Currently, most blockchains operate in isolation, making it difficult for them to communicate and share data seamlessly. Establishing standards, protocols, and frameworks for cross-chain interoperability is essential for enabling data and asset transfer between different blockchain platforms.
- *Governance and Regulation:* Blockchain's decentralized nature presents challenges in terms of governance and regulatory frameworks. Determining how decisions are made, resolving disputes, and ensuring compliance with existing regulations face huge challenges. Developing transparent and efficient

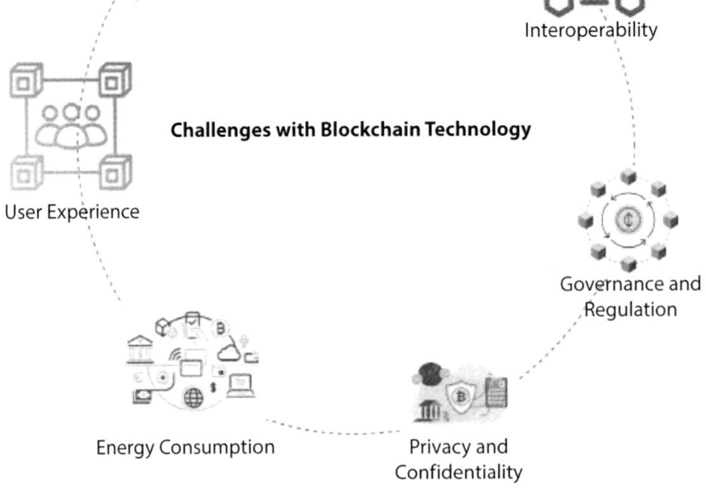

Figure 6.9 Key challenges with blockchain technology.

governance models and regulatory frameworks that balance decentralization, innovation, and legal compliance is necessary.
- *Privacy and Confidentiality:* While blockchain provides transparency and immutability, privacy and confidentiality of sensitive data remain important challenges. Finding ways to protect user privacy and secure confidential information while maintaining the benefits of transparency and auditability is a complex task. Advancements in privacy-preserving technologies, such as zero-knowledge proofs or secure multi-party computation, are being discussed to address this challenge.
- *Energy Consumption:* Blockchain's energy consumption, particularly in proof-of-work (PoW) based networks, is an issue due to its environmental impact. As blockchain adoption grows, the energy requirements of mining operations become more essential. Exploring alternative consensus mechanisms with lower energy consumption, such as proof-of-stake (PoS), or implementing energy-efficient solutions is important for sustainable blockchain technology.
- *User Experience:* Blockchain technology still faces challenges in terms of user experience and accessibility. Complex wallet setups, private key management, and unfamiliar interfaces make it difficult for non-technical users to interact with blockchain applications. Improving the user experience, simplifying key management, and providing intuitive interfaces are important for broader adoption.
- *Security and Trust:* Blockchain networks need to address security vulnerabilities, including smart contract bugs, 51% attacks, or phishing scams. Building robust security measures, conducting thorough audits, and promoting best practices in smart contract development are essential to ensure the integrity and trustworthiness of blockchain systems.

Hence, addressing these important challenges requires collaborative efforts from researchers, developers, industry leaders, and regulatory bodies. Continued research, innovation, and the implementation of scalable, interoperable, privacy-preserving, and user-friendly solutions will drive the future of blockchain technology and unlock its transformative potential.

6.12 Conclusion

In this chapter, we start with the introduction to distributed systems and distributed ledger technology highlights the transformative impact of these concepts in various industries. Distributed systems provide scalability, fault tolerance, and high availability by distributing computation and data across multiple nodes. This enables efficient handling of large-scale applications and ensures resilience against failures. On the other hand, distributed ledger technology, such as blockchain, provides a decentralized and transparent way to securely record and verify transactions or other forms of digital information. By eliminating the need for intermediaries and central authorities, distributed ledgers enhance trust, transparency, and data integrity in complex ecosystems involving multiple participants. Through the case studies discussed, we have seen how distributed systems have been successfully employed in diverse domains. E-commerce platforms, social media networks, healthcare systems, financial institutions, and IoT applications have all used distributed systems to address scalability, fault tolerance, real-time processing, and data management challenges. Additionally, the adoption of distributed ledger technology has brought huge/ rapid improvements to supply chain management, cross-border payments, decentralized applications, and even the management of sensitive data like medical records. The unique characteristics of distributed ledgers, including immutability, consensus mechanisms, and smart contracts, provide enhanced security, privacy, and auditability, fostering trust among participants. In summary, the introduction to distributed systems and distributed ledger technology has demonstrated their immense potential in revolutionizing the way we handle data, conduct transactions, and build resilient and trustworthy systems. As these technologies continue to evolve, we can expect even more innovative applications and transformative impacts across various industries, shaping the future of decentralized and distributed systems.

References

1. Zaabar, B., Cheikhrouhou, O., Jamil, F., Ammi, M., Abid, M., HealthBlock: A secure blockchain-based healthcare data management system. *Comput. Networks*, 200, 108500, 2021.
2. Mondal, S., Shafi, M., Gupta, S., Gupta, S.K., Blockchain based secure architecture for electronic healthcare record management. *GMSARN Int. J.*, 16, 4, 413–26, 2022.
3. Kumari, S. and Muthulakshmi, P., Artificial Intelligence—Blockchain Enabled Technology for Internet of Things: Research Statements, Open Issues, and Possible Applications in the Near Future, in the book: *Privacy Preservation of Genomic and Medical Data*, 2023. https://doi.org/10.1002/9781394213726.ch18
4. Ahmed, S. and Khan, M., Securing the Internet of Things (IoT): A comprehensive study on the intersection of cybersecurity, privacy, and connectivity in the IoT ecosystem, in: *AI, IoT and the Fourth Industrial Revolution Review*, vol. 13, pp. 1–17, 2023.
5. Lindsay, D., Gill, S.S., Smirnova, D., Garraghan, P., The evolution of distributed computing systems: from fundamental to new frontiers. *Computing*, 103, 8, 1859–1878, 2021.
6. Hashem, I.A.T., Yaqoob, I., Anuar, N.B., Mokhtar, S., Gani, A., Khan, S.U., The rise of "big data" on cloud computing: Review and open research issues. *Inf. Syst.*, 47, 98–115, 2015.
7. Natarajan, H., Krause, S., Gradstein, H., *Distributed ledger technology and blockchain*, 2017.
8. Efanov, D. and Roschin, P., The all-pervasiveness of the blockchain technology. *Proc. Comput. Sci.*, 123, 116–121, 2018.
9. Werbach, K., *The blockchain and the new architecture of trust*, Mit Press, 2018.
10. Dhillon, V., Metcalf, D., Hooper, M., *Blockchain enabled applications*, Apress, Berkeley, CA, 2017.
11. Mourtzis, D., Angelopoulos, J., Panopoulos, N., Blockchain Integration in the Era of Industrial Metaverse. *Appl. Sci.*, 13, 3, 1353, 2023, https://doi.org/10.3390/app13031353.
12. Aoun, A., Ilinca, A., Ghandour, M., Ibrahim, H., A review of Industry 4.0 characteristics and challenges, with potential improvements using blockchain technology. *Comput. Ind. Eng.*, 162, 107746, 2021.
13. Leng, J., Ruan, G., Jiang, P., Xu, K., Liu, Q., Zhou, X., Liu, C., Blockchain-empowered sustainable manufacturing and product lifecycle management in industry 4.0: A survey. *Renew. Sustain. Energy Rev.*, 132, 110112, 2020.
14. Bhushan, B., Sahoo, C., Sinha, P., Khamparia, A., Unification of Blockchain and Internet of Things (BIoT): requirements, working model, challenges and future directions. *Wirel. Netw.*, 27, 55–90, 2021.

15. Niranjanamurthy, M., Nithya, B.N., Jagannatha, S.J.C.C., Analysis of Blockchain technology: pros, cons and SWOT. *Cluster Comput.*, 22, 14743–14757, 2019.
16. Hughes, L., Dwivedi, Y.K., Misra, S.K., Rana, N.P., Raghavan, V., Akella, V., Blockchain research, practice and policy: Applications, benefits, limitations, emerging research themes and research agenda. *Int. J. Inf. Manage.*, 49, 114–129, 2019.
17. Lashkari, B. and Musilek, P., A comprehensive review of blockchain consensus mechanisms. *IEEE Access*, 9, 43620–43652, 2021.
18. Ali, O., Jaradat, A., Kulakli, A., Abuhalimeh, A., A comparative study: Blockchain technology utilization benefits, challenges and functionalities. *IEEE Access*, 9, 12730–12749, 2021.
19. Wu, J. and Tran, N.K., Application of blockchain technology in sustainable energy systems: An overview. *Sustainability*, 10, 9, 3067, 2018.
20. Angelis, J. and Da Silva, E.R., Blockchain adoption: A value driver perspective. *Bus. Horizons*, 62, 3, 307–314, 2019.
21. Ma, M., Shi, G., Li, F., Privacy-oriented blockchain-based distributed key management architecture for hierarchical access control in the IoT scenario. *IEEE Access*, 7, 34045–34059, 2019.
22. Toufaily, E., Zalan, T., Dhaou, S.B., A framework of blockchain technology adoption: An investigation of challenges and expected value. *Inf. Manage.*, 58, 3, 103444, 2021.
23. Zachariadis, M., Hileman, G., Scott, S.V., Governance and control in distributed ledgers: Understanding the challenges facing blockchain technology in financial services. *Inf. Organ.*, 29, 2, 105–117, 2019.
24. Biswas, B. and Gupta, R., Analysis of barriers to implement blockchain in industry and service sectors. *Comput. Ind. Eng.*, 136, 225–241, 2019.
25. Nair, M.M. and Tyagi, A.K., Chapter 11 - AI, IoT, blockchain, and cloud computing: The necessity of the future, in: *Distributed Computing to Blockchain*, R. Pandey, S. Goundar, S. Fatima (Eds.), pp. 189–206, Academic Press, 2023, ISBN 9780323961462, https://doi.org/10.1016/B978-0-323-96146-2.00001-2.
26. Tyagi, A.K., Chapter 2 - Decentralized everything: Practical use of blockchain technology in future applications, in: *Distributed Computing to Blockchain*, R. Pandey, S. Goundar, S. Fatima (Eds.), pp. 19–38, Academic Press, 2023, ISBN 9780323961462, https://doi.org/10.1016/B978-0-323-96146-2.00010-3.
27. Tyagi, A.K., Mishra, A.K., Vedavathi, N., Kakulapati, V., Sajidha, S.A., Chapter 21 Futuristic Technologies for Smart Manufacturing, in: *Research Statement and Vision for the Future*, Tyagi, A.K. (Ed.), 2023, https://doi.org/10.1002/9781394213948.ch21.
28. Tyagi, A.K., Chandrasekaran, S., Sreenath, N., Blockchain Technology:– A New Technology for Creating Distributed and Trusted Computing Environment. *2022 International Conference on Applied Artificial Intelligence and Computing (ICAAIC)*, pp. 1348–1354, 2022, doi: 10.1109/ICAAIC53929.2022.9792702.
29. Tyagi, A.K., Kumari, S., Chidambaram, N., Sharma, A., Engineering Applications of Blockchain in This Smart Era, in the book: *Enhancing Medical Imaging with Emerging Technologies,* 2024. DOI: 10.4018/979-8-3693-5261-8.ch011
30. Tyagi, A.K., Kukreja, S., Richa, Sivakumar, P., Role of Blockchain Technology in Smart Era: A Review on Possible Smart Applications, *J. Inform. Knowl. Manage.,* February 2024. https://doi.org/10.1142/S0219649224500321

7

Digital Twins Tools and Technologies in Smart Manufacturing

K. Jayashree[1]*, S. Muralidharan[2], V. Sathya[3], M. Rajakumaran[4] and C.M. Nalayini[5]

[1]Department of AI&DS, Panimalar Engineering College, Chennai, India
[2]Department of EEE, Madha Engineering College, Chennai, India
[3]Department of Computer Science and Engineering, VelTech Rangarajan Dr. Sagunthala R&D Institute of Science and Technology, Chennai, India
[4]School of Computing, Sastra University Deemed to be University, Thanjavur, India
[5]Department of Information Technology, Velammal Engineering College, Chennai, India

Abstract
The evolution of information technology has sparked the rise of ground breaking concept of DTs. Developing a DT empowers manufacturers to simulate diverse scenarios and experiment with multiple configurations without interrupting the live production line. This capability facilitates more efficient testing and fine-tuning of production processes, leading to heightened quality control and predictive maintenance. In essence, DTs serve as a crucial component in smart production, enabling manufacturers to bolster efficiency and curtail expense. By integrating DT technology, firms can streamline the operation and upkeep of their physical systems and manufacturing processes. It plays a pivotal role in enabling the firm to achieve peak profitability and enhance shareholder value. Thus this chapter discusses the DT technologies in manufacturing.

Keywords: Smart manufacturing, digital twins, applications of digital twin, digital twin in manufacturing, digital twin characteristics

7.1 Introduction

Innovations in Information Technology (IT) have disrupted conventional manufacturing practices, forming the foundation of the Fourth Industrial Revolution, also known as Industry 4.0. The convergence of technologies like the Internet of Things (IoT), Artificial Intelligence (AI), Cyber-Physical Systems (CPS), and Big Data is propelling the manufacturing sector toward smart production by enabling seamless connectivity among intellectual elements within the shop floor [14]. The growing accessibility of real-time operational data facilitated by IoT technologies, coupled with the enhanced capabilities of AI in knowledge as well as rational, serves as a catalyst in achieving a vision where physical products or processes are complemented by evolving virtual depictions throughout their life cycles [15].

**Corresponding author*: k.jayashri@gmail.com

Amit Kumar Tyagi, Shrikant Tiwari, Senthil Kumar Arumugam and Avinash Kumar Sharma (eds.) Artificial Intelligence-Enabled Digital Twin for Smart Manufacturing, (125–142) © 2024 Scrivener Publishing LLC

These depictions, termed Digital Twins (DTs), mirror real-world objects in digital form, spanning from individual components within CNC machines to entire factory setups. The manufacturing sector has transitioned from solely relying on physical-mechanical processes to embracing a new era characterized by the interplay and iterative relationship between the tangible and computerized realm [52].

Foremost companies frequently maintain substantial product inventories in warehouses, primarily concentrated in specific regions. Nevertheless, efficiently transporting these stockpiles to regions where they are urgently needed presents a formidable challenge. Despite the high level of coordination in these supply chains, they experience strain and can fail when faced with unforeseen events. DTs offer a solution by simulating abrupt fluctuations in supply. This simulation capability enables organizations to anticipate and address sudden increases or decreases in supply, aiming to fulfill production demands and streamline the entire process [12].

In the realm of manufacturing, the Digital Twin (DT) stands as a virtual embodiment of the physical product, reflecting its design, construction, and maintenance phases. It is enriched by real-time process insights and analytical data derived from precise configurations of the physical product, production systems, or machinery. This operational framework forms the core context of the DT necessary for enhancing performance. While conceptual in nature, virtual models are complemented by real-time data, serving as a digital representation of tangible physical occurrences. CAD prototypes encapsulate the DT's virtual attributes mirroring the physical counterpart in terms of fit, form, and function. However, the execution of analytics applications, defining the state and behavior of the performance-oriented DT for optimization and process enhancement, necessitates real-time operational and asset data. Achieving smart manufacturing involves automating and digitizing manufacture methods through the integration of industrialized machineries through sensors, as well as communication tools. This integration results in enhanced production efficiency and more effective management. Digitization has enabled the construction of cyberspace, representing intricate corporal info. The degree of elegance is attainable in industrialized hinges on the depth of integration between the physical world and this digital cyberspace [9].

Drawing from practical applications, manufacturers are exploring innovative business models centered on selling services rather than the products themselves. Leveraging the DT, they monitor and enhance the availability and efficiency of these services. Customers are presented with the opportunity to utilize the product or equipment while receiving comprehensive maintenance and operational optimization. This is achieved through the extrapolative as well as rigid competences embedded within the DT. In this model, the manufacturer retains ownership of the equipment and offers maintenance services driven by the DT, presenting a more streamlined and lucrative approach to conducting business.

Thus this chapter describes the applications of DT in Section 7.2, in Section 7.3 DT in manufacturing is explained, Section 7.4 discusses the related works, Section 7.5 describes the case study. Challenges and Open research are described in sections 7.6 and 7.7 respectively. Section 7.8 concludes the chapter.

7.2 Applications and Characteristics of DT

Kreger [16], highlighted many areas where DT applications are being applied. The various applications such as Manufacturing Sector, Healthcare, Automobile, Asset Management

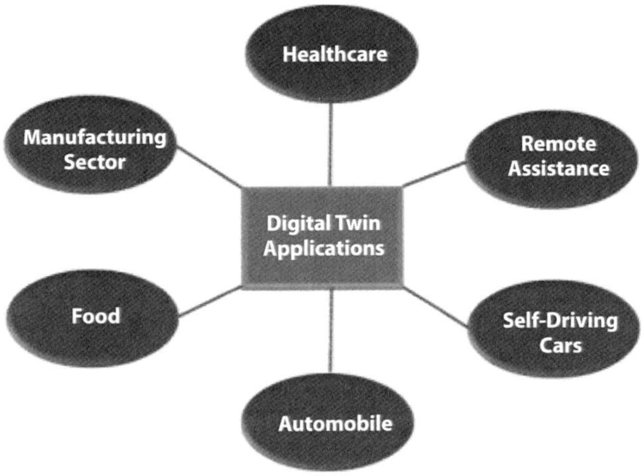

Figure 7.1 Applications of digital twin [11].

at Real-Time, Remote Assistance, Self-Driving Car and Food Industry are as shown in Figure 7.1.

Characteristic of Digital Twin

DTs exhibit numerous features, including modularity, homogenization, connectivity, and digital traceability. Embedded and connectivity systems serve as foundational elements within the DT framework due to their significant influence on the expansion and development of the digital environment. These systems play a crucial role in enabling the functionalities and growth potential of DTs, contributing to their effectiveness and adaptability within various applications and industries.

Connectivity

Connectivity stands as a pivotal distinctive within DTs. It's indispensable because the corporal artifact and its digital counterpart rely on a robust linkage facilitated by sensors, 4G/5G networks. This connection is the lifeline of a DT; without it, the essence of a DT ceases to exist. Moreover, connectivity isn't limited to just the product and its digital representation; it also extends to the relationship between the product and the customer within the supply chain. This interconnectedness ensures seamless communication and information flow, fostering a more efficient and responsive supply chain network.

Homogenization

Homogenization of diverse data bases and the ability to disentangle them into distinct physical components are enabled by DTs. Moore's Law, which observes that the computing power doubles approximately every two years, signifies that as computing capabilities escalate, the overall cost of computing diminishes over time. Essentially, this means that the marginal cost associated with implementing DTs will decrease, thereby allowing future users to conduct product tests, address prevailing issues, and even forecast future outcomes regarding targeted products. The accessibility and affordability of DT technology will progressively empower users to leverage its capabilities for a broader spectrum of applications, including predictive analytics, problem-solving, and product development.

Programmability

The digital component within a corporal artifact can be programmed using sensors, allowing for the modification and adaptation of its functionalities. DTs associated with individual products have the capability to gather performance-related data through embedded sensors. Moreover, these DTs can be reprogrammed to essentially create alternative versions or configurations of the product, such as developing a different iteration of an engine. This flexibility and adaptability are inherent strengths of DTs. By utilizing sensors and programming capabilities, manufacturers can collect real-time data, understand performance intricacies, and even simulate changes in design or functionality without altering the physical product itself. This versatility supports innovation, iterative improvements, and the development of diverse product variations within a digital framework.

Digital Tracing

DTs generate traces of data that are incredibly valuable for engineers when diagnosing and troubleshooting defective product engines. These digital traces essentially capture the entire lifecycle of a product, collecting data on its design, manufacturing, performance, and any issues encountered. Engineers rely on these traces to conduct in-depth analyses, identify anomalies or malfunctions, and pinpoint areas requiring improvement or correction. By examining these digital traces, manufacturers can gain crucial insights into the root causes of defects or inefficiencies in the final physical products. This knowledge is then used iteratively in the design and development phase to prevent similar defects from occurring in future product iterations, enhancing overall product quality and reliability. The digital traces serve as a rich resource for continuous improvement, informing and guiding the refinement of product designs to mitigate potential issues in subsequent manufacturing cycles.

Adaptability and flexibility

DTs lie in their adaptability and flexibility. They can be customized and adjusted to cater to individual needs or specific requirements. Whether it involves altering parameters, simulating changes, or even fine-tuning functionalities, DTs provide a platform where adjustments to products or systems can be made and tested virtually before implementing them in the physical realm. This adaptability allows for personalized modifications tailored to unique needs without directly impacting the physical product. Engineers and designers can experiment within the DT environment, testing various adjustments and configurations to meet specific user preferences or operational demands. This flexibility significantly reduces the risk and cost associated with making alterations directly to the physical product, offering a safe and efficient means to tailor products according to individual needs.

Context Awareness

Context-awareness sensors within DTs enable the identification and monitoring of individual product components in a given context. These sensors are instrumental in capturing detailed data about the product's surroundings, conditions, and interactions, providing a holistic view of the product's environment and its components. By leveraging these sensors, DTs gain a nuanced understanding of how individual components function within the larger system. They can detect changes, track performance metrics, and analyze interactions between various elements. This contextual awareness empowers the DT to offer insights into the behavior and condition of each component, facilitating precise diagnostics, predictive maintenance, and optimization strategies for enhanced performance and reliability.

Embeddedness

Network devices and objects are integrated into virtual environments through sensor communication with the physical environment. These network devices and objects, equipped with sensors, establish a bridge between the physical and virtual realms. Sensors embedded within these devices facilitate the collection of real-time data from the physical environment. This data is then transmitted to the DT, enabling a dynamic replication of the physical system in a virtual space. Through this communication, the DT accurately mirrors the behavior, interactions, and state of the physical components, fostering a comprehensive understanding and representation within the virtual environment. This synchronization between the physical and virtual domains is instrumental in enabling simulations, analyses, and predictive modeling within DT.

7.3 DT in Manufacturing

Societies are experiencing population growth and industrialization, integral elements of emerging economies. There's been a notable surge in the production of counterfeit and below-standard goods within our current physical environments. To counter this trend and prevent the proliferation of such products, redesigning existing product models becomes crucial. The DT solution emerges as a promising approach within the industrial sector, as it integrates various technologies, enabling more efficient service delivery and it is as shown in Figure 7.2. Achieving a digital replica of a physical product and preventing counterfeit production is feasible by enabling interaction between the physical product and its virtual counterpart [6]. Smart manufacturing seamlessly integrates industrial processes with information technology, resulting in substantial enhancements in both production effectiveness and product class. At the heart of this advancement lies DT technology, harnessing authentic production data to replicate and simulate the behaviors of physical objects in the real world [50].

Modularity, self-sufficiency, and widespread connectivity are reshaping manufacturing, ushering in novel production prospects. These include distributed manufacturing, automated order planning and execution, adaptive responses to new orders or disruptions leading to automatic production rescheduling, and enhanced decision-making support. To achieve these advancements, seamless data collection and accessibility are crucial across the entire production lifecycle. The synergy between the IoT, CPS, and DTs fulfills this essential need. CPS integrates computational abilities with physical processes, where embedded computers and networks monitor and govern physical operations, influencing computations and being influenced by them in return [18, 19].

Connected CPS can be perceived as interconnected IoT devices. When coupled with robust data analytics and artificial intelligence, they hold the potential to empower intelligent systems across diverse application domains, such as the development of smart cities [35]. The concept of CPS stands at the intersection of the physical and digital realms, setting them apart from traditional information systems. CPS operates by integrating physical and digital elements, incorporating functions like sensing, control, and networking.

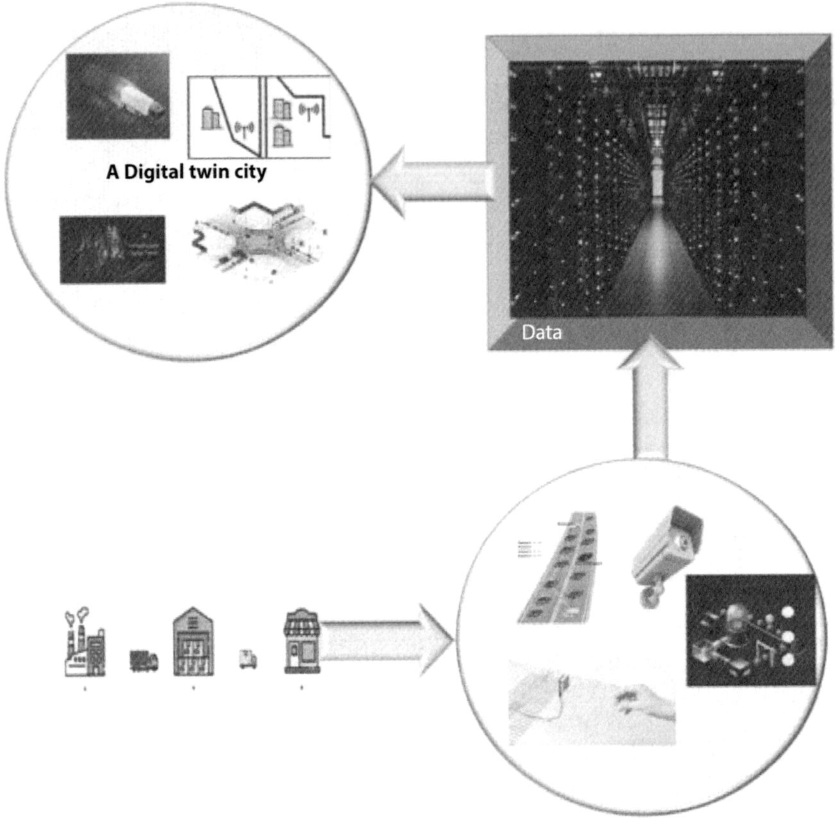

Figure 7.2 Digital twin in manufacturing.

This integration allows CPS to impact the real world directly by influencing physical processes [2]. CPS expands the scope of real-world objects by offering digital representations of physical entities. These digital descriptions directly connect with the digital counterparts of the physical world, commonly known as DTs [17].

In the realm of smart manufacturing, physical components within a factory interface with the Industrial Internet through standard cyber gateways, forming DTs within the cyber domain. Each DT represents its real-world counterpart by mirroring its physical state. The cyber domain serves as a repository for and processor of data streamed from these interconnected physical objects. This data is then harnessed for modeling, simulation, and forecasting the status of each physical element amidst varying operational conditions. The extensive utilization of smart technologies, such as Big Data Processing and AI, empowers the extraction of manufacturing insights in real-time during every phase of production. The collective intelligence pooled from locally connected factories and the cyber domain sets the stage for profound transformations in intra-business operations, inter-business collaboration, and the production paradigm [25].

In smart manufacturing, DT emerges from the fusion of real-time data sourced from physical sensors alongside CAD models and various simulation tools. This amalgamation empowers manufacturers to actively oversee and assess equipment and processes as they operate, enabling the anticipation of potential problems before they manifest.

By leveraging this information, manufacturers can implement data-driven strategies, ensuring optimized production and informed decision-making [49]. A DT has the capability to replicate the real-life performance of manufactured parts across various operational settings within virtual environments. Testing serves as a pivotal phase in component design, essential for assessing the enactment criteria of manufactured parts in line with working settings and industry standards. Through the establishment of a DT and the utilization of digital simulation and analysis, the necessity for constructing corporal archetypes can be eliminated [39]. Liu *et al.* [20] discussed that employing DT within a manufacturing site extends beyond singular parts or components; it extends to the comprehensive oversight and enhancement of an entire production line or even the entirety of the manufacturing process. From the inception and evolution of a product to its final production stages, DT serve as a tool to monitor, optimize, and refine the entirety of the manufacturing journey.

Leveraging DT for the exploration and alteration of parts significantly amplifies the optimization process in part design. Consequently, companies can harness these tools to scrutinize product designs within virtual environments, ultimately refining accuracy and elevating the quality of the resulting manufactured parts [55]. The convergence of DT technology with smart manufacturing amenities has the potential to revolutionize utilization practices, product design methodologies, and various other processes across industries [3]. The authors [54] have discussed the smart manufacturing cell is designed to undertake various specific tasks, including 1. Autonomous Cleaning of the Process System: The smart manufacturing cell operates by continuously monitoring the processing areas within the system in real time. If the system identifies an excessive accumulation of chips in a specific part, impeding the processing efficiency, it autonomously initiates a cleaning process. Through control of robots and associated execution equipment, the system effectively clears the chips. This proactive action is aimed at ensuring optimal processing quality and efficiency in production. 2. Fault Handling in Processing: The smart manufacturing cell possesses autonomous fault detection and resolution capabilities, eliminating the need for human intervention. It can identify issues like inappropriate holding, workpieces falling from their clamps, or cutter damage during processing, and addresses these concerns independently. 3. Workpiece Quality Management: Throughout the processing life cycle, smart sensors consistently monitor the workpiece quality. If any subpar quality is detected, the execution devices promptly redirect the deficient workpiece to the designated waste area. This swift action significantly enhances the overall operational efficiency of the system. 4. Processing Environment Safety and Security: Continuous monitoring ensures the safety of the smart manufacturing cell's environment. Detection of foreign objects or individuals approaching the cell prompts the system to take immediate action. It can control processing devices to slow down, stop operations, or execute other relevant reactions, ensuring the secure operation of the system.

7.4 Related Work

Continuous data monitoring informs smart manufacturing approaches, empowering companies to customize production runs more effectively to meet consumer demands [51]. A few examples of manufacturing use cases that can be achieved through upcoming Industrial

IoT and edge computing solutions such as Predictive maintenance and equipment safety, Production flow monitoring and optimization and Supply chain optimization [5].

Min *et al.* [22] delve into the optimization of production within the petrochemical industry through the exploration of DT. Lattanzi *et al.* [28] presented the review that aimed to evaluate and update the current understanding of DT concepts, specifically focusing on their practical application in smart manufacturing industries. It aimed to assess their suitability for implementation in real industrial processes and to highlight the most recent advancements in industrial methods. Jain *et al.* [13] extant a model study focusing on a DT methodology for diagnosing faults in distributed Photovoltaic Systems. Leveraging advancements in DT, the team successfully develops a real-time capable DT that is capable of accurately estimating faults associated with PV energy units.

Giacomo *et al.* [7] utilize service composition methods to coordinate DTs, aiming to devise an efficient manufacturing method plan that minimizes charges while upholding the final product's quality. Their approach extends service composition methodologies into a stochastic framework, accommodating the unpredictable behaviors of services such as DTs and the underlying manufacturing elements like machines and human resources subject to wear and tear. By addressing this uncertainty, they formulate an optimal solution by tackling a probabilistic planning problem, specifically solving a Markov decision process. This solution factors in the probabilities of breakdowns and considers the costs associated with employing particular manufacturing elements, ensuring a balanced and effective manufacturing strategy. Xiao *et al.* [47] introduced and investigated modeling concepts, methodological ideas, and theoretical frameworks centered on DT systems within strategic enterprise contexts, particularly in the realm of smart manufacturing. Shikata *et al.* [40] employed DT in the prototyping of automobiles to assess the relationship between customers' ride ease and power ingestion. Their research also investigated the impact of power-related components on overall vehicle performance.

Qi *et al.* [31] delineated a framework showcasing the utilization of a DT in manufacturing services. Their study scrutinized how manufacturers leverage diverse components of DTs, employing them as service-oriented entities within the manufacturing domain.

Rosen *et al.*, [34] discussed that the idea behind DTs involves employing advanced modeling and simulation methods to generate virtual replicas of physical systems. This facilitates precise emulation of the behaviors and traits demonstrated by the actual system, offering a wide range of potential applications across multiple domains. Integrating DTs equips operators with the capability to monitor production processes, identify irregularities within a controlled virtual environment, and bolster security protocols within process industries. Zhang *et al.*, [53] introduced a structure within the manufacturing system realm that harnesses DT technology. This framework aims to streamline engineering analysis, fostering swift design iterations, and facilitating optimization specifically tailored for crystal manufacture lines.

Chhetri *et al.*, [4] discussed the impact of digitalization on the evolution of smart manufacturing design and advancements. They introduce a methodology supported by a case study leveraging IoT's expansion to construct a DT of a manufacturing process. This approach involves IoT sensors gathering and storing data streams to circuitously highlight ancillary channel conditions. Such data aids in localizing faults and identifying the issues within manufacturing operations.

Liu *et al.* [21] applied a DT to evaluate the production method of marine diesel engines, aiming to enhance artifact value and expedite the progress phase. They devised three fundamental technologies such as establishing a real-time correlation between machining data and method strategy info, creating a DT-centered framework for assessing machining processes, and implementing process evaluations guided by DT data. Longo *et al.* [24] used DT skill towards controlling the manufacture rate as well as method eminence of industrialized methods.

Han *et al.* [11] introduced a monitoring and traceability system centered on DTs for a continuous casting machine, specifically targeting parts necessitating frequent maintenance. This system, tailored for application in steel-making enterprises, acquires real-time operational data and historical work-related information concerning machine components. The delivery of this data to operators occurs via a 3D DT model, providing a comprehensive overview of the machine's operation and maintenance status. Manufacturing works in the growth and task of a digital factory twin [1]. They suggested an eight dimensional DT model. Pargmann *et al.* [29] introduced a cloud-based monitoring system for DTs, specifically applied in the progress and discerning of wind farms. They showcased a functional prototype employing data streams and specified parameters from both methodical and commercial perspectives. This facilitated the establishment of a viable twin representing the progression of a wind farm expansion. Marrella *et al.* [27] demonstrate the use of automated planning to address unexpected and exceptional events. Specifically, planning techniques are utilized to rectify a process instance, reinstating the conditions necessary to resume the standard, manually distinct method.

Rauch and Pietrzyk [33] incorporated DTs into the industrialized method of great strengthen steel shreds. They developed an effective systematic line that comprised essential components. The evolution and advancement of IoT as well as big data have played a pivotal role in the widespread adoption of DTs in manufacturing. This adoption is attributed to their benefits in real-time processes, exceptional accuracy, traceability, and seamless incorporation. In merchandise expansion, recreation tools allows for dynamic alterations within a merchandise's core constituents, resulting in substantial cost reductions and minimized substantiation efforts. Within manufacturing, virtual environments can aid in equipment diagnostics, process simulations, and predictive analysis, effectively mitigating irregular production occurrences. Continuous gathering and analysis of data through DT models empower predictive maintenance strategies, consequently minimizing manufacturing risks [32].

Lim *et al.* [23] devised a prototype utilizing DTs for tower cranes. This approach aimed to create digital representations of possessions and environs within a vibrant setting, integrating them into a simulation system to facilitate decision-making processes. To expedite the creation of the DT, the authors proposed a comprehensive DT architecture with a three-layer technology heap. Building upon this architecture, they introduced a DT system capable of managing disruptions within complex environments by enhancing context awareness. Additionally, they applied this DT to the design of modular tower cranes engaged in transportation activities within oil and gas facilities.

Automotive companies frequently encounter challenges such as meeting deadlines due to the ever-changing demands and orders as of with their consumers. Researchers have suggested employing a web-based cyber-physical system, utilizing DT technology, to handle unusual situations within production lines in the automotive industry. This system's effectiveness was evaluated through experiments [41]. Modoni *et al.* [26] proposed a conceptual exemplary aiming to deepen comprehension of the DT concept, emphasizing the constant synchronization between an actual factory and its digital replica. This synchronization is achieved through two data streams. The first stream encompasses real-time monitored data, comprising physical variables identified by widespread sensors across the factory floor. These sensors are attached to diverse physical components and transmit data at high frequency to the digital realm. The second stream involves actions to be executed in real time or near real time within the factory. These actions constitute feedback from the digital space back to the physical factory, establishing a continuous loop of information exchange. Xu *et al.* [48] have proposed an AI-centered elucidation dedicated to fault diagnosis within smart manufacturing setups. Their work emphasizes the hurdles associated with acquiring sufficient training data to develop precise AI algorithms, a crucial element in crafting accurate DTs for novel industrialized procedures. To counter these tasks, their approach involves initially deploying a DT at the forefront. This DT learns and identifies faults while simultaneously generating essential training data. Subsequently, in the second stage, transfer learning comes into play, leveraging the accumulated training data from the initial phase of the algorithm. This innovative strategy culminates in a more refined fault-diagnosing system.

7.5 Case Study: Challenge Advisory [30]

Challenge Advisory, a firm specializing in DT Solutions for Manufacturing, undertook a project in the automobile manufacturing sector, introducing a DT Genie. The primary project objectives were to enhance yearly revenues by 15-20%, shorten manufacturing time to 12-13 hours, and ensure on-demand operational process visibility for the client.

Remarkably, within a concise two-week period involving data gathering and the implementation of the DT Genie, the results surpassed the client's outlooks significantly:

- Profit margins surged impressively by 41-54%.
- Additional monitoring features were integrated, such as daily mileage tracking and real-time engine performance monitoring

This successful implementation showcased substantial improvements across profitability, production efficiency, and operational insights, exemplifying the transformative impact of DT technology in the manufacturing landscape.

7.5.1 Case Study: Whirlpool

Whirlpool has embraced a digital thread approach, leveraging it to gather real-world data from smart associated devices aligned with product definitions. Their utilization of DT technology allows for testing low-fidelity prototype with marginal investment. This strategic implementation has yielded several advantageous outcomes for Whirlpool such as Faster innovation in the product expansion stage, enabling the company to introduce new products to the market at a higher frequency, Centralization of data facilitated by the digital thread, and fostering increased agility throughout their entire value chain and Cost reduction across product development and testing, due to simulations in the cyber space, ultimately leading to substantial decreases in overall manufacturing costs.

7.5.2 Case Study: Woodward

One of the leading producers of industrial components and controllers worldwide, Woodward sought to enhance their adaptability and decision-making capabilities. They integrated the DT concept across their design, manufacturing, and service phases, resulting in a centralized data repository that encapsulated the entirety of their end-to-end processes. This integration brought forth several advantageous outcomes for Woodward such as Enhanced Operational Visibility, Experienced improved production rates, Gained flexibility and responsiveness in their processes and Attained higher product quality while reducing manufacturing expenses.

7.6 Challenges to Implement DT [36]

Navigating the vast landscape of DT technology presents numerous challenges. The hurdles in developing a DT vary based on its scope and complexity, yet some barriers are universally encountered. Despite its immense potential, DT's infancy stage introduces complexities that span engineering, technology, and commercial domains [37]. These complexities encompass issues such as the unclear definition of DT, inadequate tools, high investment costs, data-related obstacles, absence of standardized regulations, and more.

7.6.1 Innovation of Technology

As DT remains in its nascent stage, leaving a void in understanding its potential benefits for individuals, businesses, and industries. The deficiency in technical expertise obstructs its advancement, compounded by the absence of documented successful DT implementations and accurate cost projections [38]. DT relies on a convergence of various evolving technologies like 3D simulations, IoT/IIoT, AI, big data, ML and CC. However, the developmental phase of these foundational technologies hampers the seamless progression of DT. Improving the infrastructure is essential to bolster the effectiveness of DT implementation [43].

7.6.2 Time and Cost

One of the primary hurdles hindering the full realization of DT is its substantial implementation cost. The intricate process of crafting highly detailed computer models and simulating processes demands extensive time and labor, alongside a considerable computational prowess. Consequently, the investment in DT becomes notably expensive [8].

7.6.3 Lack of Standards and Regulations

Given the abundance of DT models and architectures documented in literature, there arises a crucial necessity: establishing a unified framework for DT across industries. This framework should encompass standardized interfaces, fostering shared understanding, and ensuring uniformity. Simultaneously, it should prioritize an optimized design for data flow, facilitating easier data accessibility without conceding security measures [45]. Establishing standards for models, interfaces, protocols, and data holds paramount importance, particularly in industries like aerospace, automotive, and healthcare. This standardization plays a pivotal role in ensuring efficient third-party communication, safeguarding product and human safety, and upholding data security and integrity. In addition to technical standards, the development of standards and interoperability based on those standards is crucial to tackle the social and organizational challenges arising from digital transformation within industries [46].

7.6.4 Data Related Issues

A significant hurdle concerning data within DT lies in its convergence and ascendancy. Initiatives revolving around big data often encounter failures attributed to inadequate ascendancy and management of data. These challenges encompass issues such as identifying and accessing data, harmonizing diverse data from various sources, grappling with data of subpar quality, and encountering translation discrepancies [44]. Addressing these challenges is pivotal to ensure the success of projects reliant on big data within the realm of DT.

7.6.5 Life-Cycle Mismatching

One of the pressing concerns with DT involves products characterized by extensive life cycles, such as buildings, aircraft, ships, machinery, and even cities. The longevity of these products surpasses the lifespan of software utilized for DT design, simulation, data storage, and analysis. Consequently, there's a looming risk that the software formats may become outdated or constrained by vendor lock-in when transitioning to new software versions or alternative authoring tools in the future [10]. This predicament poses a potential challenge, as it could hinder seamless integration and continuity in leveraging DT for long-standing products and infrastructures.

7.7 Open Research

Smart manufacturing has the potential to streamline manufacturing processes and overhaul entire business operations, leading to heightened productivity, reduced costs, and minimized waste materials [31]. Soori *et al.* [42] have discussed some areas for future research in DT technology for smart manufacturing:

7.7.1 Integration with the Internet of Things (IoT)

The IoT constitutes a network of physical devices equipped with sensors, software, and connectivity, facilitating data interchange. When combined with DT technology, this integration empowers real-time observing, examining, and mechanism of manufacturing developments.

7.7.2 Integration with Industry 4.0 Technologies

Encompassing DT technology within a spectrum of Industry 4.0 innovations like IoT, AI, and machine learning amplifies the capabilities and efficiency of DTs, bolstering their role in optimizing smart manufacturing processes.

7.7.3 Multi-Scale DTs

Utilizing multi-scale DTs offers a nuanced and exhaustive comprehension of the manufacturing method, integrating various levels of intricacy from the microscopic to the macroscopic. This method facilitates the detection and resolution of issues across different scales, culminating in superior method optimization.

7.7.4 Integration with Advanced Analytics

The utilization of DT technology yields extensive data regarding manufacturing methods and tools enactment. When coupled with sophisticated analytical tools like ML and AI, this integration empowers manufacturers with profound operational insights, enabling informed decision-making.

7.7.5 Real-Time Optimization

The real-time optimization facilitated by DTs empowers manufacturers to swiftly adapt their processes to dynamic conditions, such as shifts in demand or disruptions in the supply chain. This agility enhances process efficiency, minimizes downtime, and elevates whole production.

7.7.6 Collaboration and Communication

DTs serve as a conduit for fostering collaboration and communication among various stakeholders involved in the manufacturing cycle, encompassing inventors, engineers, workers, and administrators.

7.7.7 Optimization of Supply Chain Processes

The application of DT technology extends to generating virtual depictions of supply chain networks, empowering constructors toward streamline logistics and enhances inventory management. Upcoming research endeavors might concentrate on refining intricate models for supply chain optimization through the utilization of DT tools.

7.7.8 Human–Machine Collaboration

Future exploration initiatives could delve into amalgamating DT technology with human-machine interfaces, aiming to enhance association and decision-making for a more cohesive and effective operational environment.

7.7.9 Cybersecurity

As the prevalence of DT rises within manufacturing, safeguarding these virtual replicas against cyber threats becomes paramount. Imminent research initiatives may center on devising robust cybersecurity rules and tools aimed at shielding DT from potential mischievous spasms.

7.7.10 Scalability

With escalating concerns about environmental sustainability, manufacturers are actively seeking methods to curtail their carbon footprint and diminish waste. DT technology emerges as a tool to model the ecological impact of manufacturing processes, pinpointing areas for enhancement. Its application spans diverse industries.

7.8 Conclusion

The utilization of a DT proves highly advantageous within the business landscape. It enables the prediction of current and future conditions of physical products by analyzing their digital counterparts within a virtual environment. Implementing DTs supports enterprises in enhancing product performance, improving customer satisfaction, and facilitating informed decision-making regarding operational strategies. By leveraging DTs that mirror both the product and production systems, manufacturers can significantly curtail the time and expenses linked with assembling, installing, and validating factory production systems. Moreover, the adoption of DTs in asset management commonly yields measurable advantages in maintaining field equipment. Thus this chapter discussed the challenges and the open research to further work in DTs in Manufacturing.

References

1. Azeez, N.A. and Adjekpiyede, O.O., Digital Twin Technology: A Review of Its Applications and Prominent Challenges. *Covenant J. Inf. Commun. Technol.*, 10, 1, 2022.
2. Boyes, H., Hallaq, B., Cunningham, J., Watson, T., The industrial Internet of Things (IIoT): An analysis framework. *Comput. Ind.*, 101, 1–12, 2018.
3. Cinar, Z.M., A.A.Nuhu, Zeeshan, Q., Korhan, O., DTs for Industry 4.0: A Review. *Comput. Sci.*, 193–203, 2019.
4. Chhetri, S.R., Faezi, S., Canedo, A., Faruque, M.A.A., QUILT: quality inference from living DTs in IoT-enabled manufacturing systems, in: *Proceedings of the International Conference on Internet of Things Design and Implementation - IoTDI '19*, pp. 237–248, 2019.
5. O'Connell, E., 'Brien, W.O., Bhattacharya, M., Moore, D., Penica, M., DTs: Enabling Interoperability in Smart Manufacturing. *Netw. Telecom.*, 4, 265–2784, 2023.
6. Ezekiel, L., Building Blocks for DT of Additive Manufacturing. 35–37, 2018.
7. De Giacomo, G., Favorito, M., Leotta, F., Mecella, M., Silo, L., DT composition in smart manufacturing via Markov decision Processes. *Comput. Ind.*, 149, 2023, 103916, 2023.

8. Gabor, T., Belzner, L., Kiermeier, M., Beck, M.T., Neitz, A., A simulation-based architecture for smart cyber-physical systems, in: *Proceedings of the 2016 IEEE International Conference on Autonomic Computing (ICAC)*, Wuerzburg, Germany, vol. 17–22, pp. 374–379, 2016.
9. Geng, R., Li, M., Hu, Z., Han, Z., Zheng, R., DT in smart manufacturing: remote control and virtual machining using VR and AR technologies. *Struct. Multidiscipl. Optim.*, 65, 11, 32, 2022.
10. Gagne, M.R., DTs, Another Reason to Worry about the iot and Data Security, 2020, Available online: https://irishtechnews.ie/digital-twins-iot-and-data-security/.
11. Han, Z., Li, Y., Yang, M., Yuan, Q., Ba, L., Xu, E., DT-driven 3D visualization monitoring and traceability system for general parts in continuous casting machine. *J. Adv. Mech. Des. Syst. Manuf.*, 14, 7, JAMDSM0100–JAMDSM0100, 2020.
12. Hinduja, H., Kekkar, S., Chourasia, S., Bharadwaj Chakrapani, H., Industry 4.0: DT and its Industrial Applications. *Int. J. Sci. Eng. Technol.*, 8, 4, 1–7, 2020.
13. Jain, P., Poon, J., Singh, J.P., Spanos, C., Sanders, S., Panda, S.K., A DT Approach for Fault Diagnosis in Distributed Photovoltaic System. *IEEE Trans. Power Electron.*, 1, 1, 99, 2019.
14. Negri, E., Fumagalli, L., Macchi, M., A Review of the Roles of DT in CPS-based Production Systems. *Proc. Manuf.*, 11, June, 939–948, 2017.
15. Koulamas, C. and Kalogeras, A., Cyber-physical systems and DTs in the industrial internet of things. *Comput. (Long. Beach. Calif.)*, 51, 11, 95–98, 2018.
16. Kreger, L., *Future Visions: Tales on Realization of DT*, pp. 13–17, Springer, Berlin, 20182016.
17. Kalogeras, A.P., Rivano, H., Ferrarini, L., Alexakos, C., Iova, O., Rastegarpour, S., Mbacké, A.A., Cyber physical systems and Internet of Things: Emerging paradigms on smart cities, in: *Proc. 1st Int. Conf. Societal Autom. (SA)*, pp. 1–13, 2019.
18. Lee, E.A., Cyber-physical systems-are computing foundations adequate, in: *Proc. Position Paper NSF Workshop cyber-Phys. Syst., Res. Motivat., Techn. Roadmap*, vol. 2, Citeseer, Princeton, NJ, USA, pp. 1–9, 2006.
19. Lee, E.A., Cyber physical systems: Design challenges, in: *Proc. 11th IEEE Int. Symp. Object Compon.-Oriented Real-Time Distrib. Comput. (ISORC)*, May 2008, pp. 363–369, 2008.
20. Liu, Q., Leng, J., Yan, D., Zhang, D., Wei, L., Yu, A., Zhao, R., Zhang, H., Chen, X., DT-based designing of the configuration, motion, control, and optimization model of a flow-type smart manufacturing system. *J. Manuf. Syst.*, 58, 52–64, 2021.
21. Liu, J.F., Zhou, H.G., Liu, X.J., Tian, G.Z., Wu, W.F., Cao, L.P., Wang, W., Dynamic evaluation method of machining process planning based on DT. *IEEE Access*, 7, 19312–19323, 2019.
22. Lattanzi, L., Raffaeli, R., Peruzzini, M., Pellicciari, M., DT for smart manufacturing: a review of concepts towards a practical industrial implementation. *Int. J. Comput. Integr. Manuf.*, 34, 567–597, 2021.
23. Lim, K.Y.H., Zheng, P., Chen, C.H., Huang, L., A DT-enhanced system for engineering product family design and optimization. *J. Manuf. Syst.*, 57, 82–93, 2020.
24. Longo, F., Nicoletti, L., Padovano, A., Ubiquitous knowledge empowers the smart factory: The impacts of a service-oriented DT on enterprises performance. *Ann. Rev. Control*, 47, 221–236, 2019.
25. Lu, Y., Liu, C., Wang, K., Huang, H., Xu, X., DT-driven smart manufacturing: Connotation, reference model, applications and research issues. *Robot. Comput. Integr. Manuf.*, 61, 101837, 2020, 10.1 016/j.rcim.2019.101837.
26. Modoni, G.E., Caldarola, E.G., Sacco, M., Terkaj, W., Synchronizing physical and digital factory: Benefits and technical challenges. *Proc. CIRP*, 79, 472–477, 2019.
27. Marrella, A., Mecella, M., Sardina, S., Intelligent process adaptation in the SmartPM system. *ACM Trans. Intell. Syst. Technol.*, 8, 2, 1–43, 2016.
28. Min, Q., Lu, Y., Liu, Z., Su, C., Wang, B., Machine Learning based DT Framework for Production Optimization in Petrochemical Industry. *Int. J. Inf. Manage.*, 2019.

29. Pargmann, H., Euhausen, D., Faber, R., Intelligent big data processing for wind farm monitoring and analysis based on cloud-technologies and DTs: A quantitative approach, in: *2018 IEEE 3rd International Conference on Cloud Computing and Big Data Analysis (ICCCBDA), (Chengdu)*, IEEE, pp. 233–237, 2018.
30. Bhamre, P., Significance and future scope of DT technology in advanced manufacturing. *Int. J. Manage.*, 11, 11, 1579–1588, 2020, Article ID: IJM_11_11_149.
31. Qi, Q., Tao, F., Zuo, Y., Zhao, D., DT Service towards Smart Manufacturing. *Proc. CIRP*, 72, 237–242, 2018.
32. Qiu, H., Chen, Y., Zhang, H., Yi, W., Li, Y., Evolutionary DT model with an agent-based discrete-event simulation method. *Appl. Intell.*, 53, 6, 6178–6194, 2023.
33. Rauch, L. and Pietrzyk, M., DTs as a modern approach to design of industrial processes. *J. Mach. Eng.*, 19, 1, 86–97, 2019.
34. Rosen, R., Von Wichert, G., Lo, G., Betten Hausen, K.D., About the importance of autonomy and DTs for the future of manufacturing. *IFAC-PapersOnLine*, 48, 3, 567–572, 2015.
35. Serpanos, D., The cyber-physical systems revolution. *Computer*, 51, 3, 70–73, 2018.
36. Singh, M., Fuenmayor, E., Hinchy, E.P., Qiao, Y., Murray, N., Devine, D., Review DT: Origin to Future. *Appl. Syst. Innov.*, 4, 36, 2021.
37. Singh, S., Shehab, E., Higgins, N., Fowler, K., Tomiyama, T., Fowler, C., *Challenges of DT in High. Value Manufacturing*, pp. 0148–7191, SAE Technical Paper, 2018.
38. Simchenko, N., Tsohla, S., Chyvatkin, P., IoT & DTs Concept Integration Effects on Supply Chain Strategy: Challenges and Effects. *Int. J. Sup. Chain. Mgt.*, 2019, 8, 803.
39. Shao, G., Jain, S., Laroque, C., Lee, L.H., Lendermann, P., Rose, O., DT for smart manufacturing: the simulation aspect, in: *2019 Winter Simulation Conference (WSC)*, IEEE, pp. 2085–2098, 2019.
40. Shikata, H., Yamashita, T., Arai, K., Nakano, T., Hatanaka, K., Fujikawa, H., DT environment to integrate vehicle simulation and physical verification. *SEI Tech. Rev.*, 88, 18–21, 2019.
41. Son, Y.H., Park., K.T., Lee, D., Jeon, S.W., Noh, S.D., DT-based cyber-physical system for automotive body production lines. *Int. J. Adv. Manuf. Technol.*, 115, 291–310, 2021, doi: 10.1007/s00170-021-07183-3.
42. Soori, M., Arezoo, B., Dastres, R., DT for smart manufacturing, A review. *Sustain. Manuf. Serv. Econ.*, 2, 100017, 2023.
43. Tao, F., Cheng, J., Qi, Q., Zhang, M., Zhang, H., Sui, F., DT-driven product design, manufacturing and service with big data. *Int. J. Adv. Manuf. Technol.*, 94, 3563–3576, 2018.
44. Tyagi, P., Demirkan, H., Lakes, D., The Biggest Big Data Challenges. *Anal. Mag.*, 2016.
45. Wagner, R., Schleich, B., Haefner, B., Kuhnle, A., Wartzack, S., Lanza, G., Challenges and potentials of DTs and Industry 4.0 in product design and production for high performance products. *Proc. CIRP*, 2019, 84, 88–93, 2019.
46. Wyckoff, A. and Pilat, D., *Key Issues for Digital Transformation in the G20*, OECD, Berlin, Germany, 2017.
47. Xiao, J.H., Xie, K., Chi JY, J.Y., Intelligent manufacturing, DT and strategic scenario modelling. *J. Beijing Jiaotong Univ.: Soc. Sci. Ed.*, 18, 2, 69–71, 2019.
48. Xu, Y., Sun, Y., Liu, X., Zheng, Y., A Digital-Twin-Assisted Fault Diagnosis Using Deep Transfer Learning. *IEEE Access*, 7, 19990–19999, 2019.
49. Yasin, A., Pang, T.Y., Cheng, C.T., Miletic, M., A roadmap to integrate DTs for small and medium-sized enterprises. *Appl. Sci.*, 11, 9479, 2021.
50. Yi, H. and Fan, K., Co-simulation-based DT for thermal characteristics of motorized spindle. *Int. J. Adv. Manuf. Technol.*, 1–13, 2023.
51. Zeid, A., Sundaram, S., Moghaddam, M., Kamarthi, S., Marion, T., Interoperability in smart manufacturing: Research challenges. *Machines*, 7, 21, 2019.

52. Zhang, Y.F., Dang, Z., Shan, R., Survey on current research and future trends of smart manufacturing and its key technologies. *Mech. Sci. Technol. Aeros. Eng.*, 38, 3, 329–338, 2019.
53. Zhang, H., Liu, Q., Chen, X., Zhang, D., Leng, J., A DT-based approach for designing and multi-objective optimization of hollow glass production line. *IEEE Access*, 5, 26901–26911, 2017.
54. Zhang, J., Li, P., Luo, L., DT-based smart manufacturing cell: Application Case, System Architecture and Implementation. *J. Phys. Conf. Ser.*, 1884, 2021, 012017, 2021.
55. Zheng, P. and Sivabalan, A.S., A generic tri-model-based approach for product-level digital twin development in a smart manufacturing environment. *Robot. Comput. Integr. Manuf.*, 64, 101958, 2020.

8

Blockchain Based Digital Twin for Smart Manufacturing

Amit Kumar Tyagi[1]*, Shabnam Kumari[2], Richa[3] and Utkarsh Kumar[4]

[1]Department of Fashion Technology, National Institute of Fashion Technology, New Delhi, Delhi, India
[2]Department of Computer Science, Faculty of Science and Humanities, SRM Institute of Science and Technology, Kattankulathur, Chennai, Tamilnadu, India
[3]Department of Computer Science and Engineering, Birla Institute of Technology, Mesra, Ranchi, India
[4]Technical Engineer, Energy Efficiency Services Ltd., India

Abstract

The fusion of blockchain and digital twin concepts revolutionizes smart manufacturing, leveraging blockchain's security and decentralization to augment digital twin capabilities. Digital twins, virtual replicas of physical manufacturing systems, gather real-time data to mirror their real-world counterparts. By integrating blockchain, data security and integrity are guaranteed, enhancing manufacturing optimization and efficiency. The blockchain-based digital twin for smart manufacturing provides numerous advantages, including enhanced data security, transparent data sharing among stakeholders, and a single source of truth for all parties involved. This technology enables real-time monitoring, predictive maintenance, and streamlined operations, ultimately leading to improved productivity, reduced downtime, and cost savings. As smart manufacturing continues to evolve, the blockchain-based digital twin is poised to revolutionize the industry, providing a foundation for more efficient and agile manufacturing processes. However, to realize its full potential, challenges related to interoperability, standardization, and scalability must be addressed, and industry-wide collaboration is essential. In summary, the blockchain-based digital twin for smart manufacturing holds the promise of greater efficiency, security, and transparency in manufacturing operations. Hence, by combining the strengths of blockchain and digital twin technology, manufacturers can thrive in an increasingly competitive and dynamic landscape while fostering innovation and sustainable practices.

Keywords: Blockchain technology, security, smart manufacturing, digital twin technology

8.1 Introduction to Blockchain, Digital Twin, and Smart Manufacturing

Blockchain, known for its role in cryptocurrencies like Bitcoin, is a decentralized ledger technology. It stores data in immutable blocks, each containing transactions. Consensus

**Corresponding author*: amitkrtyagi025@gmail.com

Amit Kumar Tyagi, Shrikant Tiwari, Senthil Kumar Arumugam and Avinash Kumar Sharma (eds.) Artificial Intelligence-Enabled Digital Twin for Smart Manufacturing, (143–178) © 2024 Scrivener Publishing LLC

from the network is required to alter or delete blocks, ensuring data integrity [1, 2]. Few key features of blockchain technology include:

- Decentralization: Blockchain operates peer-to-peer, removing the need for centralized authorities like banks or governments to validate transactions.
- Transparency: All transactions are visible to participants, fostering trust through high transparency.
- Security: Blockchain employs cryptographic techniques for data security, resisting fraud and hacking attempts.
- Immutability: Transactions once added are nearly impossible to alter or delete, ensuring a permanent, auditable record.
- Wide Applications: Blockchain spans various sectors from finance to supply chain management, promising secure and transparent data and transaction management, revolutionizing industries.

Digital Twin

A digital twin is a virtual copy of a physical object or system, created by combining real-time data with CAD and simulation software. It accurately mirrors its physical counterpart and finds applications in various domains like manufacturing, healthcare, and urban planning [3]. Few Key aspects of digital twins include:

- Data Integration: Digital twins incorporate data from sensors, IoT devices, and other sources to provide a real-time and accurate representation of the physical asset or system.
- Simulation and Analysis: They enable simulations, predictive modeling, and analysis of the physical counterpart, helping in optimizing performance and making informed decisions.
- Monitoring and Control: Digital twins allow for remote monitoring and control of physical assets, improving operational efficiency and reducing downtime.
- Lifecycle Management: They can be used throughout the entire lifecycle of an asset, from design and manufacturing to operation and maintenance.

Digital twins are particularly valuable in industries like manufacturing, where they can optimize production processes, monitor equipment health, and support predictive maintenance. They provide a means for organizations to gain better information, make data-driven decisions, and improve the efficiency and performance of their physical assets.

Smart Manufacturing

Smart manufacturing, synonymous with Industry 4.0, integrates advanced technologies into manufacturing to boost productivity, quality, and flexibility. Leveraging the Internet of Things (IoT), artificial intelligence (AI), automation, and data analytics, it ensures more efficient and responsive manufacturing processes [23]. Few Key components of smart manufacturing include:

- IoT and Sensors: Smart manufacturing relies on a network of sensors and IoT devices to collect real-time data from machines, products, and processes.

- Data Analytics: Data from sensors and other sources are analyzed to gain information and improve decision-making.
- Automation: Automation is used to optimize processes, reduce human error, and increase efficiency.
- Integration: Smart manufacturing integrates various systems and processes, creating a seamless flow of data and information across the production line.

Digital Twin in Smart Manufacturing

Digital twins play an important role in smart manufacturing by providing a virtual replica of the production environment, allowing for real-time monitoring and control. They can help manufacturers optimize production, reduce downtime, predict maintenance needs, and improve product quality [3].

Blockchain in Smart Manufacturing

Blockchain in smart manufacturing secures and streamlines supply chain processes, tracks material provenance, and ensures product data authenticity. It enhances transparency, trust, and traceability throughout the manufacturing process, making it a valuable tool for quality control and supply chain management [4].

In summary, blockchain, digital twins, and smart manufacturing are cutting-edge technologies with the potential to transform industries. Blockchain enhances security and transparency, digital twins enable real-time monitoring and analysis, and smart manufacturing uses these technologies to create more efficient and data-driven manufacturing processes.

8.1.1 Background of Blockchain, Digital Twin, and Smart Manufacturing

Background of Blockchain

Conceptual Origins: The concept of a blockchain-like system can be traced back to 1991 when Stuart Haber and W. Scott Stornetta introduced a cryptographically secure chain of blocks to timestamp digital documents, making them tamper-proof.

- Bitcoin's introduction in 2009, attributed to "Satoshi Nakamoto," pioneered blockchain's practical application, creating a decentralized digital currency [5].
- Ethereum's 2015 debut extended blockchain's utility beyond cryptocurrencies, introducing smart contracts for programmable agreements.
- Blockchain's adoption across industries such as finance, supply chain management, and healthcare ensured heightened security and transparency in digital transactions and records.

Background of Digital Twin

- Origin in Engineering: Digital twin concept originated from creating digital representations of physical objects for simulation, rooted in fields like CAD and simulation software [6].
- NASA's Pioneering: NASA utilized digital twin technology in the 1960s for spacecraft simulations and mission planning, showcasing early adoption.

- IoT Integration: Integration of IoT enabled advanced digital twin development by providing real-time data from sensors, facilitating virtual representations mirroring physical assets.

Background of Smart Manufacturing

- Evolution of Industrial Revolutions: Smart manufacturing aligns with the Fourth Industrial Revolution, building upon previous advancements such as mechanization, mass production, electricity, and automation.
- Industry 4.0 Concept: Coined in Germany, Industry 4.0 integrates technologies like automation, data exchange, IoT, cloud computing, and AI into manufacturing for enhanced connectivity and intelligence.
- IoT's Role: IoT's widespread adoption facilitated smart manufacturing by interconnecting machines, sensors, and devices, enabling data collection, transmission, and analysis for improved control and optimization.

In summary, blockchain, digital twin technology, and smart manufacturing have all evolved over time in response to the growing need for increased security, efficiency, and transparency in various industries. While blockchain originated with the creation of Bitcoin, digital twins developed from early simulation and modeling concepts, and smart manufacturing emerged as a response to the Fourth Industrial Revolution and the rise of IoT technologies. Each of these technologies is now poised to revolutionize how various industries operate.

8.1.2 Role of Digital Twins in Manufacturing

Digital twins play an essential role in manufacturing by providing a range of benefits that contribute to improved efficiency, productivity, and decision-making [8]. Here are some key aspects of the importance of digital twins in manufacturing (Figure 8.1):

- Real-time Monitoring and Control: Digital twins offer real-time insights into asset performance, production processes, and supply chain operations, enabling immediate response to issues and minimizing disruptions.
- Predictive Maintenance: Continuous monitoring allows digital twins to forecast maintenance needs, helping schedule proactive maintenance to prevent breakdowns and optimize asset lifespan.
- Process Optimization: Through simulations and data analysis, digital twins identify bottlenecks and inefficiencies, enabling experimentation with scenarios to optimize production and enhance operational efficiency.
- Quality Control: Real-time monitoring helps detect deviations from quality standards, ensuring only high-quality products reach customers.
- Product Development and Prototyping: Digital twins facilitate virtual prototyping, allowing testing and refinement of designs before physical prototypes are produced, saving time and resources.
- Supply Chain Optimization: Extending beyond manufacturing, digital twins track material and product movement, optimizing logistics and enhancing transparency.

Figure 8.1 Importance of Digital Twins in manufacturing.

- Energy Efficiency: Monitoring energy consumption identifies opportunities for savings, reducing costs, and supporting sustainability goals.
- Customization and Personalization: Understanding customer preferences enables manufacturers to deliver customized products at scale.
- Simulation and What-If Analysis: Digital replicas support simulation and analysis, allowing assessment of changes before implementation.
- Data-Driven Decision-Making: Analysis of digital twin data informs continuous improvement, reducing costs and increasing competitiveness.
- Continuous Improvement: Data-driven insights foster ongoing enhancements in processes and products, supporting a culture of continuous improvement.

In summary, digital twins in manufacturing provide transformative technology that enhances operational efficiency, product quality, and decision-making. They empower manufacturers to proactively manage their assets, optimize processes, reduce costs, and remain competitive in an increasingly complex and dynamic manufacturing landscape.

8.1.3 The Role of Blockchain in Smart Manufacturing

Blockchain technology plays an important role in smart manufacturing by providing various benefits related to security, transparency, and efficiency [4]. Here's how blockchain can be used in the context of smart manufacturing [31–35]:

- Supply Chain Transparency: Blockchain ensures an immutable ledger of supply chain transactions, enhancing traceability of materials and reducing counterfeiting risks.

- Product Provenance and Authentication: Tracking product lifecycle on blockchain verifies authenticity, crucial for industries like aerospace and pharmaceuticals.
- Smart Contracts: Blockchain automates contract enforcement in smart manufacturing, triggering actions like payments upon predefined conditions, streamlining procurement.
- Secure Data Sharing: Blockchain enables secure, decentralized data sharing among manufacturing stakeholders, safeguarding sensitive information.
- Quality Assurance and Certification: Storing quality data and certifications on blockchain simplifies audits and ensures compliance with industry standards.
- Asset and Inventory Management: Real-time tracking of assets and inventory on blockchain enhances transparency and reduces fraud risks.
- Intellectual Property Protection: Blockchain safeguards intellectual property rights by securely time stamping and storing records of inventions and patents.
- Interoperability: Blockchain technology can facilitate data exchange and interoperability between different systems and platforms used in smart manufacturing. It provides a standardized and secure way to integrate data from various sources, ensuring that different components of the manufacturing process work seamlessly together.
- Reduced Fraud and Errors: Blockchain's inherent security and immutability help reduce fraud and errors in manufacturing processes. Unauthorized changes to records are nearly impossible, and all transactions are transparently recorded.
- Trust and Collaboration: Blockchain fosters trust among different stakeholders in the manufacturing ecosystem. It encourages collaboration and information sharing while maintaining data integrity, helping to build stronger partnerships within the industry.

In smart manufacturing, blockchain technology enhances the security, efficiency, and transparency of operations, making it an important tool for achieving Industry 4.0 objectives. It enables manufacturers to streamline processes, build trust with stakeholders, and improve supply chain management, ultimately leading to more reliable, cost-effective, and sustainable manufacturing practices.

8.1.4 Organization of the Work

This work is summarized in 12 sections.

8.2 Issues and Challenges in Conventional Manufacturing Processes

Conventional manufacturing processes, while effective, face several issues and challenges that can impact their efficiency, cost-effectiveness, and sustainability [10, 11]. Some of the key challenges in conventional manufacturing processes include (Figure 8.2):

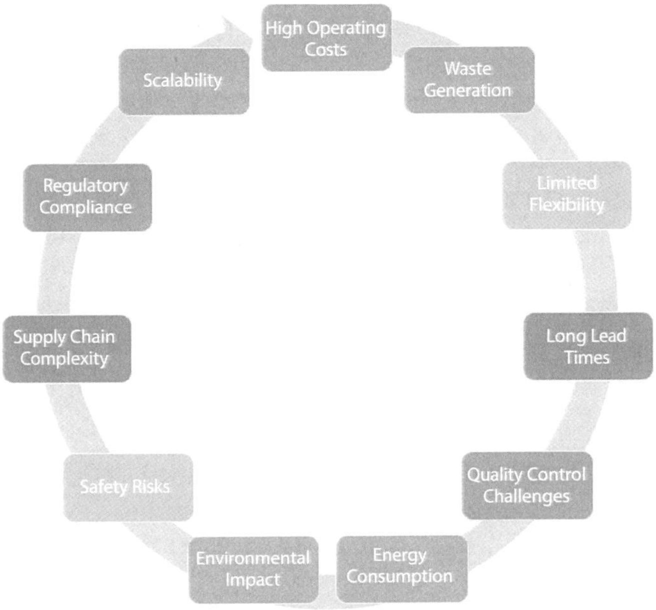

Figure 8.2 Issues and challenges in conventional manufacturing processes.

- High Operating Costs: Conventional manufacturing often involves substantial labor, energy, and material costs. The need for extensive manual labor and large quantities of raw materials can impact a company's bottom line.
- Waste Generation: Traditional manufacturing processes can generate substantial waste, including raw material waste, energy waste, and hazardous waste products. This not only contributes to environmental issues but also increases disposal costs.
- Limited Flexibility: Conventional manufacturing processes are often designed for specific products or tasks. Reconfiguring or adapting these processes to accommodate new products or changing market demands can be time-consuming and costly.
- Long Lead Times: The time required to set up and execute traditional manufacturing processes can result in long lead times. This can be problematic when businesses need to respond quickly to market changes or customer demands.
- Quality Control Challenges: Maintaining consistent product quality can be a challenge in conventional manufacturing, especially in processes where manual labor is involved. Variability in human input and machine performance can lead to quality control issues.
- Energy Consumption: Many conventional manufacturing processes are energy-intensive, contributing to a higher carbon footprint and increased operational costs. Reducing energy consumption is a key issue for sustainability.
- Environmental Impact: Conventional manufacturing can have a negative impact on the environment through the emission of greenhouse gases, air and water pollution, and the depletion of natural resources.

- Safety Risks: Manual labor in manufacturing processes can expose workers to safety risks, particularly in industries with heavy machinery and hazardous materials. Ensuring worker safety is a priority but can be challenging.
- Supply Chain Complexity: Managing complex supply chains with multiple suppliers and long lead times can be difficult, and it can lead to inefficiencies, delays, and quality issues.
- Regulatory Compliance: Compliance with local, regional, and international regulations is essential, but it can be challenging in manufacturing, particularly if products need to meet various quality and safety standards.
- Scalability: Expanding production capacity in conventional manufacturing often requires essential/ huge capital investment in new equipment, facilities, and labor. This can limit the ability to scale production quickly in response to market demands.
- Global Competition: Conventional manufacturing faces increasing competition from global markets. Companies often need to find ways to remain competitive by optimizing processes and reducing costs.

To address these challenges, many manufacturing organizations are transitioning toward more advanced and innovative approaches, such as smart manufacturing, automation, and sustainable practices. These changes aim to improve efficiency, reduce waste, enhance product quality, and align with environmental and sustainability goals.

8.2.1 Need for Digital Transformation in Manufacturing

Digital transformation in manufacturing is driven by a combination of internal and external factors, all of which underscore the need for the industry to embrace technology and adapt to a rapidly evolving landscape [12, 13]. Here are some of the key reasons why digital transformation is imperative for manufacturing:

- Competitive Advantage: Manufacturers need to remain competitive in a global market. Companies that can adopt digital technologies to streamline their operations and provide innovative products and services gain a competitive edge.
- Efficiency and Productivity: Digital technologies, including automation, IoT (Internet of Things), and data analytics, can enhance operational efficiency and productivity. These improvements reduce lead times, cut costs, and increase throughput.
- Real-time Data and Insights: Digital transformation provides manufacturers with real-time data and actionable information about their processes and products. This data-driven decision-making enables proactive management, better quality control, and predictive maintenance.
- Customization and Personalization: Digital transformation allows for more flexible and customized manufacturing processes. Manufacturers can better respond to customer demands, providing tailor-made products at scale.
- Supply Chain Optimization: Digital technologies provide end-to-end visibility and transparency in the supply chain. This improves inventory management, reduces disruptions, and ensures timely deliveries.

- Sustainability and Environmental Responsibility: Digital transformation supports sustainability efforts by reducing energy consumption, waste, and emissions. It enables manufacturers to meet environmental regulations and align with consumer demand for eco-friendly products.
- Remote Monitoring and Control: With digital transformation, manufacturers can remotely monitor and control equipment and processes. This is particularly important for distributed or global manufacturing operations.
- Predictive Maintenance: Predictive maintenance, driven by IoT and data analytics, helps manufacturers reduce unplanned downtime, extend the lifespan of equipment, and save on maintenance costs.
- Quality Assurance: Digital technologies improve quality control by enabling real-time monitoring and reporting. Manufacturers can quickly identify and rectify defects, reducing recalls and warranty costs.
- Innovation and Product Development: Digital tools support innovation and product development. Prototyping, simulations, and digital twins help manufacturers test and refine products before physical production.
- Compliance and Traceability: Digital transformation assists in maintaining compliance with industry standards and regulations. It enables end-to-end traceability for products and processes, reducing legal and reputational risks.
- Skill Enhancement: Manufacturers must invest in the skills of their workforce to embrace digital transformation fully. This contributes to employee growth and satisfaction while ensuring the organization has the talent to use digital tools effectively.
- Ecosystem Collaboration: Digital transformation encourages collaboration within the manufacturing ecosystem. It allows for integrated systems and data sharing, fostering partnerships and new business models.
- Customer-Centric Approach: Manufacturers can use digital transformation to engage with customers more directly, gather feedback, and provide better after-sales support. This customer-centric approach can build brand loyalty.
- Market Adaptation: As markets change rapidly, digital transformation enables manufacturers to adapt quickly. They can shift production, customize products, and reconfigure processes as needed to meet evolving customer demands.

In summary, digital transformation in manufacturing is essential to stay competitive, efficient, and responsive to changing market dynamics. It empowers manufacturers to harness the benefits of data, automation, and advanced technologies, ensuring their sustainability and growth in the modern manufacturing landscape.

8.3 Digital Twins and Blockchain in Manufacturing

8.3.1 Digital Twin, Types, and Applications of Digital Twins in Manufacturing

A digital twin is a virtual representation of a physical object or system used to simulate, analyze, and monitor its real-world counterpart. In manufacturing, digital twins are crucial for enhancing operational efficiency. They come in various types and find applications across different manufacturing processes [14, 15]. Here are the types and applications of digital twins in manufacturing:

Types of Digital Twins

- Product Digital Twins: These digital twins represent physical products or assets, such as machinery, equipment, and even entire production lines. They are used to monitor and optimize the performance of these assets, simulate different operating conditions, and plan for maintenance or upgrades.
- Process Digital Twins: Process digital twins model and simulate manufacturing processes and operations. They are used to optimize production workflows, monitor process parameters, and identify potential bottlenecks or inefficiencies.
- System Digital Twins: System digital twins combine product and process digital twins to model an entire manufacturing system or supply chain. They help in simulating and optimizing end-to-end operations and logistics.
- Performance Digital Twins: These digital twins focus on performance metrics and key performance indicators (KPIs) for manufacturing operations. They provide real-time insights and analytics for decision-making and continuous improvement.
- Physics-Based Digital Twins: Physics-based digital twins use mathematical models to simulate the behavior of physical systems. They are commonly used for complex and important manufacturing processes, such as fluid dynamics in chemical manufacturing.

Applications of Digital Twins in Manufacturing

- Design and Prototyping: Product digital twins are used in the design and prototyping phase to simulate and evaluate product performance, which can help in identifying and rectifying design flaws early in the development cycle.
- Manufacturing Process Optimization: Process digital twins allow manufacturers to simulate and optimize production processes. This includes improving manufacturing workflows, reducing cycle times, and increasing overall efficiency.
- Quality Control: Digital twins can be used to monitor and control product quality in real-time. They help identify defects and variations during the manufacturing process, reducing waste and ensuring consistent quality.
- Predictive Maintenance: By monitoring equipment and machinery, product digital twins can predict when maintenance is needed. This prevents unplanned downtime and extends the lifespan of manufacturing assets.
- Supply Chain Management: System digital twins are valuable for supply chain management, allowing manufacturers to optimize logistics, inventory management, and distribution to ensure the timely delivery of materials and products.

- Energy Efficiency: Digital twins can help reduce energy consumption by optimizing the usage of machines and systems. This is important for both cost savings and environmental sustainability.
- Inventory Management: System digital twins can track inventory levels in real-time and help manufacturers reduce excess inventory while ensuring that materials are available when needed.
- Customization and Personalization: In industries like automotive and consumer goods, digital twins enable mass customization by simulating how different configurations affect the final product, allowing for personalized providing.
- Safety and Risk Management: Digital twins can simulate and analyze potential safety hazards and risks within manufacturing operations. This aids in developing safety protocols and mitigating risks.
- Data Analysis and Decision Support: Performance digital twins provide data-driven insights, facilitating better decision-making for manufacturers. Real-time analytics enable immediate responses to changing conditions.

Digital twins are transforming manufacturing by providing a comprehensive view of products, processes, and systems. Their applications extend across the entire manufacturing lifecycle, from design and prototyping to operations and maintenance. As technology advances and the Internet of Things (IoT) becomes more integrated into manufacturing, digital twins are expected to play an even more important role in optimizing operations and enhancing efficiency (Table 8.1).

8.3.2 Benefits and Key Use Cases of Digital Twin in Manufacturing

Digital twins offer numerous advantages in manufacturing by creating virtual representations of physical assets, processes, and products. These benefits enhance operations, decision-making, and product development [16–17]. Here are some key advantages of using digital twins in manufacturing:

Benefits

- Real-time Monitoring and Control: Digital twins enable real-time monitoring of physical assets and processes, facilitating quick decision-making and response to deviations.
- Predictive Maintenance: Continuous monitoring predicts maintenance needs, reducing downtime and optimizing maintenance schedules.
- Process Optimization: Simulation and analysis aid in optimizing manufacturing processes, enhancing efficiency, and reducing waste.
- Quality Control: Real-time data helps maintain consistent product quality by promptly identifying defects or variations.
- Product Development: Digital twins assist in prototyping and simulating new product designs, saving time and resources.
- Inventory Management: Real-time visibility into inventory levels and movement enhances inventory management and supply chain operations.

Table 8.1 Digital Twin, types and applications of Digital Twins in manufacturing.

Digital Twin, Types and Applications of Digital Twins in Manufacturing		
	Types of Digital Twins	Process Digital Twins
		Product Digital Twins
		System Digital Twins
		Performance Digital Twins
		Physics-Based Digital Twins
	Applications of Digital Twins in Manufacturing	Design and Prototyping
		Manufacturing Process Optimization
		Quality Control
		Predictive Maintenance
		Supply Chain Management
		Energy Efficiency
		Inventory Management
		Customization and Personalization
		Safety and Risk Management

- Energy Efficiency: Monitoring energy consumption identifies opportunities for savings, reducing costs and environmental impact.
- Customization and Personalization: Tailoring production processes enables manufacturers to offer customized products at scale.
- Remote Monitoring: Remote monitoring facilitates managing operations from anywhere, valuable for global manufacturing setups.
- Simulation and What-If Analysis: Virtual replicas aid in simulating scenarios and assessing changes before implementation, reducing risks.

Key Use Cases

- Smart Factory: Digital twins replicate entire manufacturing facilities, enabling monitoring and optimization for maximum efficiency.
- Equipment Health Monitoring: Digital twins monitor individual machine performance for predictive maintenance, reducing downtime.
- Product Lifecycle Management: Digital twins are used across product lifecycles, from design to maintenance.
- Supply Chain Management: End-to-end visibility and traceability in the supply chain optimize logistics and reduce delays.

- Quality Assurance: Digital twins ensure consistent product quality by monitoring processes in real-time
- Sustainable Manufacturing: Digital twins monitor energy consumption and waste, contributing to eco-friendly practices.
- Custom Manufacturing: For businesses that need to provide customized or personalized products, digital twins enable more flexible and efficient production processes to meet specific customer demands.
- Operational Resilience: Digital twins enhance the resilience of manufacturing operations by providing a virtual backup and enabling quick response to unforeseen disruptions or issues.

In summary, digital twins are a valuable technology in manufacturing, providing numerous benefits for monitoring, optimizing, and innovating manufacturing processes and products. They are increasingly becoming an integral part of Industry 4.0 initiatives in the manufacturing sector (Table 8.2).

Table 8.2 Benefits and key use cases of Digital Twin in manufacturing.

Benefits and Key Use Cases of Digital twin in manufacturing		
	Benefits	Real-time Monitoring and Control
		Predictive Maintenance
		Process Optimization
		Quality Control
		Product Development
		Inventory Management
		Energy Efficiency
		Customization and Personalization
		Remote Monitoring
		Simulation and What-If Analysis
	Key Use Cases	Smart Factory
		Equipment Health Monitoring
		Product Lifecycle Management
		Supply Chain Management
		Quality Assurance
		Sustainable Manufacturing
		Custom Manufacturing
		Operational Resilience

8.3.3 Blockchain Components: Blocks, Transactions, and Smart Contracts in Smart Manufacturing

In the context of smart manufacturing and blockchain technology, understanding the components of blockchain, including blocks, transactions, and smart contracts, is important [4]. Here's an explanation of these components:

Blocks

- Blockchain Structure: A chain of blocks, each containing transaction data, ensures data immutability and integrity.
- Block Components: Blocks consist of a header containing metadata (timestamp, previous block reference), and transactions representing various manufacturing activities (assembly, quality checks, maintenance, supply chain movements).

Transactions

- Transactions: Essential records of actions or data added to a block in a blockchain.
- Smart Manufacturing Usage: Transactions in smart manufacturing encompass sensor data, quality control, material movements, and more.
- Cryptographic Signing: Transactions are signed cryptographically for data integrity and security.
- Immutability: Once confirmed by consensus, transactions become immutable, making blockchain highly resistant to tampering or fraud.

Smart Contracts

- Smart Contracts: Self-executing agreements programmed to enforce specified terms automatically.
- Smart Manufacturing Usage: Automates production and supply chain processes.
- Transparency: Written in code, stored on blockchain, and accessible to relevant parties.
- Automated Actions: Triggered based on predefined conditions/events, like ordering replacement parts or executing quality control checks.
- Efficiency and Transparency: Reduce intermediaries, and manual interventions, improving process efficiency and transparency.

By integrating blockchain technology with smart manufacturing, companies can enhance transparency, traceability, and security within their operations. The use of blocks, transactions, and smart contracts allows for the creation of an immutable and trustworthy ledger of manufacturing data, helping to improve process efficiency and quality control while reducing the risk of fraud or errors.

8.3.4 Security and Data Integrity Issues in Implementing Digital Twin and Blockchain in Smart Manufacturing

Security and data integrity are important issues when implementing digital twin and blockchain technologies in smart manufacturing. While these technologies provide numerous advantages, they also introduce potential vulnerabilities and risks [4, 17]. Here are some of the key security and data integrity issues in the context of smart manufacturing (Table 8.3):

Security Issues

- Cybersecurity Threats: Smart manufacturing faces risks like hacking, malware, ransomware, and DDoS attacks, exacerbated by interconnected Industry 4.0 systems.
- Unauthorized Access: Breaches into sensitive manufacturing data or control systems can cause disruptions, IP theft, or physical asset damage.
- Data Breaches: Compromised manufacturing data, trade secrets, or confidential information can harm a company's competitiveness and reputation.

Table 8.3 Security and data integrity issues in digital and blockchain in smart manufacturing.

Security and Data Integrity Issues in Digital and Blockchain in Smart manufacturing		
	Security Issues	Cyber security Threats
		Unauthorized Access
		Data Breaches
		Supply Chain Vulnerabilities
		Identity and Access Management
	Data Integrity Issues	Data Tampering
		Chain of Custody
		Counterfeiting and Product Quality
		Interoperability
	Issues and Mitigations	Security Protocols
		Regular Updates and Patching
		Security Awareness
		Third-Party Vendors
		Data Validation and Verification
		Chain of Trust
		Blockchain Consensus Mechanisms
		Incident Response Plan

- Supply Chain Vulnerabilities: Interconnected supply chains are prone to disruptions and cyber-attacks, leading to delays, increased costs, and security risks.
- Identity and Access Management: Strong authentication and access controls are crucial to prevent unauthorized access to systems and data, minimizing security breaches.

Data Integrity Issues

- Data Tampering: Ensuring the integrity of manufacturing data is essential. Without proper security measures, data can be tampered with, leading to unreliable records and decisions.
- Chain of Custody: In a supply chain, it's important to maintain a clear and trustworthy chain of custody for materials and products. Any gaps or inconsistencies in this chain can lead to data inaccuracies and potential fraud.
- Counterfeiting and Product Quality: Digital and blockchain technologies are used to combat counterfeiting and maintain product quality. If these systems are compromised, it can have a major impact on product authenticity and customer safety.
- Interoperability: Ensuring that data from various sources and systems is integrated and standardized can be challenging. Inconsistent data formats or lack of data interoperability can hinder data integrity.

Issues and Mitigations

- Security Protocols: Employ robust security measures like encryption, multi-factor authentication, and intrusion detection systems for digital and blockchain systems.
- Regular Updates and Patching: Keep software, firmware, and systems current with security patches to mitigate vulnerabilities.
- Security Awareness: Educate staff and partners on security best practices and digital technology risks to minimize human-related security threats.
- Third-Party Vendors: Conduct thorough security assessments when integrating third-party systems or services to ensure they meet security standards.
- Data Validation and Verification: Implement processes for validating and verifying data to maintain accuracy and prevent tampering.
- Chain of Trust: Establish traceability and data integrity throughout the supply chain with a robust chain of trust for data and products.
- Blockchain Consensus Mechanisms: Choose appropriate consensus mechanisms, considering security, decentralization, and energy efficiency.
- Incident Response Plan: Develop a comprehensive incident response plan to address security breaches or data integrity issues effectively.

In smart manufacturing, addressing security and data integrity concerns is vital to safeguard sensitive information, maintain system reliability, and ensure supply chain integrity.

8.4 Synergy of Blockchain and Digital Twins for Smart Manufacturing

The synergy of blockchain and digital twins in smart manufacturing can provide a powerful combination of technologies that enhances transparency, security, and efficiency across the entire manufacturing ecosystem [18, 19]. Here's how the integration of blockchain and digital twins can benefit smart manufacturing (Figure 8.3):

- Data Integrity and Trustworthiness: Blockchain guarantees the immutability of data in digital twins, ensuring a trusted record of transactions and changes.
- Provenance and Traceability: Digital twins offer detailed representations of assets and processes, with blockchain ensuring traceability and provenance crucial for supply chain management and regulatory compliance.
- Secure Data Sharing: Blockchain enables secure and permissioned data sharing among stakeholders, facilitating seamless collaboration without compromising security.
- Smart Contracts for Automation: Blockchain-based smart contracts automate processes and trigger actions within digital twins, such as scheduling maintenance or ordering replacement parts.
- Quality Control and Monitoring: Digital twins monitor and simulate manufacturing processes while blockchain ensures authenticity and accuracy of recorded data, enhancing quality control.
- Supply Chain Optimization: Digital twins represent the entire supply chain, with blockchain integration enabling transparent recording of each step, facilitating tracking, optimization, and risk mitigation.
- Predictive Maintenance: Digital twins provide real-time information about the condition of assets and equipment, while blockchain maintains an immutable record of maintenance history and actions taken. This combination supports predictive maintenance strategies, reducing downtime and maintenance costs.
- Interoperability and Standardization: Blockchain can act as a standardized layer for data and transactions between various digital twin implementations, promoting interoperability and data consistency across the manufacturing ecosystem.
- Energy Efficiency and Sustainability: Digital twins can model and optimize energy consumption, while blockchain records and verifies the use of sustainable energy sources. This combination helps manufacturers meet environmental goals and reduce energy costs.
- Regulatory Compliance: Blockchain's transparency and auditability make it easier for manufacturers to comply with industry regulations and report on compliance with confidence, especially in highly regulated sectors like pharmaceuticals and aerospace.

In summary, the integration of blockchain and digital twins in smart manufacturing creates a secure and transparent digital environment that ensures data integrity, automates processes, and supports the entire manufacturing lifecycle. This synergy enhances

Figure 8.3 Synergy of blockchain and Digital Twins for smart manufacturing.

efficiency, quality control, traceability, and sustainability, positioning smart manufacturing to meet the demands of Industry 4.0 and beyond.

8.5 AI, Blockchain, IoT and Other Emerging Technologies: Role in Smart Manufacturing

Smart manufacturing, often associated with Industry 4.0, uses a variety of emerging technologies, including artificial intelligence (AI), blockchain, the Internet of Things (IoT), and others, to transform traditional manufacturing processes into efficient, data-driven, and interconnected ecosystems [20]. Here's a look at the roles of these technologies in smart manufacturing:

Artificial Intelligence (AI)

- Predictive Maintenance: AI analyzes data from sensors and equipment to predict when maintenance is needed, reducing unplanned downtime.
- Quality Control: AI-driven image and data analysis can quickly detect defects and anomalies in real time, ensuring product quality.
- Process Optimization: Machine learning algorithms optimize production processes by identifying areas for improvement and making real-time adjustments.

Blockchain

- Supply Chain Transparency: Blockchain fosters transparency and traceability in supply chains, curbing fraud and verifying product/material authenticity.
- Data Integrity: Blockchain offers an immutable ledger, safeguarding data integrity and deterring tampering.

- Smart Contracts: Automation via smart contracts streamlines operations by automating processes and actions based on preset conditions, minimizing intermediaries.

Internet of Things (IoT)

- Real-time Monitoring: IoT sensors and devices provide real-time data on machine performance, energy consumption, and environmental conditions.
- Asset Management: IoT enables the tracking and management of assets, optimizing their usage and maintenance.
- Inventory Control: IoT devices in warehouses and along the supply chain improve inventory management, reducing costs and minimizing waste.

Robotics and Automation

- Manufacturing Robots: Robots and automated systems perform repetitive and dangerous tasks with precision, increasing efficiency and worker safety.
- Cobots (Collaborative Robots): These robots work alongside human workers, enhancing productivity and flexibility in manufacturing processes.

Edge Computing

- Real-time Processing: Edge computing facilitates data processing at or near the data source, minimizing latency and enabling real-time decision-making in smart manufacturing.

3D Printing and Additive Manufacturing

- Prototyping: 3D printing allows for rapid prototyping and product development, reducing time to market.
- Customization: It enables on-demand and customized production, catering to individual customer needs.

Augmented Reality (AR) and Virtual Reality (VR)

- Training and Maintenance: AR and VR provide immersive training and maintenance guidance for workers, reducing errors and improving efficiency.
- Product Design: Designers can use AR and VR for virtual prototyping and product visualization.

Cyber Security

- Data Protection: As smart manufacturing relies on interconnected systems and data sharing, robust cyber security measures are essential to protect sensitive data and systems from cyber threats.

Cloud Computing

- Data Storage and Processing: Cloud-based solutions store and process large amounts of data generated in smart manufacturing, providing scalability and accessibility.

Sustainability Technologies

- Renewable Energy Integration: Smart manufacturing incorporates renewable energy sources and energy-efficient practices to reduce the environmental impact.
- Waste Reduction: IoT and data analytics help minimize waste and support sustainable manufacturing practices.

The integration of these emerging technologies in smart manufacturing results in improved efficiency, reduced costs, enhanced quality control, greater sustainability, and the ability to respond quickly to changing market demands. It also provides a foundation for the evolution of manufacturing in an increasingly digital and connected world.

8.6 Key Technologies for Blockchain-Based Digital Twins

Blockchain-based digital twins synergize the advantages of digital twins and blockchain, bolstering data security, transparency, and trustworthiness across industries like manufacturing, supply chain management, healthcare, and beyond [21]. Here are the key technologies involved in creating blockchain-based digital twins (Figure 8.4):

- Blockchain Technology: Acts as the foundation for securely managing digital twin data, ensuring integrity, immutability, and transparency through components like blocks, transactions, smart contracts, consensus algorithms, and nodes.

Figure 8.4 Key Technologies for blockchain-based Digital Twins.

- Digital Twins: Virtual replicas of physical objects, processes, or systems capturing real-time IoT sensor data, simulations, and more, utilized for individual assets, production processes, or entire ecosystems.
- IoT (Internet of Things): Provides real-time data from physical assets, integrated with blockchain to ensure digital twin accuracy.
- Data Analytics and AI: Process and analyze data from digital twins and IoT sensors, identifying patterns, anomalies, and optimizing processes.
- Smart Contracts: Self-executing contracts automating and enforcing agreements, triggering actions like maintenance requests or quality checks in blockchain-based digital twins.
- Interoperability Standards: Facilitate effective communication and data exchange between different digital twin systems and blockchain platforms, enabling collaboration among stakeholders.
- Cyber security Protocols: Essential security measures like encryption, access control, and identity management safeguard data and systems involved in blockchain-based digital twins.
- Consensus Mechanisms: Validate and add transactions to the blockchain, ensuring security and integrity through mechanisms like proof of work (PoW) and proof of stake (PoS).
- Data Visualization and User Interfaces: User-friendly interfaces and visualization tools enable stakeholders to interact with and interpret digital twin data for informed decision-making.
- Cloud Computing: Utilized for storing, processing, and accessing data related to blockchain-based digital twins, offering scalability, accessibility, and redundancy.

The combination of these technologies in blockchain-based digital twins provides a secure, transparent, and reliable way to represent physical assets and processes in a digital form. This technology has applications in various industries, from optimizing manufacturing operations to ensuring the authenticity and traceability of products in supply chains.

8.6.1 Integration of Emerging Technologies with Existing Manufacturing Systems

The integration of emerging technologies with existing manufacturing systems is essential for driving digital transformation and ensuring that traditional manufacturing processes remain competitive, efficient, and adaptable to changing market demands [22]. Here are key steps and considerations for effectively integrating emerging technologies into existing manufacturing systems:

Assessment and Strategy Development

- Begin with a thorough assessment of your current manufacturing systems, processes, and infrastructure.
- Define your digital transformation objectives, such as optimizing productivity, minimizing expenses, elevating product standards, or facilitating personalization.

Identify Suitable Technologies

- Identify emerging technologies that align with your strategic goals and address specific pain points or opportunities in your manufacturing operations.
- Consider technologies like IoT, AI, blockchain, digital twins, robotics, and 3D printing based on their relevance to your industry and use cases.

Data Connectivity and Integration

- Ensure that existing machines, sensors, and equipment can connect and share data with the new technologies. This may require the installation of IoT sensors or retrofitting existing machinery.
- Implement middleware or integration platforms to enable data exchange between old and new systems.

Interoperability and Standards

- Use industry standards and open protocols to ensure interoperability between different technologies and systems.
- Standardization facilitates the seamless exchange of data and communication among diverse components.

Pilot Projects

- Start with small-scale pilot projects to test and validate the integration of emerging technologies with existing systems.
- Evaluate the impact on operations, costs, and quality before scaling up.

Change Management

- Prepare employees for the transition by providing training and creating a culture that embraces new technologies.
- Encourage collaboration and cross-functional teams to ensure that everyone understands the benefits of the changes.

Security Measures

- Implement vigorous cyber security actions to guard the integrity and confidentiality of data across the integrated systems.
- This includes encryption, access controls, and regular security audits.

Scalability and Flexibility

- Design your integrated systems to be scalable and flexible to accommodate future growth and technological advancements.
- Be prepared to adapt to new emerging technologies as they become relevant.

Data Analytics and AI

- Use data analytics and AI to gain information from the integrated systems and optimize manufacturing processes.
- AI can help identify patterns, anomalies, and opportunities for process improvement.

Continuous Improvement

- Monitor the performance of the integrated systems and gather feedback from users.
- Continuously assess and refine your strategies to maximize the benefits of emerging technologies.

Collaboration with Suppliers and Partners

- Collaborate with suppliers and partners to ensure that their systems align with the integrated technologies, promoting end-to-end transparency and efficiency in the supply chain.

ROI Assessment

- Measure and evaluate the return on investment (ROI) to assess the impact of the integration and make data-driven decisions about future investments.

The successful integration of emerging technologies with existing manufacturing systems can enhance efficiency, competitiveness, and innovation within your organization. It allows you to adapt to evolving market dynamics and positions your manufacturing operations for long-term success in the digital age.

8.7 Applications of Blockchain-Based Digital Twins in Smart Manufacturing

Blockchain-based digital twins offer practical applications in smart manufacturing, providing benefits such as enhanced transparency, data integrity, and automation [2]. Here are some key applications of blockchain-based digital twins in the context of smart manufacturing:

Supply Chain Management: Blockchain-based digital twins ensure supply chain visibility and traceability, fostering transparency and trust among stakeholders by tracking the movement of materials and products (Figure 8.5).

- Provenance and Authenticity: Manufacturers can use digital twins and blockchain to verify the authenticity and provenance of products, ensuring that they meet quality standards and are not counterfeit.

Figure 8.5 Applications of blockchain-based Digital Twins in smart manufacturing.

- Quality Control and Assurance: Digital twins monitor production processes in real time, and blockchain records the data securely. This combination allows for immediate detection of defects or deviations, improving overall product quality and reducing waste.
- Predictive Maintenance: Smart manufacturing facilities use digital twins for real-time asset monitoring. When sensor data indicates maintenance is needed, a smart contract on the blockchain can automatically schedule maintenance, improving equipment reliability and reducing downtime.
- Energy Efficiency: Digital twins model and optimize energy consumption, while blockchain verifies the use of sustainable energy sources. This combination supports energy efficiency and sustainability goals.
- Smart Contracts for Automation: Blockchain smart contracts automate manufacturing tasks like order processing, payment settlements, and inventory management, streamlining operations and reducing manual interventions.
- Process Optimization: Data analytics and AI can analyze data from digital twins to identify process inefficiencies and areas for improvement. Smart contracts on the blockchain can then execute optimizations in real time.
- Warranty and Maintenance Records: Blockchain-based digital twins store maintenance and warranty records securely. This ensures that manufacturers can access and validate warranty claims and maintenance histories, reducing disputes and enhancing customer satisfaction.
- Inventory Management: Digital twins provide real-time information about inventory levels and movements. Blockchain ensures that inventory data is accurate and secure, allowing for optimized inventory management and reduced carrying costs.
- Regulatory Compliance: Blockchain-based digital twins help manufacturers comply with industry standards and regulations by providing a transparent and auditable record of processes and quality control.
- Product Lifecycle Management: Throughout the product lifecycle, digital twins and blockchain can track design changes, production processes, and usage history, improving product development and support.

- Customization and Personalization: Digital twins can be used to tailor production processes to individual customer requirements, creating customized and personalized products at scale.
- Interoperability and Data Sharing: Blockchain ensures that data is shared securely among different stakeholders in the manufacturing ecosystem, including suppliers, manufacturers, and customers, facilitating efficient collaboration.
- Sustainability Reporting: Blockchain-based digital twins record and verify sustainability practices and environmental data, enabling manufacturers to meet sustainability reporting requirements and demonstrate eco-friendly operations.

These applications demonstrate how blockchain-based digital twins can enhance efficiency, quality, transparency, and sustainability in smart manufacturing, creating a more resilient and competitive manufacturing ecosystem.

8.8 Security and Data Privacy in Smart Manufacturing - In General

Security and data privacy are important issues in smart manufacturing to protect sensitive information, maintain the integrity of operations, and ensure compliance with privacy regulations [24]. Here are key aspects to consider for security and data privacy in smart manufacturing:

- Network Security: Safeguard manufacturing networks with firewalls, intrusion detection, and encryption to protect data in transit.
- Access Control: Employ strict access measures, like strong authentication and role-based control, limiting access to authorized personnel.
- Data Encryption: Encrypt data both at rest and in transit, using robust protocols to prevent unauthorized access.
- Data Integrity: Ensure data integrity with checksums, digital signatures, and hash functions to detect and prevent tampering.
- IoT Device Security: Secure IoT devices with updates, strong passwords, and vulnerability assessments to prevent breaches.
- Cyber security Training: Train employees to recognize and respond to threats, reducing human error risks.
- Incident Response Plan: Develop a response plan for prompt mitigation of cyber security incidents.
- Secure Data Storage: Protect stored data with encryption and access controls, and back it up regularly.
- Vendor and Supply Chain Security: Assess and collaborate with trusted partners who meet security standards.
- Patch Management: Keep systems up to date with security patches to mitigate vulnerabilities.
- Privacy by Design: Incorporate privacy considerations from the outset of system design.
- Data Minimization: Collect only necessary data to minimize privacy risks.

- Compliance with Regulations: Adhere to relevant privacy regulations, such as GDPR or CCPA.
- Data Ownership and Sharing: Clearly define data ownership and sharing agreements, implementing access controls and encryption.
- Secure Remote Access: Use secure VPNs and multi-factor authentication for remote system access.
- Supply Chain Security: Monitor supplier security practices to prevent supply chain vulnerabilities.
- Security Audits and Penetration Testing: Regularly audit systems and conduct penetration tests to identify weaknesses.

Continuous vigilance, proactive strategies, and adherence to regulations are essential in ensuring security and data privacy in smart manufacturing, mitigating risks and protecting operations and sensitive data.

8.8.1 Cyber Security in the Smart Factory

Cyber security in the smart factory is important to protect against potential cyber threats and vulnerabilities that can disrupt operations, compromise sensitive data, and face safety risks [25]. Here are key cyber security issues for smart factories:

- Network Segmentation: Segment the network to isolate important systems from non-important ones. This limits lateral movement for attackers.
- Firewalls and Intrusion Detection Systems (IDS): Employ firewalls to filter traffic and IDS to monitor for suspicious activities and potential threats.
- Access Control: Implement strong access control measures with role-based access permissions to limit access to authorized personnel only.
- Endpoint Security: Secure IoT devices and endpoints with up-to-date firmware, regular patch management, and strong passwords.
- Security Updates: Ensure all systems and software are regularly updated with the latest security patches to address vulnerabilities.
- Data Encryption: Encrypt data both at rest and in transit to safeguard sensitive information from unauthorized access.
- Data Backups: Regularly back up critical data to prevent loss in the event of a cyber-incident.
- Security Policies and Training: Establish and enforce security policies while providing cyber security training to employees to enhance awareness and reduce errors.
- Incident Response Plan: Develop a comprehensive plan to promptly address and mitigate cyber security incidents, including data breaches and cyber-attacks.
- Third-Party Vendor Assessment: Assess the cyber security practices of third-party vendors and partners connected to smart factory systems.
- Physical Security: Implement measures to restrict access to essential infrastructure and equipment.
- Audit and Monitoring: Continuously monitor network traffic, system logs, and device behavior to detect anomalies and potential threats.

- Privacy by Design: Integrate privacy and data protection into smart manufacturing system design from the outset.
- Supply Chain Security: Evaluate supplier security practices and ensure compliance with security standards to mitigate supply chain risks.
- Cyber security Culture: Foster a culture of cyber security awareness and responsibility across all levels of the organization.
- Regulatory Compliance: Ensure compliance with relevant industry regulations and privacy laws such as GDPR or CCPA.
- Zero Trust Security: Adopt a zero trust security model to verify all users and devices before granting access, regardless of network status.
- Penetration Testing: Conduct regular penetration testing to identify vulnerabilities and weaknesses in smart factory cyber security defenses.
- Secure Remote Access: Implement secure VPNs and multi-factor authentication for remote connections to smart factory systems.
- Security Awareness Training: Train employees and contractors on cyber security best practices, phishing detection, and how to respond to security incidents.

Smart factories can be attractive targets for cyber-attacks due to their reliance on interconnected systems and IoT devices. Therefore, a proactive and multi-layered approach to cyber security is important to protect operations and maintain data integrity. Regular risk assessments and security audits are essential to stay ahead of emerging threats and vulnerabilities.

8.9 Case Studies

8.9.1 Blockchain for Smart Manufacturing in the Automotive Industry

Blockchain technology has several promising applications in the automotive industry, particularly in the context of smart manufacturing [26, 27]. Here are some ways in which blockchain can benefit smart manufacturing in the automotive sector:

- Supply Chain Transparency: Blockchain can provide end-to-end transparency in the automotive supply chain. It helps in tracking the movement of parts and materials from suppliers to manufacturers, ensuring authenticity and quality control.
- Provenance and Authenticity: Blockchain can verify the authenticity and provenance of important components and spare parts. This is important for ensuring that counterfeit or substandard parts do not enter the supply chain.
- Quality Control: Smart manufacturing in the automotive industry benefits from real-time quality control. Blockchain can securely record data related to quality checks and inspections, providing a tamper-proof record of the production process.
- Predictive Maintenance: Blockchain can be integrated with IoT sensors to monitor the condition of machinery and vehicles in real time. When

maintenance is required, smart contracts on the blockchain can automatically schedule maintenance or order replacement parts, reducing downtime.
- Inventory Management: The automotive industry relies on efficient inventory management. Blockchain can streamline this process by securely tracking and updating inventory levels across the supply chain.
- Smart Contracts for Automation: Smart contracts on the blockchain can automate various aspects of production and supply chain management, such as order processing, payment settlements, and compliance with quality standards.
- Warranty and Maintenance Records: Blockchain can securely store and manage warranty information and maintenance records, enabling manufacturers and customers to access and validate warranty claims and maintenance histories.
- Data Sharing with Partners: Blockchain provides a secure platform for sharing data with manufacturing partners, suppliers, dealers, and customers. This facilitates seamless collaboration and information sharing.
- Regulatory Compliance: The automotive industry is subject to various regulations and quality standards. Blockchain can help manufacturers demonstrate compliance by providing an auditable record of processes and quality control.
- Product Lifecycle Management: Blockchain can track design changes, production processes, and usage history throughout a vehicle's lifecycle. This data can enhance product development and support.
- Customization and Personalization: Digital twins and blockchain can enable the customization and personalization of vehicles, catering to individual customer preferences.
- Sustainability Reporting: Blockchain can record and verify sustainability practices and environmental data, helping automotive manufacturers meet sustainability reporting requirements.

In summary, blockchain technology enhances transparency, data integrity, and security in the smart manufacturing processes of the automotive industry. By integrating blockchain with digital twins, IoT sensors, and smart contracts, automotive manufacturers can optimize production, improve quality control, reduce operational costs, and ensure regulatory compliance. Blockchain plays an essential role in making smart manufacturing in the automotive industry more efficient, reliable, and resilient.

8.9.2 Digital Twins in Pharmaceutical Manufacturing

Digital twins have gained prominence in pharmaceutical manufacturing for their potential to improve product quality, efficiency, and compliance with regulatory requirements. A digital twin is a virtual representation of a physical manufacturing process, allowing for real-time monitoring, analysis, and optimization. Here's how digital twins are utilized in pharmaceutical manufacturing:

- Process Simulation and Optimization: Digital twins simulate the entire drug manufacturing process, helping identify potential bottlenecks, inefficiencies, and areas for optimization. By running "what-if" scenarios, manufacturers can refine processes before implementation.

- Real-time Monitoring and Control: Digital twins enable real-time monitoring of equipment, variables, and quality metrics within the manufacturing process. This real-time data allows for immediate adjustments to maintain product quality and consistency.
- Quality Control and Assurance: Digital twins incorporate quality control checkpoints, ensuring that products meet regulatory and quality standards. Any deviations from the expected quality parameters trigger alerts and corrective actions.
- Predictive Maintenance: IoT sensors integrated into digital twins continuously monitor equipment condition. Predictive maintenance algorithms can identify maintenance needs before equipment failures occur, reducing downtime and ensuring uninterrupted production.
- Regulatory Compliance: Digital twins can document and demonstrate compliance with pharmaceutical regulations, ensuring that processes are followed correctly and that data is securely recorded.
- Rapid Scale-up and Tech Transfer: When bringing a new drug to market or transferring a process to a different facility, digital twins facilitate faster scale-up and technology transfer by providing a virtual platform for testing and optimization.
- Batch Record and Documentation: Digital twins automatically create and maintain batch records, reducing the risk of human errors and simplifying documentation for regulatory compliance.
- Product Lifecycle Management: Digital twins track the entire lifecycle of a pharmaceutical product, including changes in formulations, manufacturing processes, and quality parameters, which is essential for post-approval changes and regulatory filings.
- Customization and Personalization: Digital twins can accommodate customized pharmaceutical products by adjusting process parameters according to individual patient needs, such as personalized medications.
- Supply Chain Visibility: Digital twins extend beyond manufacturing to include the entire supply chain. They track raw materials, production, distribution, and delivery, ensuring end-to-end visibility and traceability.
- Data Analytics and Artificial Intelligence: Advanced data analytics and AI can be integrated into digital twins to identify trends, anomalies, and opportunities for process improvement.
- Energy Efficiency and Sustainability: Digital twins model energy consumption and environmental impact, allowing manufacturers to optimize energy use and reduce their carbon footprint.

Digital twins in pharmaceutical manufacturing provide a comprehensive, real-time view of processes, enabling manufacturers to improve quality control, reduce operational costs, and expedite the development and scale-up of new drugs. They also play an important role in ensuring compliance with regulatory requirements, which is essential in the pharmaceutical industry.

8.9.3 Blockchain and Digital Twin in Aerospace Manufacturing

The integration of blockchain and digital twin technology in aerospace manufacturing provides a range of benefits, including enhanced transparency, traceability, and security in a highly regulated and safety-important industry. Here's how blockchain and digital twins are utilized in aerospace manufacturing:

- Supply Chain Management: Blockchain provides end-to-end visibility and traceability in the aerospace supply chain. It helps track the movement of parts, components, and materials from suppliers to manufacturers, ensuring authenticity and quality control.
- Provenance and Authenticity: Blockchain verifies the authenticity and provenance of important components and spare parts, ensuring that counterfeit or substandard parts do not enter the supply chain.
- Quality Control: Digital twins monitor production processes in real time, and blockchain records the data securely. This combination allows for immediate detection of defects or deviations, improving overall product quality and reducing waste.
- Predictive Maintenance: Blockchain can be integrated with IoT sensors and digital twins to monitor the condition of aircraft components and equipment in real time. When maintenance is required, smart contracts on the blockchain can automatically schedule maintenance or order replacement parts, reducing downtime and enhancing safety.
- Inventory Management: The aerospace industry relies on efficient inventory management. Blockchain can streamline this process by securely tracking and updating inventory levels across the supply chain.
- Smart Contracts for Automation: Smart contracts on the blockchain can automate various aspects of production and supply chain management, such as order processing, payment settlements, and compliance with quality standards.
- Warranty and Maintenance Records: Blockchain can securely store and manage warranty information and maintenance records, enabling manufacturers and customers to access and validate warranty claims and maintenance histories.
- Data Sharing with Partners: Blockchain provides a secure platform for sharing data with manufacturing partners, suppliers, maintenance providers, and regulatory agencies, facilitating seamless collaboration and information sharing.
- Regulatory Compliance: The aerospace industry is subject to strict regulatory requirements. Blockchain can help manufacturers demonstrate compliance by providing an auditable record of processes and quality control.
- Product Lifecycle Management: Blockchain and digital twins can track design changes, production processes, and usage history throughout an aircraft's lifecycle. This data can enhance product development, support, and regulatory reporting.
- Customization and Personalization: Digital twins and blockchain can enable customization and personalization of aircraft components, catering to individual customer preferences and specific mission requirements.

- Security and Data Integrity: Blockchain ensures the integrity and security of data recorded in digital twins, providing a trusted and tamper-proof record of all transactions and changes in the aerospace manufacturing process.

In the aerospace manufacturing sector, safety, quality, and regulatory compliance are of utmost importance. The combination of blockchain and digital twins enhances the industry's ability to maintain transparency, quality control, and traceability throughout the entire product lifecycle, contributing to safer and more reliable aerospace products. Additionally, it helps in complying with stringent regulatory requirements and increasing overall operational efficiency.

8.10 Future Research Towards Integration of AI and Blockchain for Autonomous Manufacturing

The fusion of AI and blockchain for autonomous manufacturing has the potential to transform manufacturing processes significantly. Though progress has been made in this field, there remains ample opportunity for future research and development.

Here are some key areas [28, 29] where further research can be conducted:

- AI-Enabled Predictive Maintenance: Develop AI algorithms for predictive maintenance in manufacturing, complemented by blockchain for secure maintenance history records, reducing downtime.
- Supply Chain Transparency and Traceability: Utilize blockchain to ensure transparency and traceability across the supply chain, augmented by AI for real-time data analysis.
- Smart Contracts for Manufacturing Agreements: Implement blockchain-based smart contracts to automate manufacturing agreements, with AI monitoring and enforcement for compliance.
- Quality Control and Defect Detection: Deploy AI-driven quality control systems using computer vision and machine learning, integrated with blockchain for tamper-proof documentation of defects.
- Production Optimization: Research AI-driven real-time production optimization, leveraging blockchain for secure recording of parameter adjustments.
- Data Privacy and Security: Explore methods for ensuring data privacy with AI and blockchain, incorporating techniques like homomorphic encryption and multi-party computation.
- Energy Efficiency: Investigate AI and blockchain applications for optimizing energy consumption in manufacturing, potentially through incentivized smart contracts.
- Interoperability and Standards: Establish industry standards for seamless integration of AI and blockchain in manufacturing systems and processes.

- Scalability: Research how to make the integration of AI and blockchain more scalable for large manufacturing ecosystems. This includes finding solutions for handling a high volume of transactions and data.
- Regulatory Compliance: Address the regulatory challenges associated with AI and blockchain in manufacturing. Research ways to ensure compliance with data protection, safety, and industry-specific regulations.
- Human-Machine Collaboration: Investigate how humans and AI-driven autonomous systems can collaborate in a manufacturing environment. Develop methods for efficient interaction and decision-making between human workers and AI-driven machines.
- Robustness and Resilience: Study methods to make AI and blockchain systems more resilient to cyber-attacks and system failures, ensuring the continuous and secure operation of manufacturing processes.
- Environmental Sustainability: Discuss how AI and blockchain can be used to monitor and reduce the environmental impact of manufacturing processes, including resource consumption, emissions, and waste management (Figure 8.6).

Collaboration among researchers, industry, and government is crucial for advancing AI and blockchain integration in manufacturing. This promises efficiency, cost reduction, and transparency, driving future research and development.

8.11 Sustainability and Environmental Impact via Smart Manufacturing

Smart manufacturing, leveraging IoT, AI, and data analytics, is important for addressing sustainability and reducing the environmental footprint of manufacturing processes,

Figure 8.6 Future research towards integration of AI and blockchain.

aligning with contemporary priorities [7, 9, 30, 31, 34]. Here are several ways in which smart manufacturing can contribute to sustainability and minimize environmental impact:

- Energy Efficiency: Smart manufacturing systems can continuously monitor and optimize energy consumption. This includes the use of AI algorithms to adjust machine operations, lighting, and heating/cooling systems to minimize energy usage while maintaining productivity.
- Resource Optimization: AI-driven predictive analytics can help manufacturers better forecast demand, enabling them to optimize the use of raw materials and reduce waste. Manufacturers can also implement recycling and reusing strategies for materials within their processes.
- Reducing Emissions: Smart manufacturing systems can monitor emissions and provide real-time feedback on air and water quality. AI can be used to minimize emissions through better process control and identifying opportunities for cleaner energy sources.
- Supply Chain Efficiency: Improved supply chain management through smart manufacturing can lead to fewer emissions and less waste by reducing unnecessary transportation, improving inventory management, and minimizing overproduction.
- Lean Manufacturing: Smart manufacturing embraces lean principles, which focus on eliminating waste and maximizing value. This leads to more efficient processes, reduced resource consumption, and fewer environmental impacts.
- Quality Control: AI-based quality control systems can detect defects and variations in real-time, reducing the need for rework and waste. This improves product quality while minimizing material waste.
- Waste Reduction: Smart manufacturing can incorporate real-time monitoring of waste generation and implement strategies to minimize it. This includes tracking and optimizing the disposal of hazardous waste and promoting recycling and responsible waste management practices.
- Environmental Monitoring: IoT sensors can be used to continuously monitor environmental conditions and report data to regulatory authorities. This ensures compliance with environmental regulations and enables early response to potential issues.
- Life Cycle Assessment: Smart manufacturing can integrate life cycle assessment (LCA) tools to evaluate the environmental impact of products and processes from design to disposal. This helps identify opportunities for improvement.
- Sustainable Materials: Smart manufacturing can facilitate the use of sustainable and eco-friendly materials in the production process. This includes sourcing materials from suppliers with green practices and promoting circular economy principles.
- Optimized Production Scheduling: AI-driven scheduling can optimize production workflows, reducing idle time and energy consumption. This leads to better resource utilization and lower environmental impact.
- Water Management: Smart manufacturing can monitor and optimize water consumption and wastewater treatment, helping to reduce water waste and minimize the environmental impact of industrial discharges.

- Employee Awareness: Smart manufacturing can provide real-time information to employees about the environmental impact of their work, fostering a culture of sustainability and encouraging responsible behavior.
- Compliance and Reporting: Smart manufacturing can streamline the collection and reporting of environmental data, making it easier for companies to meet regulatory requirements and demonstrate their commitment to sustainability.

In summary, the integration of smart manufacturing practices with a strong focus on sustainability and environmental impact reduction can result in more efficient, eco-friendly, and competitive manufacturing processes. It also demonstrates a commitment to corporate social responsibility and can lead to improved brand reputation and consumer trust.

8.12 Conclusion

Blockchain-based digital twins in smart manufacturing promise revolutionary changes by enhancing transparency, data security, and traceability, leading to more efficient, sustainable, and agile operations. In conclusion, several key points can be highlighted:

- Trust and Transparency: Blockchain's immutable ledger ensures data integrity, fostering trust among stakeholders in collaborative manufacturing environments.
- Data Integrity and Provenance: Blockchain records data provenance, crucial for traceability, quality control, and compliance in manufacturing.
- Smart Contracts and Automation: Automation via smart contracts reduces manual intervention, improving efficiency and minimizing errors in manufacturing processes.
- Real-time Data Sharing: Blockchain-based digital twins enable real-time data sharing and collaboration among manufacturers, suppliers, and regulatory bodies.
- Supply Chain Optimization: Real-time visibility into supply chain movements reduces inefficiencies and environmental impacts.
- Security and Privacy: Blockchain enhances data security and privacy, safeguarding intellectual property and sensitive manufacturing data.
- Compliance and Reporting: Simplified data collection and reporting streamline regulatory compliance and support sustainability efforts.
- Efficient Maintenance: AI-driven predictive maintenance powered by blockchain reduces downtime and enhances product quality.
- Environmental Sustainability: Real-time monitoring and optimization contribute to reducing energy consumption and waste in manufacturing.
- Scalability and Interoperability: Standardized protocols ensure scalability and interoperability of blockchain-based digital twins across the manufacturing sector.

Blockchain-based digital twins offer transformative potential in manufacturing, enhancing transparency, efficiency, security, and sustainability. Continued research and development will further refine and expand their application, paving the way for a more resilient, environmentally responsible, and competitive manufacturing sector. Embracing this technology positions manufacturers to thrive amidst evolving industry demands and gain a competitive advantage in the smart manufacturing landscape.

References

1. Tao, F., Zhang, Y., Cheng, Y., Ren, J., Wang, D., Qi, Q., Li, P., Digital twin and blockchain enhanced smart manufacturing service collaboration and management. *J. Manuf. Syst.*, 62, 903–9145, 2022.
2. Shen, W., Hu, T., Zhang, C., Ma, S., Secure sharing of big digital twin data for smart manufacturing based on blockchain. *J. Manuf. Syst.*, 61, 338–3505, 2021.
3. Li, L., Lei, B., Mao, C., Digital twin in smart manufacturing. *J. Ind. Inf. Integr.*, 26, 1002895, 2022.
4. Leng, J., Ye, S., Zhou, M., Zhao, J.L., Liu, Q., Guo, W., Cao, W., Fu, L., Blockchain-secured smart manufacturing in industry 4.0: A survey. *IEEE Trans. Syst. Man Cybern.: Syst.*, 51, 1, 237–252, 2020.
5. Nakamoto, S., Bitcoin: A peer-to-peer electronic cash system. *Decentralized business Rev.*, 2008.
6. Uhlenkamp, J.-F., Hribernik, K., Wellsandt, S., Thoben, K.-D., Digital Twin Applications: A first systemization of their dimensions, in: *2019 IEEE International Conference on Engineering, Technology and Innovation (ICE/ITMC)*, IEEE, pp. 1–8, 2019.
7. Mittal, S., Khan, M.A., Romero, D., Wuest, T., Smart manufacturing: Characteristics, technologies and enabling factors. *Proceedings of the Institution of Mechanical Engineers, Part B: Journal of Engineering Manufacture*, vol. 233, pp. 1342–1361, 2019.
8. Shen, T. and Li, B., Digital twins in additive manufacturing: a state-of-the-art review. *Int. J. Adv. Manuf. Technol.*, 1–30, 2024.
9. Tofail, S.A.M., Koumoulos, E.P., Bandyopadhyay, A., Bose, S., O'Donoghue, L., Charitidis, C., Additive manufacturing: scientific and technological challenges, market uptake and opportunities. *Mater. Today*, 21, 1, 22–37, 2018.
10. Garetti, M. and Taisch, M., Sustainable manufacturing: trends and research challenges. *Prod. Plan. Control.*, 23, 2-3, 83–104, 2012.
11. Jones, M.D., Hutcheson, S., Camba, J.D., Past, present, and future barriers to digital transformation in manufacturing: A review. *J. Manuf. Syst.*, 60, 936–9485, 2021.
12. Liere-Netheler, K., Packmohr, S., Vogelsang, K., Drivers of digital transformation in manufacturing, 2018.
13. Enders, M.R. and Hoßbach, N., Dimensions of digital twin applications-a literature review, 2019.
14. Ruzsa, C., Digital twin technology-external data resources in creating the model and classification of different digital twin types in manufacturing. *Proc. Manuf.*, 54, 209–2155, 2021.
15. Onaji, I., Tiwari, D., Soulatiantork, P., Song, B., Tiwari, A., Digital twin in manufacturing: Conceptual framework and case studies. *Int. J. Comput. Integr. Manuf.*, 35, 8, 831–858, 2022.
16. Raj, P. and Surianarayanan, C., Digital twin: the industry use cases, in: *Advances in computers*, vol. 117, pp. 285–320, Elsevier, 2020.
17. Juma, M., Alattar, F., Touqan, B., Securing Big Data Integrity for Industrial IoT in Smart Manufacturing Based on the Trusted Consortium Blockchain (TCB). *IoT*, 4, 1, 27–55, 2023.
18. Sahal, R., Alsamhi, S.H., Brown, K.N., O'shea, D., McCarthy, C., Guizani, M., Blockchain-empowered digital twins collaboration: Smart transportation use case. *Machines*, 9, 9, 193, 2021.
19. Yaqoob, I., Salah, K., Uddin, M., Jayaraman, R., Omar, M., Imran, M., Blockchain for digital twins: Recent advances and future research challenges. *IEEE Netw.*, 34, 5, 290–298, 2020.

20. Shahbazi, Z. and Byun, Y.-C., Integration of blockchain, IoT and machine learning for multi-stage quality control and enhancing security in smart manufacturing. *Sensors*, 21, 4, 1467, 2021.
21. Suhail, S., Hussain, R., Jurdak, R., Oracevic, A., Salah, K., Hong, C.S., Matulevičius, R., Blockchain-based digital twins: research trends, issues, and future challenges. *ACM Comput. Surv.*, 54, 11s, 1–34, 2022.
22. Paez, O., Dewees, J., Genaidy, A., Tuncel, S., Karwowski, W., Zurada, J., The lean manufacturing enterprise: An emerging sociotechnological system integration. *Hum. Factors Ergon. Manuf. Serv. Ind.*, 14, 3, 285–306, 2004.
23. Shen, W., Hu, T., Zhang, C., Ma, S., Secure sharing of big digital twin data for smart manufacturing based on blockchain. *J. Manuf. Syst.*, 61, 338–3505, 2021.
24. Tuptuk, N. and Hailes, S., Security of smart manufacturing systems. *J. Manuf. Syst.*, 47, 93–1065, 2018.
25. Maggi, F., Balduzzi, M., Vosseler, R., Rösler, M., Quadrini, W., Tavola, G., Pogliani, M., Quarta, D., Zanero, S., Smart factory security: A case study on a modular smart manufacturing system. *Proc. Comput. Sci.*, 180, 666–6755, 2021.
26. Shahbazi, Z. and Byun, Y.-C., Smart manufacturing real-time analysis based on blockchain and machine learning approaches. *Appl. Sci.*, 11, 8, 3535, 2021.
27. Sharma, P.K., Kumar, N., Park, J.H., Blockchain-based distributed framework for automotive industry in a smart city. *IEEE Trans. Ind. Inf.*, 15, 7, 4197–4205, 2018.
28. Patel, D., Sahu, C.K., Rai, R., Security in modern manufacturing systems: integrating blockchain in artificial intelligence-assisted manufacturing. *Int. J. Prod. Res.*, 1–31, 2023.
29. Leng, J., Zhong, Y., Lin, Z., Xu, K., Mourtzis, D., Zhou, X., Zheng, P., Liu, Q., Zhao, J.L., Shen, W., Towards resilience in Industry 5.0: A decentralized autonomous manufacturing paradigm. *J. Manuf. Syst.*, 71, 95–114, 2023.
30. Huang, A., Triebe, M., Li, Z., Wu, H., Joung, B.G., Sutherland, J.W., A review of research on smart manufacturing in support of environmental sustainability. *Int. J. Sustain. Manuf.*, 5, 2-4, 132–163, 2022.
31. Tyagi, A.K. and Richa, 2023, Digital Twin Technology: Opportunities and Challenges for Smart Era's Applications. *Proceedings of the 2023 Fifteenth International Conference on Contemporary Computing (IC3-2023). Association for Computing Machinery*, New York, NY, USA, pp. 328–336, https://doi.org/10.1145/3607947.3608015.
32. Tyagi, A.K. and Richa, Smart Manufacturing Using Internet of Things, Artificial Intelligence, and Digital Twin Technology, in: *Global Perspectives on Robotics and Autonomous Systems: Development and Applications*, 2023.
33. Tyagi, A.K., Fernandez, T.F., Mishra, S., Kumari, S., Intelligent Automation Systems at the Core of Industry 4.0, in: *Intelligent Systems Design and Applications. ISDA 2020. Advances in Intelligent Systems and Computing*, vol. 1351, A. Abraham, V. Piuri, N. Gandhi, P. Siarry, A. Kaklauskas, A. Madureira (Eds.), Springer, Cham, 2021, https://doi.org/10.1007/978-3-030-71187-0_1.
34. Tyagi, A.K., Chandrasekaran, S., Sreenath, N., Blockchain Technology:– A New Technology for Creating Distributed and Trusted Computing Environment. *2022 International Conference on Applied Artificial Intelligence and Computing (ICAAIC)*, pp. 1348–1354, 2022, doi: 10.1109/ICAAIC53929.2022.9792702.
35. Nair, M.M., Tyagi, A.K., Sreenath, N., The Future with Industry 4.0 at the Core of Society 5.0: Open Issues, Future Opportunities and Challenges. *2021 International Conference on Computer Communication and Informatics (ICCCI)*, pp. 1–7, 2021, doi: 10.1109/ICCCI50826.2021.9402498.

9

Blockchain for Internet of Things and Machine Learning-Based Automated Sectors

Amit Kumar Tyagi[1]*, Shabnam Kumari[2] and Shrikant Tiwari[3]

[1]Department of Fashion Technology, National Institute of Fashion Technology, New Delhi, Delhi, India
[2]Department of Computer Science, Faculty of Science and Humanities, SRM Institute of Science and Technology, Kattankulathur, Chennai, Tamil Nadu, India
[3]School of Computing Science and Engineering (SCSE), Galgotias University, Greater Noida, Uttar Pradesh, India

Abstract

The convergence of Blockchain technology with the Internet of Things (IoT) and Machine Learning (ML) heralds a transformative era for automated sectors. This chapter discusses the integration of Blockchain to fortify the security, transparency, and efficiency of data transactions within IoT networks and the application of ML algorithms in automating decision-making processes. The convergence of Blockchain and IoT addresses inherent challenges of data integrity and security in the large ecosystem of interconnected devices. By using distributed ledger principles, this synergy ensures tamper-resistant record-keeping, secure data sharing, and streamlined communication protocols. The result is a robust foundation for reliable and trustful interactions between IoT devices, mitigating vulnerabilities, and enhancing the overall resilience of automated systems. Furthermore, the integration of ML algorithms into automated sectors opens avenues for intelligent decision-making, predictive analytics, and optimization. This chapter investigates how Blockchain, as a secure and transparent data layer, facilitates the training and deployment of ML models by providing a trustworthy data source and an immutable audit trail for model outcomes. In this chapter, real-world applications and case studies will be examined to illustrate the impact of this triad – Blockchain, IoT, and ML – on diverse sectors such as smart cities, industrial automation, and healthcare. We also discuss challenges and opportunities arising from this integration, including scalability, interoperability, and ethical issue. By keeping light on the symbiotic relationship between Blockchain, IoT, and ML in automated sectors, this chapter aims to contribute to a deeper understanding of the technological advancements shaping the future of intelligent, secure, and efficient automated systems.

Keywords: Blockchain, Internet of Things (IoT), machine learning (ML), automated sectors, data integrity, predictive maintenance

**Corresponding author*: amitkrtyagi025@gmail.com

Amit Kumar Tyagi, Shrikant Tiwari, Senthil Kumar Arumugam and Avinash Kumar Sharma (eds.) Artificial Intelligence-Enabled Digital Twin for Smart Manufacturing, (179–214) © 2024 Scrivener Publishing LLC

9.1 Introduction

Blockchain technology, coupled with the Internet of Things (IoT) and machine learning, has the potential to revolutionize various sectors by enabling secure, decentralized, and intelligent automation [1]. The Internet of Things (IoT) refers to the network of interconnected physical devices, sensors, and actuators that collect and exchange data. IoT enables the seamless integration of the physical and digital worlds, providing valuable information and enabling automation in various sectors such as manufacturing, logistics, healthcare, agriculture, and smart cities [2]. Machine learning, a subset of artificial intelligence, empowers systems to learn from data and make intelligent predictions or decisions without explicit programming. Machine learning algorithms analyze large datasets, discover patterns, and improve their performance over time, enhancing automation capabilities. Blockchain technology, known for its decentralized and immutable nature, provides a secure and transparent framework for data storage, verification, and transaction execution [3]. By combining blockchain with IoT and machine learning, several advantages can be realized which is shown in the Figure 9.1.

Data Security and Integrity: Blockchain ensures the security and integrity of IoT data. Each transaction or data point is cryptographically linked and recorded in a transparent and tamper-proof manner, reducing the risk of data manipulation or unauthorized access. This feature is particularly important in sectors where data accuracy and security are paramount, such as healthcare and supply chain management.

Decentralized and Trustless Systems: Blockchain eliminates the need for intermediaries by enabling peer-to-peer interactions. This decentralization enhances the efficiency and reliability of IoT networks by reducing single points of failure and improving trust among participants. It also enables automated transactions and interactions between devices without relying on a centralized authority.

Smart Contracts for Automation: Smart contracts, self-executing contracts with predefined rules and conditions, can automate processes in IoT environments. By integrating machine learning algorithms into smart contracts, intelligent automation can be achieved. For example, in supply chain management, smart contracts can automatically trigger actions based on predefined conditions, such as product delivery, inventory management, and quality control.

Data Monetization and Sharing: Blockchain technology allows individuals and organizations to securely monetize and share IoT data. By enabling direct peer-to-peer transactions and establishing transparent ownership and usage rights, blockchain facilitates data marketplaces and incentivizes data sharing. This opens up opportunities for new business models and collaborations while preserving data privacy and control.

Improved Machine Learning Models: Blockchain can enhance the quality and robustness of machine learning models. By considering decentralized data sources from IoT devices

Figure 9.1 Key aspects related to blockchain.

and other participants, machine learning algorithms can access a larger and more diverse dataset, leading to improved accuracy and performance. Blockchain also enables the sharing and validation of machine learning models in a trusted and transparent manner.

In summary, the integration of blockchain with the Internet of Things and machine learning has the potential to transform automated sectors. This combination provides enhanced data security, decentralized automation, smart contract-driven processes, data monetization opportunities, and improved machine learning models. As these technologies continue to evolve and mature, we can expect rapid/ huge advancements in various industries, leading to more efficient, secure, and intelligent automated systems.

9.2 Evolution Variants and Architecture of Internet of Things

The Internet of Things (IoT) has evolved rapidly over the years, and today it encompasses a wide range of devices, technologies, and applications [4–7]. Let us discuss the evolution, variants, and architecture of IoT is shown in the Figure 9.2, Figure 9.3, and Figure 9.4.

Evolution of IoT

a) *Early Stage:* In the early stages, IoT was focused on machine-to-machine (M2M) communication and industrial applications. It involved connecting machines and sensors to monitor and control various processes.

b) *Consumer IoT:* The next phase brought IoT into the consumer domain, with the proliferation of connected devices such as smart home appliances, wearables, and personal assistants. Consumer IoT aimed to enhance convenience, automation, and personalization in everyday life.

c) *Industrial IoT (IIoT):* IIoT focuses on using IoT technologies and data analytics to optimize industrial processes, improve operational efficiency, and enable predictive maintenance. It involves connecting industrial equipment, machines, and sensors for real-time monitoring and control.

d) *IoT Ecosystems:* The evolution of IoT has led to the emergence of IoT ecosystems, where various devices, platforms, and services interoperate to create integrated solutions. These ecosystems foster collaboration among different stakeholders and enable seamless connectivity and interoperability.

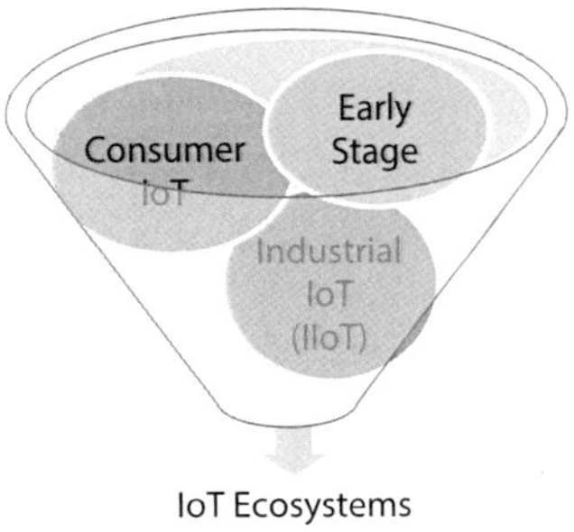

Figure 9.2 Key aspects related to blockchain.

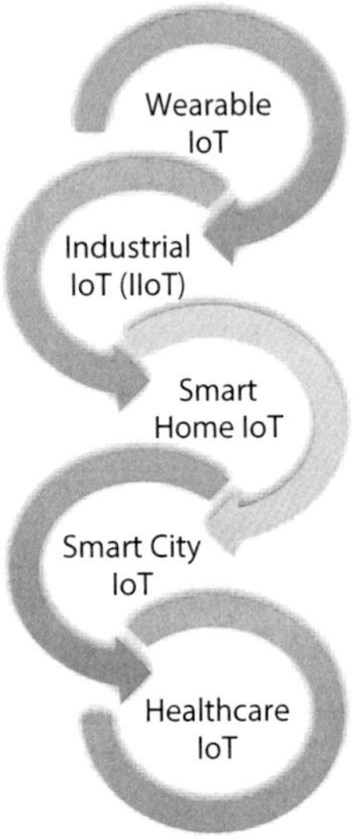

Figure 9.3 Key aspects related to blockchain.

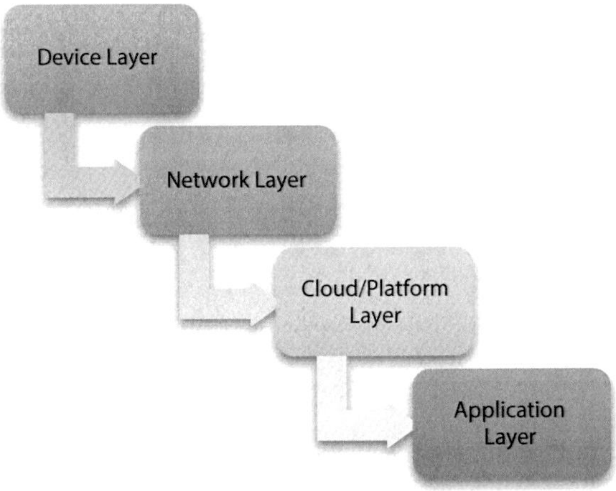

Figure 9.4 Key aspects related to blockchain.

Variants of IoT

a) *Wearable IoT:* Wearable devices, such as smartwatches, fitness trackers, and medical monitoring devices, form a different variant of IoT. These devices collect and transmit data related to health, fitness, and personal activities.
b) *Industrial IoT (IIoT):* IIoT focuses on the integration of sensors, machines, and industrial systems to enhance productivity, efficiency, and safety in industrial sectors like manufacturing, energy, transportation, and agriculture.
c) *Smart Home IoT:* Smart home IoT refers to the networked ecosystem of devices and systems within a residential setting. It enables automation and

control of various aspects of home life, including lighting, security, temperature, entertainment, and appliances.

d) *Smart City IoT:* Smart city IoT involves the deployment of connected sensors, devices, and infrastructure to optimize urban services and improve quality of life. It encompasses areas like transportation, energy management, waste management, public safety, and citizen engagement.

e) *Healthcare IoT:* Healthcare IoT focuses on considering connected devices and data analytics to improve patient care, remote monitoring, medical diagnostics, and treatment. It includes wearable health trackers, remote patient monitoring systems, and smart medical devices.

IoT Architecture

a) *Device Layer:* The device layer consists of physical devices or sensors that collect data from the environment. These devices can range from simple sensors to complex embedded systems with connectivity capabilities.

b) *Network Layer:* The network layer provides connectivity and communication infrastructure for IoT devices. It includes wireless technologies like Wi-Fi, Bluetooth, Zigbee, cellular networks, and emerging technologies like NB-IoT and LoRaWAN.

c) *Cloud/Platform Layer:* The cloud/platform layer is responsible for data storage, processing, and analysis. It includes cloud-based platforms that manage and analyze the large amount of data generated by IoT devices.

d) *Application Layer:* The application layer represents the end-user applications that utilize IoT data and provide services or information. It includes various domains such as smart homes, healthcare applications, industrial automation, and smart city services.

In summary, IoT has evolved from early M2M applications to encompass a wide range of domains and use cases. Variants of IoT include wearable IoT, industrial IoT, smart home IoT, smart city IoT, and healthcare IoT. The architecture of IoT involves the device layer, network layer, cloud/platform layer, and application layer. As IoT continues to advance, we can expect further integration, innovation, and transformative impacts across industries and sectors.

9.3 Evolution, Variants, and Architecture Machine Learning

Machine learning (ML) has also experienced important evolution, resulting in various variants and architectural approaches [7, 8]. Let us discuss the evolution, variants, and architecture of machine learning is shown in the Figure 9.5, Figure 9.6, and Figure 9.7.

Evolution of Machine Learning

a) *Traditional Machine Learning:* Initially, machine learning focused on traditional algorithms such as linear regression, decision trees, and support vector

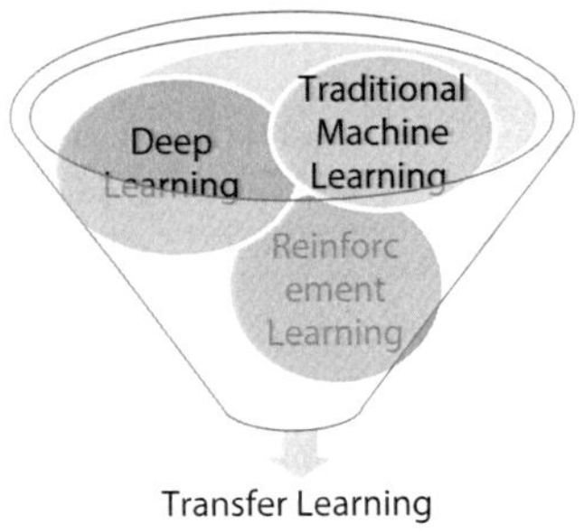

Figure 9.5 Key aspects related to blockchain.

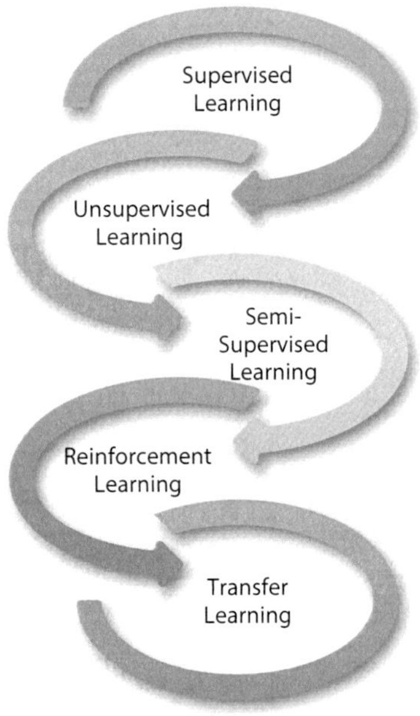

Figure 9.6 Key aspects related to blockchain.

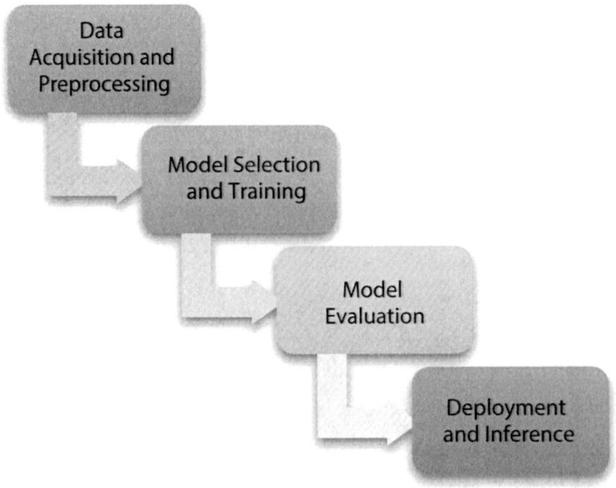

Figure 9.7 Key aspects related to blockchain.

machines. These algorithms were applied to structured datasets and primarily relied on manual feature engineering.

b) *Deep Learning:* The emergence of deep learning revolutionized machine learning by introducing neural networks with multiple layers. Deep learning algorithms, such as convolutional neural networks (CNNs) and recurrent neural networks (RNNs), excelled in tasks involving image recognition, natural language processing, and speech recognition.
c) *Reinforcement Learning:* Reinforcement learning involves training models to make sequential decisions by interacting with an environment. It has been successfully applied in domains like robotics, game playing, and autonomous systems.
d) *Transfer Learning:* Transfer learning enables the use of pre-trained models on one task to improve performance on a related task. It allows models to use knowledge gained from one domain to enhance learning in another domain, even with limited data.

Variants of Machine Learning

a) *Supervised Learning:* Supervised learning involves training models on labeled datasets, where the input data and corresponding output labels are provided. The models learn to predict or classify new, unseen data based on the labeled examples.
b) *Unsupervised Learning:* Unsupervised learning aims to discover patterns, structures, or relationships in unlabeled datasets. It includes techniques like clustering, dimensionality reduction, and generative models.
c) *Semi-Supervised Learning:* This variant combines supervised and unsupervised learning by using a small amount of labeled data along with a larger set of unlabeled data for training. It uses the structure in the unlabeled data to improve model performance.
d) *Reinforcement Learning:* Reinforcement learning involves training an agent to learn optimal actions in an environment by receiving rewards or punishments based on its actions. The agent discusses the environment and learns to maximize cumulative rewards over time.
e) *Transfer Learning:* Transfer learning allows models to transfer knowledge from one task or domain to another, accelerating learning and improving performance on new tasks with limited labeled data.

Machine Learning Architecture

a) *Data Acquisition and Preprocessing:* In this stage, data is collected, cleaned, and prepared for training the machine learning model. It involves tasks like data cleaning, feature engineering, and splitting the data into training and testing sets.
b) *Model Selection and Training:* In this phase, the appropriate machine learning algorithm or model is selected based on the task and data characteristics.

The selected model is trained using the labeled or unlabeled data, and model parameters are adjusted to minimize the prediction error.

c) *Model Evaluation:* The trained model is evaluated on a separate test set to assess its performance and generalization capabilities. Various metrics, such as accuracy, precision, recall, and F1-score, are used to evaluate the model's performance.
d) *Deployment and Inference:* Once the model is trained and evaluated, it can be deployed to make predictions or perform inference on new, unseen data. The model takes input data and generates predictions or decisions based on the learned patterns.

In summary, machine learning has evolved from traditional algorithms to deep learning and reinforcement learning techniques. Variants of machine learning include supervised, unsupervised, semi-supervised, reinforcement, and transfer learning. The machine learning architecture involves data acquisition and preprocessing, model selection and training, model evaluation, and deployment/inference. As machine learning continues to advance, we can expect further advancements in algorithms, architectures, and applications, driving innovation across industries.

9.4 Blockchain for Internet of Things and Machine Learning

Blockchain technology, when combined with the Internet of Things (IoT) and machine learning, has the potential to unlock new possibilities and address important challenges in various domains [9, 10]. Let us discuss how blockchain can enhance IoT and machine learning applications which is shown in the Figure 9.8.

Data Integrity and Security: Blockchain provides a decentralized and tamper-proof ledger that ensures the integrity and security of IoT data. IoT devices can securely store data on the blockchain, making it resistant to unauthorized modifications or tampering. This

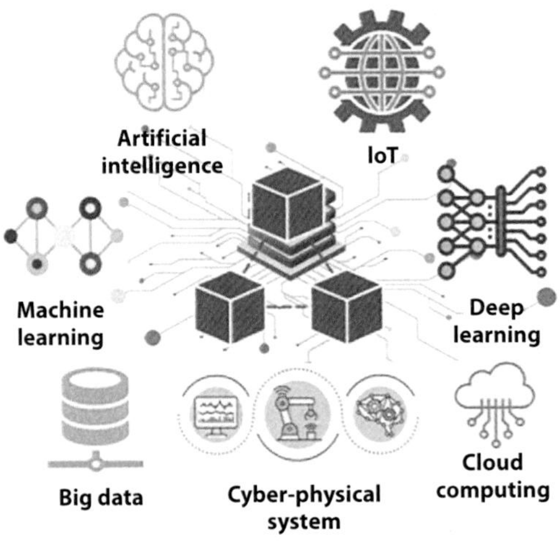

Figure 9.8 Blockchain for Internet of Things and machine learning [11].

ensures the reliability and trustworthiness of the data, which is important in sensitive applications like healthcare, supply chain management, and smart contracts.

Immutable Audit Trail: Blockchain enables the creation of an immutable audit trail for IoT data. Every transaction and data update is recorded in a transparent and verifiable manner, allowing for traceability and accountability. This audit trail is particularly valuable in scenarios where multiple stakeholders need to access and verify the history of IoT data, such as in regulatory compliance or product provenance.

Data Sharing and Monetization: Blockchain facilitates secure and transparent data sharing and monetization in IoT ecosystems. With the use of smart contracts, IoT devices can autonomously negotiate and trade data directly with other devices or entities. This opens up new possibilities for data marketplaces, where individuals and organizations can securely share and monetize their IoT data while maintaining control over data ownership and privacy.

Enhanced Interoperability: Blockchain can improve interoperability between different IoT devices, platforms, and protocols. By establishing a standardized framework for data exchange and communication, blockchain enables seamless integration and interoperability across heterogeneous IoT systems. This simplifies the development of complex IoT solutions and facilitates collaboration among various stakeholders.

Machine Learning Model Security and Transparency: Blockchain can enhance the security and transparency of machine learning models. By storing the models and their training data on the blockchain, the integrity and ownership of the models can be verified. This ensures that the models have not been tampered with and allows for transparent auditing and accountability. It also enables collaborative machine learning, where multiple parties can contribute data and models securely and immutably.

Decentralized and Trustless Orchestration: Blockchain enables decentralized orchestration of IoT devices and machine learning processes. Smart contracts can automate interactions and transactions between devices and enable decentralized decision-making. This reduces the need for intermediaries and centralized authorities, promoting trustless interactions and enhancing the efficiency and autonomy of IoT systems.

Scalability and Performance: Blockchain technology is continuously evolving to address scalability and performance challenges. Innovations like sharding, off-chain solutions, and consensus algorithms are being developed to improve transaction throughput and reduce latency. These advancements will enable blockchain to handle the massive scale of IoT data and machine learning computations.

In summary, the integration of blockchain with the Internet of Things and machine learning brings numerous benefits, including enhanced data integrity, security, interoperability, data sharing, and transparency. By using blockchain technology, IoT and machine learning applications can be made more reliable, secure, and efficient, paving the way for innovative solutions in various domains such as healthcare, supply chain, smart cities, and industrial automation.

9.5 Blockchain-Based Learning Automated Analytics Platforms

Blockchain-based learning automated analytics platforms use the power of blockchain technology to enhance learning analytics and provide advanced educational solutions. These platforms combine the benefits of blockchain, machine learning, and analytics to improve the efficiency, transparency, and personalization of the learning process [12]. Here are the key aspects and advantages of blockchain-based learning automated analytics platforms is shown in the Figure 9.9.

Data Security and Privacy: Blockchain ensures the security and integrity of learner data. Student records, achievements, and other educational data can be securely stored on the blockchain, preventing unauthorized access, tampering, or data loss. Blockchain's decentralized nature and cryptographic techniques provide robust data protection and privacy.

Transparent and Immutable Records: Blockchain maintains a transparent and immutable record of learner achievements, certifications, and credentials. This enables reliable verification of educational accomplishments, eliminating the need for manual verification processes. Employers and educational institutions can easily validate a learner's credentials, reducing fraud and enhancing trust.

Personalized Learning Paths: Machine learning algorithms integrated into blockchain-based platforms can analyze learner data and generate personalized learning paths. By tracking learner progress, behavior, and preferences, the platform can recommend tailored content, activities, and assessments, ensuring a customized and engaging learning experience for each student.

Enhanced Collaboration and Peer Review: Blockchain-based platforms facilitate peer-to-peer collaboration and assessment. Smart contracts can automate and verify peer reviews, ensuring fair evaluation and feedback. Learners can engage in decentralized learning communities, share resources, and collaborate on projects while maintaining transparency and trust in the evaluation process.

Figure 9.9 Blockchain-based learning automated analytics platforms.

Microcredentialing and Badges: Blockchain-based platforms support microcredentialing, enabling learners to earn and store digital badges or certifications directly on the blockchain. These badges can represent specific skills, achievements, or competencies, making them easily shareable and transferable across educational institutions and employers.

Data Analytics: By using machine learning algorithms, blockchain-based learning platforms can analyze large amounts of educational data to generate actionable information. This information can help educators and institutions identify trends, personalize instruction, optimize curriculum, and improve learning outcomes.

Tokenized Incentives and Gamification: Blockchain-based platforms can introduce tokenized incentives and gamification elements to motivate learner engagement and achievement. Learners can earn tokens or digital assets for completing tasks, reaching milestones, or demonstrating mastery. These tokens can be exchanged for tangible rewards or used to unlock advanced learning opportunities.

Lifelong Learning and Credential Portability: Blockchain-based platforms enable lifelong learning and portability of credentials. Learners can store their educational achievements and credentials on the blockchain, maintaining a lifelong record of their learning journey. This facilitates the transfer of credentials across educational institutions, employers, and borders, empowering learners with greater mobility and recognition.

In summary, blockchain-based learning automated analytics platforms provide secure data storage, transparency, personalized learning paths, collaboration, microcredentialing, data analytics, gamification, and lifelong learning capabilities. By harnessing the power of blockchain and machine learning, these platforms have the potential to revolutionize education by providing efficient, personalized, and globally recognized educational experiences.

9.6 Blockchain Inclusion in Internet of Things Architecture and Machine Learning

Blockchain technology can be integrated into the architecture of the Internet of Things (IoT) and machine learning systems to enhance their capabilities and address various challenges [13]. Here's how blockchain can be included in the architecture of IoT and machine learning:

Internet of Things (IoT) Architecture with Blockchain

a) *Device Integration:* IoT devices can be equipped with blockchain capabilities, allowing them to securely interact with the blockchain network. This ensures the authenticity and integrity of data collected from IoT devices, as well as secure communication and transactions between devices.
b) *Data Integrity and Security:* Blockchain provides a decentralized and immutable ledger, ensuring the integrity and security of IoT data. IoT devices can store data on the blockchain, preventing unauthorized modifications or tampering. This enhances data reliability and trustworthiness.
c) *Decentralized Governance:* Blockchain enables decentralized governance models for IoT networks, eliminating the need for a central authority. Consensus

mechanisms ensure agreement on data validity and enforce rules for device interactions, enhancing trust and reducing dependency on intermediaries.

d) *Smart Contracts:* Smart contracts can be deployed on the blockchain to automate interactions between IoT devices. These self-executing contracts facilitate secure and transparent transactions, allowing devices to autonomously negotiate and exchange data or services.

e) *Data Sharing and Monetization:* Blockchain enables secure and transparent data sharing and monetization in IoT ecosystems. With the use of smart contracts, IoT devices can negotiate data sharing agreements and receive compensation for providing valuable data to other entities.

Machine Learning Architecture with Blockchain

a) *Data Provenance and Integrity:* Blockchain can ensure the integrity and provenance of training data used in machine learning models. By recording the data acquisition and processing steps on the blockchain, the history, and authenticity of the training data can be verified, improving model reliability and accountability.

b) *Model Security and Transparency:* Blockchain can store machine learning models and their training processes, providing a transparent and auditable record. This ensures that the models have not been tampered with and enables verification of their integrity and fairness.

c) *Federated Learning and Data Collaboration:* Blockchain facilitates secure and privacy-preserving collaboration in federated learning scenarios. Multiple parties can contribute their data and models to collectively train a shared model without exposing sensitive data. Blockchain ensures the fairness and transparency of the collaboration process.

d) *Incentive Mechanisms:* Blockchain-based incentive mechanisms, such as tokenized rewards or data exchange, can encourage data sharing and participation in machine learning collaborations. Participants can be rewarded for contributing data or computational resources, fostering a collaborative and incentivized machine learning ecosystem.

Hence, integrating blockchain into the architecture of IoT and machine learning systems brings benefits such as data integrity, security, transparency, decentralized governance, and incentivized collaboration. It enhances the trustworthiness, efficiency, and privacy of these systems, enabling innovative applications in various domains like supply chain management, healthcare, smart cities, and personalized learning.

9.7 Features Benefits Limitations Applications and Challenges of Internet of Things

In this section, we will discuss Features, Benefits, Limitations, Applications, and Challenges of Internet of Things in detail as:

Features of Internet of Things (IoT)

a) *Connectivity:* IoT enables the connection of physical devices and objects through the internet, allowing them to communicate and share data.
b) *Sensing and Data Collection:* IoT devices have sensors and actuators that collect data from the physical environment and transmit it to other devices or systems.
c) *Data Processing and Analytics:* IoT systems employ data processing and analytics techniques to extract meaningful information from the collected data.
d) *Automation and Control:* IoT devices can automate processes and control physical objects based on collected data and predefined rules.
e) *Remote Monitoring and Management:* IoT enables remote monitoring and management of devices and systems, enhancing efficiency and convenience.
f) *Scalability:* IoT systems can scale to accommodate a large number of connected devices, enabling widespread deployment in various applications.

Benefits of Internet of Things (IoT)

a) *Improved Efficiency and Productivity:* IoT systems optimize operations, automate tasks, and enable real-time monitoring, leading to increased efficiency and productivity.
b) *Enhanced Decision-Making:* IoT provides real-time data and valuable insights, enabling informed and data-driven decision-making.
c) *Cost Savings:* IoT helps optimize resource utilization, reduce operational costs, and minimize maintenance expenses.
d) *Improved Safety and Security:* IoT enables remote monitoring of assets, enhances safety protocols, and facilitates proactive security measures.
e) *Enhanced Customer Experience:* IoT enables personalized and context-aware experiences, improving customer satisfaction.
f) *Environmental Sustainability:* IoT can contribute to environmental sustainability through optimized resource consumption, energy management, and waste reduction.

Limitations of Internet of Things (IoT)

a) *Security and Privacy Risks:* IoT devices and networks can be vulnerable to security breaches and privacy issues, requiring robust security measures.
b) *Interoperability Challenges:* Different IoT devices and platforms may use different protocols and standards, posing challenges for seamless integration and interoperability.
c) *Scalability and Network Congestion:* As the number of connected devices increases, IoT systems may face challenges related to scalability and network congestion.
d) *Complexity and Technical Challenges:* Developing and managing IoT systems can be complex, requiring expertise in multiple domains such as networking, data management, and device integration.

e) *Data Overload and Quality:* The massive amount of data generated by IoT devices can overwhelm systems and face challenges for data storage, processing, and ensuring data quality.

Applications of Internet of Things (IoT)

a) *Smart Home Automation:* IoT devices enable automation and control of home appliances, security systems, and energy management.
b) *Industrial Automation and Manufacturing:* IoT optimizes industrial processes, monitors equipment, and enhances production efficiency.
c) *Healthcare and Remote Monitoring:* IoT facilitates remote patient monitoring, wearable health devices, and real-time healthcare analytics.
d) *Smart Cities and Urban Planning:* IoT enhances urban infrastructure, transportation systems, and energy management in smart city initiatives.
e) *Supply Chain and Logistics:* IoT enables real-time tracking, monitoring, and optimization of supply chain operations.
f) *Agriculture and Environmental Monitoring:* IoT helps optimize agricultural processes, monitor crop health, and enable efficient resource management.

Challenges of Internet of Things (IoT)

a) *Security and Privacy:* IoT systems are vulnerable to security breaches and privacy threats, requiring robust security measures.
b) *Interoperability:* Ensuring seamless integration and interoperability among different IoT devices and platforms remains a challenge.
c) *Standardization:* The lack of standardized protocols and frameworks faces challenges for widespread adoption and compatibility.
d) *Data Management and Analytics:* Handling and processing large volumes of IoT data, extracting meaningful information, and ensuring data quality are challenges.
e) *Power Consumption:* IoT devices often rely on batteries and need to optimize power consumption for long-term operation.
f) *Regulatory and Ethical Issues:* IoT raises issues about data privacy, ownership, ethical implications, and regulatory frameworks that need to be addressed.

9.8 Features Benefits Limitations Applications and Challenges of Machine Learning

Machine Learning (ML) is a subfield of artificial intelligence (AI) that focuses on the development of algorithms and models that enable computers to learn and make predictions or decisions without being explicitly programmed [14]. Here are the key features, benefits, limitations, applications, and challenges of machine learning:

Features of Machine Learning

a) Automated Learning: ML algorithms can automatically learn patterns and relationships from data, eliminating the need for manual rule-based programming.
b) Adaptability: ML models can adapt and improve their performance over time as they receive new data.
c) Handling Complex Data: ML algorithms can handle and process large and complex datasets, including unstructured data like text, images, and videos.
d) Decision Making: ML models can make predictions, classifications, and decisions based on patterns and information learned from training data.

Benefits of Machine Learning

a) *Automation and Efficiency:* ML automates repetitive tasks and processes, improving efficiency and productivity.
b) *Accurate Predictions:* ML models can analyze large amount of data to make accurate predictions, enabling better decision-making.
c) *Personalization:* ML algorithms can personalize user experiences and recommendations based on individual preferences and behaviors.
d) *Continuous Improvement:* ML models can continually learn and adapt to new data, improving their performance over time.

Limitations of Machine Learning

a) *Data Dependency:* ML models heavily rely on high-quality and representative training data. Insufficient or biased data can lead to inaccurate or biased results.
b) *Interpretability:* Complex ML models can be difficult to interpret, making it challenging to understand and explain the reasoning behind their decisions.
c) *Overfitting:* ML models may become too specialized to the training data, resulting in poor generalization to new, unseen data.
d) *Computational Resources:* Training complex ML models can require huge computational resources and time.

Applications of Machine Learning

a) *Image and Speech Recognition:* ML is used in applications like facial recognition, object detection, and speech-to-text conversion.
b) *Natural Language Processing:* ML algorithms power chatbots, language translation, sentiment analysis, and text summarization.
c) *Recommendation Systems:* ML models enable personalized recommendations in e-commerce, streaming services, and content platforms.
d) *Fraud Detection:* ML can identify patterns and anomalies to detect fraudulent transactions and activities in finance and cybersecurity.

Challenges of Machine Learning

a) Data Privacy and Ethics: ML raises issues about the privacy and ethical use of personal data, requiring careful handling and regulation.
b) Bias and Fairness: ML models can inherit biases present in the training data, leading to unfair or discriminatory outcomes.
c) Scalability: Scaling ML models to handle increasing amounts of data and users can be challenging, requiring distributed computing infrastructure.
d) Model Interpretability: Explaining the decisions and behavior of complex ML models remains a challenge, particularly in important applications such as healthcare and autonomous systems.

In summary, machine learning provides automated learning, adaptability, and handling of complex data, leading to benefits such as automation, accurate predictions, personalization, and continuous improvement. However, limitations like data dependency, interpretability issues, overfitting, and resource requirements exist. ML finds applications in various fields like image recognition, natural language processing, recommendation systems, and fraud detection. Challenges include data privacy, bias and fairness issues, scalability, and model interpretability.

9.9 Physical, Network, Software Attacks in Internet of Things and Machine Learning-Based Applications

Internet of Things (IoT) and Machine Learning (ML) based applications are susceptible to various types of attacks, including physical, network, and software attacks [15]. Here is an overview of each of these attack categories:

Physical Attacks

a) *Tampering:* Physical attackers can gain access to IoT devices or ML systems and tamper with their hardware or software components, altering their functionality or extracting sensitive information.
b) *Side-Channel Attacks:* Attackers exploit unintended information leakage from physical characteristics like power consumption, electromagnetic emissions, or timing behavior to extract sensitive data or gain unauthorized access.

Network Attacks

a) *Eavesdropping:* Attackers intercept and monitor network communications between IoT devices or ML systems, potentially gaining access to sensitive data or information.
b) *Man-in-the-Middle (MitM) Attacks:* Attackers position themselves between the communication path of IoT devices or ML systems, intercepting and altering the exchanged data or impersonating legitimate entities.

c) *Denial of Service (DoS) Attacks:* Attackers flood IoT devices or ML systems with a high volume of requests, overwhelming their resources and causing a disruption in their normal operations.

Software Attacks

a) *Malware Injection:* Attackers exploit vulnerabilities in IoT device firmware or ML system software to inject malicious code that can disrupt the device's functionality, gain unauthorized access, or steal sensitive information.
b) *Data Poisoning:* Attackers manipulate the training data used by ML models to introduce biased or malicious information, leading to inaccurate predictions or compromised decision-making.
c) *Model Inversion Attacks:* Attackers exploit the responses of ML models to extract sensitive information used during the training phase, such as confidential data or personal attributes of individuals.

Note that these attacks can be interconnected and may involve multiple stages or combinations. For example, a physical attack on an IoT device can lead to the compromise of network communications, enabling subsequent software attacks.

To mitigate these risks, IoT and ML applications should incorporate security measures such as:

a) Ensuring strong physical security for devices and systems, including tamper-resistant components and secure storage of sensitive information.
b) Implementing encryption and authentication mechanisms to protect network communications and prevent unauthorized access.
c) Regularly updating software and firmware to address known vulnerabilities and applying security patches promptly.
d) Implementing intrusion detection and prevention systems to identify and mitigate network attacks.
e) Employing anomaly detection techniques and robust data validation methods to detect and mitigate data poisoning attacks.
f) Implementing privacy-preserving techniques in ML models, such as differential privacy, to protect sensitive training data and prevent model inversion attacks.
g) Continuous monitoring, threat modeling, and adherence to security best practices are important to ensure the robustness and integrity of IoT and ML-based applications in the face of evolving security threats.

9.10 Countermeasures for Raised Challenges and Reliability of Blockchain in Machine Learning/Artificial Intelligence

This section will discuss several countermeasures for raised challenges in Machine Learning and Artificial Intelligence, as:

Data Privacy and Ethics

a) Implement strong data governance practices, including data anonymization and encryption, to protect user privacy.
b) Adhere to ethical guidelines and regulations when collecting, storing, and using personal data.
c) Employ techniques like federated learning or secure multi-party computation to train ML models on distributed data without sharing raw data.

Bias and Fairness

a) Carefully curate and preprocess training data to mitigate bias.
b) Regularly evaluate and audit ML models for fairness and bias, using techniques like fairness metrics and bias-correction algorithms.
c) Diversify the data sources and include diverse perspectives when creating training datasets.

Scalability

a) Utilize distributed computing and parallel processing techniques to handle large-scale data and computational demands.
b) Employ cloud-based infrastructure or distributed systems to scale ML and AI applications efficiently.

Model Interpretability

a) Use interpretable ML models like decision trees or rule-based systems for applications where explainability is important.
b) Employ techniques such as model-agnostic interpretability methods, feature importance analysis, or rule extraction algorithms to gain information into complex ML models.

9.10.1 Reliability of Blockchain in Machine Learning/Artificial Intelligence

Blockchain technology can enhance the reliability and security of ML and AI systems in the following ways [16, 17]:

Data Integrity and Traceability: Blockchain provides a tamper-resistant and transparent ledger, ensuring the integrity of training data and preventing unauthorized modifications. Smart contracts on the blockchain can define rules and conditions for data sharing and usage, ensuring proper data governance.

Decentralization and Trust: By utilizing a decentralized network, blockchain can reduce the reliance on a single central authority, promoting trust and transparency in ML and AI systems. Peer-to-peer consensus mechanisms ensure agreement on the state of the blockchain, reducing the risk of data manipulation or fraud.

Auditing and Accountability: The immutability of blockchain records allows for easier auditing and tracking of ML model training, data sources, and decision-making processes.

Smart contracts can enforce accountability and compliance, ensuring adherence to predefined rules and regulations.

Data Sharing and Collaboration: Blockchain can facilitate secure and controlled data sharing among multiple parties, enabling collaborative ML and AI initiatives while maintaining data privacy and ownership.

However, note that blockchain technology may introduce certain limitations, such as scalability constraints, computational overhead, and the need for consensus mechanisms, which can impact the performance of ML and AI applications. Selective consideration should be given to the specific requirements of the application to determine the suitability of blockchain as a reliable solution.

9.11 Countermeasures for Raised Challenges and Reliability of Blockchain in IoT

This section will discuss few countermeasures for challenges in Internet of Things (IoT)-

Data Privacy and Security

a) Implement strong encryption and authentication mechanisms to protect data transmitted between IoT devices and backend systems.
b) Apply secure communication protocols like Transport Layer Security (TLS) or Datagram Transport Layer Security (DTLS) to ensure data integrity and confidentiality.
c) Regularly update firmware and software to address known vulnerabilities and apply security patches promptly.
d) Implement secure access controls and user authentication mechanisms to prevent unauthorized access to IoT devices and data.

Network Security

a) Employ network segmentation and isolation techniques to prevent lateral movement of attackers within IoT networks.
b) Implement intrusion detection and prevention systems (IDPS) to monitor and detect suspicious activities or network anomalies.
c) Use strong network firewalls and access control mechanisms to restrict unauthorized access to IoT devices and networks.
d) Conduct regular security audits and vulnerability assessments to identify and address potential weaknesses.

Device Authentication and Authorization

a) Implement robust authentication mechanisms such as digital certificates or secure tokens to ensure that only authorized devices can connect to IoT networks.

b) Employ two-factor authentication or multi-factor authentication to enhance the security of device authentication.
c) Apply strict access controls and role-based permissions to limit device capabilities and prevent unauthorized actions.

Reliability of Blockchain in IoT
Data Integrity and Trust

a) Blockchain provides a decentralized and tamper-resistant ledger, ensuring the integrity and immutability of IoT data.
b) By recording IoT data transactions on the blockchain, trust is established among multiple parties, reducing the reliance on centralized authorities.

Distributed Consensus

a) Blockchain's consensus mechanisms ensure agreement on the state of the blockchain, enhancing the reliability and fault tolerance of IoT networks.
b) Consensus algorithms such as Proof of Work (PoW) or Proof of Stake (PoS) ensure that majority agreement is reached before adding new data to the blockchain.

Transparency and Auditing

a) Blockchain's transparent nature allows for easy auditing and verification of IoT data transactions, enabling better accountability and traceability.
b) Smart contracts on the blockchain can automate auditing processes and enforce predefined rules and conditions for data transactions.

Secure Data Sharing

a) Blockchain enables secure and controlled data sharing among multiple IoT devices and stakeholders.
b) Through blockchain-based smart contracts, data sharing permissions and access controls can be defined, ensuring privacy and data ownership.

Hence, while blockchain technology provides benefits in terms of data integrity and trust, it may introduce challenges related to scalability, latency, and energy consumption in IoT environments. Evaluating the specific requirements and constraints of an IoT application is important in determining the suitability and reliability of blockchain implementation.

9.12 Adoption of Blockchain with Internet of Things Systems

The adoption of blockchain technology with Internet of Things (IoT) systems (refer Figure 9.10) provides several potential benefits and use cases [18]. Here are some key aspects of adopting blockchain with IoT systems:

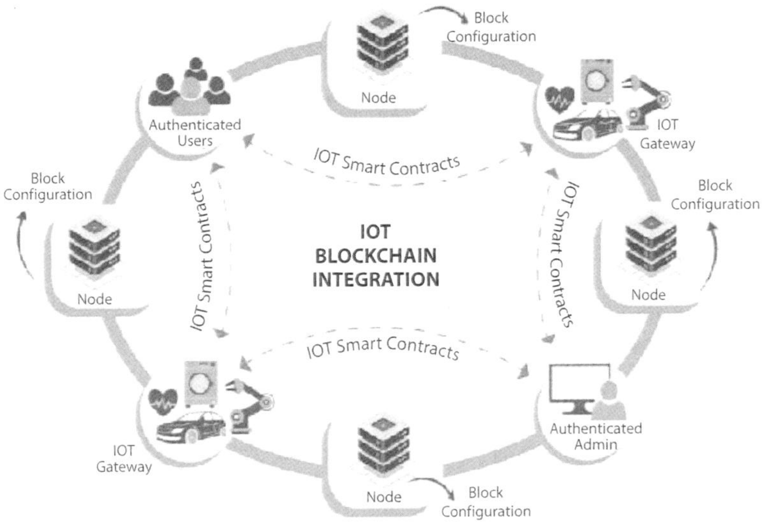

Figure 9.10 IoT blockchain integration [19].

Data Integrity and Trust: Blockchain can provide a tamper-resistant and transparent ledger for recording and verifying IoT data transactions. This ensures the integrity and immutability of IoT data, reducing the need for centralized trust authorities.

Secure Data Sharing: Blockchain enables secure and controlled data sharing among multiple IoT devices and stakeholders. It allows for direct peer-to-peer data exchanges, eliminating the need for intermediaries and reducing the risk of data breaches or unauthorized access.

Distributed Consensus: Blockchain's distributed consensus mechanisms ensure agreement on the state of the blockchain among multiple nodes in a decentralized IoT network. This enhances the reliability, fault tolerance, and resilience of IoT systems.

Improved Security: Blockchain can enhance the security of IoT systems by providing a tamper-resistant platform for storing and validating device identities, access permissions, and software/firmware updates. It can help prevent unauthorized access, data manipulation, and device spoofing.

Smart Contracts and Automation: Blockchain-based smart contracts can automate and enforce predefined rules and conditions for IoT transactions, such as data sharing, access controls, and payments. This allows for automated, trustless interactions between IoT devices and stakeholders.

Supply Chain Management: Blockchain can be used in IoT-based supply chain management to provide end-to-end visibility, transparency, and traceability of products, components, and transactions. This helps reduce fraud, counterfeiting, and ensures compliance with regulatory standards.

Energy Management and Microgrids: Blockchain can enable peer-to-peer energy transactions and decentralized energy management in IoT-enabled microgrids. It allows for transparent energy trading, secure metering, and efficient allocation of energy resources.

Data Monetization and Ownership: Blockchain technology can enable individuals to have better control over their personal data generated by IoT devices. It empowers users to selectively share and monetize their data while maintaining privacy and ownership rights.

Hence despite the potential benefits, we need to consider the challenges of implementing blockchain in IoT systems, including scalability, latency, energy consumption, interoperability, and the complexity of integrating blockchain with existing IoT infrastructure. Careful evaluation, proof-of-concept projects, and considering specific use cases are essential for successful adoption of blockchain in IoT systems.

9.13 Adoption of Blockchain with Machine Learning-Based Computing Environment

The adoption of blockchain technology in machine learning (ML) based computing environments provides several potential benefits and use cases [20]. Here are some key aspects of adopting blockchain with ML-based computing environments is shown in the Figure 9.11.

Data Security and Privacy: Blockchain can enhance the security and privacy of ML datasets and models. By storing data on a tamper-resistant and decentralized ledger, it reduces the risk of data breaches, unauthorized access, and manipulation. Users can maintain control over their data while allowing secure and auditable sharing with ML algorithms.

Trusted Data Sharing and Collaboration: Blockchain enables secure and controlled sharing of ML datasets and models among multiple parties. It allows for transparent verification of data sources, ensuring the authenticity and integrity of shared data. This promotes collaboration, data exchange, and the development of more accurate and robust ML models.

Data Provenance and Auditing: Blockchain provides an immutable record of data transactions, enabling transparent tracking of data provenance and lineage. This helps in auditing and verifying the origin, quality, and history of ML datasets, ensuring trust and transparency in the data used for training ML models.

Model Governance and Intellectual Property: Blockchain can facilitate the management and governance of ML models and intellectual property rights. Smart contracts on the blockchain can define rules for model ownership, usage rights, and royalties, enabling secure and automated enforcement of intellectual property agreements.

Federated Learning and Collaboration: Blockchain can be used to facilitate federated learning, where ML models are trained collaboratively across multiple distributed devices or organizations without sharing raw data. Blockchain's consensus mechanisms and privacy-preserving techniques ensure fairness, transparency, and security in federated learning scenarios.

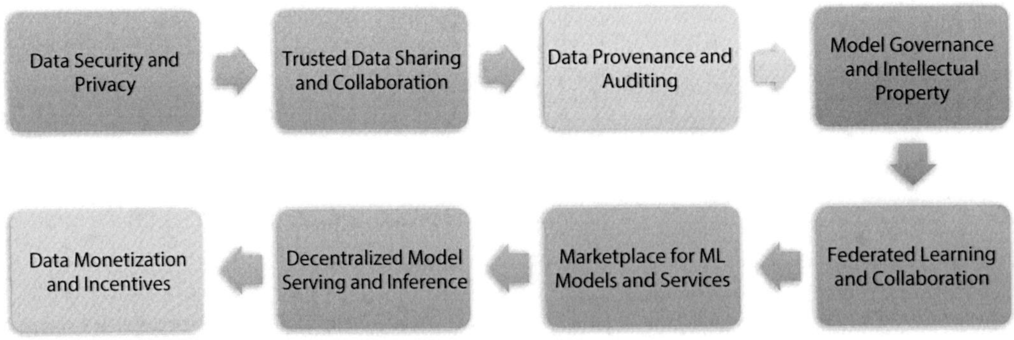

Figure 9.11 Machine learning blockchain integration.

Marketplace for ML Models and Services: Blockchain-based platforms can create decentralized marketplaces for ML models and services, connecting ML developers, data providers, and consumers. Smart contracts enable transparent and automated transactions, ensuring trust, and facilitating a fair and efficient marketplace.

Decentralized Model Serving and Inference: Blockchain can enable decentralized and trustless model serving and inference. ML models can be deployed on the blockchain network, allowing users to request model predictions without relying on a central server, enhancing availability and reliability.

Data Monetization and Incentives: Blockchain enables individuals to have more control over their data and participate in data marketplaces, where they can monetize their data contributions to ML models. Blockchain-based tokens and smart contracts can facilitate fair and transparent data monetization schemes.

While there are potential benefits, we need to consider the challenges of adopting blockchain in ML-based computing environments, including scalability, computational overhead, latency, interoperability, and the complexity of integrating blockchain with existing ML infrastructure. Careful evaluation, proof-of-concept projects, and considering specific use cases are essential for successful adoption of blockchain in ML-based computing environments.

9.14 Blockchain- and Machine Learning-Based Solutions for Big Data Challenges

Blockchain and machine learning (ML) can be used together to address various challenges associated with big data [21]. Here are some solutions that use the combination of blockchain and ML:

Data Quality and Trust

a) ML algorithms can be used to analyze and validate data on the blockchain, ensuring its quality and accuracy.
b) Blockchain provides an immutable ledger that ensures the integrity and transparency of data, establishing trust in the data sources.

Data Sharing and Collaboration

a) Blockchain enables secure and controlled data sharing among multiple parties by defining access permissions and data usage rules through smart contracts.
b) ML algorithms can use shared data on the blockchain to enhance their training and prediction capabilities, resulting in more accurate and robust models.

Data Privacy and Security

a) Blockchain can facilitate the storage and sharing of encrypted data, ensuring privacy while maintaining transparency.

b) ML techniques like federated learning or homomorphic encryption can be employed to process data on the blockchain without exposing sensitive information.

Data Monetization and Incentives

a) Blockchain-based tokens and smart contracts can create economic systems that reward data contributors for sharing their data with ML models.
b) ML algorithms can analyze and extract valuable information from big data, enabling better monetization opportunities for data providers.

Scalability and Performance

a) ML algorithms can be used to optimize blockchain operations and consensus mechanisms, improving scalability and reducing computational overhead.
b) Blockchain can help distribute the processing and storage of big data across a network of nodes, enhancing performance for large-scale data sets.

Fraud Detection and Anomaly Detection

a) ML algorithms can analyze data stored on the blockchain to identify patterns and detect fraudulent activities or anomalies.
b) Blockchain provides a transparent and auditable record of data transactions, enabling better fraud detection and mitigation.

Data Governance and Compliance

a) Blockchain can establish a decentralized and auditable data governance framework, ensuring compliance with regulations and data usage policies.
b) ML algorithms can be used to monitor and enforce compliance rules on the blockchain, ensuring data integrity and adherence to privacy requirements.

These solutions demonstrate how the combination of blockchain and ML can address big data challenges by enhancing data quality, privacy, security, collaboration, and monetization while improving scalability, fraud detection, and compliance. However, we need to carefully evaluate the specific requirements and constraints of each use case to determine the suitability and effectiveness of these solutions.

9.15 Blockchain-Enabled Internet of Things (IoTs) Platforms for Automation in Intelligent Transportation Systems

Blockchain-enabled Internet of Things (IoT) platforms provide automation capabilities in Intelligent Transportation Systems (ITS). By combining blockchain technology with IoT devices, these platforms enhance the efficiency, security, and reliability of transportation systems [22]. Few key benefits of blockchain-enabled IoT platforms in ITS automation include is shown in the Figure 9.12.

Enhanced Data Security: Blockchain's decentralized and immutable nature provides robust security for IoT devices in transportation systems. It ensures secure data transmission, storage, and access, reducing the risk of data breaches and tampering.

Improved Trust and Transparency: Blockchain technology enables a transparent and auditable system for tracking and validating transactions and data exchanges in IoT-enabled transportation. This enhances trust between various stakeholders, such as vehicle owners, manufacturers, service providers, and regulatory bodies.

Seamless Interoperability: IoT devices integrated with blockchain platforms can communicate and interact seamlessly, regardless of their underlying protocols and technologies. This promotes interoperability among different transportation systems and devices, leading to more efficient and coordinated operations.

Smart Contracts and Automation: Blockchain-enabled IoT platforms can utilize smart contracts, which are self-executing agreements with predefined rules and conditions. Smart contracts automate processes and transactions within transportation systems, such as toll collection, insurance claims, and maintenance scheduling, eliminating the need for intermediaries and reducing costs.

Real-time Data Analytics: IoT devices generate large amount of data in transportation systems. By integrating blockchain, data can be securely shared, stored, and analyzed in real-time. This enables actionable information for optimizing traffic management, predicting maintenance needs, and improving overall operational efficiency.

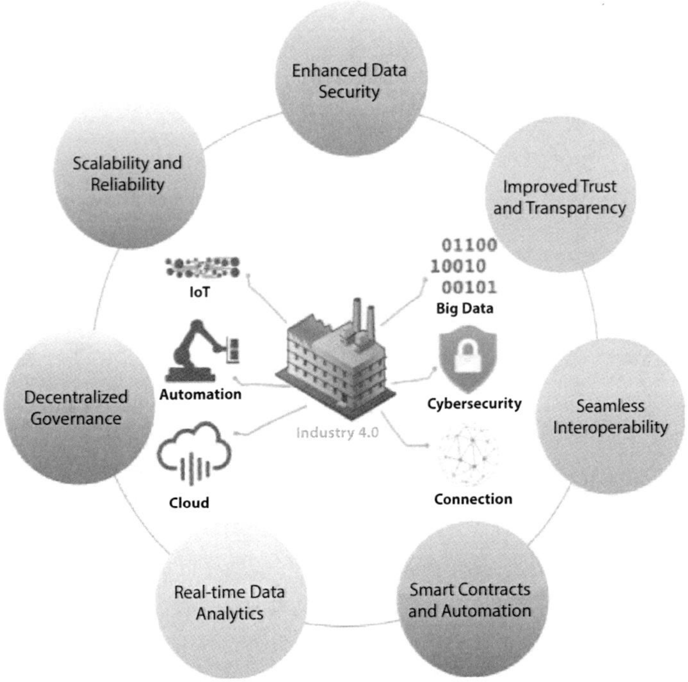

Figure 9.12 Key benefits of blockchain-enabled IoT platforms in ITS automation.

Decentralized Governance: Blockchain technology allows for decentralized governance models in IoT platforms, where decision-making processes are distributed among participants. This can facilitate more democratic and inclusive approaches to managing transportation systems, involving stakeholders from various domains.

Scalability and Reliability: Blockchain technology enables distributed data storage and processing, enhancing scalability and fault-tolerance in IoT platforms for intelligent transportation systems. This ensures that the system can handle a large number of devices, transactions, and data streams without compromising performance or reliability.

In summary, blockchain-enabled IoT platforms provide a foundation for automation and optimization in Intelligent Transportation Systems, leading to enhanced efficiency, security, and coordination among various stakeholders.

9.16 Blockchain-Enabled Internet of Things (IoTs) Platforms for Automation in Software Development

Blockchain-enabled Internet of Things (IoT) platforms have the potential to automate and streamline various aspects of software development processes. By integrating blockchain technology with IoT devices, these platforms provide several benefits for automation in software development is shown in the Figure 9.13 [23, 24].

Secure Code Management: Blockchain provides a secure and tamper-resistant platform for managing code repositories. Developers can store their code on the blockchain, ensuring

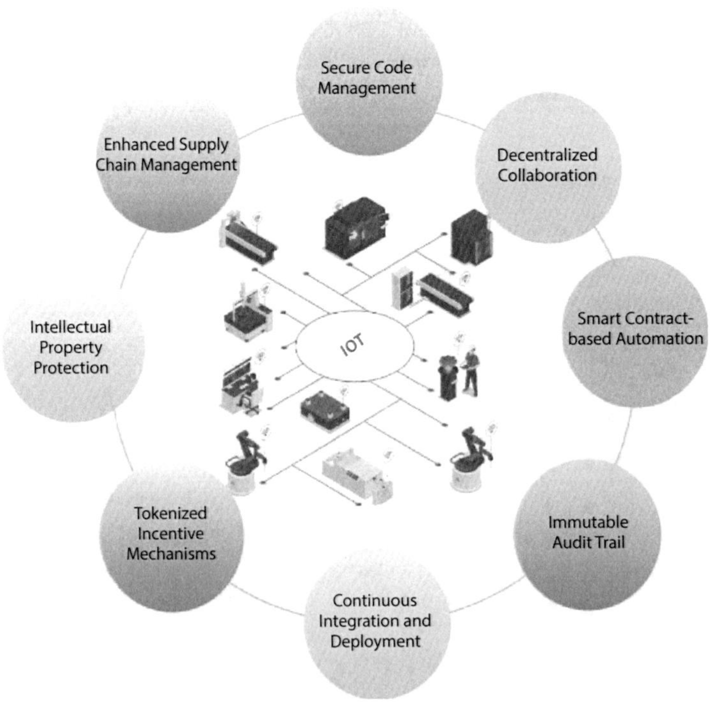

Figure 9.13 Benefits for automation in software development.

its integrity and traceability. This enhances security and reduces the risk of unauthorized modifications or intellectual property theft.

Decentralized Collaboration: Blockchain-based IoT platforms enable decentralized collaboration among developers. They can securely share code, collaborate on projects, and maintain a transparent record of contributions and changes. This promotes efficient teamwork and eliminates the need for centralized project management systems.

Smart Contract-based Automation: Smart contracts, powered by blockchain technology, can automate various aspects of software development processes. These self-executing contracts can define and enforce predefined rules and conditions, such as code testing, deployment, and payment terms. This reduces manual intervention and ensures compliance with agreed-upon processes.

Immutable Audit Trail: Blockchain's immutable nature provides an auditable record of all actions and changes made during software development. This audit trail can be valuable for regulatory compliance, debugging, and resolving disputes. Developers can easily track the history of code modifications, ensuring accountability and transparency.

Continuous Integration and Deployment (CI/CD): Blockchain-enabled IoT platforms can facilitate automated continuous integration and deployment processes. Smart contracts can trigger automatic code testing, quality assurance, and deployment to production environments. This streamlines the software delivery pipeline, reduces manual effort, and improves the speed and efficiency of software releases.

Tokenized Incentive Mechanisms: Blockchain-based IoT platforms can introduce tokenized incentive mechanisms to encourage developer participation and contribution. Developers can be rewarded with tokens for their code contributions, bug fixes, or peer reviews. This gamification element can foster collaboration, innovation, and quality improvements in software development.

Intellectual Property Protection: Blockchain technology can assist in protecting intellectual property rights in software development. By storing timestamps and digital fingerprints of code on the blockchain, developers can establish proof of ownership and originality. This can be important in cases of copyright infringement or disputes over code ownership.

Enhanced Supply Chain Management: Blockchain-enabled IoT platforms can automate supply chain management in software development. They can track and verify the origin and authenticity of software components, licenses, and dependencies. This ensures the integrity and security of software supply chains, reducing the risk of malware or unauthorized modifications.

In summary, blockchain-enabled IoT platforms provide automation capabilities in software development processes. They enhance security, collaboration, auditability, and efficiency, enabling streamlined and reliable software delivery.

9.17 Blockchain-Enabled Internet of Things (IoTs) Platforms for Automation in Protecting Systems

Blockchain-enabled Internet of Things (IoT) platforms provide automation capabilities for protecting systems against various threats and vulnerabilities. By integrating blockchain technology with IoT devices, these platforms provide enhanced security and automation features to safeguard systems [25]. Here are some ways in which blockchain-enabled IoT platforms can automate system protection is shown in the Figure 9.14.

Device Authentication: Blockchain-based identity management enables automated device authentication in IoT systems. Each device has a unique digital identity stored on the blockchain, allowing secure and automated verification of device authenticity before granting access to the system.

Secure Data Transmission: IoT devices connected to a blockchain platform can utilize cryptographic techniques to secure data transmission. Automation ensures that data is encrypted, digitally signed, and transmitted securely, protecting against unauthorized access, tampering, or eavesdropping.

Distributed Threat Intelligence: Blockchain-enabled IoT platforms facilitate the automated sharing of threat intelligence among devices and systems. Information about known threats, vulnerabilities, and attack patterns can be shared securely and in real-time through the blockchain, enabling automated defense mechanisms and timely responses to emerging threats.

Intrusion Detection and Prevention: Blockchain-based IoT platforms can automate intrusion detection and prevention systems (IDPS). Smart contracts can analyze incoming data streams, identify abnormal patterns or malicious behavior, and trigger automated responses such as isolating compromised devices or blocking suspicious network traffic.

Immutable Audit Trail: Blockchain's immutability ensures the creation of an immutable audit trail of system activities and events. Automated logging and timestamping of security-related events on the blockchain enable efficient forensic analysis, incident response, and compliance auditing.

Decentralized Security Governance: Blockchain technology enables decentralized security governance models in IoT platforms. Automated consensus mechanisms and smart contracts can automate security policies, access control, and authorization processes. This reduces reliance on central authorities and enhances the resilience of the system against single points of failure.

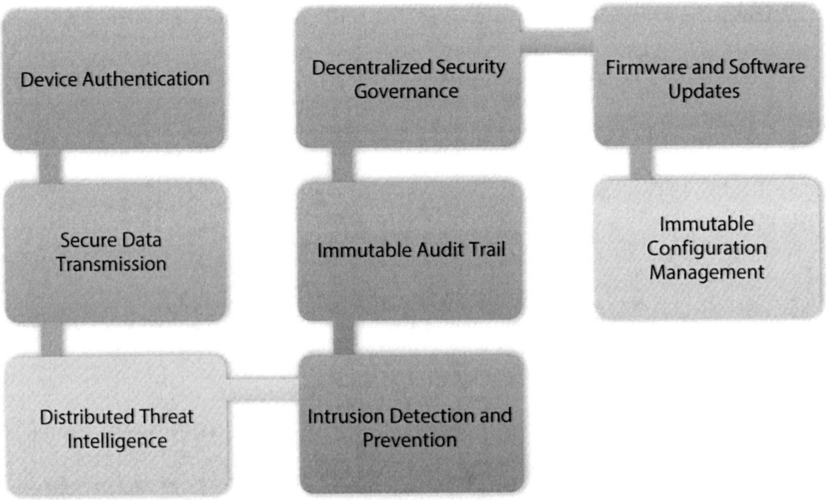

Figure 9.14 Blockchain-enabled IoT platforms automate system protection.

Firmware and Software Updates: Blockchain-based IoT platforms can automate the secure distribution and verification of firmware and software updates. Smart contracts can ensure that updates originate from trusted sources, are cryptographically signed, and are propagated securely to IoT devices. Automated update mechanisms help prevent vulnerabilities from outdated software versions.

Immutable Configuration Management: Blockchain technology can automate configuration management in IoT systems. Device configurations and settings can be securely stored on the blockchain, ensuring tamper-resistant and auditable records of system configurations. Automation can enforce configuration policies and detect unauthorized changes.

By using blockchain-enabled IoT platforms, automation can enhance the protection of systems against cybersecurity threats, ensuring the integrity, confidentiality, and availability of IoT devices and data.

9.18 Issues and Challenges Towards Blockchain Based IoT–ML Applications

While blockchain-based Internet of Things (IoT) and Machine Learning (ML) applications provide numerous benefits, they also face several issues and challenges that need to be addressed [26]. Some of the key challenges are:

Scalability: Blockchain technology faces scalability challenges when it comes to processing a high volume of IoT data and ML computations. The transaction processing speed and block size limitations can hinder the real-time processing requirements of IoT-ML applications.

Data Privacy and Confidentiality: Blockchain's inherent transparency faces challenges for preserving data privacy and confidentiality in IoT-ML applications. As blockchain maintains a public ledger of all transactions, ensuring the privacy of sensitive IoT and ML data becomes important.

Energy Consumption: Blockchain networks, particularly those using proof-of-work consensus algorithms, require huge computational power and energy consumption. This can be a challenge for resource-constrained IoT devices that have limited processing capabilities and battery life.

Integration Complexity: Integrating blockchain with existing IoT and ML systems can be complex. Adapting legacy systems to support blockchain technology and ensuring interoperability between different platforms and protocols can face challenges during implementation.

Governance and Consensus: Establishing governance models and consensus mechanisms that align with the requirements of IoT-ML applications can be challenging. Achieving consensus among numerous IoT devices and ML models, ensuring fairness, and maintaining a decentralized decision-making process are key issues.

Standardization: Lack of standardized protocols and frameworks for blockchain-based IoT-ML applications hampers interoperability and hinders seamless integration between different devices and platforms. Standardization efforts are needed to address compatibility issues and promote widespread adoption.

Data Quality and Reliability: Blockchain relies on the accuracy and reliability of data input. In IoT-ML applications, data quality can be compromised due to various factors such as sensor errors, data tampering, or malicious attacks. Ensuring the integrity and reliability of data becomes important for accurate ML model training and decision-making.

Regulatory and Legal Challenges: Blockchain-based IoT-ML applications may face regulatory and legal challenges related to data ownership, privacy, and compliance with industry-specific regulations. The decentralized nature of blockchain and the complexities of data handling require careful consideration of legal and regulatory frameworks.

Cost and Efficiency: Implementing and maintaining blockchain infrastructure can be costly, especially for resource-constrained IoT devices. Striking a balance between the benefits of blockchain-enabled IoT-ML applications and the associated costs is a challenge that needs to be addressed.

Education and Awareness: Lack of awareness and understanding about the potential benefits and challenges of blockchain-based IoT-ML applications can impede their adoption. Education and training programs are essential to empower developers and stakeholders to effectively utilize these technologies.

Hence, addressing these challenges requires collaborative efforts from researchers, developers, policymakers, and industry stakeholders. Overcoming these hurdles can unlock the full potential of blockchain-based IoT-ML applications and drive innovation in various domains.

9.19 Open Challenges of Blockchain, Internet of Things, and Machine Learning Integration

The integration of blockchain, Internet of Things (IoT), and Machine Learning (ML) presents several open challenges that researchers and practitioners are actively working to address [27]. Some of the key challenges included in the Figure 9.15.

Scalability: The combination of blockchain, IoT, and ML introduces scalability challenges. Processing a large volume of IoT data, training ML models, and maintaining a distributed ledger can be computationally intensive. Developing scalable solutions that can handle the increasing size and complexity of data is a priority.

Data Privacy and Security: Preserving data privacy and security is important in integrated systems. While blockchain provides inherent data immutability, ensuring confidentiality and privacy of sensitive IoT and ML data remains a challenge. Developing robust encryption, access control mechanisms, and privacy-enhancing techniques is essential.

Interoperability and Standards: Achieving interoperability and standardization among diverse IoT devices, ML algorithms, and blockchain platforms is an essential challenge. Creating common protocols, data formats, and frameworks that enable seamless communication, data exchange, and integration across different technologies is a research focus.

Real-time Processing and Latency: Many IoT-ML applications require real-time processing and low latency. However, blockchain's consensus mechanisms and distributed nature can introduce delays in transaction validation and data processing. Developing efficient consensus algorithms and optimizing data handling for real-time requirements is a challenge.

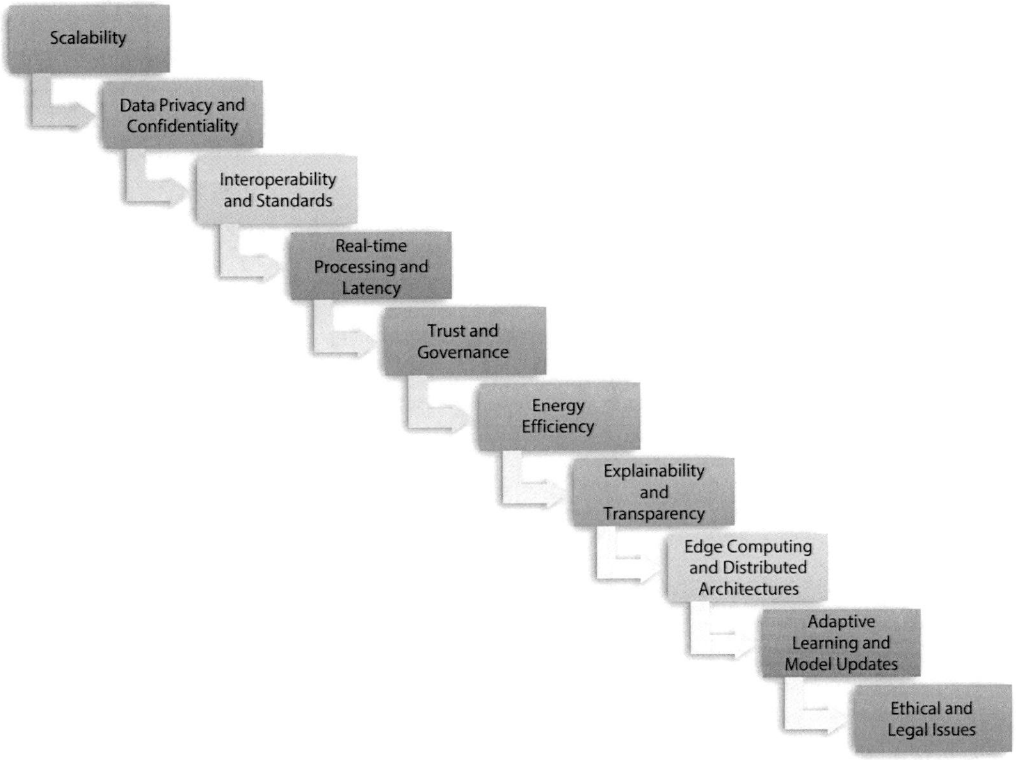

Figure 9.15 Issues and challenges towards blockchain based IoT–ML applications.

Trust and Governance: Establishing trust among IoT devices, ML models, and blockchain networks is important. Ensuring the integrity of data, validating the accuracy of ML models, and designing governance mechanisms that maintain fairness and transparency face challenges. Determining trust metrics and governance models is an ongoing research area.

Energy Efficiency: Blockchain networks, especially those relying on proof-of-work consensus, are energy-intensive. When integrated with IoT devices, energy consumption becomes an issue. Exploring energy-efficient consensus algorithms and optimizing resource usage in integrated systems is important for sustainability.

Explainability and Transparency: ML models integrated with blockchain may lack transparency and explainability, which can be problematic for important decision-making processes. Developing techniques to interpret and explain ML models' decisions and ensuring transparency in the integration are open challenges.

Edge Computing and Distributed Architectures: Many IoT devices operate at the edge of the network, where computational resources are limited. Designing efficient and distributed architectures that use edge computing for ML model inference and blockchain consensus can enhance performance and reduce reliance on centralized components.

Adaptive Learning and Model Updates: IoT environments are dynamic, and data distributions can change over time. Ensuring ML models adapt to changing IoT data patterns and enabling automated model updates face challenges. Developing techniques for adaptive learning and efficient model updates in decentralized environments is an active area of research.

Ethical and Legal Issues: The integration of blockchain, IoT, and ML raises ethical and legal challenges related to data ownership, privacy, bias, and accountability. These issues require developing frameworks and guidelines that ensure responsible and ethical use of integrated technologies.

Hence, these open challenges require interdisciplinary research, collaboration, and innovation. Overcoming these difficulties will follow the way for the effective integration of blockchain, IoT, and ML, enabling new applications and driving advancements in various industries.

9.20 Future Research Opportunities for Blockchain, Internet of Things, and Machine Learning Integration

The integration of blockchain, Internet of Things (IoT), and Machine Learning (ML) provides exciting research opportunities that can shape the future of various industries [28–35]. Some potential research directions include:

Privacy-Preserving Techniques: Developing advanced privacy-preserving techniques that allow for secure and confidential data sharing and processing in integrated systems. This involves exploring cryptographic methods, differential privacy, and federated learning approaches to protect sensitive data while maintaining the benefits of blockchain, IoT, and ML.

Scalable and Efficient Consensus Mechanisms: Designing consensus algorithms tailored for IoT-ML applications to address scalability and energy efficiency challenges. Developing lightweight consensus protocols, such as proof-of-stake or practical Byzantine fault tolerance, can enhance the scalability of blockchain networks integrated with IoT devices and ML models.

Federated Learning in Decentralized Environments: Investigating federated learning approaches that consider blockchain as a secure and decentralized framework for aggregating and sharing ML models trained on distributed IoT devices. Research can focus on optimizing federated learning algorithms for heterogeneous and resource-constrained IoT networks.

Trust and Security Management: Exploring novel approaches for establishing and maintaining trust among IoT devices, ML models, and blockchain networks. This includes investigating reputation systems, secure identity management, and decentralized governance mechanisms to ensure the integrity, accountability, and reliability of integrated systems.

Explainable and Transparent AI: Developing techniques to enhance the explainability and transparency of ML models integrated with blockchain and IoT. This involves research on interpretable ML algorithms, transparent decision-making processes, and methods for capturing and auditing model training and inference in decentralized environments.

Energy-Efficient Edge Computing: Investigating energy-efficient computing architectures that combine edge computing with blockchain and ML. Research can focus on optimizing ML model inference and training at the edge, reducing energy consumption, and using edge resources for blockchain consensus.

Standardization and Interoperability: Contributing to the development of industry standards and protocols that enable seamless interoperability between different blockchain, IoT, and ML technologies. Research can focus on creating common data formats, communication protocols, and interoperability frameworks to promote widespread adoption and integration.

Real-time Analytics and Decision-making: Exploring real-time analytics techniques that consider blockchain and ML to enable rapid data processing, insights generation, and automated decision-making in IoT environments. This involves developing algorithms and frameworks for real-time data convergence, anomaly detection, and dynamic decision-making in integrated systems.

Sustainable and Green Computing: Investigating the environmental impact of blockchain, IoT, and ML integration and developing sustainable computing approaches. Research can focus on energy-efficient consensus mechanisms, green data centers, and optimization techniques to minimize the carbon footprint of integrated systems.

Domain-specific Applications: Exploring the integration of blockchain, IoT, and ML in specific domains such as healthcare, supply chain, energy, smart cities, and agriculture. Research can focus on addressing domain-specific challenges, developing use cases, and evaluating the benefits and limitations of integrated systems in different industries.

Hence, these research opportunities have the potential to advance the integration of blockchain, IoT, and ML, leading to innovative applications, improved system performance, and enhanced trust and security in future technology ecosystems.

9.21 Conclusion

The convergence of blockchain technology with the Internet of Things (IoT) and machine learning has the potential to revolutionize various sectors, enabling secure, decentralized, and automated solutions. Through the analysis of case studies, several key findings emerge:

a) Blockchain provides a robust and tamper-proof framework for securing IoT networks and devices. By storing transactional data in an immutable ledger, it ensures data integrity, mitigates the risks of cyber-attacks, and establishes trust in the IoT ecosystem.
b) Blockchain facilitates trusted data exchange and interoperability between different IoT devices and platforms. It enables seamless integration and communication among disparate devices, ensuring transparent and reliable data sharing.
c) Supply chain management benefits hugely from blockchain implementation. It enables end-to-end traceability, verifying the authenticity and provenance of IoT-enabled products. Transparency in the supply chain is enhanced, reducing fraud and counterfeiting.
d) Machine learning algorithms integrated with blockchain technology enhance automated decision-making processes. In sectors like predictive maintenance and intelligent routing, machine learning enables efficient and optimized operations, while blockchain ensures the integrity and transparency of the decision-making process.
e) Blockchain enables the creation of decentralized marketplaces for IoT devices and services, eliminating intermediaries and facilitating peer-to-peer

transactions. This disintermediation fosters cost savings, efficiency, and greater user control over their IoT assets.

f) Privacy and data ownership are important issues in blockchain-based IoT ecosystems. The combination of blockchain and machine learning should adhere to data protection regulations, provide user consent mechanisms, and discuss approaches for data monetization that prioritize individual privacy.
g) Scalability remains a challenge for large-scale IoT deployments utilizing blockchain technology. Efforts to optimize blockchain consensus mechanisms and implement off-chain scaling solutions are important to meet the demands of IoT data volumes and transaction throughput.
h) The integration of blockchain and machine learning is driving innovation in sectors such as autonomous vehicles and energy grids. Blockchain enables secure data sharing and communication, while machine learning algorithms enhance decision-making for real-time optimization and efficiency.
i) Healthcare systems can benefit from blockchain and machine learning by securely sharing and analyzing medical IoT data. This facilitates accurate diagnoses, personalized treatment plans, and efficient healthcare delivery while prioritizing patient privacy and data security.

In summary, the combination of blockchain, IoT, and machine learning holds immense potential for creating secure, automated, and decentralized systems in various sectors. While challenges remain, ongoing research, technological advancements, and industry collaborations are paving the way for transformative applications that improve efficiency, transparency, and trust in automated sectors.

References

1. Singh, S.K., Rathore, S., Park, J.H., Block IoT Intelligence: A blockchain-enabled intelligent IoT architecture with artificial intelligence. *Future Gener. Comput. Syst.*, 110, 721–743, 2020.
2. Vermesan, O. and Friess, P. (Eds.), *Internet of things: converging technologies for smart environments and integrated ecosystems*, River publishers, 2013.
3. Politou, E., Casino, F., Alepis, E., Patsakis, C., Blockchain mutability: Challenges and proposed solutions. *IEEE Trans. Emerging Top. Comput.*, 9, 4, 1972–1986, 2019.
4. Wortmann, F. and Flüchter, K., Internet of things: technology and value added. *Bus. Inf. Syst. Eng.*, 57, 221–224, 2015.
5. Gubbi, J., Buyya, R., Marusic, S., Palaniswami, M., Internet of Things (IoT): A vision, architectural elements, and future directions. *Future Gener. Comput. Syst.*, 29, 7, 1645–1660, 2013.
6. Asghari, P., Rahmani, A.M., Javadi, H.H.S., Internet of Things applications: A systematic review. *Comput. Netw.*, 148, 241–261, 2019.
7. Boutaba, R., Salahuddin, M.A., Limam, N., Ayoubi, S., Shahriar, N., Estrada-Solano, F., Caicedo, O.M., A comprehensive survey on machine learning for networking: evolution, applications and research opportunities. *J. Internet Serv. Appl.*, 9, 1, 1–99, 2018.
8. Rodrigues, T.K., Suto, K., Nishiyama, H., Liu, J., Kato, N., Machine learning meets computation and communication control in evolving edge and cloud: Challenges and future perspective. *IEEE Commun. Surv. Tutorials*, 22, 1, 38–67, 2019.

9. Ferrag, M.A., Derdour, M., Mukherjee, M., Derhab, A., Maglaras, L., Janicke, H., Blockchain technologies for the internet of things: Research issues and challenges. *IEEE Internet Things J.*, 6, 2, 2188–2204, 2018.
10. Mohanta, B.K., Jena, D., Satapathy, U., Patnaik, S., Survey on IoT security: Challenges and solution using machine learning, artificial intelligence and blockchain technology. *Internet Things*, 11, 100227, 2020.
11. Ullah, N., Mugahed Al-Rahmi, W., Alzahrani, A., II, Alfarraj, O., Alblehai, F.M., Blockchain technology adoption in smart learning environments. *Sustainability*, 13, 4, 1801, 2021.
12. Ahmed, I., Zhang, Y., Jeon, G., Lin, W., Khosravi, M.R., Qi, L., A blockchain- and artificial intelligence-enabled smart IoT framework for sustainable city. *Int. J. Intell. Syst., Wiley*, 37, 9, 6493–6507, 2022, doi: 10.1002/int.22852.
13. Torky, M. and Hassanein, A.E., Integrating blockchain and the internet of things in precision agriculture: Analysis, opportunities, and challenges. *Comput. Electron. Agric.*, 178, 105476, 2020.
14. Cioffi, R., Travaglioni, M., Piscitelli, G., Petrillo, A., De Felice, F., Artificial intelligence and machine learning applications in smart production: Progress, trends, and directions. *Sustainability*, 12, 2, 492, 2020.
15. Al-Garadi, M.A., Mohamed, A., Al-Ali, A.K., Du, X., Ali, I., Guizani, M., A survey of machine and deep learning methods for internet of things (IoT) security. *IEEE Commun. Surv. Tutorials*, 22, 3, 1646–1685, 2020.
16. Atlam, H.F., Azad, M.A., Alzahrani, A.G., Wills, G., A Review of Blockchain in Internet of Things and AI. *Big Data Cogn. Comput.*, 4, 4, 28, 2020.
17. Zhang, C., Li, W., Luo, Y., Hu, Y., AIT: An AI-enabled trust management system for vehicular networks using blockchain technology. *IEEE Internet Things J.*, 8, 5, 3157–3169, 2020.
18. Ferrag, M.A., Derdour, M., Mukherjee, M., Derhab, A., Maglaras, L., Janicke, H., Blockchain technologies for the internet of things: Research issues and challenges. *IEEE Internet Things J.*, 6, 2, 2188–2204, 2018.
19. Mahlous, A.R. and Ara, A., The Adoption of Blockchain Technology in IoT: An Insight View. *2020 6th Conference on Data Science and Machine Learning Applications (CDMA)*, pp. 100–105, 2020.
20. Singh, S., Rathore, S., Alfarraj, O., Tolba, A., Yoon, B., A framework for privacy-preservation of IoT healthcare data using Federated Learning and blockchain technology. *Future Gener. Comput. Syst.*, 129, 380–388, 2022.
21. Waheed, N., He, X., Ikram, M., Usman, M., Hashmi, S.S., Usman, M., Security and privacy in IoT using machine learning and blockchain: Threats and countermeasures. *ACM Comput. Surv.*, 53, 6, 1–37, 2020.
22. Alsharari, N., Integrating blockchain technology with internet of things to efficiency. *Int. J. Innov. Technol. Manage.*, 1, 2, 01–13, 2021.
23. Bahga, A. and Madisetti, V.K., Blockchain platform for industrial internet of things. *J. Software Eng. Appl.*, 9, 10, 533–546, 2016.
24. Panarello, A., Tapas, N., Merlino, G., Longo, F., Puliafito, A., Blockchain and iot integration: A systematic survey. *Sensors*, 18, 8, 2575, 2018.
25. Da Xu, L., Lu, Y., Li, L., Embedding blockchain technology into IoT for security: A survey. *IEEE Internet Things J.*, 8, 13, 10452–10473, 2021.
26. Viriyasitavat, W., Da Xu, L., Bi, Z., Hoonsopon, D., Blockchain technology for applications in internet of things—mapping from system design perspective. *IEEE Internet Things J.*, 6, 5, 8155–8168, 2019.
27. Nižetić, S., Šolić, P., Gonzalez-De, D.L.D., II, Patrono, L., Internet of Things (IoT): Opportunities, issues and challenges towards a smart and sustainable future. *J. Clean. Prod.*, 274, 122877, 2020.

28. Khalil, R.A., Saeed, N., Masood, M., Fard, Y.M., Alouini, M.S., Al-Naffouri, T.Y., Deep learning in the industrial internet of things: Potentials, challenges, and emerging applications. *IEEE Internet Things J.*, 8, 14, 11016–11040, 2021.
29. Singh, R., Tyagi, A.K., Arumugam, S.K., Imagining the Sustainable Future With Industry 6.0: A Smarter Pathway for Modern Society and Manufacturing Industries, in: *Machine Learning Algorithms Using Scikit and TensorFlow Environments*, IGI Global, 2024, doi: 10.4018/978-1-6684-8531-6.ch016.
30. Meghna, M.N. and Tyagi, A.K., Chapter 11 - AI, IoT, blockchain, and cloud computing: The necessity of the future, in: *Distributed Computing to Blockchain*, R. Pandey, S. Goundar, S. Fatima (Eds.), pp. 189–206, Academic Press, 2023, ISBN 9780323961462, https://doi.org/10.1016/B978-0-323-96146-2.00001-2.
31. Tyagi, A.K., Chapter 2 - Decentralized everything: Practical use of blockchain technology in future applications, in: *Distributed Computing to Blockchain*, R. Pandey, S. Goundar, S. Fatima (Eds.), pp. 19–38, Academic Press, 2023, ISBN 9780323961462, https://doi.org/10.1016/B978-0-323-96146-2.00010-3.
32. Tyagi, A.K. and Richa, Smart Manufacturing Using Internet of Things, Artificial Intelligence, and Digital Twin Technology, in: *Global Perspectives on Robotics and Autonomous Systems: Development and Applications*, 2023.
33. Tyagi, A.K., Lakshmi Priya, R., Mishra, A.K., Balamurugan, G., Industry 5.0: Potentials, Issues, Opportunities, and Challenges for Society 5.0, in: *Privacy Preservation of Genomic and Medical Data*, vol. 17, pp. 409–32, 2023 Oct.
34. Tyagi, A.K., Fernandez, T.F., Mishra, S., Kumari, S., Intelligent Automation Systems at the Core of Industry 4.0, in: *Intelligent Systems Design and Applications. ISDA 2020. Advances in Intelligent Systems and Computing*, vol. 1351, A. Abraham, V. Piuri, N. Gandhi, P. Siarry, A. Kaklauskas, A. Madureira (Eds.), Springer, Cham, 2021, https://doi.org/10.1007/978-3-030-71187-0_1.
35. Tyagi, A.K., Mishra, A.K., Vedavathi, N., Kakulapati, V., Sajidha, S.A., Chapter 21 Futuristic Technologies for Smart Manufacturing, in: *Research Statement and Vision for the Future*, Tyagi, A.K. (Ed.), 2023, https://doi.org/10.1002/9781394213948.ch21.

10

An Enhanced Threat Detection Model to Assist Supply Chain Management Using Artificial Intelligence

Ambika N.

Dept. of Computer Science and Applications, St. Francis College, Bangalore, India

Abstract

The previous framework combines computer algorithms and physical processes to create a comprehensive system. Risk on Screen Character is a potential threat actor or hacker who may attempt to exploit vulnerabilities in the system. Tactics, Techniques, and Procedures (TTP) are strategies and methods used by threat actors to manipulate cybersecurity weaknesses in the supply chain. Cybersecurity in the Supply Chain (CSC) secures the information and processes within the supply chain to ensure the organization's security and operational goals are met. CSC Requirements are the security requirements and measures used to protect the supply chain from cyber threats. Attack Entity could refer to the threat actor or entity attempting to compromise the supply chain's cybersecurity. Digital Incident Report is created in response to a cybersecurity incident and can impact the organization's goals related to inbound and outbound supply chains. The probability of a cyber attack developing into a significant threat is determined by the information and intelligence gathered about the threat actor's capabilities and intentions. The proposal causes the hash key utilizing the Merkle tree concept creating the trail datasets, and stores them. It improves the system by enhancing security by 8.99% and early threat detection by 7.02%.

Keywords: Supply chain management, security, cyber systems, artificial intelligence, hashing, early threat detection, datasets

10.1 Introduction

Supply chain management (SCM) (Basu and Wright 2010) (Boyer and Verma 2009) is a systematic approach to planning, designing, implementing, and controlling the flow of materials, information, and finances as they move from supplier to manufacturer to wholesaler to retailer to the end customer. The primary goal of supply chain management is to optimize the overall efficiency and effectiveness of this process, ensuring that products or services are delivered to the right place, at the right time, and the appropriate cost while maintaining high quality and customer satisfaction.

Email: ambika.nagaraj76@gmail.com

Amit Kumar Tyagi, Shrikant Tiwari, Senthil Kumar Arumugam and Avinash Kumar Sharma (eds.) Artificial Intelligence-Enabled Digital Twin for Smart Manufacturing, (215–232)

Effective supply chain management [1, 2] can lead to reduced costs, improved product quality, faster delivery times, and increased customer satisfaction. It plays a crucial role in the success of businesses in various industries, from manufacturing to retail to service providers, and has become even more critical in today's global and interconnected business environment.

The cyber-physical approach is analysis and physical procedures. It creates a whole technique. The joining of innovations has improved business forms (Ballou 2007) speed up, and decreased appropriation costs. The expanded interdependencies among different store network partners have brought numerous difficulties – an absence of outsider review instruments and falling digital dangers. It has prompted assaults, for example, the control of the plan determinations, changes, and control during dispersion. Key elements for efficient working are –

- *Association with dealers* – The execution of slender methods depends on extended haul and solid associations with dealers. It distinguishes significant drivers of partners' joint efforts.
- *Preparation* – It advances the utilization of joint effort in the production network through preparing drives to serve the providers.
- *Assets* – The organizations utilize master counseling to execute these practices, so assets for counseling are fundamental.
- *Closeness* – It is a significant element for convenient conveyance. Figure 10.1 portrays the preliminary framework of the supply chain framework.

Supply chain management with artificial intelligence and cybersecurity refers to the strategic use of AI technologies and robust cybersecurity measures to enhance the efficiency, resilience, and security of supply chain operations. This approach combines the benefits of AI-driven optimization with comprehensive security measures to protect sensitive data and critical supply chain processes. It involves:

- Data Protection: Implementing encryption, access controls, and data privacy measures to secure sensitive information, such as customer data, intellectual property, and supplier contracts.

The study [3] is a sophisticated and technologically advanced framework for safeguarding digital security, confidentiality, and industrial-commercial discretion. It presents a comprehensive approach to ensuring the safety, privacy, and honesty of system communications within the industry 4.0 standard. It leverages advanced technologies like neural networks, homomorphic encryption, and blockchain to provide a robust defense against security threats and breaches. When one digital instrument wishes to connect with another, the framework's smart security mechanism is triggered. The framework employs a network traffic control system designed to identify and sense irregularities during communication. In the initial stage, the features of system gridlock are extracted. These features are essential data points for analyzing network behavior. The extracted network traffic features serve as input to a neural grid. This neural network is robotically created through the neural framework Exploration method, which helps optimize the network's architecture for the specific task. The neural network prototypical is initially trained on a host server using initial data. This training process helps the model learn and improve its ability to detect anomalies.

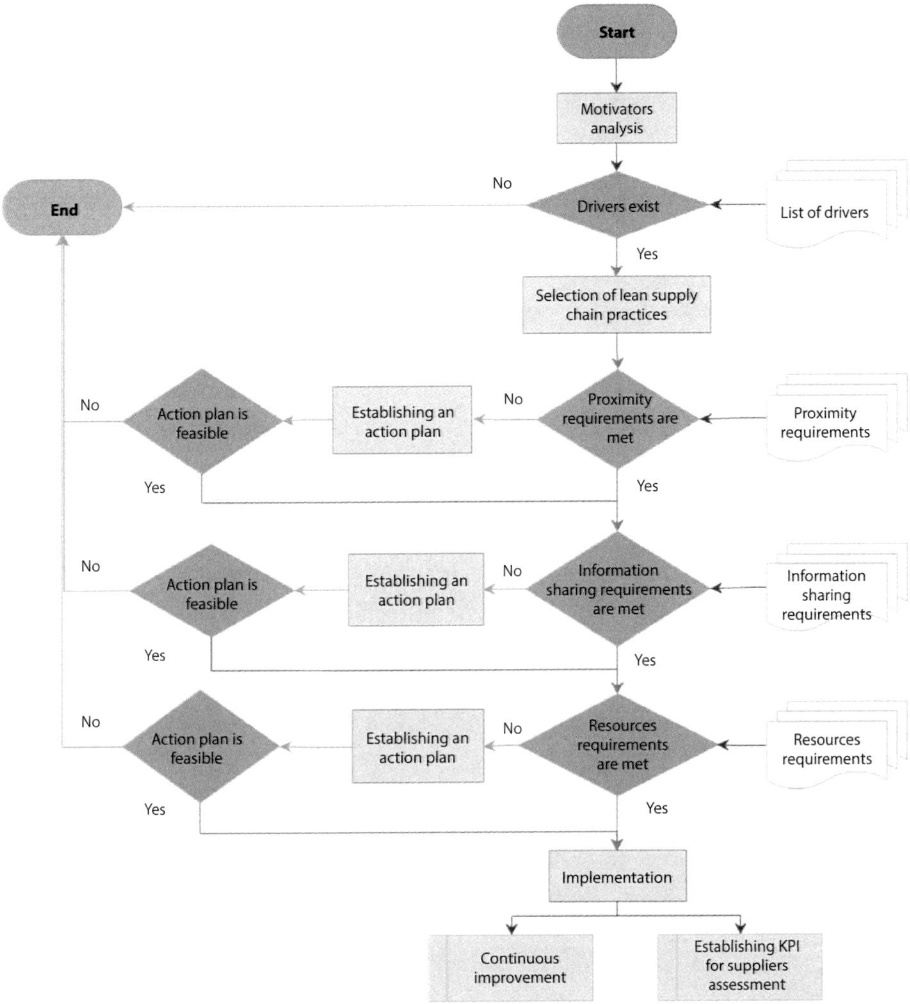

Figure 10.1 A preliminary framework for supply chain management [9].

After training, the model is encrypted using homomorphic encryption. The encrypted model is transmitted via a blockchain to nodes within the network that will utilize it. Blockchain technology ensures secure and transparent data transmission. The recipient nodes take the encrypted model and enhance it by incorporating the data, which they have at their disposal. The improved model is then encrypted and returned through the blockchain to the host server. The framework aggregates the various improved models, and a weighted average is computed using the Grid Search Weighted Average Ensemble method. It helps identify the most effective model. The final selected model is returned via a blockchain medium to the final nodes, ensuring that all nodes in the network have access to the best-performing security model. This framework is tailored for applications in the context of Industry 4.0, which represents the fourth industrial revolution, characterized by automation, data exchange, and IoT technologies [4, 5]. It integrates advanced communication technologies and security measures to protect critical data and ensure the secure and efficient operation of Industry 4.0 systems.

- Threat Detection: Employing AI-driven cybersecurity solutions to detect and respond to cyber threats in real-time, protecting supply chain systems from potential breaches and disruptions.

The work [6] described involves an integration of Cyber Threat Intelligence (CTI) and Machine Learning (ML) within a systematic process to enhance cybersecurity. This approach uses inbound and outbound chains of vulnerabilities, enabling organizations (specifically the CSC organization) to identify potential system weaknesses. The approach combines the power of CTI and ML techniques to bolster cybersecurity efforts. CTI focuses on gathering and analyzing threat data, while ML is employed to predict potential threats. The approach considers vulnerabilities in both inbound (external) and outbound (internal) aspects of an organization's network and systems. This comprehensive analysis ensures that the CSC organization can address a range of potential system flaws. The CTI process is utilized to collect and analyze data related to cybersecurity threats. This process involves monitoring and collecting information about potential threats from various sources. Machine Learning techniques are applied to develop predictive models for identifying threats. These ML techniques use classification algorithms to analyze and learn from datasets. This learning process is crucial for improving the accuracy of threat detection and predictive analytics. The CTI lifecycle process provides input parameters to aid in the identification of known threats. In contrast, ML generates output parameters that enable the prediction of known and unknown threats, offering a forward-looking perspective on cybersecurity.

- Resilience Planning: Develop contingency plans and redundancy strategies to ensure supply chain operations can continue in the event of a cybersecurity incident or other disruptions.

The study [7] aims to create a shared understanding and approach to enhancing cybersecurity within the energy sector. It implies that various actors within the sector are working together to address common cybersecurity challenges. It employs a collaborative framework to facilitate the co-production of cybersecurity guidance. This approach likely involves different stakeholders within the energy sector working together to develop solutions and best practices for cybersecurity. A questionnaire is created to collect information. It is used to assess the cybersecurity practices and readiness of energy sector participants. It is based on an analysis of the types of questions that energy operators commonly pose to their suppliers. Resource workers acknowledge that cybersecurity is a significant danger that needs to be managed. This recognition suggests an awareness of the importance of cybersecurity for maintaining resilient energy systems. Energy providers may be required to validate their cybersecurity organization practices to comply with regulatory requirements. It suggests that there are likely legal and regulatory mandates in place to ensure cybersecurity within the energy sector. The collection of dealers and workers has proposed a "Code of Practice and Partnership model" as an introductory method to ensuring cybersecurity within the source system. It involves establishing a set of best practices and guidelines that all participants in the energy sector should follow to enhance cybersecurity. The partnership model implies cooperation among stakeholders to achieve these goals.

- Security Audits: Conduct regular security audits and assessments to identify vulnerabilities and areas of improvement within the supply chain and address them proactively.

The work [8] is a framework or model for conducting cybersecurity audits, either internally or externally, within various types of organizations, including enterprises and non-profit organizations. It is used for conducting cybersecurity audits within the organization (internal audits) and by external audit teams. It implies that it is versatile and can be used by organizations to assess their cybersecurity controls, or by external auditors to evaluate an organization's cybersecurity posture. It provides flexibility in terms of the scope of the audit. It can be used for full-scale cybersecurity audits that cover all 18 domains or for selective audits where specific domains are chosen for control verification and improvement. It allows organizations to focus on areas where they have specific concerns or priorities. It is structured into 18 domains. It is hierarchical structure for a more granular assessment of cybersecurity controls. It employs checklists to evaluate sub-controls within the various domains and sub-domains. Each checklist is associated with a specific domain and sub-domain. It enables auditors to systematically assess and verify the effectiveness of controls.

The previous structures [10] combine computer algorithms and physical processes to create a comprehensive system. Risk on Screen Character is a potential threat actor or hacker who may attempt to exploit vulnerabilities in the system. Tactics, Techniques, and Procedures (TTP) are strategies and methods used by threat actors to manipulate cybersecurity weaknesses in the supply chain. Cybersecurity (N. Ambika 2021) (N. Ambika 2020) in the Supply Chain (CSC) secures the information and processes within the supply chain to ensure the organization's security and operational goals are met. CSC Requirements are the security requirements and measures used to protect the supply chain from cyber threats [11]. Attack Entity could refer to the threat actor or entity attempting to compromise the supply chain's cybersecurity. Digital Incident Report is created in response to a cybersecurity incident and can impact the organization's goals related to inbound and outbound supply chains. The probability of a cyber attack developing into a significant threat is determined by the information and intelligence gathered about the threat actor's capabilities and intentions. Data protection experts employ proxy hosts to examine the matter. It recognizes a possible danger in the cyber framework or grid defects that permit pirates to earn unlicensed permits to stock administration. It infiltrates into them.

The study [12] discusses the deceptive nature of spyware, when it disguises itself as a potentially unwanted application, and how malicious actors may use such software to gather information about users. Additionally, it touches on the activities of hackers who engage in corporate espionage by investigating the rankings and employees of companies. The context in which this passage is found would provide a clearer understanding of its significance. Spyware appears to present its software programs in a way that suggests they provide helpful features and can be integrated with end-user authorization arrangements. When masquerading as a PUA (Potentially Unwanted Application), discreetly collects information about a user's activities and may install itself on a system without the user's knowledge or consent. Spyware gathers information about what a user does on their device or computer. It highlights the stealthy nature of spyware when it infiltrates a system without the user's awareness or consent. It appears to be involved in investigating the rankings or statuses of companies and their employees. It suggests some form of cyber-espionage or unauthorized access to corporate information.

It is a detailed process [13] involving technology, authentication, and management, possibly in the context of investments or transactions. It refers to a person or entity involved in an investment process. Stripe Design initiates the design process, which could involve structuring, possibly related to investments. Distinctive Recognition uniquely identifies something or someone in the process. Cryptographically Inscribed Naming indicates that some form of cryptographic technology is used to encode or secure names or information. Blockchain technology secure and transparent record-keeping. Patch Authentication involves verifying the authenticity of a patch or component, possibly related to the transaction or investment. Patch Organization suggests a system in place to manage and arrange these patches. Patch Industrialization makes these patches suitable for mass production or use. Synchronization Assistance is a need for synchronization or alignment in the process. The vendor verifies or authenticates something, possibly related to the transaction or investment. A digital profile is a transaction that is created and recorded. Patch Confirmation involves verifying that the patch or component meets the required standards or criteria. The patch is placed or integrated into a test environment or context. Automation of the Tester Team implies that the testing process is automated, possibly using technology and tools. If any issues or impacts occur, the operation is halted, and the system reverts to a previous stable state.

Some protection parameters can be implemented to make it a strong network. The study implements the Merkle tree concept to enhance security in the network by 8.99% compared to (Mylrea and Gourisetti 2018). The work has six sections. Background follows the introduction section. The third segment summarizes the previous contributions. The fourth section explains the proposed model. The fifth segment details the analysis investigation. The work concludes in the sixth segment.

10.2 Background

A Merkle tree has a sequence of information packets., where the value of an interior parent is designated using the hash text of its offspring. Its assessment of an end vertex allocates the explicit hash text of the connected facts packet. The hash process meets the preimage-resistance property. It locates the preimage of the provided hash content. The highest depth from the last node to the parent is at $\lceil \log 2n \rceil$ for "n" information packets. The Merkle tree works as a certified information framework for efficient confirmation of the online scope. Figure 10.2 represents the same.

10.3 Literature Survey

It is a detailed process [13] involving technology, authentication, and management, possibly in the context of investments or transactions. It refers to a person or entity involved in an investment process. Stripe Design initiates the design process, which could involve structuring, possibly related to investments. Distinctive Recognition uniquely identifies something or someone in the process. Cryptographically Inscribed Naming indicates that some form of cryptographic technology is used to encode or secure names or information. Blockchain

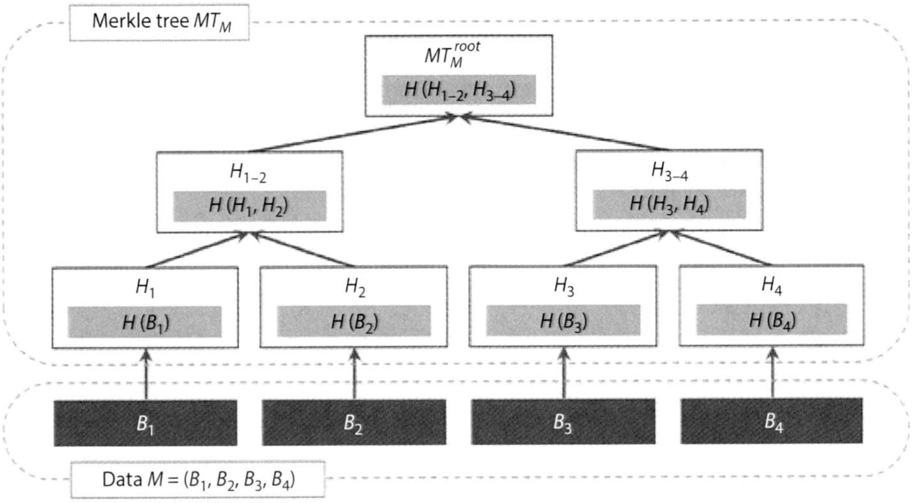

Figure 10.2 Merkle tree [14].

technology secure and transparent record-keeping. Patch Authentication involves verifying the authenticity of a patch or component, possibly related to the transaction or investment. Patch Organization suggests a system in place to manage and arrange these patches. Patch Industrialization makes these patches suitable for mass production or use. Synchronization Assistance is a need for synchronization or alignment in the process. The vendor verifies or authenticates something, possibly related to the transaction or investment. A digital profile is a transaction that is created and recorded. Patch Confirmation involves verifying that the patch or component meets the required standards or criteria. The patch is placed or integrated into a test environment or context. Automation of the Tester Team implies that the testing process is automated, possibly using technology and tools. If any issues or impacts occur, the operation is halted, and the system reverts to a previous stable state.

The work [15] is a collaborative concept that combines additive manufacturing production methods with the cyber world. It is a cooperative approach that allows for the assessment of various risks and operations associated with this integration. It plays a pivotal role in identifying and executing solutions to address these risks and demonstrates security gaps. A robust design program is used to create a 3D representation of the final product. This design step can be time-consuming and may require iterations. It has a set of research software tools known for limited feature research, computational fluid dynamic analysis, and stabilization. The computer-Aided Design (CAD) record appears to go through a process from the prototype stage to producing a physical component using a 3D printer. The approach is described as a sub-component of a reserve chain, which could indicate its role within a broader context or process.

The approach [16] details how enterprise entities, where essential to the supply chain, may be traced. Every supply chain participant must develop an internal smart manufacturing system including an extension to upload vital control data to the blockchain. A specific stakeholder is the focus of the tiered structure. Its foundation is the three-tier architecture of IIoT. The processes used by the companies are indicated by the Process Layer. The procedures and model allow the development of AI/ML models as well as root cause analysis. To carry out the previously established procedures and satisfy client needs, resources are needed. Connecting to databases in the data layer and resources in the resource layer is the responsibility of the edge layer. The data layer facilitates local or cloud-hosted database storage. Key Performance Indicators (KPIs) offer an all-encompassing perspective of a business. The Advanced Function Layer includes all of the features needed to manage production, inventory, quality, and maintenance. The required context for the data is provided by Smart Layer.

Maersk, a global shipping firm, is the example company. Maersk uses TradeLens for its international operations. HyperLedger Fabric blockchain technology [17] powers the open API that allows users to access TradeLens, a blockchain application built on IBM cloud infrastructure. Maersk shall arrange for a cargo ship to travel a suitable route from the harbor of beginning to the harbor of terminus following the receipt of an order notice and request for a shipment. The gatherings involved (such as the funders, the dealer, and the port) are informed about the shipping schedule, route, and necessary documentation. The primary focus of the machine learning-integrated blockchain technology is marine risk assessment identification. The computational and numerical examines of large information about before allocated courses and ancient risk regions form the basis of the machine learning method for this detection. A cycle of statistical and computational analysis is applied to fresh data in that algorithm. Better forecasts will come from this cycle, sometimes referred to as learning from experience or improving via it.

This suggestion suggests a safe e-wellbeing architecture (Sefira) [18], which integrates an advanced chronological blockchain to overcome the previously mentioned challenges. The major element of this paradigm was the authenticity and truthfulness of electrical wellbeing annals, which was heavily stressed. The electric wellbeing record is meant to be preserved by the architecture without the need for outside assistance. The two most important parts of Sefira are the Java-implemented context-based access control (CBAC) of a smart contract and the context-based Merkle tree (CBMT). A decentralized network of electronic health data and information on access restrictions is built into smart contracts. After the blockchain's mining mechanism is completed, medical researchers use the anonymized data to pay the blockchain storage provider for their labor. Linking the hash chain to its form and a strong hash function that attacks both examining the input model and testing with any data to attempt to decode the contents are the two most challenging difficulties when utilizing a regular blockchain.

This paper [19] suggests a system that levitates the blockchain and effectively carries out commercial operations across the agricultural supply chain to control crop price and traceability. The proposed framework method eliminates the need for intermediaries, centralized authority, and transaction records while improving effective science and safety with high integrity and reliability. Every transaction is first recorded, then it is linked to a decentralized network and stored in the blockchain's immutable ledger. This guarantees a strong, trustworthy, and efficient environment for the supply chain with a high level of traceability and transparency. There are four separate stages, all of which are interdependent and necessary for the other phases to function. The farmer, whose details include name, address, phone number, crop name, and crop selling price (FRP). It is the first block on the blockchain. All data is entered into the smart contract, which generates the hash code using the sha256 technique. The crop price information, which is set by the government or a particular organization, is contained in the dealer's login registration. Subdealers can buy goods from dealers at government-set fixed pricing. The consumer, who completes the product purchase, is the last link in the chain.

The previous structures [10] combine computer algorithms and physical processes to create a comprehensive system. Risk on Screen Character is a potential threat actor or hacker who may attempt to exploit vulnerabilities in the system. Tactics, Techniques, and Procedures (TTP) are strategies and methods used by threat actors to manipulate cybersecurity weaknesses in the supply chain. Cybersecurity in the Supply Chain (CSC) secures the information and processes within the supply chain to ensure the organization's security and operational goals are met. CSC Requirements are the security requirements and measures used to protect the supply chain from cyber threats. Attack Entity could refer to the threat actor or entity attempting to compromise the supply chain's cybersecurity. Digital Incident Report is created in response to a cybersecurity incident and can impact the organization's goals related to inbound and outbound supply chains. The probability of a cyber attack developing into a significant threat is determined by the information and intelligence gathered about the threat actor's capabilities and intentions.

Improved techniques to lower the volume of data produced by the system are covered in this paper [20]. The suggested approach uses the header verification mechanism for transactions to minimize the amount of data that is processed and its increase. This approach uses a Merkle tree to check just the header file information during the transaction. This system can create a blockchain by storing every transaction made by a node. As a result, every node can confirm legitimate transactions using simple header data, without retaining the entirety of the blockchain's data. The plan's anticipated result is a marked slowdown in the blockchain's data expansion.

The work [21] is a blockchain-secured platform for production security and monitoring. This platform tracks and stores data through IoT-connected sensors. The olive tree production process is observed using this technique. Monitoring everything that comes into and goes out of our olive tree production—from fertilizers, pesticides, and fortifiers to olives, pruning, and other materials—is the aim. The decentralized nature of the blockchain allows for transparent, safe, and unchangeable monitoring. A dashboard makes it easier for us to draw attention to the changes and helps farmers with their work. A Raspberry Pi 4 platform will be utilized for embedding the prototype.

The technology under development [22] aims to teach cars how to transport commodities securely and accident-free from producer to consumer. Reinforcement learning, a machine learning algorithm, and artificial intelligence were combined with blockchain technology. The heuristic search strategy was specifically designed to train the self-driving car for this supply chain in the proposed system. A Reinforcement Learning (RL) computation efficiently examines its state to gather information and attempts to use the knowledge obtained to make decisions or predictions. Every logistics starting with the distribution from every state is recorded on the blockchain. Every state takes action and receives rewards from the relevant destination state.

The project [23] demonstrates how supply chain management (SCM) and developing technology integration are extended for product lifecycle management (PLM) in the component manufacture of boxed, flat sheet metal parts for railcar manufacturing. Because the ledger can be searched and examined using key-value data produced by the system, it is utilized for PDM. The construction of railcars, manufacturing suppliers, the fabrication, and assembly of boxed flat sheet metal components, the disassembly of a railcar, and the recycling company at EOL for scrap metal parts are all depicted in SCM, which provides an overview of part design requirements and inventory management. In supply chain management, data provenance is crucial, and the Hyperledger Fabric multi-level architecture makes this possible. Combining blockchain technology with QR codes is a useful way to transfer, store, and embed data for non-electronic devices. The use of blockchain technology in traceability makes it possible to choose high-security levels for a QR code.

The system [24] is composed of two primary modules: a machine learning-based medicine recommendation system for customers and a blockchain-based drug supply chain management system. The drug supply chain management system (Basu & Wright, 2010), which can continually track and monitor the medication distribution process in the smart pharmaceutical business, is implemented in the first module utilizing Hyperledger fabrics. On the other side, the machine learning module uses the N-gram and LightGBM models to suggest the best or highest-rated medications to the pharmaceutical industry's clients. The UCI, an open-source machine learning repository, has a well-known drug reviews dataset that has been made freely available for training these models. The blockchain-based system stores information about patients, physicians, hospitals, pharmacies, manufacturers, distributors, and other system users. The medication, raw material, order, and record repositories comprise the DSCMR systems' assets. Every system user has access to a client application-based front-end via which they can conveniently complete transactions and interact with the blockchain network. Using the client application, all pertinent system participants may monitor the medication delivery status during the whole DSCMR process. Conversely, a data storage pool—also known as a stored-off blockchain—is a distinct data library.

10.4 Methodology Adopted

The work [10] emphasizes the need to assess and address potential threats and vulnerabilities in the supply chain to protect the organization's objectives and maintain the security of inbound and outbound supply chains.

- Digital Frameworks: These are systems that combine computer algorithms and physical processes to create a comprehensive system.
- Risk on Screen Character: This appears to refer to a potential threat actor or hacker who may attempt to exploit vulnerabilities in the system.
- Tactics, Techniques, and Procedures (TTP): These are strategies and methods used by threat actors to manipulate cybersecurity weaknesses in the supply chain.
- Cybersecurity in the Supply Chain (CSC): The practice secures the information and processes within the supply chain to ensure the organization's security and operational goals are met.
- CSC Requirements: These are the security requirements and measures used to protect the supply chain from cyber threats.
- Attack Entity: This could refer to the threat actor or entity attempting to compromise the supply chain's cybersecurity.
- Digital Incident Report: This document is created in response to a cybersecurity incident and can impact the organization's goals related to inbound and outbound supply chains.

- Interrelationship: This term implies a connection or relationship between various elements, such as threat actors, TTP, risks, and threats, in understanding the impact on the CSC.
- Likelihood of Threat Becoming a Risk: The probability of a cyber attack (Ambika N., 2020) developing into a significant threat is determined by the information and intelligence gathered about the threat actor's capabilities and intentions.

Drawback

- Retaining and using the same credential is hackable and not reliable.

The suggestion makes an improvement to the previous work [10] by enhancing security [25] into the system. It does the following steps-

- Phase 1 – The goal of the system is recognized according to the previous work. The three activities are as follows-
    - Activity 1 – The activity identifies the stakeholders of the procedure. This happens in two scenarios-
        - The products are assembled to generate the final outcome. Each product is supplied by the stakeholder. The same has an assigned hash key. This hash key is generated using the algorithm 1. Table 10.1 represents the same. The various hash keys are assembled using Merkle tree concept. The same is represented in algorithm 2. Table 10.2 represents the same. The recommendation maintains the hash values in the database. This is compared when similar products are manufactured. This system improves reliability in the system.

The datasets are created and used for comparison purpose. The datasets contain the data relevant to the product. Table 10.3 represents the same.

The procedure uses Algorithm 3 to map it with the hash keys at every level. Hence there is early detection of any attack. The artificial algorithm uses the trial dataset to compare with intake. Table 10.4 represents the same.

The rest of the procedure follows the same as previous work [10].

Table 10.1 Algorithm to generate the hash key.

Step 1 – Accept the identification of the stockholder (24 bits)
Step 2 – Insert '1' in the even position of the input (output – 48 bits)
Step 3 – XOR the odd position bits with even position bits (output -24 bits)
Step 4 – Apply Right circular shift
Step 5 – Divide the output into two halves (12 bits each)
Step 6 – First half bits are arranged according in the odd positions (considering 24-bit positions)
Step 7 – Second half bits are arranged in even positions. (Final output)

Table 10.2 Algorithm to generate the final hash key.

Step 1 – Take the hash key from the Table 10.1 (4 inputs are considered assuming that the final product is the assembly of 4 components) (4*24 bits)
Step 2 – First and third component hash key undergoes Right shift.
Step 3 – Second and fourth component undergo Left shift.
Step 4 – First component hash key is XORED with second component (output – 24 bits)
Step 5 – Third and fourth component hash keys are XORED (output – 24 bits)
Step 6 – The output of step 4 and step 5 are XORED to get the final output (output – 12 bits)

Table 10.3 Dataset information.

Final product derivative	List of stockholders	Relevant hash key and intermediate hash keys	Final hash key
P_1	S_1, S_2	H_1, H_2	H_{1-2}
P_2	S_1, S_2, S_3, S_4	$H_1, H_2, H_{1-2}, H_{3-4}$	$H_{1-2-3-4}$

Table 10.4 Algorithm to use trial datasets for comparison.

Step 1 – Accept the input parameters (all the relevant stockholders of the product)
Step 2 – The relevant hash keys are generated (every time)
Step 3 – Compare the intermediate hash keys with the trial datasets.
Step 4 – If any discrepancies, send the alert message to the respective device.
Step 5 – After successful completion, send the acknowledgement to the respective device. (In case of inconsistency, the procedure is halted and reported; in case of normal execution, the procedure is completed and acknowledged)

10.5 Analysis of the Work

The previous work [10] combine computer algorithms and physical processes to create a comprehensive system. Risk on Screen Character is a potential threat actor or hacker who may attempt to exploit vulnerabilities in the system. Tactics, Techniques, and Procedures (TTP) are strategies and methods used by threat actors to manipulate cybersecurity weaknesses in the supply chain. Cybersecurity [26] in the Supply Chain (CSC) (Ballou, 2007) secures the information and processes within the supply chain to ensure the organization's security and operational goals are met. CSC Requirements are the security requirements and measures used to protect the supply chain from cyber threats (Howard, 2009). Attack Entity could refer to the threat actor or entity attempting to compromise the supply chain's (Boyer & Verma, 2009) cybersecurity. Digital Incident Report is created in response to a cybersecurity incident and can impact the organization's goals related to inbound and outbound supply chains. The probability of a cyber attack developing into a significant threat is determined by the information and intelligence gathered about the threat actor's capabilities and intentions.

The venture specialist (Mylrea and Gourisetti, 2018) begins the system of the striping plan. The dealer acknowledges and records the cryptographically engraved naming with the impeding innovation. The fix goes through verification. The owner capacities fix association initially. The work goes through fix industrialization and synchronization help. The seller confirms the investigating history. He records the advanced profile. The fix affirms and inserts in a preliminary foundation. The framework mechanizes the analyzer group. The activity is halted and withdrawn to the previous consistent exhibition (if any connection makes sway). The framework upgrades unwavering quality into the framework.

The work is simulated using NS2. Table 10.5 jolts down the simulation details.

- *Security* – The suggestion generates the hash key using Merkle tree concept. It enhances security by 8.99%. Figure 10.3 represents the same.

Table 10.5 Simulation details.

Parameters	Description
Area of the network	200 m*200 m
Number of products manufactured	10
Number of different kinds of products manufactured	3 (2 + 2 + 6)
Number of components of product 1	4
Number of components of product 2	2
Number of components of product 3	4
Number of bits to represent the component	24 bits
Total simulation time	60 ms

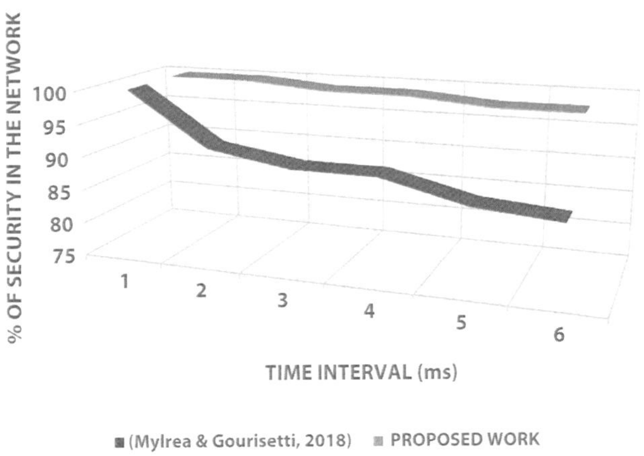

Figure 10.3 Security comparison in the network.

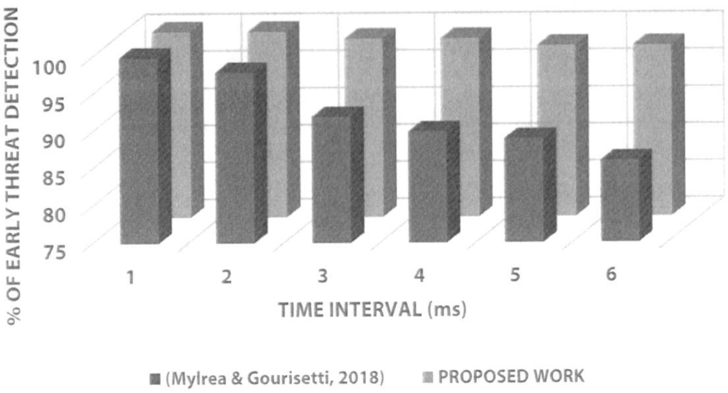

Figure 10.4 Early threat detection.

- *Early threat detection* – The procedure creates the trail datasets and stores. This is made comparison with intake, its intermediate and final hash keys. Hence the threat is detected at an early stage by 7.02% compared to previous work [13]. Figure 10.4 represents the same.

10.6 Future Work

The future of Supply Chain Management (SCM) with Artificial Intelligence (AI) is promising and transformative. AI technologies are revolutionizing supply chains operation, optimizing processes, enhancing decision-making, and driving efficiency.

- Predictive Analytics and Demand Forecasting: AI-powered predictive analytics can analyze vast amounts of data to predict demand trends, enabling companies to optimize inventory levels, production, and distribution. It leads to reduced costs and improved customer satisfaction through better stock availability.
- Optimized Supply Chain Planning: AI algorithms can optimize supply chain planning by considering various factors such as production schedules, transportation, inventory levels, and demand forecasts. This results in better allocation of resources and efficient scheduling.
- Smart Inventory Management: AI can optimize inventory levels by considering demand patterns, seasonality, and lead times. It reduces excess inventory and minimizes stockouts, ultimately saving costs and improving customer service.
- AI-Driven Logistics and Transportation: AI can optimize logistics and transportation through route optimization, real-time tracking, and predictive maintenance of vehicles. It improves efficiency, reduces fuel consumption, and minimizes transit times.
- Enhanced Supplier Relationship Management: AI can analyze supplier performance data, predict supplier behavior, and provide insights for better

negotiation and relationship management. It ensures a reliable supply of materials and reduces disruptions.

- Blockchain Integration for Transparency and Traceability: Integrating AI with blockchain technology can enhance transparency and traceability in the supply chain. It is crucial for verifying the authenticity of products and ensuring compliance with regulations.
- Chatbots and Customer Service: AI-powered chatbots can handle customer inquiries, order tracking, and issue resolution, improving customer experience and reducing the workload on customer service teams.
- Robotic Process Automation (RPA) for Operations: RPA can automate repetitive and rule-based tasks within the supply chain, such as data entry and invoice processing. It saves time and reduces errors, leading to greater efficiency.
- AI-Enabled Risk Management: AI can analyze external factors like geopolitical events, natural disasters, or economic shifts to predict and mitigate potential risks to the supply chain.
- Sustainability and Green Supply Chain: AI can assist in designing and optimizing environmentally friendly supply chain strategies, focusing on reducing carbon footprint, waste, and energy consumption.

10.7 Conclusion

Supply chain management is where the components are assembled to derive the final product. Industry 4.0 has brought Internet acting like an interface between the end-user and the factory setup. The happenings in the industrial setup can be viewed and communicated to the user's device. The user can give instruction from a remote location.

The previous structure combines computer algorithms and physical processes to create a comprehensive system. Risk on Screen Character is a potential threat actor or hacker who may attempt to exploit vulnerabilities in the system. Tactics, Techniques, and Procedures (TTP) are strategies and methods used by threat actors to manipulate cybersecurity weaknesses in the supply chain. Cybersecurity in the Supply Chain (CSC) secures the information and processes within the supply chain to ensure the organization's security and operational goals are met. CSC Requirements are the security requirements and measures used to protect the supply chain from cyber threats. Attack Entity could refer to the threat actor or entity attempting to compromise the supply chain's cybersecurity. Digital Incident Report is created in response to a cybersecurity incident and can impact the organization's goals related to inbound and outbound supply chains. The probability of a cyber attack (Ambika N., 2021) developing into a significant threat is determined by the information and intelligence gathered about the threat actor's capabilities and intentions.

The suggestion uses hash key generation on the stockholders' details followed by the assembly to derive the end-product. A trial dataset created with various commodities are compared using the artificial algorithm. The proposal enhances security by 8.99% and early threat detection by 7.02% compared to previous work.

References

1. Asante, M., Epiphaniou, G., Maple, C., Al-Khateeb, H., Bottarelli, M., Ghafoor, K.Z., Distributed ledger technologies in supply chain security management: A comprehensive survey. *IEEE Trans. Eng. Manage.*, 70, 2, 713–739, 2021.
2. Vyas, N., Beije, A., Krishnamachari, B., *Blockchain and the supply chain: concepts, strategies and practical applications*, Kogan Page Publishers, London, England, 2019.
3. Demertzis, K., Iliadis, L., Pimenidis, E., Tziritas, N., Koziri, M., Kikiras, P., Tonkin, M., Federated blockchained supply chain management: A CyberSecurity and privacy framework, in: *17th IFIP WG 12.5 International Conference*, Hersonissos, Crete, Greece, 2021.
4. Nagaraj, A., *Introduction to Sensors in IoT and Cloud Computing Applications*, Bentham Science Publishers., UAE, 2021.
5. Nagaraj, A., Adapting Blockchain for Energy Constrained IoT in Healthcare Environment, in: *Sustainable and Advanced Applications of Blockchain in Smart Computational Technologies*, p. 103, CRC press, Boca Raton, Florida, 2022.
6. Yeboah-Ofori, A., Islam, S., Lee, S.W., Shamszaman, Z.U., Muhammad, K., Altaf, M., Al-Rakhami, M.S., Cyber threat predictive analytics for improving cyber supply chain security. *IEEE Access*, 9, 94318–94337, 2021.
7. Wallis, T. and Dorey, P., Implementing Partnerships in Energy Supply Chain Cybersecurity Resilience. *Energies*, 16, 4, 1868, 2023.
8. Sabillon, R., Serra-Ruiz, J., Cavaller, V., Cano, J., A comprehensive cybersecurity audit model to improve cybersecurity assurance: The cybersecurity audit model (CSAM), in: *International Conference on Information Systems and Computer Science (INCISCOS)*, Quito, Ecuador, 2017.
9. Ibn El Farouk, I., Moufad, I., Frichi, Y., Arif, J., Jawab, F., Proposing a Supply Chain Collaboration Framework for Synchronous Flow Implementation in the Automotive Industry: A Moroccan Case Study. *Information*, 11, 431, 2020.
10. Yeboah-Ofori, A. and Islam, S., Cyber Security Threat Modeling for Supply Chain Organizational Environments. *Future Internet*, 11, 63, 1–25, 2019.
11. Scholz, R.W., Digital threat and vulnerability management: the SVIDT method. *Sustainability*, 9, 4, 554, 2017.
12. Egele, M., Kruegel, C., Kirda, E., Yin, H., Song, D., Dynamic spyware analysis. *USENIX Annual Technical Conference*, Berkeley, CA. United States, 2007.
13. Mylrea, M. and Gourisetti, S.N.G., Blockchain for supply chain cybersecurity, optimization and compliance, in: *Resilience Week (RWS)*, Denver, CO, USA, 2018.
14. Koo, D., Shin, Y., Yun, J., Hur, J., Improving Security and Reliability in Merkle Tree-Based Online Data Authentication with Leakage Resilience. *Appl. Sci.*, 8, 2532, 2018.
15. Gupta, N., Tiwari, A., Bukkapatnam, S.T., Karri, R., Additive manufacturing cyber-physical system: Supply chain cybersecurity and risks. *IEEE Access*, 8, 47322–47333, 2020.
16. Karadgi, S., Kulkarni, V., Doddamani, S., Traceable and intelligent supply chain based on blockchain and artificial intelligence. *J. Phys. Conf. Ser.*, 2070, 1, 012158, 2021.
17. Wong, S., Yeung, J.-K.-W., Lau, Y.-Y., So, J., Technical Sustainability of Cloud-Based Blockchain Integrated with Machine Learning for Supply Chain Management. *Sustainability*, 13, 15, 8270, 2021.
18. Sharma, V., Ramirez-Asis, E., Ahmad, A.J., Silva-Zapata, M., Alvarado-Tolentino, J., Kumar, H., Krah, D., On the Internet of Things, Blockchain Technology for Supply Chain Management (IoT). *Wireless Commun. Mobile Comput.*, 2022.
19. Chiranjeevi, K., Tripathi, M.K., Maktedar, D.D., Block chain technology in agriculture product supply chain, in: *International Conference on Artificial Intelligence and Smart Systems (ICAIS)*, Coimbatore, India, 2021.

20. Amin, M.R., Zuhairi, M.F., Saadat, M.N., Enhanced Blockchain transaction: a case of food supply chain management. *J. Eng. Appl. Sci.*, 15, 1, 99–106, 2020.
21. Ghorbel, O., Frikha, T., Alabdali, R., Ayadi, R., Abbas Elmasry, M., Blockchain-Based Supply Chain System for Olive Fields Using WSNs. *Comput. Intell. Neurosci.*, 2022.
22. Ahamed, N.N. and Karthikeyan, P., A reinforcement learning integrated in heuristic search method for self-driving vehicle using blockchain in supply chain management. *Int. J. Intell. Networks*, 1, 92–101, 2020.
23. Matenga, A.E. and Mpofu, K., Blockchain-based Product Lifecycle Management using Supply Chain Management for Railcar Remanufacturing, in: *30th CIRP Life Cycle Engineering Conference*, New Brunswick, New Jersey, 2023.
24. Abbas, K., Afaq, M., Ahmed Khan, T., Song, W.-C., A Blockchain and Machine Learning-Based Drug Supply Chain Management and Recommendation System for Smart Pharmaceutical Industry. *Electronics*, 9, 5, 852, 2020.
25. Tyagi, A.K., *Handbook of Research on Technical, Privacy, and Security Challenges in a Modern World*, IGI Global, US, 2022.
26. Deshmukh, A., Patil, D.S., Soni, G., Tyagi, A.K., Cyber Security: New Realities for Industry 4.0 and Society 5.0, in: *Handbook of Research on Quantum Computing for Smart Environments*, pp. 299–325, IGI Global, US, 2023.
27. Ballou, R.H., *Business logistics/supply chain management: planning, organizing, and controlling the supply chain*, Pearson Education, India, 2007.
28. Howard, R., *Cyberfraud: tactics, techniques and procedures*, Auerbach Publications, Boca Raton, Fla., United States, 2009.
29. Boyer, K. and Verma, R., *Operations and Supply Chain Management for the 21st Century*, Cengage Learning., Boston, MA, 2009.
30. Ambika, N., An Improved Solution to Tackle Cyber Attacks, in: *Advanced Controllers for Smart Cities*, pp. 15–23, Cham., Springer, 2021.
31. Ambika, N., Improved Methodology to Detect Advanced Persistent Threat Attacks, in: *Quantum Cryptography and the Future of Cyber Security*, pp. 184–202, IGI Global, US, 2020.
32. Basu, R. and Wright, J.N., *Total supply chain management*, Routledge, England, UK, 2010.

11

Role of AI and Digital Twin in Smart Manufacturing

M. Anand[1*], T. M. Sheeba[2] and C. Fancy[1]

[1]*Department of Networking and Communications, Faculty of Engineering and Technology, SRM Institute of Science and Technology, Kattankulathur, Chennai, India*
[2]*Department of Computer Applications, Faculty of Science and Humanities, SRM Institute of Science and Technology, Kattankulathur, Chennai, India*

Abstract
Recent years we have seen the emergence of several technologies that are critical to the development of the Industrial IoT and smart manufacturing. These include next-generation material science, advanced robotics, cyber-physical systems, big data, advanced analytics, artificial intelligence (AI) and machine learning (ML), operational intelligence and generative design for additive manufacturing. Manufacturers are thinking about new business models based on real-world implementations, in which they offer services instead of products and use the digital twin to optimize the product's performance and availability. In addition to full maintenance and operational optimization established on the digital twin's predictive and prescriptive capabilities, customers are provided the usage of the product or equipment. As a more profitable and manageable business model, the manufacturer retains ownership of the equipment while offering the maintenance service established on a digital twin.

Keywords: Artificial intelligence, digital twins, smart manufacturing

11.1 Introduction

Recent years have seen the emergence of several technologies that are critical to the development of the Industrial IoT and smart manufacturing. These include next-generation material science, advanced robotics, cyber-physical systems, big data, advanced analytics, artificial intelligence (AI) and machine learning (ML), operational intelligence and generative design for additive manufacturing. Manufacturers are thinking about new business models based on real-world implementations, in which they offer services instead of products and use the digital twin to optimize the product's performance and availability. In addition to full maintenance and operational optimization established on the digital twin's predictive and prescriptive capabilities, customers are provided the usage of the product or equipment. As a more profitable and manageable business model, the manufacturer retains

**Corresponding author*: manandinbox@gmail.com

Amit Kumar Tyagi, Shrikant Tiwari, Senthil Kumar Arumugam and Avinash Kumar Sharma (eds.) Artificial Intelligence-Enabled Digital Twin for Smart Manufacturing, (233–248) © 2024 Scrivener Publishing LLC

ownership of the equipment while offering the maintenance service established on a digital twin.

A digital twin in manufacturing is a computerized model of a physical product that is created, built, and maintained. It is enhanced by real-time process data and analytics that are derived from precise arrangements of the product, production processes, or equipment. This is essentially the digital twin's operating context, which is required to enable performance optimization. Real-time and operational data is a computer depiction of actual tangible events, whereas virtual models are conceptual in nature. AI-driven digital twins have become indispensable in the wake of smart manufacturing. These are digital representations of physical systems made with sensor and Internet of Things device data. They enable for virtual interaction and alteration, which is frequently quicker, safer, and more convenient than in the actual world. They mimic physical items or systems in real time. Using sensor data, digital twins update their models continually to match the status of their physical counterparts. Their two-way, dynamic information flow distinguishes them from static digital models. Digital twins that have been enhanced by AI are now able to suggest activities in the real world. In order to handle problems like plant outages or bottlenecks, this helps businesses to make well-informed decisions and swift process adjustments. In order to preserve, optimize, and simplify the underlying physical process, a method known as "digital twinning" entails building a virtual replica of any actual thing. Digital twins may be utilized to view the asset, track-changes, understand and improve asset execute throughout the product lifetime inspection process. Furthermore, the data collected from digital twins may provide the whole product lifetime and the processes required to oversee the supply chain, improve part manufacturing workflows, and maintain product quality [1]. He continued by defining digital twinning as comprising three primary components in Article [2]: a virtual twin, its consistent physical twin, and a data flow-cycle that moves data and operations back and forth between the two physical twins. All that a virtual twin is an algorithm that, given certain input values, produces the same output as its corresponding physical counterpart, mimicking its behavior. It is mainly believed of as a module of the smart manufacturing process, even though it may be used in any area, including construction, education, business, transportation, power and electronics, human and healthcare, sports, networking, and communications. The digital twin's principal purpose was to track the craft's structural evolution over time as a maintenance tool. Next, it was virtual as a whole twin in- order- to mimic its entire life-cycle and anticipate its recital [3]. Subsequently, as more sectors sought to make their operations more dynamic, adaptive, and clever depending on the operating environment, digital twinning began to gain traction. Due to its ability to predict failure, reduce production costs, monitor resources in real time, and facilitate the identification of product defects, this technology is in higher demand globally. As a result, digital twinning became the most popular technical trend [4]. Numerous surveys that highlight the most recent developments in digital twinning research across a range of disciplines have been published. Wanasinghe *et al.* [5] emphasized the state-of-the-art in digital twinning within the oil and gas industry. Lu *et al.* [6] and Cimino *et al.* [7] examined the reference models, applications, and research problems that are now present in manufacturing. In order to achieve smart manufacturing, Qi and Tao [8] focused on the importance of data and digital twinning. Tao *et al.* [9] review patents linked to DT in several sectors. Rasheed *et al.* also study the modeling viewpoint of digital twin [10].

11.2 Digital Twins

With the use of digital twin technology, technological and physical assets may be virtually simulated and maintained while remaining unchanged, all thanks to the data contained in a physical object. The Digital Twin system must use techniques like Artificial Intelligence (AI), Machine Learning, and Neural Networks to continuously learn and adapt its mode of operation depending on inputs and updates. Digital twins and other technologies, including the Internet of Things (IoT), have made data integration and cyber-physical interaction more advantageous. As an enabling technology, digital twins are anticipated to be integrated into industry 4.0 for broader coverage, particularly in the manufacturing, healthcare, and smart city IoT sectors to enhance safety, reduce costs, and create new products. For technicians, digital twin technology—which is typically readily available—is a critical tool for tracking, evaluating, and projecting the potential of their assets. IoT and AI technologies coexist alongside digital twin technology, which means they face similar difficulties. Three categories are often used to categorize digital twins. They are digital twin instance (DTI), which is the digital twin of each separate instance of the product once it is manufactured, digital twin aggregate (DTA), which is connected to its physical counterpart for the remaining years of the physical counterpart's life, and digital twin prototype (DTP), which is made up of the designs, analyses, and processes that realize a physical product. Use cases drive the specific data that is included in the digital twins. Since the digital twin is a conceptual construct, it's possible that other programs store the real data and information. The idea of digital twinning is showed in Figure 11.1.

11.3 Digital Twin for Smart Manufacturing

Comprehensive explanations of the integration and optimization of innovative digital technologies for smart manufacturing are found in Emerging Approaches and Applications. Throughout the examination of the product lifetime, digital twins may be used to visualize the asset, track-changes, comprehend, and optimize asset-performance. Additionally, the full lifespan of a product and its operations may be obtained from the data gathered from the digital twin, which can be used to achieve the supply chain, optimize part manufacturing workflows, and control product worth. Smart manufacturing can minimize time to market by utilizing digital twins to create and evaluate production processes in virtual environments prior to manufacturing. With the use of digital twins, complete simulation platforms that evaluate and simulate product acts in terms of analysis and produced part change may be provided. Three key components of smart manufacturing are: 1) gathering data from production cells using a range of sensors; 2) managing the collected data and 3) exchanging data with various devices and computers. Data is gathered from a real

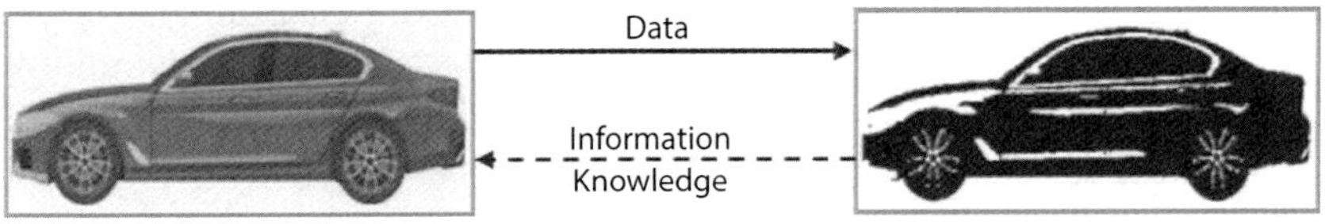

Figure 11.1 Digital twinning concept.

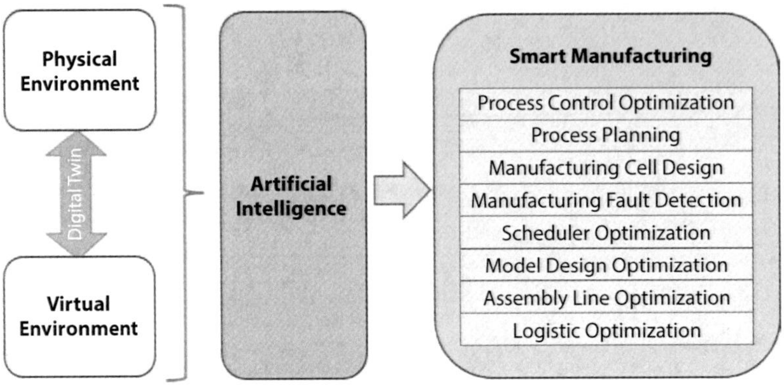

Figure 11.2 Smart manufacturing using artificial intelligence.

manufacturing cell or its equivalent simulated cell in a DT environment. With AI techniques, such data may also be used for defect diagnostics, an efficient assembly line, manufacturing process optimization, etc. Figure 11.2 shows the digital twinning method for smart manufacturing that is based on AI and ML. Three crucial phases were delineated for smart manufacturing: (1) Create networks to identify issues (2) Create platforms for modeling, sharing, and invention and (3) Implement smart manufacturing regulations. Among the digital twin's features are assembly line optimization, logistic optimization, scheduler optimization, model design optimization, manufacturing cell design, manufacturing fault detection, process control optimization, and process planning.

11.3.1 Networks to Define Problems

A data-driven paradigm known as the Digital Twin Network (DTN) uses network-related data to create a digital twin of a network. Usually, real-world networks, specialized network test beds, or network simulation tools are the sources of this data. Although the data may have a variety of characteristics, it must specify the network environment or phenomena that the network operator wishes to replicate. The DTN's reference architecture is shown in Figure 11.3. The DT, which essentially includes a network model that imitates the real network, is the main element of the design. A set of network-related data is fed into this model, which outputs a set of performance measures. Furthermore, a network optimizer may be used by some Traffic Engineering tools to identify the best solutions for various optimization issues.

The ML-based network model is housed in what might be thought of as a "black box," the DTN. This box generates several network-related performance measures and accepts various parameters as inputs, such as traffic, topology, routing, and scheduling policies. The environment and problem determine the inputs and results. The DTN's performance measurements are visible to the network operator in real-time as they adjust the input parameter values. This is done without the requirement of performing costly simulations or instrumenting the real-world network. The operator is able to test any input value, even if it might interrupt network services, because the DTN is an accurate replica of the real network. This is a result of the DTN being operated in a secure setting apart from the real-world network. The DTN's output performance metrics can take many different forms, depending on what the network operator requires.

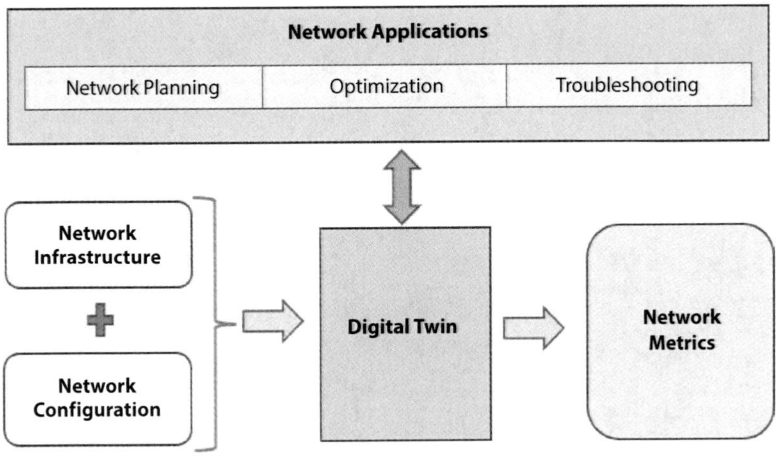

Figure 11.3 Digital twin network architecture.

11.3.2 Production Monitoring and Modification Using Digital Twin

The idea of the "digital twin" maybe used to raise the standard of whole value chains, production procedures, or the products themselves. Real-time production process monitoring is made possible in the part manufacture process by making a digital twin for a particular machine or manufacturing line. Manufacturers can find places where manufacturing processes can be adjusted or optimized to increase productivity and product value by evaluating data from the digital twin. Digital twins can be used watch vitality usage during the part-manufacturing process and identify possibilities for cost-cutting. Digital twins can also be used to accomplish continuous improvement at industrial locations [11]. Industrial activities may be optimized using digital twins without losing money or effort. Businesses may make data-driven choices to boost productivity, save expenses, and increase efficiency by using the potential of advanced analytics and machine learning [12]. The approach may be highly beneficial for producers across various sectors, since it can eliminate the need for expensive physical testing and provide real-time quality forecasts [13]. Digital twins can support to enhance present industrial-processes by utilizing complex simulation depend on real data collected by Internet of Things sensor [14]. To model the effects of manufacturing process modifications, such changing assembly line layouts or machine settings, on production output and efficiency, one can create a digital twin of a production-line. By doing this, businesses may save time and money by determining which process modifications are the most beneficial before putting them into practice. Production teams may use the digital twin to analyze several data sources and decrease the amount of faulty products in the process of producing parts, which will enhance productivity and save costs. Industries are able to increase output while reducing idle time.

11.3.3 Smart Manufacturing Policies

The smallest implementation unit in a smart factory for carrying out production operations is the smart manufacturing cell. Typically, it is a sophisticated cyber-physical system made up of sensors, robots, logistics gear, and processing equipment with the goal of guaranteeing

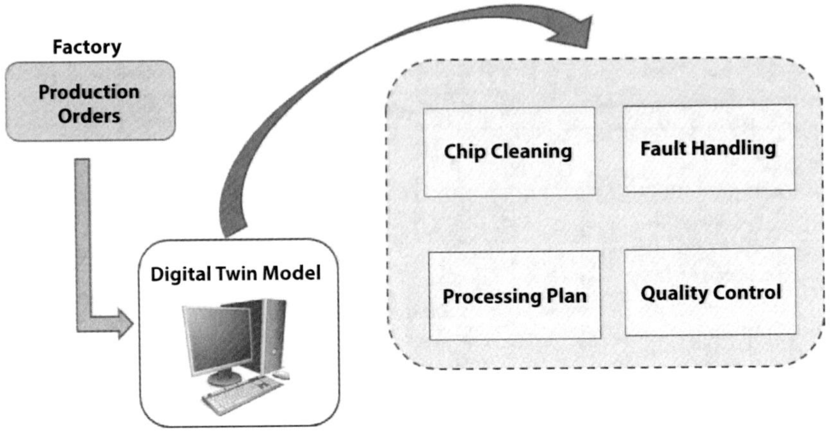

Figure 11.4 Application scenarios of smart manufacturing cell.

safe, effective, and orderly processing. Figure 11.4 displays the smart manufacturing cell's application possibilities. An autonomous system having cognitive functions including perception, decision-making, and communication is the smart manufacturing cell. It has the flexibility to react to changes in the physical environment as well as production orders supplied by MES and ERP. Real-time remote monitoring of on-site data is possible, and digital space may also be used for other tasks such processing optimization, simulation, and configuration. The subsequent instances delineate many distinct functions of the intelligent manufacturing cell: 1) Process system cleaning carried out autonomously. The process system's processing sections are continuously observed by the smart manufacturing cell. In order to maintain the value and efficiency of the processing and manufacturing, the system cleans the chips by operating the robot and other relevant execution equipment when it notices that there are too many chips gathered in one area, which impacts the processing. 2) Poor processing management. Without human assistance, the smart manufacturing cell detects and automatically corrects processing errors such incorrect clamping, workpieces falling out of clamps, cutter damage, etc. 3) Workpiece quality management Smart sensors keep an eye on the workpiece's quality throughout the whole processing life cycle. The system's overall operational efficiency is increased when the execution devices transfer the unqualified workpiece straight to the trash area when the quality is found to be inadequate. 4) Security and safety in the processing area Security surveillance is conducted in the smart manufacturing cell environment. In order to maintain the system's safety, the system may recognize information when humans or foreign objects approach the smart manufacturing cell and can then direct the processing machines to halt, slow down, and perform other relevant activities.

11.4 Artificial Intelligence Enabled with Digital Twin

One factor contributing to the latest advancements in Industry 4.0 is artificial intelligence (AI). Businesses are concentrating on increasing product uniformity, productivity, and cutting operational costs. They hope to do this through a cooperative relationship between humans and robotics. Hyperconnected manufacturing processes in smart-trades rely on various parameters that link with to each other through AI automation systems that imprisonment and understand all kinds of data. Intelligent automation technologies have the power to fundamentally change how modern manufacturing is carried out. Artificial Intelligence furnishes related data to facilitate decision-making and notify each of potential issues. Based on their desire to incorporate them into their equipment, industries will utilize AI to interpret data supplied via linked machines and Internet of things (IoT) devices. Among the digital twin's features are assembly line optimization, logistic optimization, scheduler optimization, model design optimization, manufacturing cell design, manufacturing fault detection, process control optimization, and process planning.

Methods for AI-DTS that are essential to a Digital Twin System's optimization. Creating models, updating models, modeling, data analytics, predictive analytics and decision making are some of these strategies. Therefore, the digital twin notion cannot be fully realized without the integration of AI and associated subjects like data science, machine learning, and statistics. Since DT's design offers a clear platform to highlight the variety of contributions AI can make, it is crucial to apply AI for a DTS. Six distinct AI approaches may be identified from the Digital Twin System's structural architecture. These methods are referred to as AI for Digital Twin Systems, 1) data analytics; 2) optimization, 3) modeling; 4) predictive analytics; 5) data analytics; 6) decision making. For each AI-DTS, we present a brief discussion in the following. Picture 1's visualization makes clear that the physical item to be modeled might originate from a variety of fields, such as manufacturing, city development, sustainability, engineering, or weather research. Based on this, a computer simulation that captures the key features of the physical model is used to construct a digital twin that corresponds to a dynamical system. This often entails the use of supplementary data to estimate, the restrictions of the digital twin. We refer to this process of optimization as 1.AI, or optimization. Crucially, there is an additional AI-based optimization stage that differs from the first. The second optimization process ensures that the digital twin and its physical counterpart operate in sync. This results in the digital twin's updating mechanism being optimized. 2. Optimization; see to Figure 11.5 for more information. To avoid misperception among the optimization processes of model development and model updating, we deliver information about their use. This highlights another issue that will be covered in more depth below, which is that certain AI methods are not exclusive to AI-DTS. In this instance, a DTS can benefit from optimization approaches in several ways. Generative modeling provides the third AI engagement. Generative adversarial networks (GANs) are one instance of this [15–17]. Generally speaking, data with characteristics gleaned from massive amounts of data may be produced using a GAN. A GAN may occasionally be used in place of or in addition to a conventional simulation model, such as one that is based on Boolean networks, differential equations, or agent-based models [18, 19]. This implies that artificial intelligence (AI) can both create and enhance simulation models.

Since both methods are used for data analysis, the 4th and 5th AI applications could initially appear to be comparable to one another. The data sources for this analysis, however, come from completely different sources. In particular, it is possible to discriminate between two data sources: the digital twin and external data. Because of this, there are two different kinds of data analysis that we can identify: predictive analytics and data analytics. It ought to be evident that an analysis's interpretation is significantly influenced by the data's source.

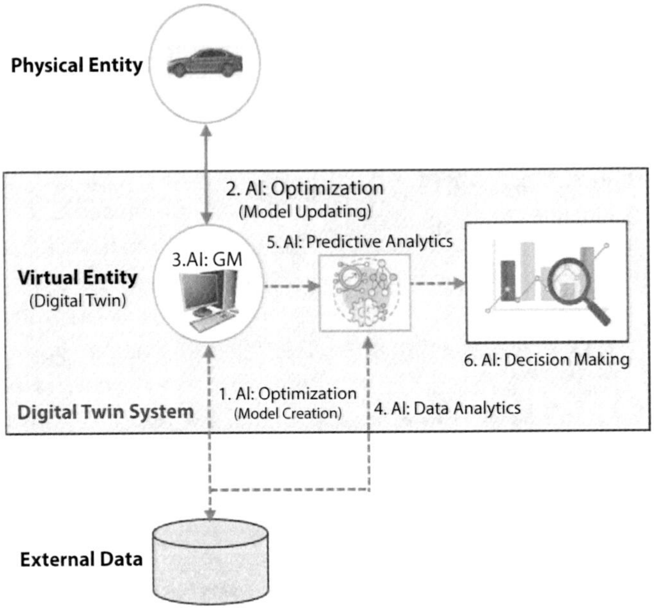

Figure 11.5 Digital twin system with an interface to a physical and external data.

Complementary techniques are offered by AI predictive analytics, and the approach we shall discuss next aims to combine them. In particular, decision making, which is the sixth application of AI, is utilized to make decisions and to summarize each unique outcome that has been attained thus far. The final result of a DTS is produced by integrating everything and creating a quantitative or qualitative summary. Plotting and charting visualizations are examples of quantitative summaries. These figures may even be interactive, enabling the examination of outcomes, depending on how these visualizations are presented. Because the outcomes of a DTS may be complicated, such an-interactive study is extremely beneficial, especially for the utilization of non-AI specialists. A dashboard would be one specific use for this kind of interactive visualization. AI 2023, 4 724 Several strategies, such as multitask learning, multi-label learning, or transfer-learning, can be used for the quantitative integration [20–22]. The underlying models of deep learning may be trained for a range of tasks, such as regression and classification, thanks to these learning paradigms [23]. Furthermore, promising ways for condition making under ambiguity have been demonstrated using graphical models [24]. Given that a DTS's decision-making power is essential and arguably the most vital component of the system as a whole, it is expected that this issue will garner a lot of attention in the years to come. AI methods that have enormous promise to advance sustainability, climate research, and health. Obstacles to Health Generative adversarial networks (GANs) present a potential approach for the health sciences to tackle the problems associated with sparse data while studying health-related problems that are frequently encountered in clinical and medical research. Generally speaking, generative adversarial networks, or GANs, are a class of artificial intelligence models that are made up of two primary parts: a generator and a discriminator. Generating new data that is similar to a given training dataset is the aim of a GAN. The generator is in charge of producing fresh samples, such pictures that replicate the features and patterns seen in the training set.

In order to generate more realistic samples, it first generates random noise and then gradually refines it. The generator's objective is to produce samples that the discriminator cannot tell apart from actual samples. These days, picture production from a variety of medical imaging technique, such as X-ray, MRIs, and CT-scan, is done using generative adversarial networks (GANs) [25–27]. These imaging technologies are frequently used in hospitals and other therapeutic settings for a variety of objectives, including prognoses, treatment planning, and diagnosis. They are clinically relevant. Therefore, the progression of generative adversarial networks (GANs) and its integration with a digital twin system hold significant possible for the development of precision and personalized medicine [28, 29]. The virtual testing of medical theories is another significant field with implications for AI [30]. Traditionally, clinical studies are required to prove a medicinal drug's safety, effectiveness, and to rule out any possible adverse effects. The evidence-based information from these studies is essential for regulatory bodies to assess whether a medicine is suitable for general use. Alternatively, virtual testing of patient digital twin models might be utilized to investigate "What-If" scenarios pertaining to the dispensing of medication. This is just one more instance of how AI may support decision-making. Difficulties in Climate Science Another new area of application for digital twins is in climate research, where sophisticated computation systems are combined with observational data to simulate and forecast intricate weather events [31]. Weather change adaptation and catastrophe risk management for extreme weather are major topics for the European Destination Earth (DestinE) effort [32]. Nonetheless, there are further fields where a digital twin might be useful, such as forecasting global surface temperatures and El Niño/Southern Oscillation (ENSO) conditions [33–36]. AI 2023, 4 725 The assessment of model predictions is one of these difficulties' concerns. For example, the lack of controlled laboratory settings makes it difficult to evaluate forecasts pertaining to the Earth's global surface temperature [37], which includes averages for both land and sea surface temperatures. This means that methods other than those used in a typical laboratory environment are needed to acquire summarized measurements [38]. Additional instances included the use of AI and statistical techniques, which are necessary to improve our understanding of temporal discontinuities, missing data, and limited spatial coverage. Restricted Spatial Coverage: Measurements from different parts of the Earth's surface-region are needed to estimate the global surface temperature. Nevertheless, there may be variations in the geographical coverage of temperature monitoring stations, which makes it difficult to provide accurate and thorough data. Absence of Information and Temporal Breaks: Errors in data collection, equipment failures, or missing observations can leave gaps in temperature records that make it difficult to estimate the global surface temperature. Furthermore, it might be difficult to create consistent and trustworthy temperature records when there are temporal discontinuities brought on by modifications to measurement procedures or station relocations. *Obstacles to Sustainability.* Sustainability is a multidisciplinary field that aims to promote social justice, environmental preservation, and economic viability. It entails making the transition to greener and renewable energy sources, as well as boosting energy efficiency and reducing pollution—all essential elements of energy sustainability [39]. This also covers the creation and uptake of renewable energy sources, such wind and solar energy [40]. Renewable energy sources have the drawback of occasionally being erratic and weather-dependent. For instance, wind energy may fluctuate greatly in a matter of minutes, while solar energy is not accessible at night [41, 42]. The fluctuations in energy production might provide difficulties in guaranteeing a steady and dependable power source. Furthermore, significant infrastructure designs and grid adjustments are needed for the deployment and integration of these technologies in order to properly handle the variable output of renewable energy sources. AI can assist in this case, for instance, by offering highly developed predictive analytics [43, 44]. AI might also be very useful in optimizing the updating procedure of a digital twin for wind-power generation. These wind farm or individual wind turbine digital twin models are based on real-time data collected from weather forecasting systems or sensors installed on the wind turbines themselves [45]. By using this data, the digital twin gives an interactive and dynamics represent of the wind-energy model, allowing operators to continually monitor and assess its recital. But maintaining the digital twin current with the most recent data is a big task. Here's when AI methods can be useful. Data from several sources may be automatically gathered, integrated, and processed with the use of AI algorithms. Large amounts of real-time data may be handled by these algorithms with efficiency, guaranteeing that the digital twin is correct and represents the wind energy system's current status. AI becomes an essential component of higher order computation when it is combined with DT as a service to the physical-system, allowing DT to anticipate, optimize, and make decisions instantly. However, able-to perform these duties, the DT must be dependable, accurate and independent. Changes in the environment and in operations have an effect on the gradual degradation of physical systems.

11.5 AI and DT Lifecycle

Creating an entire model that accurately represents the real system is usually difficult and errors in modeling or synchronization may happen. It implies that there could be problems with the model that the DT represents, and that it has to be revised and examined before being put into use. Thus, it is important to evaluate the virtual system confidence and ensure that it meets the original prospects prior to deploying it. Physical systems are exposed to a range of conditions and traits that change them throughout time. As a result, the behavior of the DT system need to be able to adjust to changes and continuously represent the state of a physical model. Moreover, it must to be dynamic and monitor modifications to the physical model. If we follow the DT through its lifespan, we may distinguish between two crucial phases: the DT's design and exploitation. a) Design: this stage is primarily concerned with building models and collecting and storing data; b) Exploitation: this stage is further divided into three stages, which are as follows: Setup: by seeing the system operate, the DT's settings are changed; Run: The DT is prepared to function and carry out the anticipated tasks when the parameters have been verified; Updating and maintaining the model is necessary because disturbances might happen and the model's parameters can change. Phase of setup To ensure that the virtual model exactly represents the physical model, the parameters of the latter must be adjusted during the validation and correction phase. Figure 11.6 provides a schematic depiction of this phase. The fact that the setup phase is started in this case by a human operator is noteworthy since the goal is to reduce the difference between the virtual and real twins, and the exact value of the DT parameters is now unknown. Execute phase The DT model should be able-to recreate the behavior of a physical models exactly as it is since it is a dynamic system with elements set during the run phase. The model has to keep an eye on how well the nominal behavior

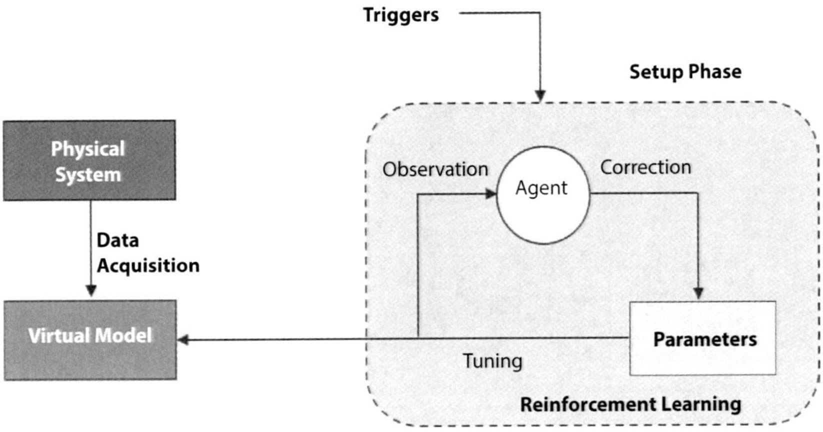

Figure 11.6 Setup phase.

determined by the models matches the physical twin's observations. During this phase, it is anticipated that AI approaches like supervised or unsupervised learning would identify any unforeseen patterns that may be present on the distribution of the combined data, as seen in Figure 11.7. Such abnormalities in data are typically found using algorithms like Principal Components, DBSCAN, Local Outlier Factor, and Isolation Forest. It should be mentioned that this stage does not directly involve human competence. But this accuracy is outside the purview of this contribution; instead, its experience is often applied to the analysis of the anomalies displayed by the AI tool.

Phase of maintenance, The human operator must first determine the severity of any divergence in the DT's behavior that the unsupervised learning algorithms have found. The worker initiates the maintenance phase, wherein the DT elements undergo further evaluation, if necessary. Because the elementer has changed is necessary to keep the latter or not, as seen in Figure 11.8, the tuning parameter that the human operator activates in this instance differs from the first step. Nonetheless, this phase's features are mostly comparable to those of the setup phase, and the same methods may likely be applied most of the time.

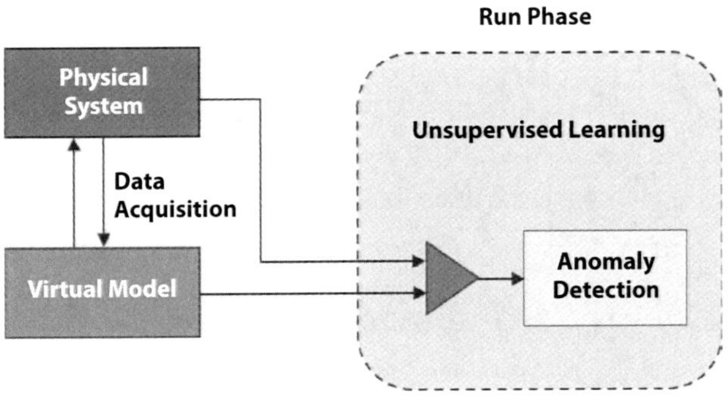

Figure 11.7 Run phase.

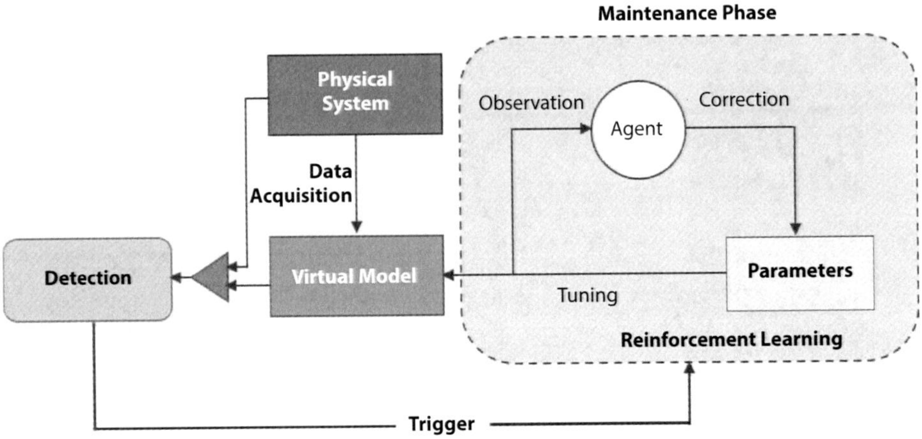

Figure 11.8 Maintenance phase.

11.6 AI-Enabled Digital Twins in Manufacturing

Artificial intelligence (AI)-driven digital twins have become a vital piece of equipment for smart manufacturing. Digital twins are digital replicas of real-world systems that use data from sensors and Internet of Things (IoT) to imitate physical objects and systems in real time [46]. A digital twin is not the same as a digital model that was made for simulation purposes during the design-phase. During the design process, simulations are quite useful for exploring different design choices. On the other hand, idealized states are represented by digital models, not the real physical states of systems. Moreover, models produced using reverse-engineering techniques are not digital twins. The physical system is unaffected by them. A digital twin replicates the present state of the physical system by continuously updating its model using information from sensors. The digital twin receive the data from the physical system and vice versa. Digital twins vary from digital models in that they have a bidirectional information flow. For digital twins to be effective in actual production, they need, in summary, possess the following qualities: Quantification of Uncertainty: The predicted system state is always subject to uncertainty when sensors are utilized. As a result, it is impossible to make wise judgments without taking this uncertainty into consideration. For the digital twin to be able to provide suggestions with the right degrees of confidence, it must be able to measure uncertainty. Performance Prediction: Given the current state of the system, digital twins ought to be able to forecast future behavior or performance. For the purpose of evaluating the potential effects of various future events and accounting for uncertainties in the system state, digital twins must run simulations in real time. Physical System Monitoring and Control: Digital twins should be able to track the operation of physical production systems in real time and give the controllers in charge of the process useful data.

11.7 Digital Twins Used in Manufacturing

Task planners and schedulers can make choices depending on the condition of the production system with the help of digital twins. 1) Digital twins track the performance and state of machinery and equipment in real time to forecast when repairs will be required. 2) Real-time quality control and flaw identification are carried out using digital twins. 3) Digital twins can be utilized to identify the issue and suggest the appropriate remedial measures when the system reaches an incorrect state. 4) Digital twins can pinpoint areas for optimization or development by examining process data. 5) To guarantee adherence to pertinent rules, digital twins can offer a comprehensive record of processing or operational circumstances. 6) Real-time manufacturing processes optimization is made possible by digital twins, which facilitate the on-demand creation of customized goods.

11.8 AI Impacting Digital Twins

Digital twin technology's capabilities are being enhanced by AI more and more. A robotic surface finishing cell's sensors may create a model of a component within the cell, potentially doing away with the need for pricey part-holding apparatuses. AI is required because tool paths must take into account uncertainty in component models generated by sensors in order to completely avoid the chance of collision. Hoses and cables that are attached to the tool are typically necessary for robots that do surface finishing. These limbs might limit the robot's range of motion. Building models of every peripheral that is installed on the robot is made possible by a digital twin. By using AI, the system is able to determine the robot's motion restrictions by estimating the states of its associated peripherals. Digital twins can make use of AI-based health management and prognostics to make sure that the beginning of unfavorable occurrences can be automatically identified. For instance, a digital twin can identify the reason behind quick tool wear in robotic finishing and implement preventative actions by using force and vision data.

11.9 Organizations Succeed Deploying Digital Twins

There are still a few measures manufacturing organizations need to do in order to properly use digital twins. a) Data Quality: The digital twin receives a lot of data from sensors. It is important for organizations to implement suitable steps to ensure that sensors consistently generate precise data. The data may be denoised and sent in the appropriate format using AI technologies. b) Scalable System Integration: The implementation of digital twins requires system integration. The underlying technologies are evolving quickly, and new sources of data are being introduced on a regular basis. It is possible to prevent system integration from becoming a bottleneck by using a scalable system integration approach. Using the right middleware in conjunction with a modular strategy will provide scalable solutions. c) Sturdy Computational Infrastructure: In order to handle data, digital twins need a lot of processing power. It is necessary to have a strong computing infrastructure with sufficient redundancy and capability. d) Workforce: The implementation of digital twins necessitates a workforce with a specialized skill set, which calls for a combination of using solution providers' services and upskilling the current workforce. e) Long-Term Sustainability: To guarantee that they provide precise forecasts, digital twins need to undergo routine upkeep and improvements. For the long-term sustainability of digital twins, organizations should create a sustainability strategy. One helpful strategy for preventing surprises is a subscription model offered by solution vendors. f) Privacy and Cybersecurity Concerns: As connectivity rises, so do the hazards to cybersecurity. Privacy issues are also raised by the use and storage of data. Modern cybersecurity solutions must be put in place to address these issues. Additionally, conducting frequent security audits and data encryption may be necessary to ensure regulatory compliance with privacy problems. When manufacturing companies start using digital twins, they will need to handle these issues with a comprehensive approach. By doing this, companies improve their ability to participate in the new era of smart manufacturing, which is characterized by decreased prices, less mistakes, improved quality, and increased performance.

Summary

An electronic version of a physical model known as a "digital twin," is used to monitor, control, and enhance its real-world counterpart. In smart manufacturing, a digital twin can be used to form a simulated representation of a production line or plant. With the use of this model, producers may mimic and optimize their manufacture process, identified possible problem and areas for development, and make knowledgeable conditions about the distribution of resources and the timing of output. By guaranteeing flawless product or process functioning, promoting operational excellence, and increasing time and cost efficiency, the use of digital twins may dramatically increase a company's added value. Through the use of data and advanced analytics, digital twins may help organizations achieve increased productivity, better decision-making, and operational optimization. In order to increase efficacy in part manufacturing, the use of digital twins may also be used to expedite production processes and reduce downtime. Reduction in expenses, enhanced quality control, predictive maintenance, and increased productivity and efficiency are some advantages of employing digital twins in smart manufacturing. All things considered, the real-time monitoring, optimization, and cooperation that digital twin technology offers the ability to bring about will transform smart manufacturing.

References

1. Soori, M., Arezoo, B., Dastres, R., Digital twin for smart manufacturing, A review. *Sustain. Manuf. Serv. Econ.*, 2, 2023.
2. Grieves, M., Digital twin: Manufacturing excellence through virtual factory replication. *White Paper*, 1, 1–7, 2014.
3. Tuegel, E.J., Ingraffea, A.R., Eason, T.G., Spottswood, S.M., Reengineering aircraft structural life prediction using a digital twin. *Int. J. Aerosp. Eng.*, 2011, 1–14, Aug. 2011.
4. Cearley, D., Burke, B., Smith, D., Jones, N., Chandrasekaran, A., Lu, C., Top 10 strategic technology trends for 2020, in: *Gartner, Stamford, CT, USA, Tech. Rep.*, 2019.
5. Wanasinghe, T.R., Wroblewski, L., Petersen, B.K., Gosine, R.G., James, L.A., De Silva, O., Mann, G.K., II, Warrian, P.J., Digital twin for the oil and gas industry: Overview, research trends, opportunities, and challenges. *IEEE Access*, 8, 104175–104197, 2020.
6. Lu, Y., Liu, C., Wang, K., II, Huang, H., Xu, X., Digital twin driven smart manufacturing: Connotation, reference model, applications and research issues. *Robot. Comput. Integr. Manuf.*, 61, Feb. 2020, Art. no. 101837.

7. Cimino, C., Negri, E., Fumagalli, L., Review of digital twin applications in manufacturing. *Comput. Ind.*, 113, Dec. 2019, Art. no. 103130.
8. Qi, Q. and Tao, F., Digital twin and big data towards smart manufacturing and industry 4.0: 360 degree comparison. *IEEE Access*, 6, 3585–3593, 2018.
9. Tao, F., Zhang, H., Liu, A., Nee, A.Y.C., Digital twin in industry: State-of-the-art. *IEEE Trans. Ind. Informat.*, 15, 4, 2405–2415, Apr. 2019.
10. Rasheed, A., San, O., Kvamsdal, T., Digital twin: Values, challenges and enablers from a modeling perspective. *IEEE Access*, 8, 21980–22012, 2020.
11. Ding, K., Chan, F.T., Zhang, X., Zhou, G., Zhang, F., Defining a digital twin-based cyber-physical production system for autonomous manufacturing in smart shop floors. *Int. J. Prod. Res.*, 57, 6315–6334, 2019.
12. Lo, C., Chen, C., Zhong, R.Y., A review of digital twin in product design and development. *Adv. Eng. Inf.*, 48, 101297, 2021.
13. Guo, H., Chen, M., Mohamed, K., Qu, T., Wang, S., Li, J., A digital twin-based flexible cellular manufacturing for optimization of air conditioner line. *J. Manuf. Syst.*, 58, 65–78, 2021.
14. Sun, X., Bao, J., Li, J., Zhang, Y., Liu, S., Zhou, B., A digital twin-driven approach for the assembly-commissioning of high precision products. *Robot. Comput. Integr. Manuf.*, 61, 101839, 2020.
15. Gui, J., Sun, Z., Wen, Y., Tao, D., Ye, J.A., review on generative adversarial networks: Algorithms, theory, and applications. *IEEE Trans. Knowl. Data Eng.*, 35, 3313–3332, 2021.
16. Deekshetha, H.R. and Tyagi, A.K., Chapter 16 - Automated and intelligent systems for next-generation-based smart applications, in: *Data Science for Genomics*, A.K. Tyagi and A. Abraham (Eds.), pp. 265–276, Academic Press, 2023, https://doi.org/10.1016/B978-0-323-98352-5.00019-7.
17. Tyagi, A.K. and Richa, Smart Manufacturing Using Internet of Things, Artificial Intelligence, and Digital Twin Technology, in: *Global Perspectives on Robotics and Autonomous Systems: Development and Applications*, 2023, doi: 10.4018/978-1-6684-7791-5.ch008.
18. Area, I., Fernández, F.J., Nieto, J.J., Tojo, F.A.F., Concept and solution of digital twin based on a Stieltjes differential equation. *Math. Methods Appl. Sci.*, 45, 7451–7465, 2022.
19. Barat, S., Parchure, R., Darak, S., Kulkarni, V., Paranjape, A., Gajrani, M., Yadav, A., Kulkarni, V., An agent-based digital twin for exploring localized non-pharmaceutical interventions to control covid-19 pandemic. *Trans. Indian Natl. Acad. Eng.*, 6, 323–353, 2021.
20. Caruana, R., Multitask learning. *Mach. Learn.*, 28, 41–75, 1997.
21. Bashath, S., Perera, N., Tripathi, S., Manjang, K., Dehmer, M., Emmert-Streib, F., A data-centric review of deep transfer learning with applications to text data. *Inf. Sci.*, 585, 498–528, 2022.
22. Gibaja, E. and Ventura, S., Multi-label learning: A review of the state of the art and ongoing research. *Wiley Interdiscip. Rev. Data Min. Knowl. Discov.*, 4, 411–444, 2014, [CrossRef].
23. Emmert-Streib, F., Yang, Z., Feng, H., Tripathi, S., Dehmer, M., An introductory review of deep learning for prediction models with big data. *Front. Artif. Intell.*, 3, 4, 2020, [CrossRef] [PubMed].
24. Kapteyn, M.G., Pretorius, J.V., Willcox, K.E., A probabilistic graphical model foundation for enabling predictive digital twins at scale. *Nat. Comput. Sci.*, 1, 337–347, 2021, [CrossRef].
25. Madani, A., Moradi, M., Karargyris, A., Syeda-Mahmood, T., Semi-supervised learning with generative adversarial networks for chest X-ray classification with ability of data domain adaptation, in: *Proceedings of the 2018 IEEE 15th International Symposium on Biomedical Imaging (ISBI 2018)*, Washington, DC, USA, 4–7 April 2018, IEEE, New York, NY, USA, pp. 1038–1042, 2018.
26. Jiang, Y., Chen, H., Loew, M., Ko, H., COVID-19 CT image synthesis with a conditional generative adversarial network. *IEEE J. Biomed. Health Inform.*, 25, 441–452, 2020, [CrossRef] [PubMed].

27. Zhao, J., Huang, J., Zhi, D., Yan, W., Ma, X., Yang, X., Li, X., Ke, Q., Jiang, T., Calhoun, V.D. *et al.*, Functional network connectivity (FNC)-based generative adversarial network (GAN) and its applications in classification of mental disorders. *J. Neurosci. Methods*, 341, 108756, 2020, [CrossRef] [PubMed].
28. Tian, Q., Price, N.D., Hood, L., Systems cancer medicine: Towards realization of predictive, preventive, personalized and participatory (P4) medicine. *J. Intern. Med.*, 271, 111–121, 2012, [CrossRef].
29. Chan, I.S. and Ginsburg, G.S., Personalized Medicine: Progress and Promise. *Annu. Rev. Genom. Hum. Genet.*, 12, 217–244, 2011, [CrossRef].
30. An, G. and Cockrell, C., Drug development digital twins for drug discovery, testing and repurposing: A schema for requirements and development. *Front. Syst. Biol.*, 2, 928387, 2022, [CrossRef].
31. Voosen, P., Europe builds' digital twin'of Earth to hone climate forecasts. *Science*, 370, 16, 2020, [CrossRef] [PubMed].
32. Destination Earth—A digital twin in support of climate services. *Clim. Serv.*, 30, 100394, 2023, [CrossRef].
33. Ham, Y.G., Kim, J.H., Luo, J.J., Deep learning for multi-year ENSO forecasts. *Nature*, 573, 568–572, 2019, [CrossRef] [PubMed].
34. Lean, J.L. and Rind, D.H., How will Earth's surface temperature change in future decades? *Geophys. Res. Lett.*, 36, 15708, 2009, [CrossRef].
35. Cifuentes, J., Marulanda, G., Bello, A., Reneses, J., Air temperature forecasting using machine learning techniques: A review. *Energies*, 13, 4215, 2020, [CrossRef].
36. Taylor, J. and Feng, M., A deep learning model for forecasting global monthly mean sea surface temperature anomalies. *Front. Clim.*, 4, 178, 2022, [CrossRef].
37. Hansen, J., Ruedy, R., Sato, M., Lo, K., Global surface temperature change. *Rev. Geophys.*, 48, RG4004, 2010, [CrossRef].
38. Niu, G.Y., Yang, Z.L., Mitchell, K.E., Chen, F., Ek, M.B., Barlage, M., Kumar, A., Manning, K., Niyogi, D., Rosero, E. *et al.*, The community Noah land surface model with multiparameterization options (Noah-MP): 1. Model description and evaluation with local-scale measurements. *J. Geophys. Res. Atmos.*, 116, D12109, 2011, [CrossRef].
39. Ahmad, T., Zhang, D., Huang, C., Zhang, H., Dai, N., Song, Y., Chen, H., Artificial intelligence in sustainable energy industry: Status Quo, challenges and opportunities. *J. Clean. Prod.*, 289, 125834, 2021, [CrossRef].
40. Qazi, A., Hussain, F., Rahim, N.A., Hardaker, G., Alghazzawi, D., Shaban, K., Haruna, K., Towards sustainable energy: A systematic review of renewable energy sources, technologies, and public opinions. *IEEE Access*, 7, 63837–63851, 2019.
41. Milan, P., Wächter, M., Peinke, J., Turbulent character of wind energy. *Phys. Rev. Lett.*, 110, 138701, 2013, [CrossRef] [PubMed].
42. Anvari, M., Lohmann, G., Wächter, M., Milan, P., Lorenz, E., Heinemann, D., Tabar, M.R.R., Peinke, J., Short term fluctuations of wind and solar power systems. *New J. Phys.*, 18, 063027, 2016, [CrossRef].
43. Liu, Z., Jiang, P., Zhang, L., Niu, X., A combined forecasting model for time series: Application to short-term wind speed forecasting. *Appl. Energy*, 259, 114137, 2020, [CrossRef].
44. Sharadga, H., Hajimirza, S., Balog, R.S., Time series forecasting of solar power generation for large-scale photovoltaic plants. *Renew. Energy*, 150, 797–807, 2020, [CrossRef].
45. Wang, M., Wang, C., Hnydiuk-Stefan, A., Feng, S., Atilla, I., Li, Z., Recent progress on reliability analysis of offshore wind turbine support structures considering digital twin solutions. *Ocean. Eng.*, 232, 109168, 2021, [CrossRef].
46. https://www.forbes.com/sites/forbestechcouncil/2023/11/27/what-to-know-about-ai-enabled-digital-twins-in-manufacturing/?sh=6c3038a82f86

12

Data Analytics and Visualization in Smart Manufacturing Using AI-Based Digital Twins

M. Sivakumar[1*], M. Maranco[1], N. Krishnaraj[1] and U. Srinivasulu Reddy[2]

[1]*Department of Networking and Communications, Faculty of Engineering and Technology, SRM Institute of Science and Technology, Kattankulathur, Tamil Nadu, India*
[2]*Machine Learning and Data Analytics Lab, Centre of Excellence in Artificial Intelligence, Department of Computer Applications, National Institute of Technology, Tiruchirappalli, India*

Abstract

Smart manufacturing encompasses the use of cutting-edge technologies, including IoT (Internet of Things), AI (Artificial Intelligence), and data analytics, to optimize production procedures, enhance effectiveness, minimize expenses, and elevate total productivity. A crucial element of smart manufacturing is the notion of digital twins, which refer to virtual duplicates of tangible assets, procedures, or systems. These digital replicas are generated by utilizing real-time data obtained from sensors integrated into tangible assets, and are employed for the purposes of monitoring, analysis, prediction, and optimization. The primary issue lies in integrating data from many sources throughout the production ecosystem while ensuring data quality. Scalable analytics and visualization solutions are necessary to manage the substantial amount of data produced by sensors and other devices in smart manufacturing environments. Manufacturing processes can be intricately intricate, encompassing a multitude of interconnected systems and variables. Scientists are utilizing AI algorithms to examine data from digital replicas for the purpose of predictive maintenance. This allows manufacturers to foresee equipment malfunctions and plan maintenance in advance, thus reducing the amount of time that operations are halted. Manufacturers are utilizing AI-based digital twins to simulate and enhance industrial processes, allowing them to discover areas of congestion, optimize the usage of resources, and enhance overall productivity. Artificial intelligence algorithms are being utilized to analyze data from digital replicas of physical objects, known as digital twins, in order to promptly identify any flaws or imperfections, thereby guaranteeing the quality of the product and minimizing unnecessary waste. Ongoing progress in AI algorithms, namely in deep learning and reinforcement learning, will provide more precise forecasts and enhancements using data derived from digital twins. The significance of edge computing technologies will grow as they enable immediate analysis and visualization of data produced by sensors located at the network's periphery, thereby diminishing the need for low latency and high bandwidth.

Keywords: Smart manufacturing, artificial intelligence, digital twins, data analytics, deep learning

**Corresponding author*: siva.mcs@gmail.com

Amit Kumar Tyagi, Shrikant Tiwari, Senthil Kumar Arumugam and Avinash Kumar Sharma (eds.) Artificial Intelligence-Enabled Digital Twin for Smart Manufacturing, (249–278) © 2024 Scrivener Publishing LLC

12.1 Introduction

Smart manufacturing incorporates cutting-edge technology, including the Internet of Things (IoT), artificial intelligence (AI), and data analytics, to optimize industrial processes and improve efficiency. A fundamental element of intelligent manufacturing is the notion of digital twins (Tyagi & Richa, 2023), which are virtual duplicates of tangible resources. Data analytics approaches applied in smart manufacturing through AI-based digital twins have a wide range of applications and significant effects. AI-powered digital replicas assess live sensor data and past maintenance records to forecast equipment malfunctions in advance. Manufacturers can proactively schedule maintenance, minimize downtime, and prevent expensive unplanned shutdowns by detecting early signals of degradation or irregularities in machine behavior. Utilizing data analytics techniques on digital twins allows for the continuous monitoring of production processes and the assessment of product quality in real-time. Manufacturers can utilize sensor data and production parameters to analyze and identify deviations from quality standards, as well as potential faults, at an early stage in the manufacturing process. This helps to minimize waste and maintain consistent product quality. AI-powered digital replicas replicate and enhance manufacturing procedures to enhance effectiveness, productivity, and resource allocation (Xia *et al.*, 2023). Manufacturers may achieve optimal production schedules, decrease energy usage, and streamline operations for maximum throughput by analyzing data from digital twins and discovering opportunities for optimization. Applying data analytics approaches to digital twins enhances the efficiency of supply chain operations, encompassing inventory management, demand forecasting, and logistics planning. Manufacturers can enhance overall supply chain efficiency by utilizing digital twins and external data sources to analyze and manage inventory levels, predict demand variations, and optimize transportation routes, resulting in cost reduction.

AI-powered digital replicas enable virtual prototyping and product design by replicating the performance of a product under different circumstances (Mihai *et al.*, 2022). Manufacturers can enhance product design and optimization by analyzing data from digital twins and conducting virtual trials. This iterative process enables them to meet performance criteria, shorten time-to-market, and expedite innovation cycles. Applying data analytics approaches to digital twins enhances the efficiency of resource allocation and capacity planning in manufacturing plants. Manufacturers can enhance production schedules, distribute resources effectively, and optimize production capacity by scrutinizing production data and demand estimates. This enables them to satisfy consumer demand while saving expenses. AI-powered digital replicas assess data on energy use and production factors to maximize energy utilization and minimize ecological footprint. Manufacturers can decrease carbon emissions, enhance sustainability, and meet regulatory requirements by identifying energy-intensive procedures and applying energy-saving strategies. Applying data analytics approaches to digital twins enables the real-time monitoring and provision of decision assistance for manufacturing operations. Operators and managers may enhance production results by utilizing interactive dashboards to visualize key performance indicators (KPIs) and performance metrics. This allows them to make informed decisions, discover patterns, and take timely measures. In summary, the utilization of data analytics approaches in smart manufacturing using AI-based digital twins allows producers to attain enhanced efficiency, productivity, and adaptability in the current competitive market. Manufacturers may exploit the potential of data analytics and AI-driven insights to discover fresh prospects for innovation, expansion, and sustainability.

Visualization approaches are essential in smart manufacturing using AI-based digital twins since they offer clear and practical insights into intricate manufacturing processes (Wang *et al.*, 2024). Visualization dashboards present up-to-date information from AI-powered digital replicas, enabling operators to visually and easily track key performance indicators (KPIs), equipment conditions, and production metrics. Real-time feedback on production performance is provided through interactive charts, graphs, and gauges. This allows operators to swiftly spot issues, trends, and anomalies and promptly take necessary steps. Visualization tools facilitate the detection of anomalies and underlying causes in industrial processes by visually representing patterns, trends, and correlations within the data. Heatmaps, scatter plots, and time-series charts are used to emphasize deviations from typical operating conditions. This allows engineers to examine the root causes and apply remedial actions to avoid future occurrences. Visualization tools display predictive maintenance data produced by AI-powered digital replicas, including equipment health scores, forecasts of remaining useful life, and recommendations for maintenance. Predictive maintenance dashboards present information on the likelihood of equipment failure, timetables for maintenance, and evaluations of risk. This allows maintenance teams to prioritize jobs, allocate resources effectively, and optimize maintenance processes.

Visualization techniques are employed to display simulation findings and optimization recommendations generated by AI-based digital twins, with the aim of enhancing industrial processes. 3D models, animations, and virtual reality (VR) representations replicate production operations, material flows, and resource consumption, enabling engineers to detect bottlenecks, inefficiencies, and possibilities for enhancement. Visualization tools facilitate "what-if" scenario research, enabling engineers to investigate different process setups, layouts, and parameters in order to enhance performance and save expenses. Visualization dashboards display quality control data produced by AI-powered digital replicas, including defect rates, quality measures, and inspection outcomes. Pareto charts, histograms, and defect maps are graphical tools that display defects categorized by kind, location, and frequency. These tools allow quality control teams to detect patterns, determine underlying causes, and pinpoint areas that require enhancement. Utilizing real-time defect tracking and visualization aids in diminishing faults, enhancing product quality, and augmenting client satisfaction. Visualization techniques are used to represent supply chain data and logistics information that is created by AI-powered digital replicas, known as digital twins. This includes visualizing inventory levels, demand predictions, and transit routes. Supply chain dashboards exhibit supply chain performance indicators, such as rates of order fulfillment, turnover of inventory, and durations of lead times, allowing supply chain managers to detect inefficiencies, hazards, and possibilities for enhancement. Geographic maps, network diagrams, and flowcharts provide visual representations of supply chain networks, allowing managers to optimize transportation routes, warehouse placements, and inventory quantities in order to save expenses and enhance operational effectiveness. In summary, the utilization of visualization approaches in smart manufacturing through the implementation of AI-based digital twins allows manufacturers to acquire more profound understanding, make well-informed choices, and enhance performance throughout the entirety of the production process. Visualization technologies enable operators, engineers, and managers to promote continuous improvement, increase efficiency, and achieve operational excellence by presenting visual representations of intricate data and insights.

12.2 Smart Manufacturing and Digital Twins

Smart manufacturing is the integration of cutting-edge technologies, data-driven analysis, and intelligent automation to enhance manufacturing processes and achieve operational excellence (Leng *et al.*, 2021). The fundamental objective of smart manufacturing is to optimize efficiency, productivity, and flexibility throughout the whole production process.

12.2.1 Key Concepts of Smart Manufacturing

Smart manufacturing is an innovative method of production that utilizes cutting-edge technologies and data-driven analysis to enhance manufacturing processes, enhance operational efficiency, and foster innovation along the whole value chain, as depicted in Figure 12.1.

- *Interconnectivity:* Smart manufacturing utilizes the Internet of Things (IoT) to establish connections between physical objects, sensors, and machinery in the production setting. This network of interconnected systems allows for the continuous monitoring, gathering of information, and communication among different parts of the production ecosystem.
- *Data Analytics:* Data analytics plays a crucial role in smart manufacturing by analyzing large volumes of data collected from sensors, production equipment, and other sources. Manufacturers can employ advanced analytics techniques such as machine learning and predictive analytics to acquire actionable insights into production processes, identify patterns, and improve performance.
- *Automation and Robotics:* Automation and robotics technologies are employed to automate repetitive tasks, streamline manufacturing processes, and enhance operational efficiency in smart industrial environments. Advanced robotics systems, such as collaborative robots (cobots) and autonomous guided vehicles (AGVs), work together with human operators to do tasks in a safe and efficient way (Tyagi *et al.*, 2023).

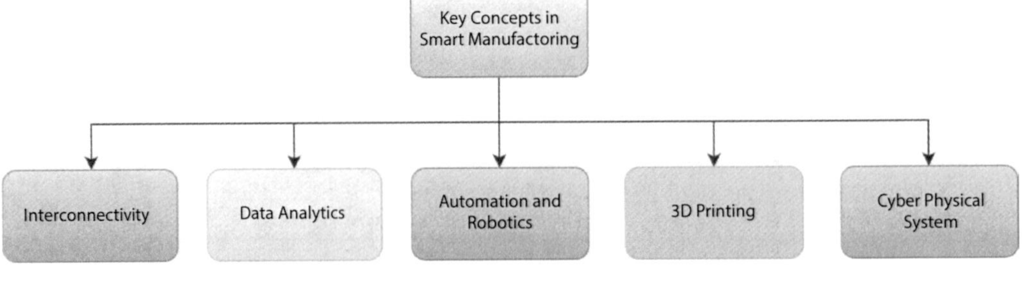

Figure 12.1 Key concepts of smart manufacturing.

- *Additive Manufacturing (3D Printing):* 3D printing and other additive manufacturing technologies enable the rapid production, customization, and fabrication of complex parts and components. Additive manufacturing, through the elimination of traditional manufacturing methods, offers more flexibility, cost-effectiveness, and design independence in the field of smart manufacturing.
- *Cyber-Physical Systems:* Smart manufacturing integrates cyber-physical systems (CPS) that establish a connection between physical processes and digital technologies. CPS combine sensors, actuators, and control systems with computer intelligence to actively monitor, assess, and optimize industrial operations in real time.

12.2.2 Role of Digital Twins in Optimizing Manufacturing Operations

Digital twins are virtual replicas of physical assets, processes, or systems that accurately mimic their real-world counterparts within a digital setting (Friederich *et al.*, 2022). Within the realm of smart manufacturing, digital twins are crucial for optimizing production operations due to the following reasons:

- *Simulation and Modeling:* Digital twins simulate and model manufacturing processes, enabling engineers to test different scenarios, optimize parameters, and predict outcomes before implementation in the physical world. By simulating production workflows, material flows, and resource utilization, digital twins help identify inefficiencies, bottlenecks, and opportunities for improvement.
- *Real-Time Monitoring and Control:* Digital twins enable the continuous monitoring and management of industrial activities through the collection of data from sensors, production equipment, and Internet of Things (IoT) devices. Digital twins facilitate predictive maintenance, quality control, and performance optimization by evaluating real-time data streams. This capability empowers manufacturers to promptly adapt to changing conditions and minimize potential hazards.
- *Predictive Analytics and Maintenance:* Digital twins utilize data analytics methodologies to forecast equipment malfunctions, anticipate maintenance requirements, and optimize maintenance timetables. Digital twins utilize historical data, sensor readings, and equipment health indicators to detect early indications of deterioration or abnormalities in machine performance. This allows for preventive maintenance and reduces the amount of time that the unit is out of service.
- *Performance Optimization:* Digital twins enhance industrial performance by analyzing production data, identifying areas for optimization, and providing recommendations for improvement. By simulating production scenarios and monitoring performance indicators, digital twins help improve production schedules, resource allocation, and workflow efficiency to increase throughput and decrease costs.

Smart manufacturing concepts utilize cutting-edge technologies and data-driven analysis to optimize manufacturing operations along the whole value chain. Digital twins are essential in smart manufacturing since they create virtual models of physical assets and processes. This allows for the modeling, prediction, and optimization of manufacturing activities, leading to enhanced efficiency, productivity, and competitiveness.

12.2.3 Benefits of Digital Twins

Digital twins offer a range of benefits in monitoring, analysis, prediction, and optimization of manufacturing processes:

- *Real-time Monitoring:* Digital twins facilitate the continuous monitoring of manufacturing processes, allowing producers to monitor key performance indicators (KPIs), equipment status, and output metrics in real-time. Manufacturers can promptly intervene and take corrective steps by visually analyzing data obtained from sensors, IoT devices, and production equipment. This enables them to detect errors, trends, and anomalies in real-time.
- *Predictive Maintenance:* Digital twins utilize historical data, sensor readings, and machine learning algorithms to forecast equipment malfunctions in advance. Digital twins can proactively schedule maintenance, limit downtime, and prevent costly unplanned shutdowns by evaluating patterns and trends in equipment performance to detect early symptoms of degradation or anomalies.
- *Process Optimization:* Digital twins enable the simulation and optimization of production processes, enabling firms to detect and address inefficiencies, bottlenecks, and areas for improvement (Min *et al.*, 2019). Digital twins utilize production data analysis and scenario simulation to enhance production schedules, resource allocation, and workflow efficiency, hence maximizing throughput and minimizing costs.
- *Quality Control:* Digital twins provide instantaneous monitoring and analysis of product quality, enabling manufacturers to promptly identify flaws, deviations, and non-conformities at the initial stages of the manufacturing process. Digital twins utilize sensor data, production parameters, and quality measures to guarantee product uniformity, adherence to quality standards, and customer contentment.
- *Supply Chain Optimization:* Digital twins provide a visual representation and enhance the efficiency of supply chain operations, encompassing tasks such as managing inventory, predicting demand, and planning logistics. Digital twins utilize supply chain data analysis and scenario simulations to optimize inventory levels, predict demand variations, and enhance transportation routes for cost reduction and efficiency enhancement.
- *Energy Management:* Digital twins utilize energy consumption data and production characteristics to monitor and optimize energy usage, hence minimizing environmental effect. Digital twins contribute to cost reduction, carbon emission minimization, and sustainability improvement by recognizing energy-intensive operations and applying energy-saving strategies.

- *Decision Support:* Digital twins offer decision support tools and insights to operators, engineers, and managers, facilitating well-informed decision-making and strategic planning. Digital twins enable stakeholders in industrial operations to make data-driven decisions and achieve continuous improvement by visualizing key performance indicators, performance measurements, and optimization recommendations.

12.3 Data Collection and Integration

Data collection and integration are essential procedures in the data management lifecycle. They encompass the acquisition and merging of data from many sources to establish a cohesive dataset for the purposes of analysis, reporting, and decision-making.

12.3.1 Sensor Data Acquisition

The procedure of gathering sensor data from manufacturing equipment and gadgets encompasses multiple stages to guarantee precise and dependable data acquisition. Below is a summary of the standard procedure illustrated in Figure 12.2.

- *Identify Sensor Requirements:* To begin, it is essential to determine the specific types of sensors required for monitoring the pertinent characteristics or variables inside the manufacturing process. These may encompass temperature sensors, pressure sensors, flow meters, vibration sensors, proximity sensors, and other similar devices. The choice of sensors is contingent upon the particular demands of the manufacturing process and the variables that need to be monitored.
- *Sensor Installation:* After choosing the sensors, they must be properly put on the production equipment or devices in suitable positions to accurately capture the needed data. Accurate installation is crucial to guarantee the sensors' effective and dependable measurement of the variables of interest.
- *Sensor Calibration:* Following the installation, it is imperative to calibrate the sensors in order to guarantee precision and uniformity in the results. Calibration entails the process of aligning the sensor readings with established reference values or standards. This aids in the elimination of any systematic inaccuracies or gradual changes in the sensor values as time progresses.

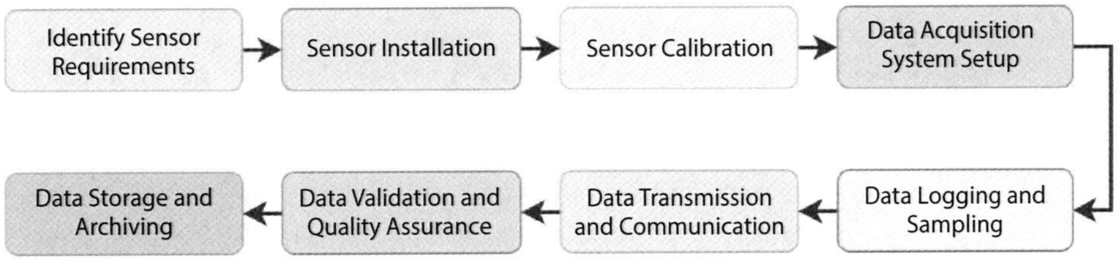

Figure 12.2 Overall process of sensor data acquisition.

- *Data Acquisition System Setup:* Subsequently, a data gathering system is established to gather and retain the sensor data. This system generally comprises of data acquisition hardware, such as data loggers or programmable logic controllers, along with software designed for data logging and display. The data acquisition system can be either linked with the control system of the manufacturing equipment or operated separately.
- *Data Logging and Sampling:* The data collection system gathers sensor data at consistent intervals, referred to as the sampling frequency or rate. The sampling frequency may vary depending on the production process needs and the monitored variables. Data logging entails the act of saving the gathered sensor data in a database or data repository for subsequent analysis.
- *Data Transmission and Communication:* Occasionally, it may be necessary to transfer sensor data instantaneously to a central monitoring system or control center for prompt analysis and decision-making. To do this, it is necessary to build dependable communication channels, such as wired or wireless networks, in order to communicate the data collected by the sensors to the data gathering system.
- *Data Validation and Quality Assurance:* After the collection of sensor data, it goes through validation and quality assurance procedures to verify its accuracy, comprehensiveness, and dependability. This task entails the identification and rectification of any inaccuracies or irregularities in the data, such as outliers, missing numbers, or measurement problems.
- *Data Storage and Archiving:* Ultimately, the sensor data is preserved and cataloged for subsequent consultation, examination, and documentation. Adhering to appropriate data storage techniques is crucial for preserving data integrity, safeguarding against security breaches, and complying with regulatory mandates.

12.3.2 Types of Sensors

- *Temperature Sensors*: These sensors measure temperature variations in the manufacturing environment and are crucial for processes where temperature control is essential, such as heat treatment or cooling processes.
- *Pressure Sensors:* Pressure sensors monitor variations in pressure levels within manufacturing equipment, which is vital for ensuring safety and efficiency in processes like hydraulic systems or pneumatic machinery.
- *Flow Sensors:* Flow sensors measure the rate of fluid flow through pipes or channels, commonly used in applications such as monitoring liquid or gas flow rates in manufacturing processes.
- *Vibration Sensors:* Vibration sensors detect mechanical vibrations in equipment, helping to identify potential faults or abnormalities in rotating machinery, motors, or bearings.
- *Proximity Sensors:* Proximity sensors detect the presence or absence of objects within a certain range, often used in automated manufacturing systems for object detection or position sensing.

- *Level Sensors:* Level sensors measure the level of liquids, solids, or powders in tanks or containers, essential for inventory management, batching, or material handling processes.
- *Humidity Sensors:* Humidity sensors measure the moisture content in the air or within materials, critical for processes where humidity control is necessary, such as in cleanrooms or humidity-sensitive manufacturing environments.

12.3.3 Data Transmission Protocols

- *Modbus:* A widely used protocol for serial communication between industrial devices, Modbus facilitates data exchange between sensors, controllers, and supervisory systems in manufacturing environments.
- *OPC UA (Unified Architecture):* OPC UA is a standardized protocol for secure and reliable data exchange in industrial automation and control systems, supporting interoperability between diverse devices and platforms.
- *Ethernet/IP:* Ethernet/IP is an industrial Ethernet protocol used for real-time communication and control in manufacturing environments, enabling high-speed data transmission and seamless integration with Ethernet-based networks.
- *Profibus:* Profibus is a fieldbus communication protocol commonly used in process automation and control systems, providing reliable data exchange between sensors, actuators, and control devices.
- *WirelessHART:* WirelessHART is a wireless communication protocol based on the HART (Highway Addressable Remote Transducer) protocol, designed for industrial applications where wired communication is impractical or costly.

12.3.4 Challenges Associated with Sensor Data Acquisition

- *Data Quality:* Ensuring the accuracy, reliability, and consistency of sensor data can be challenging due to factors such as sensor drift, calibration errors, environmental conditions, or electromagnetic interference.
- *Data Integration:* Integrating data from heterogeneous sensors and devices into a unified data acquisition system can be complex, requiring compatibility between different communication protocols, data formats, and hardware interfaces.
- *Data Security:* Protecting sensor data from unauthorized access, manipulation, or cyber-attacks is crucial for maintaining data integrity, confidentiality, and availability in manufacturing environments where sensitive information is transmitted and stored.
- *Scalability:* Scaling up sensor networks to accommodate large-scale manufacturing operations may pose challenges in terms of network infrastructure, bandwidth requirements, and data processing capabilities, particularly in distributed or remote locations.
- *Interference and Noise:* Interference from electromagnetic fields, electrical noise, or other sources can distort sensor readings and affect data accuracy,

requiring noise filtering or shielding measures to mitigate their impact on data acquisition.

- *Power Supply:* Providing reliable power sources for sensors, especially in remote or harsh environments, can be challenging, requiring solutions such as battery-powered sensors, solar panels, or energy harvesting techniques to ensure continuous operation.
- *Cost:* The cost of sensors, data acquisition systems, and communication infrastructure can be significant, particularly for deploying large-scale sensor networks or upgrading existing systems to support advanced data acquisition capabilities.

12.3.5 Data Integration and Preprocessing

The process of integrating data from numerous sources within the manufacturing ecosystem entails consolidating data from diverse sources, including sensors, equipment, databases, and enterprise systems, into a cohesive data infrastructure. This infrastructure is then used for the purposes of analysis, visualization, and decision-making. There are multiple techniques available for incorporating data from various sources into the industrial ecosystem:

- Data Standardization: Standardizing data formats, schemas, and protocols helps ensure compatibility and consistency across diverse data sources. Adopting industry-standard formats such as XML, JSON, or CSV facilitates data exchange and interoperability between different systems and applications (Tyagi *et al.*, 2023).
- Data Aggregation: Aggregating data from multiple sources involves consolidating data into a centralized repository or data warehouse for storage and analysis. This approach simplifies data management and enables comprehensive analysis of the entire manufacturing ecosystem.
- Data Transformation: Transforming data involves converting data from different formats or structures into a common format or schema to facilitate integration. This may involve data normalization, cleansing, or enrichment to ensure data consistency and quality across diverse sources.
- APIs and Web Services: Application Programming Interfaces (APIs) and web services provide standardized interfaces for accessing and exchanging data between different systems and applications. Using APIs allows seamless integration between disparate systems, enabling data sharing and collaboration across the manufacturing ecosystem.
- Middleware Platforms: Middleware platforms act as intermediaries between different systems and applications, facilitating data integration, communication, and interoperability. These platforms provide tools and services for data transformation, routing, and orchestration, enabling seamless integration of diverse data sources.
- Data Integration Tools: Data integration tools and platforms offer capabilities for extracting, transforming, and loading (ETL) data from diverse sources into a centralized data repository. These tools provide graphical interfaces,

workflow automation, and data mapping functionalities to streamline the integration process.

- Event-Driven Architecture: Event-driven architecture (EDA) enables real-time data integration and processing by reacting to events or triggers generated by diverse data sources. This approach allows for dynamic, event-driven integration of data streams from sensors, equipment, and other sources, enabling timely insights and responses to changing conditions.
- Cloud-Based Integration: Cloud-based integration platforms provide scalable and flexible solutions for integrating data from diverse sources across distributed environments. These platforms offer cloud-native services, such as data lakes, data warehouses, and integration-as-a-service (iPaaS), to facilitate seamless integration and analysis of manufacturing data.
- Semantic Interoperability: Semantic interoperability involves defining common data models, ontologies, and vocabularies to ensure shared understanding and interpretation of data across different systems and domains. By establishing semantic interoperability standards, manufacturers can enable seamless integration and interoperability of diverse data sources within the manufacturing ecosystem.
- Federated Data Architectures: Federated data architectures enable distributed data integration by maintaining data sovereignty and autonomy across different systems and organizations. This approach allows for decentralized data management while enabling seamless access and integration of data from diverse sources within the manufacturing ecosystem.

12.3.6 Techniques Commonly Used for each of these Preprocessing Tasks

- **Data Cleaning**
  - Handling Missing Values: *Imputation* involves substituting missing values with a statistical measure, such as the mean, median, or mode. *Deletion* eliminates rows or columns containing missing values if they are not significant or if there is an excessive number of missing values. *Interpolation* involves estimating missing values by utilizing neighboring data points through techniques such as linear interpolation or spline interpolation.
  - Outlier Detection and Removal: *Statistical Methods* detect and eliminate outliers by utilizing statistical measurements such as Z-score, standard deviation, or percentiles. *Visualization* utilizes data distributions and scatterplots to visually detect outliers and eliminate them depending on subject expertise.
  - Noise Removal: *Smoothing Techniques* applies smoothing filters such as moving averages or median filters to remove noise from time-series or signal data. *Clustering* identifies and remove noisy data points by clustering similar data points together and removing outliers.
- **Data Normalization**
  - Min-Max Scaling: Normalize the data by subtracting the minimum value and dividing by the range to achieve a fixed range, such as [0, 1].

- Z-Score Standardization: Standardize the data by subtracting the mean and dividing by the standard deviation to transform it into a standard normal distribution.
- Robust Scaling: Scale the data using robust statistics like median and interquartile range to mitigate the influence of outliers.
- Normalization to Unit Vector: Scale each feature to have a unit norm, which is useful for algorithms that rely on the magnitude of feature vectors, such as clustering algorithms or distance-based methods.

- **Feature Extraction**
  - Principal Component Analysis (PCA): Diminish the dimensionality of the dataset by converting it into a fresh collection of orthogonal variables (principal components) that capture the utmost volatility in the data.
  - Linear Discriminant Analysis (LDA): - Find linear combinations of features that best separate different classes or groups in the data while maximizing class separability.
  - Feature Selection: - Use statistical tests (e.g., chi-square test, ANOVA) or algorithms (e.g., recursive feature elimination) to select the most relevant features based on their importance or predictive power.
  - Text Processing Techniques: *Tokenization* refers to the process of dividing text data into individual tokens, which might be either words or phrases. *Stemming and Lemmatization* employ techniques to transform words into their base form in order to standardize text data. The *Bag-of-Words (BoW)* technique involves transforming text data into numerical vectors by considering the frequency of words in documents. The *TF-IDF (Term Frequency-Inverse Document Frequency)* method assigns weights to words in documents based on their frequency across texts, hence determining their relevance.
  - Feature Engineering: Generate novel functionalities by employing mathematical operations, merging existing attributes, or extracting domain-specific data.

12.4 AI-Based Analytics

AI-based analytics, also known as artificial intelligence (AI) analytics, refers to the utilization of artificial intelligence and machine learning techniques to examine large volumes of data, extract significant insights, and offer data-driven predictions or decisions. This technique employs advanced algorithms and computer models to uncover patterns, trends, and relationships within complex information, enabling organizations to get deep insights and make more informed decisions.

12.4.1 Predictive Maintenance

AI algorithms analyze data from digital twins to predict equipment failures and schedule maintenance proactively through a combination of data processing, pattern recognition, and predictive modeling techniques (Li *et al.*, 2014). Here's an overview of the process:

- *Data Collection:* The first step involves collecting data from sensors embedded in physical equipment and devices, as well as virtual sensors within the digital twin environment. This data may include operational parameters, environmental conditions, performance metrics, maintenance logs, and historical records.
- *Data Integration and Preprocessing:* The collected data is integrated and preprocessed to ensure consistency, completeness, and quality. This may involve tasks such as data cleaning, normalization, outlier detection, and feature engineering to prepare the data for analysis.
- *Feature Extraction and Selection:* The data is analyzed to identify significant attributes and patterns that are associated with the condition and functioning of the equipment. Feature selection approaches can be utilized to discover the most useful and predictive features for modeling.
- *Model Training:* AI algorithms, such as machine learning and deep learning models, are trained using historical data from the digital twin to learn patterns and relationships between input features and equipment failures. Supervised learning algorithms, such as classification or regression, are commonly used for predictive maintenance tasks.
- *Failure Prediction:* After the models have been trained, they are implemented to assess real-time data that is being sent from the digital twin environment. The algorithms utilize this data to generate forecasts on future equipment failures or decline in performance, relying on acquired patterns and correlations.
- *Probabilistic Risk Assessment:* Predictive maintenance algorithms evaluate the likelihood of equipment failures happening within a specific period, including criteria such equipment condition, operating conditions, environmental factors, and past failure patterns.
- *Maintenance Scheduling:* Based on the predicted risk of failure, maintenance activities are scheduled proactively to address potential issues before they escalate into costly downtime or breakdowns. Maintenance scheduling algorithms optimize the timing and prioritization of maintenance tasks to minimize disruption to operations and maximize equipment availability.
- *Feedback Loop and Model Updating:* Predictive maintenance systems incorporate a feedback loop to continuously monitor equipment performance and collect new data from the digital twin environment. This data is used to update and refine the predictive models over time, improving their accuracy and effectiveness.

Manufacturers can utilize AI algorithms to examine data from digital twins, enabling them to obtain valuable insights on the condition and performance of equipment. This allows for more precise predictions of probable failures and the ability to proactively schedule maintenance tasks, resulting in improved asset reliability and productivity optimization. Implementing this proactive maintenance strategy aids in decreasing operational downtime, minimizing expenses, and prolonging the lifespan of vital equipment and infrastructure.

12.4.2 Machine Learning and Deep Learning Techniques

Predictive maintenance in smart manufacturing employs a range of machine learning (ML) and deep learning (DL) methods to anticipate equipment malfunctions and enhance maintenance timetables (Kotsiopoulos *et al.*, 2021). Below are many frequently employed methodologies:

- **Machine Learning Techniques**
  - *Regression Analysis*: Regression models, such as linear regression or logistic regression, are used to predict continuous or categorical outcomes based on input features. They are suitable for tasks like estimating remaining useful life (RUL) or predicting the likelihood of equipment failure within a specified time frame.
  - *Classification Algorithms:* Classification algorithms, such as support vector machines (SVM), decision trees, random forests, and k-nearest neighbors (KNN), are used to categorize the health states of equipment (e.g., normal, degraded, failed) using sensor data and operating characteristics.
  - *Time Series Forecasting:* Time series forecasting methods, such as autoregressive integrated moving average (ARIMA), exponential smoothing, and seasonal decomposition, are employed to anticipate future values of sensor readings or performance measures across time. This allows for the timely identification of abnormalities or deterioration patterns.
  - *Anomaly Detection:* Anomaly detection algorithms, such as isolation forest, one-class SVM, or autoencoders, identify unusual patterns or outliers in sensor data that may indicate equipment faults, deviations from normal operation, or impending failures.

- **Deep Learning Techniques**
  - *Recurrent Neural Networks (RNNs):* Recurrent Neural Networks (RNNs), namely Long Short-Term Memory (LSTM) and Gated Recurrent Unit (GRU), are highly efficient at analyzing sequential data. They excel in tasks involving time-series sensor readings or maintenance logs, since they can effectively capture temporal dependencies. This enables them to make accurate predictions regarding future equipment states or failures.
  - *Convolutional Neural Networks (CNNs):* Convolutional neural networks (CNNs) are frequently employed for predictive maintenance jobs that involve analyzing images, specifically for identifying faults or abnormalities during visual inspections of equipment components. Additionally, they can be utilized to examine spatial data obtained from sensors, such as vibrations or audio waves.
  - *Auto-encoders:* Auto-encoders are models for unsupervised learning that aim to reconstruct input data while minimizing information loss. These methods can be employed to acquire knowledge about features, decrease the number of dimensions, and discover abnormalities by reconstructing data from sensors and detecting departures from typical patterns.

- *Generative Adversarial Networks (GANs):* GANs comprise a pair of neural networks, namely a generator and a discriminator, which engage in a competitive process to produce authentic data samples. GANs are a useful tool for generating artificial sensor data that can be used to train predictive maintenance models in situations when there is a lack of or insufficient real-world data.

- **Hybrid Models**
  - *Ensemble Methods:* Ensemble learning techniques, such as bagging, boosting, and stacking, combine multiple base models to improve predictive performance and robustness. They are used to mitigate overfitting, reduce prediction variance, and enhance the accuracy of predictive maintenance models.
  - *Hybrid Approaches:* Hybrid models combine physics-based modeling with data-driven ML techniques to leverage domain knowledge and capture complex interactions between equipment components. These models integrate sensor data with physical models of equipment behavior to improve predictive accuracy and interpretability.

By leveraging these ML and DL techniques, organizations can develop advanced predictive maintenance solutions that enable proactive identification of equipment failures, optimization of maintenance schedules, and maximization of asset reliability and performance in smart manufacturing environments.

12.4.3 Process Optimization

AI-based analytics are applied to digital twins in smart manufacturing to optimize processes, improve resource utilization, and enhance efficiency through data-driven insights and predictive capabilities (Ding *et al.*, 2020). Here's how AI-based analytics are utilized in conjunction with digital twins:

- *Real-Time Monitoring and Control:* AI systems process live data streams from physical equipment and devices, along with virtual sensors in the digital twin environment. These algorithms identify patterns, abnormalities, and deviations from anticipated behavior, enabling prompt intervention and remedial measures to enhance process efficiency and avoid interruptions.
- *Predictive Maintenance:* AI-based predictive maintenance models analyze historical and real-time data from digital twins to forecast equipment failures and schedule maintenance proactively. By predicting potential issues before they occur, organizations can minimize unplanned downtime, reduce maintenance costs, and extend the lifespan of critical assets.
- *Process Optimization:* AI algorithms analyze data from digital twins to identify inefficiencies, bottlenecks, and optimization opportunities within manufacturing processes. These insights enable organizations to streamline

workflows, optimize production schedules, and improve resource allocation to maximize productivity and minimize waste.

- *Quality Control and Defect Detection:* AI-based analytics analyze sensor data and image data from digital twins to detect defects, anomalies, and quality issues in real-time. By identifying quality deviations early in the production process, organizations can take corrective actions to reduce scrap, rework, and product defects, ensuring higher product quality and customer satisfaction.
- *Supply Chain Optimization:* AI algorithms analyze data from digital twins to optimize supply chain operations, including inventory management, demand forecasting, and logistics planning. By leveraging AI-based analytics, organizations can optimize inventory levels, reduce lead times, and improve supply chain resilience, ensuring timely delivery of materials and components to support manufacturing operations.
- *Energy Management and Sustainability:* AI-based analytics analyze energy consumption data from digital twins to identify opportunities for energy efficiency improvements and cost savings. By optimizing energy usage, organizations can reduce operational costs, minimize environmental impact, and enhance sustainability across manufacturing facilities.
- *Product Lifecycle Management:* AI algorithms analyze data from digital twins throughout the product lifecycle, from design and prototyping to production and maintenance. These insights enable organizations to optimize product design, improve manufacturing processes, and enhance product reliability and performance over time.

Overall, AI-based analytics applied to digital twins enable organizations to gain valuable insights into manufacturing processes, optimize resource utilization, and enhance operational efficiency, leading to improved productivity, quality, and competitiveness in the smart manufacturing landscape.

12.4.4 Optimization Algorithms

In smart manufacturing environments, optimization algorithms are used to improve various aspects of operations, such as production scheduling, resource allocation, inventory management, and logistics planning (Hu *et al.*, 2019). Here are some commonly used optimization algorithms and their implementation in smart manufacturing as shown in Figure 12.3.

- *Linear Programming (LP):* Linear Programming (LP) is a mathematical method employed to optimize a linear objective function while adhering to linear equality and inequality constraints. LP, or linear programming, is employed in smart manufacturing to optimize production planning, scheduling, and resource allocation. This involves establishing the most efficient production schedules, workforce assignments, and inventory levels in order to save costs or maximize throughput.
- *Integer Programming (IP):* Integer programming (IP) expands upon linear programming by allowing choice variables to be restricted to integer values. This capability enables the representation of discrete decision variables and

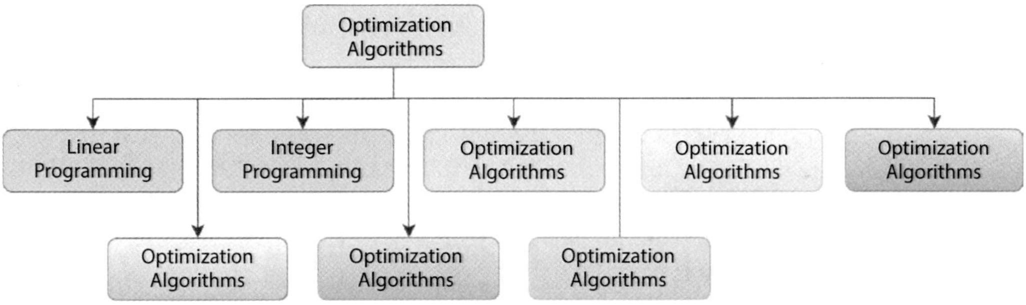

Figure 12.3 Optimization algorithms.

the solution of combinatorial optimization problems. IP, or integer programming, is utilized in smart manufacturing for several activities like production scheduling with discrete work assignments, facility location and layout optimization, and equipment selection and configuration.

- *Mixed-Integer Linear Programming (MILP):* MILP combines elements of linear and integer programming, allowing for the optimization of models with both continuous and discrete decision variables. In smart manufacturing, MILP is applied to complex optimization problems, such as production planning with discrete production quantities, resource constraints, and setup costs.
- *Genetic Algorithms (GA):* Genetic Algorithm (GA) is an optimization technique that draws inspiration from the principles of natural selection and evolution. Smart manufacturing use genetic algorithms (GA) for several purposes, including production scheduling, layout optimization, and parameter tuning in machine learning models.
- *Particle Swarm Optimization (PSO):* PSO is an optimization technique that is based on the collective behavior of birds flocking or fish schooling. PSO, or Particle Swarm Optimization, is utilized in smart manufacturing to address many challenges, including optimizing facility layout, scheduling production, and optimizing the supply chain.
- *Simulated Annealing (SA):* Simulated Annealing (SA) is a stochastic optimization method that draws inspiration from the annealing process in metallurgy. In metallurgy, materials are heated and gradually cooled to reduce imperfections. Smart manufacturing use SA for many tasks like production scheduling, facility layout optimization, and parameter tuning in machine learning models.
- *Tabu Search (TS):* TS is a metaheuristic optimization technique that systematically searches the search space by iteratively examining solutions and keeping track of previously visited solutions to prevent redundant visits. Within the realm of intelligent manufacturing, TS is utilized to address issues such as the arrangement of production schedules, allocation of resources, and optimization of the supply chain.
- *Ant Colony Optimization (ACO):* The ACO algorithm is a form of optimization that draws inspiration from the foraging behavior of ants. In this algorithm, ants communicate and locate the most efficient paths to food sources by depositing pheromone trails. ACO, or Ant Colony Optimization, is

employed in smart manufacturing for several tasks including truck routing, scheduling, and work shop optimization.
- *Reinforcement Learning (RL):* Reinforcement Learning (RL) is a type of machine learning in which an intelligent agent learns how to interact with its environment in order to maximize the total rewards it receives over a period of time. Reinforcement Learning (RL) is utilized in smart manufacturing for many tasks like dynamic scheduling, resource allocation, and control of autonomous systems.

These optimization algorithms are used in combination with mathematical models, simulation techniques, and domain-specific constraints to address complex optimization problems and improve efficiency, productivity, and competitiveness in smart manufacturing environments.

12.5 Visualization Techniques

Visualization techniques refer to methods and tools used to represent data visually, allowing users to explore, understand, and communicate insights from complex datasets. Visualization techniques aim to transform raw data into graphical representations that are intuitive, informative, and actionable.

12.5.1 Dashboard Design

Creating effective dashboards to visualize manufacturing data involves adhering to design principles and best practices that ensure clarity, usability, and actionable insights (Jwo *et al.*, 2021). Here are some key design principles and best practices for creating manufacturing dashboards:

- *Understand User Needs:* Identify the target audience for the dashboard, including operators, engineers, managers, and executives, and understand their specific information needs and goals. Tailor the dashboard design to address the requirements of different user roles, providing relevant metrics, KPIs, and visualizations that support their decision-making processes.
- *Focus on Key Metrics:* Prioritize the display of key performance indicators (KPIs) and metrics that are critical to monitoring manufacturing operations and achieving business objectives. Avoid cluttering the dashboard with unnecessary information and focus on presenting the most relevant and actionable insights to users.
- *Use Clear and Intuitive Visualizations:* Choose appropriate visualization types (e.g., line charts, bar charts, pie charts) that effectively communicate the underlying data patterns and relationships. Ensure that visualizations are clear, concise, and easy to interpret, using labels, legends, and annotations to provide context and guidance to users.
- *Provide Context and Drill-Down Capability:* Include contextual information, such as historical trends, benchmarks, and targets, to help users interpret

the data and understand its significance. Enable drill-down functionality that allows users to explore detailed information and underlying data sources behind summary metrics and visualizations.

- *Ensure Responsiveness and Interactivity:* Design dashboards that are responsive and adaptable to different screen sizes and devices, including desktops, laptops, tablets, and mobile phones. Incorporate interactive elements, such as filters, sliders, and selectors that allow users to customize views, explore data from different perspectives, and perform ad-hoc analysis.
- *Maintain Consistency and Simplicity:* Maintain a consistent visual style, layout, and color scheme across the dashboard to enhance usability and user experience. Avoid unnecessary complexity and visual clutter by simplifying the design, removing redundant elements, and focusing on the most important information.
- *Provide Real-Time Updates:* Incorporate real-time data updates and streaming capabilities to ensure that the dashboard reflects the latest information and changes in manufacturing operations. Implement automated data refresh mechanisms and alerts to notify users of significant events or deviations from expected performance.
- *Include Contextual Information:* Provide contextual information, such as equipment status, production schedules, and downtime reasons, alongside performance metrics to help users understand the factors influencing manufacturing performance. Use annotations, callouts, and tooltips to provide additional context and insights within the dashboard interface.
- *Iterate and Gather Feedback:* Continuously iterate on the dashboard design based on user feedback, usability testing, and performance evaluations to address usability issues and improve effectiveness. Solicit input from stakeholders and end-users throughout the design process to ensure that the dashboard meets their needs and expectations.

In manufacturing, key performance indicators (KPIs), data visualization techniques, and interactive dashboard features play crucial roles in monitoring and optimizing operations. Here's an overview of each:

12.5.2 Key Performance Indicators (KPIs)

Key Performance Indicators (KPIs) are measurable parameters utilized to assess the achievement or effectiveness of an organization, department, process, project, or individual in relation to certain objectives, goals, or targets. Key Performance Indicators (KPIs) are crucial for tracking advancements, pinpointing locations in need of enhancement, and making well-informed choices to propel organizational triumph.

- *Overall Equipment Effectiveness (OEE):* Measures the performance, availability, and quality of equipment, providing insights into equipment utilization and efficiency.
- *Downtime:* Tracks the amount of time equipment is offline or not in operation, helping to identify causes of downtime and prioritize maintenance efforts.

- *Production Yield:* Calculates the ratio of good-quality output to total output, indicating the efficiency and effectiveness of the manufacturing process.
- *Cycle Time:* Measures the time it takes to complete one production cycle or unit, helping to identify bottlenecks and optimize production throughput.
- *Scrap and Rework Rate:* Tracks the percentage of defective or non-conforming products, highlighting areas for quality improvement and waste reduction.
- *Inventory Levels:* Monitors the amount of raw materials, work-in-progress, and finished goods in inventory, ensuring optimal inventory management and minimizing stockouts or overstock situations.
- *Energy Consumption:* Measures the energy usage of manufacturing operations, helping to identify opportunities for energy efficiency improvements and cost savings.
- *Maintenance Costs:* Tracks the cost of preventive and corrective maintenance activities, helping to assess the effectiveness of maintenance strategies and control maintenance expenses.

12.5.3 Data Visualization Techniques

Data visualization techniques are methods used to represent data visually, making complex information more accessible, understandable, and actionable.

- *Line Charts:* Display trends and variations in KPIs over time, such as production output, equipment downtime, or energy consumption.
- *Bar Charts:* Compare performance metrics across different categories, such as comparing OEE for different production lines or comparing scrap rates for different product types.
- *Pie Charts:* Show the distribution of a total value into its component parts, such as the proportion of production downtime attributed to different downtime reasons.
- *Heatmaps:* Visualize complex data patterns and correlations, such as analyzing equipment performance across different shifts or identifying hotspots of energy consumption.
- *Gauges:* Provide a visual representation of a single KPI or metric, such as displaying OEE as a percentage or equipment uptime as a fraction.
- *Stacked Area Charts:* Illustrate the cumulative contribution of different factors to a total value, such as showing the breakdown of production downtime by downtime reason over time.
- *Scatter Plots:* Explore relationships between two variables, such as analyzing the correlation between equipment age and maintenance costs.
- *Histograms:* Display the distribution of a continuous variable, such as analyzing the distribution of cycle times or defect sizes in production.

12.5.4 Interactive Dashboard Features

Interactive dashboards are powerful tools for data visualization and analysis, allowing users to explore and interact with data dynamically (Holjevac *et al.*, 2020).

- *Filtering:* Allow users to filter data based on specific criteria, such as date ranges, production lines, or product categories.
- *Drill-Down:* Enable users to drill down from summary KPIs to detailed data and underlying metrics for further analysis.
- *Hover-over Tooltips:* Provide additional context and details about data points or visualizations when users hover over them with their cursor.
- *Data Brushing:* Allow users to select and interact with data points or regions of interest in visualizations, facilitating exploratory analysis and comparison.
- *Linked Visualizations:* Create connections between different visualizations so that interactions in one visualization update or filter data in related visualizations.
- *Dynamic Updates:* Automatically refresh data and visualizations at predefined intervals to ensure users have access to the latest information.
- *Alerts and Notifications:* Trigger alerts or notifications when KPIs or metrics exceed predefined thresholds or when significant events occur.
- *Export and Share:* Enable users to export data or visualizations in various formats (e.g., CSV, PDF) and share dashboards with colleagues or stakeholders.

12.5.5 Augmented Reality (AR) and Virtual Reality (VR)

Augmented Reality (AR) and Virtual Reality (VR) technologies provide immersive visualization capabilities that improve the comprehension and engagement with digital twins in intelligent manufacturing settings (Büttner *et al.*, 2017). Below are the different applications of augmented reality (AR) and virtual reality (VR) technologies for seeing digital twins in immersive environments.

- *Design and Prototyping*: Engineers and designers can use AR and VR to visualize digital twins of products and manufacturing facilities during the design and prototyping stages. This allows them to assess the spatial layout, ergonomics, and functionality of equipment and production processes in a virtual environment before physical implementation.
- *Training and Simulation:* AR and VR simulations provide realistic training environments for operators, technicians, and maintenance personnel to familiarize themselves with equipment operation, maintenance procedures, and emergency scenarios. Users can interact with virtual replicas of equipment and practice troubleshooting procedures in a safe and controlled environment.
- *Maintenance and Repair:* AR technologies overlay digital twin data, such as equipment status, sensor readings, and maintenance instructions, onto the physical environment in real-time. Maintenance technicians wearing AR glasses or headsets can visualize equipment health indicators, diagnostic information, and step-by-step repair instructions overlaid onto the physical equipment, facilitating faster and more accurate maintenance tasks.
- *Remote Assistance:* AR enables remote experts to provide real-time guidance and support to on-site personnel by overlaying annotations, instructions, and annotations onto the live video feed from AR-enabled devices. This allows

experts to remotely diagnose problems, provide troubleshooting advice, and guide maintenance procedures, reducing the need for travel and minimizing downtime.
- *Data Visualization and Analysis:* VR environments provide immersive data visualization capabilities for exploring complex datasets and digital twin representations. Users can navigate through 3D models of manufacturing facilities, equipment layouts, and process flows, interact with data visualizations, and gain insights into operational performance and optimization opportunities.
- *Collaborative Decision-Making:* AR and VR technologies facilitate remote collaboration and decision-making by enabling many stakeholders to perceive and interact with digital replicas in shared virtual worlds. This promotes cross-functional collaboration, the exchange of knowledge, and the establishment of agreement among teams engaged in the design, planning, and implementation of manufacturing processes.
- *Customer Engagement and Sales:* VR experiences allow customers to virtually explore and interact with digital twins of products, manufacturing facilities, and production processes. This immersive visualization enhances customer engagement, facilitates product demonstrations, and supports sales and marketing efforts by providing a compelling and interactive way to showcase capabilities and features.

In smart manufacturing environments, the use of AR and VR technologies greatly improves the visualization of digital twins. This allows for immersive experiences in various areas like as design, training, maintenance, collaboration, and customer interaction. Through the utilization of these technologies, organizations can enhance efficiency, productivity, and decision-making throughout the whole product lifecycle. Augmented Reality (AR) and Virtual Reality (VR) are extensively used in training, maintenance, and design review processes across several industries such as manufacturing, construction, healthcare, and automotive (Kaplan *et al.*, 2021; Brown *et al.*, 2023; Nee *et al.*, 2012). Here are many precise applications:

- **Training**
  - *Equipment Operation*: AR and VR simulations can provide realistic training environments for operators to learn how to operate complex machinery, equipment, and systems. Users can practice tasks such as equipment setup, calibration, and operation in a safe and controlled virtual environment.
  - *Safety Training:* AR and VR can simulate hazardous or high-risk scenarios, such as fire emergencies, chemical spills, or electrical hazards, to train personnel on safety protocols, evacuation procedures, and emergency response techniques.
  - *Soft Skills Training:* VR simulations can be used to train soft skills, such as communication, teamwork, and leadership, by simulating interactive scenarios and social interactions in virtual environments.
  - *Onboarding and Familiarization:* New employees can use AR and VR to familiarize themselves with the workplace environment, equipment layout, and standard operating procedures before starting their job roles, improving efficiency and reducing the learning curve.

- **Maintenance**
  - *Remote Assistance:* Augmented reality (AR) allows maintenance technicians to obtain immediate guidance and assistance from remote specialists by superimposing instructions, notes, and digital twin data onto the physical surroundings. This enhances the effectiveness of problem-solving, decreases the amount of time that systems are not functioning, and eliminates the necessity for in-person visits.
  - *Interactive Manuals:* AR applications can provide interactive maintenance manuals and guides that overlay step-by-step instructions, diagrams, and videos onto the physical equipment, assisting technicians in performing routine maintenance tasks and repairs.
  - *Training Simulations:* Virtual reality simulations enable technicians to practice diagnostic procedures, equipment inspections, and repair jobs in a virtual environment, replicating maintenance scenarios and processes. This allows them to gain experience and proficiency before carrying out these duties on real equipment.
  - *Predictive Maintenance Visualization:* AR and VR can visualize data from digital twins and predictive maintenance models, such as equipment health indicators, sensor readings, and fault predictions, enabling maintenance technicians to monitor equipment status and plan maintenance activities proactively.

- **Design Review Processes**
  - *Collaborative Design Reviews*: VR environments enable multidisciplinary teams to collaborate and review 3D models, CAD drawings, and design prototypes in immersive virtual spaces. Participants can interact with the models, annotate design changes, and make real-time decisions in a collaborative setting.
  - *Spatial Visualization:* AR and VR provide spatial visualization capabilities that allow designers and engineers to assess the scale, proportion, and spatial relationships of design elements, components, and assemblies in a virtual environment.
  - *Ergonomic Analysis:* VR simulations can simulate human interactions with products and environments to evaluate ergonomics, human factors, and user experience. Designers can assess factors such as reachability, visibility, and comfort to optimize product design and usability.
  - *Conceptual Design Exploration:* VR environments enable designers to explore multiple design concepts, configurations, and alternatives in a virtual space, facilitating rapid iteration and decision-making in the design process.

12.6 Case Studies and Applications

Here, Some useful case studies and real-world applications of data analytics and visualization in smart manufacturing using AI-based digital twins as given below.

12.6.1 Case Study 1: Siemens Digital Twin for Automotive Manufacturing

Siemens employed a digital twin solution in car manufacture, generating virtual duplicates of production lines to model and enhance manufacturing processes (Annanth *et al.*, 2021). The digital twin utilizes real-time data from sensors, IoT devices, and industrial equipment to actively monitor performance, identify anomalies, and forecast equipment malfunctions. Data analytics methodologies were utilized to examine past and current data, detect areas for improvement, and enhance operational effectiveness and quality. The visualization tools included interactive dashboards and 3D simulations for visualizing industrial activities, analyzing performance data, and optimizing production schedules.

12.6.2 Case Study 2: GE Aviation's Predix Platform for Aircraft Engines

GE Aviation created a digital twin platform named Predix specifically for aviation engines. This platform utilizes AI-driven analytics to oversee engine performance, forecast maintenance requirements, and enhance fuel efficiency (Bouzidi *et al.*, 2020). The digital twin utilizes sensor data, flight data, maintenance records, and environmental conditions to generate virtual replicas of aircraft engines and evaluate their condition and operational state. The data analytics algorithms examined the engine data to detect any abnormalities, identify patterns of performance decline, and forecast upcoming breakdowns. This allowed for preventative maintenance measures to be implemented. The utilization of visualization techniques offers user-friendly dashboards and interactive visualizations for the purpose of monitoring engine health, analyzing performance parameters, and optimizing maintenance plans to enhance reliability and achieve cost savings.

12.6.3 Case Study 3: ABB Ability™ Digital Twin for Industrial Plants

ABB created the ABB Ability™ Digital Twin platform for industrial facilities, incorporating AI-driven analytics and visualization capabilities to enhance plant operations and maximize asset performance (Platenius-Mohr *et al.*, 2019). The digital twin integrates real-time data from sensors, control systems, and production processes to generate virtual reproductions of industrial assets and systems. The operational data was examined by machine learning algorithms to detect inefficiencies, optimize process parameters, and forecast equipment breakdowns. This allowed for proactive maintenance and performance optimization. The visualization tools offered encompassed immersive 3D models, interactive dashboards, and augmented reality overlays. These techniques were employed to visually represent plant operations, track essential performance indicators, and aid plant operators and maintenance teams in making informed decisions.

12.6.4 Case Study 4: Bosch Rexroth's IoT Gateway and Analytics Solution

Bosch Rexroth developed an IoT gateway and analytics solution for smart manufacturing, which integrated AI-based analytics and visualization tools to optimize production processes and enhance equipment reliability (Madasamy *et al.*, 2021). The IoT gateway collected data from sensors, actuators, and control systems installed on manufacturing equipment to monitor performance, detect anomalies, and predict maintenance needs.

Data analytics algorithms analyzed sensor data to identify patterns, trends, and correlations related to equipment health, energy consumption, and production efficiency. Visualization tools provided real-time dashboards, trend charts, and heatmaps to visualize manufacturing data, analyze performance metrics, and optimize production workflows for improved productivity and cost savings.

12.6.5 Insights Based on the Use Cases and Industry Trends

The implementation of digital twin solutions entails the integration of diverse technologies, including IoT sensors, data analytics platforms, AI algorithms, and visualization tools, into pre-existing manufacturing processes. Adapting digital twin solutions is frequently essential to fulfill certain industry demands, manufacturing settings, and corporate goals. Successful implementation requires vital collaboration across cross-functional teams, encompassing engineers, data scientists, IT professionals, and domain specialists. Digital twins provide the continuous monitoring, analysis, and improvement of production processes, resulting in heightened productivity, less downtime, and improved resource allocation. Utilizing AI-driven analytics on digital replicas enables firms to anticipate equipment malfunctions, reduce unexpected periods of inactivity, and prolong the lifespan of assets through proactive maintenance. Visualization solutions offer user-friendly dashboards, interactive visualizations, and augmented reality overlays that empower stakeholders to make informed decisions based on data, optimize production schedules, and promptly adapt to evolving market demands. Organizations can attain substantial cost savings and operational efficiencies by optimizing processes, minimizing maintenance costs, and enhancing asset performance.

Ensuring data accuracy, consistency, and interoperability across disparate systems and data sources is a common challenge in digital twin projects. Organizations need to invest in data governance, data integration, and data quality management practices. As digital twin implementations scale to encompass larger production facilities or more complex systems, managing the scalability and complexity of the underlying technologies becomes increasingly important. Modular and flexible architectures are essential to accommodate evolving requirements. Adopting digital twin technologies often requires cultural and organizational changes to embrace data-driven decision-making, collaboration across departments, and continuous improvement initiatives. Change management strategies and employee training programs are critical for successful adoption. Protecting sensitive data, intellectual property, and proprietary information from cybersecurity threats and unauthorized access is a paramount concern in digital twin projects. Implementing robust security measures, encryption protocols, and access controls is essential to mitigate risks. Digital twin projects should be viewed as iterative processes that evolve over time through continuous improvement and feedback loops. Organizations should leverage lessons learned from previous implementations to refine their strategies, optimize performance, and achieve greater business value.

12.7 Challenges and Future Directions

Challenges and future directions in data analytics and visualization for smart manufacturing using AI-based digital twins revolve around addressing technical, organizational, and societal challenges while leveraging emerging technologies and trends (Rathore *et al.*, 2021).

12.7.1 Challenges

Ensuring the quality, consistency, and interoperability of data from heterogeneous sources, including sensors, IoT devices, and legacy systems, remains a significant challenge. Scaling digital twin implementations to encompass larger production facilities, more complex systems, and interconnected supply chains presents scalability and complexity challenges. Ensuring the interpretability, transparency, and trustworthiness of AI models used in digital twin applications is essential for gaining user acceptance, regulatory compliance, and stakeholder trust. Protecting sensitive data, intellectual property, and proprietary information from cybersecurity threats, data breaches, and unauthorized access is a paramount concern in smart manufacturing environments. Enhancing human-machine interaction, collaboration, and decision-making in digital twin environments requires intuitive visualization tools, user-friendly interfaces, and collaborative platforms. Addressing ethical, societal, and regulatory implications of AI-based analytics and digital twin technologies, including concerns about job displacement, algorithmic bias, and data privacy. Embracing continuous innovation, adaptation, and agility to keep pace with rapid technological advancements, market dynamics, and evolving customer needs

12.7.2 Future Direction

The implementation of digital twin solutions necessitates the integration of diverse technologies, including IoT sensors, data analytics platforms, AI algorithms, and visualization tools, into pre-existing manufacturing processes. Personalization of digital twin solutions is frequently essential to fulfill certain industry demands, manufacturing settings, and company goals. Successful implementation requires collaboration among cross-functional teams, comprising engineers, data scientists, IT professionals, and domain specialists. Digital twins provide the continuous monitoring, analysis, and optimization of production processes, resulting in heightened productivity, less downtime, and improved resource use. Utilizing AI-powered analytics on digital twins facilitates preventative maintenance, enabling firms to anticipate equipment malfunctions, reduce unexpected periods of inactivity, and prolong the lifespan of assets. Visualization technologies offer user-friendly interfaces, dynamic visual representations, and augmented reality overlays that empower stakeholders to make informed decisions based on data, improve production schedules, and promptly adapt to shifting market needs. Organizations can realize substantial cost savings and operational efficiencies by streamlining processes, decreasing maintenance costs, and enhancing asset performance.

12.7.3 Emerging Trends

Emerging trends such as edge computing, federated learning, and advanced visualization techniques have the potential to revolutionize the future of manufacturing by enabling real-time data processing, distributed intelligence, and immersive analytics capabilities.

- *Edge Computing:* Edge computing brings computing resources closer to the data source, enabling real-time processing, analysis, and decision-making at the edge of the network. In manufacturing, edge computing allows for localized data processing and analytics within industrial equipment, sensors, and devices, reducing latency, bandwidth usage, and dependence on centralized infrastructure. Edge computing enables predictive maintenance, anomaly detection, and optimization algorithms to run directly on manufacturing equipment, leading to faster response times, improved reliability, and reduced downtime. Edge computing also supports edge AI applications, such as machine learning inference and pattern recognition, enabling intelligent automation, quality control, and adaptive manufacturing processes.
- *Federated Learning:* Federated learning is a decentralized machine learning approach where model training is performed collaboratively across multiple edge devices or data sources without centralized data aggregation. In manufacturing, federated learning enables collaborative model training using data from distributed sensors, machines, and production facilities while preserving data privacy and security. Manufacturers can leverage federated learning to train predictive maintenance models, quality control algorithms, and production optimization models using data from diverse sources without exposing sensitive information or proprietary knowledge. Federated learning promotes data sovereignty, compliance with data regulations, and trust among stakeholders, fostering collaboration and knowledge sharing while preserving data ownership and privacy.
- *Advanced Visualization Techniques:* Advanced visualization techniques, such as augmented reality (AR), virtual reality (VR), and mixed reality (MR), provide immersive and interactive ways to visualize, analyze, and interact with manufacturing data and digital twins. In manufacturing, AR and VR technologies enable technicians, engineers, and operators to visualize equipment layouts, assembly instructions, and process flows in 3D space, improving comprehension, decision-making, and training effectiveness. Advanced visualization techniques enhance situational awareness, spatial understanding, and human-machine interaction in manufacturing environments, facilitating intuitive data exploration, troubleshooting, and collaborative decision-making. AR and VR applications also support remote assistance, training simulations, and design reviews, enabling distributed teams to collaborate, learn, and innovate in virtual environments, regardless of geographical location.

The convergence of edge computing, federated learning, and advanced visualization techniques holds tremendous potential to transform manufacturing operations, enabling intelligent, agile, and resilient production systems that can adapt to changing market demands, optimize resource utilization, and drive innovation in the digital age. As these technologies continue to mature and proliferate, manufacturers will increasingly leverage them to unlock new opportunities for efficiency, competitiveness, and sustainability in the global marketplace.

12.8 Conclusion

Data analytics and visualization in smart manufacturing using AI-based digital twins hold immense potential for revolutionizing manufacturing operations. Addressing challenges such as data integration, scalability, complexity, real-time analytics, and security will be crucial for realizing this potential. Current research results demonstrate promising applications in predictive maintenance, process optimization, quality control, and supply chain management, while future research directions include advancements in AI algorithms, edge computing, interoperability standards, cybersecurity, and human-machine collaboration.

References

1. Annanth, V.K., Abinash, M., Rao, L.B., Intelligent manufacturing in the context of industry 4.0: A case study of siemens industry. *J. Phys. Conf. Ser.*, IOP Publishing, 1969, 1, 012019, 2021, July.
2. Bouzidi, Z., Terrissa, L.S., Zerhouni, N., Ayad, S., An efficient cloud prognostic approach for aircraft engines fleet trending. *Int. J. Comput. Appl.*, *42*, 5, 514–529, 2020.
3. Brown, C., Hicks, J., Rinaudo, C.H., Burch, R., The use of augmented reality and virtual reality in ergonomic applications for education, aviation, and maintenance. *Ergon. Des.*, *31*, 4, 23–31, 2023.
4. Büttner, S., Mucha, H., Funk, M., Kosch, T., Aehnelt, M., Robert, S., Röcker, C., The design space of augmented and virtual reality applications for assistive environments in manufacturing: a visual approach, in: *Proceedings of the 10th International Conference on PErvasive Technologies Related to Assistive Environments*, 2017, June, pp. 433–440.
5. Ding, H., Gao, R.X., Isaksson, A.J., Landers, R.G., Parisini, T., Yuan, Y., State of AI-based monitoring in smart manufacturing and introduction to focused section. *IEEE/ASME Trans. Mechatron.*, *25*, 5, 2143–2154, 2020.
6. Friederich, J., Francis, D.P., Lazarova-Molnar, S., Mohamed, N., A framework for data-driven digital twins of smart manufacturing systems. *Comput. Ind.*, *136*, 103586, 2022.
7. Holjevac, M. and Jakopec, T., Web application dashboards as a tool for data visualization and enrichment, in: *2020 43rd International Convention on Information, Communication and Electronic Technology (MIPRO)*, 2020, September, pp. 1740–1745, IEEE.
8. Hu, Y., Zhu, F., Zhang, L., Lui, Y., Wang, Z., Scheduling of manufacturers based on chaos optimization algorithm in cloud manufacturing. *Rob. Comput. Integr. Manuf.*, *58*, 13–20, 2019.
9. Jwo, J.S., Lin, C.S., Lee, C.H., An interactive dashboard using a virtual assistant for visualizing smart manufacturing. *Mob. Inf. Syst.*, 1–9, 20212021.
10. Kaplan, A.D., Cruit, J., Endsley, M., Beers, S.M., Sawyer, B.D., Hancock, P.A., The effects of virtual reality, augmented reality, and mixed reality as training enhancement methods: A meta-analysis. *Hum. Factors*, *63*, 4, 706–726, 2021.
11. Kotsiopoulos, T., Sarigiannidis, P., Ioannidis, D., Tzovaras, D., Machine learning and deep learning in smart manufacturing: The smart grid paradigm. *Comput. Sci. Rev.*, *40*, 100341, 2021.
12. Leng, J., Wang, D., Shen, W., Li, X., Liu, Q., Chen, X., Digital twins-based smart manufacturing system design in Industry 4.0: A review. *J. Manuf. Syst.*, *60*, 119–137, 2021.
13. Li, H., Parikh, D., He, Q., Qian, B., Li, Z., Fang, D., Hampapur, A., Improving rail network velocity: A machine learning approach to predictive maintenance. *Transp. Res. Part C Emerging Technol.*, *45*, 17–26, 2014.

14. Madasamy, N., Aishvarya, B., Kumar, K., Vinith, K., Android application for accessing bosch rexroth PLC. *Int. J. Res. Eng. Sci. Manage.*, *3*, 4, 2020.
15. Mihai, S., Yaqoob, M., Hung, D.V., Davis, W., Towakel, P., Raza, M., Nguyen, H.X., Digital twins: A survey on enabling technologies, challenges, trends and future prospects. *IEEE Commun. Surv. Tutorials*, 2022.
16. Min, Q., Lu, Y., Liu, Z., Su, C., Wang, B., Machine learning based digital twin framework for production optimization in petrochemical industry. *Int. J. Inf. Manage.*, *49*, 502–519, 2019.
17. Nee, A.Y., Ong, S.K., Chryssolouris, G., Mourtzis, D., Augmented reality applications in design and manufacturing. *CIRP Ann.*, *61*, 2, 657–679, 2012.
18. Platenius-Mohr, M., Malakuti, S., Grüner, S., Goldschmidt, T., Interoperable digital twins in iiot systems by transformation of information models: A case study with asset administration shell, in: *Proceedings of the 9th International Conference on the Internet of Things*, 2019, October, pp. 1–8.
19. Rathore, M.M., Shah, S.A., Shukla, D., Bentafat, E., Bakiras, S., The role of ai, machine learning, and big data in digital twinning: A systematic literature review, challenges, and opportunities. *IEEE Access*, *9*, 32030–32052, 2021.
20. Tyagi, A.K., Fernandez, T.F., Mishra, S., Kumari, S., Intelligent Automation Systems at the Core of Industry 4.0, in: *Intelligent Systems Design and Applications. ISDA 2020. Advances in Intelligent Systems and Computing*, vol. 1351, A. Abraham, V. Piuri, N. Gandhi, P. Siarry, A. Kaklauskas, A. Madureira (Eds.), Springer, Cham, 2021, https://doi.org/10.1007/978-3-030-71187-0_1.
21. Tyagi, A.K., Lakshmi Priya, R., Mishra, A.K., Balamurugan, G., Industry 5.0: Potentials, Issues, Opportunities, and Challenges for Society 5.0, in: *Privacy Preservation of Genomic and Medical*, pp. 409–32, 2023.
22. Tyagi, A.K. and Richa, Digital Twin Technology: Opportunities and Challenges for Smart Era's Applications, in: *Proceedings of the 2023 Fifteenth International Conference on Contemporary Computing (IC3-2023)*, Association for Computing Machinery, New York, NY, USA, pp. 328–336, 2023, https://doi.org/10.1145/3607947.3608015.
23. Wang, B., Zhou, H., Li, X., Yang, G., Zheng, P., Song, C., Wang, L., Human Digital Twin in the context of Industry 5.0. *Rob. Comput. Integr. Manuf.*, *85*, 102626, 2024.
24. Xia, D., Shi, J., Wan, K., Wan, J., Martínez-García, M., Guan, X., Digital Twin and Artificial Intelligence for Intelligent Planning and Energy-Efficient Deployment of 6G Networks in Smart Factories. *IEEE Wireless Commun.*, *30*, 3, 171–179, 2023.

Part 3

ISSUES AND CHALLENGES TOWARDS AI AND DIGITAL TWIN-BASED SMART MANUFACTURING

13

The IoT of Robotics: The Frontier of Automation

G. Rithik[1] and A. Kathirvel[2*]

[1]AS Level, The Lords' International School, Chennai, Tamil Nadu, India
[2]Department of Computer Science and Engineering, Panimalar Engineering College, Chennai, Tamil Nadu, India

Abstract
The idea of the Internet of Things (IoT) is quickly developing and influencing new advancements in a variety of application domains, including the Internet of Mobile Things (IoMT), Autonomous Internet of Things (A-IoT) and the Internet of Robotic Things (IoRT), among others. The IoT influence presents new design and implementation challenges in a variety of fields, including seamless platform integration, context-based cognitive network integration, new mobile sensor/actuator network paradigms, architectural domains for smart farming, infrastructure, healthcare, agriculture, business, and commerce. Applications for automation in the Internet of Robotic Things (IoRT) are numerous and are developing quickly. IoRT blends the strength of robots and the Internet of Things (IoT), resulting in creative solutions for a range of sectors. As we ensure the authenticity of the content in this introduction, we shall investigate the wide range of IoRT automation applications. IoRT automation refers to a broad range of endeavors that use connected gadgets, sensors, and autonomous machinery to improve production, efficiency, and safety across a variety of industries. As technology develops, the range of IoRT automated applications also keeps growing. It has the ability to transform industries, enhance the standard of life, and tackle difficult problems. This overview offers a glimpse into the intriguing and diverse world of IoRT automation.

Keywords: A-IoT, IoMT, IoRT, automation and Industry 4.0

13.1 Internet of Robotic Things Automation

13.1.1 Artificial Intelligence

The realm of IoT and autonomous operations has shown great potential with the emergence of IoRT technology. Communication-centered robots that connect to wireless sensors and other network resources are becoming increasingly popular in the world of robotics. These robots can easily integrate with wired or IoT networks, allowing for the seamless utilization of IoRT's autonomous functions in this field. The IoRT technology has shown promising

**Corresponding author*: ayyakathir@gmail.com

Amit Kumar Tyagi, Shrikant Tiwari, Senthil Kumar Arumugam and Avinash Kumar Sharma (eds.) Artificial Intelligence-Enabled Digital Twin for Smart Manufacturing, (281–304) © 2024 Scrivener Publishing LLC

results in the domains of IoT and autonomous operations. The latest trend in robotics is centered around.

Communication-oriented robots that can connect with sensors and other network resources wirelessly. These robots can integrate with either wired or IoT networks, and the self-sufficient operations of the IoRT technology can be seamlessly leveraged in this industry. The incorporation of cutting-edge sensing, actuating, communication, and computing technologies elevates the original concept of IoT to new heights. This results in enhanced operational efficiency, enables businesses to uncover fresh prospects, and empowers them to predict potential hazards. These advancements present unprecedented opportunities for both consumers and providers of IoT and robotics solutions as shown in the Figure 13.1.

The Technical Committee on Networked Robots of the IEEE Robotics and Automation Society defines two types of networked robots:

- Remote-operated robots
  Networked systems make it simple for human managers to issue directives and get feedback, boosting the availability of crucial resources for study, instruction, and public awareness. This connectivity between robots and humans has a significant impact on shared responsibilities, including teleoperation and human-robot interaction. Additionally, the ability to reprogram and adapt robots on the network is affected. These technologies are now widely available for remote meeting assistance and telepresence medical equipment. Cloud robotic systems allow robots to exchange knowledge and learn from each other, while cloud infrastructure provides elastic resources to help robots overcome limitations in networked robotics.
- Automated robots
  In highly advanced systems, robots and sensors are able to communicate with one another with minimal human intervention. This network of sensors

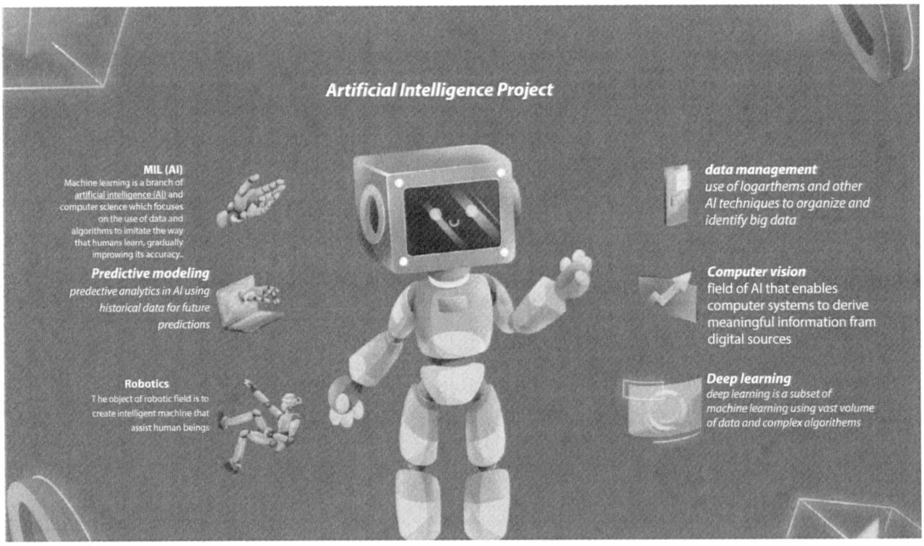

Figure 13.1 Applications of IoT.

expands the robots' sensing abilities, enabling them to interact over long distances and plan their actions accordingly. To prolong the lifespan and usefulness of the sensor network, robots can deploy, repair, and maintain it. However, a significant challenge in networked robots is establishing a scientific foundation that connects communication in order to control and enable new capabilities. Updating functionality and operations, whether remotely or locally, typically require specialized knowledge and longer maintenance periods in closed systems without open interfaces or communication channels.

By utilizing open interfaces, security and efficiency can be guaranteed through this approach. For networked robots to share information among themselves and with powerful workstations used for intensive offline processing, wireless networks are essential. These technologies encounter a number of technological obstacles, including installation, inspection, safety, the localization of operations sensor and actuators fusion, networking noise, trustworthiness, congestion, latency, stability, and range and power constraints.

The Internet of Robotic Things is a complex field that requires careful consideration of these issues. Fortunately, new hardware, software, and communication standards are emerging, providing new capabilities and possibilities.

IoT technologies and applications are fundamentally changing people's and society's perspectives on technology and business. It is vital to incorporate new IoT concepts, designs, and techniques into the conception and creation of open IoT platforms in order to support future advancements where IoT services and infrastructure will collide with artificial and autonomous system technologies. New apps, increased functionality, new business models, and investment opportunities will all be made possible by this [1].

13.2 Converging Sensing/Actuating Information Network

There exists a wide variety of products that differ in their level of intricacy, sensing and acting capabilities, communication, processing power, intelligence, mobility, and platform compatibility. When we use the term "robotic," we are referring to a specific subset of these products that are highly advanced, intelligent, and self-governing. These robotic devices rely on both edge computing and cloud IoT platforms, and incorporate cutting-edge robotics and artificial intelligence techniques to function.

In order to provide specialized solutions for certain jobs, the Internet of Robotic Things (IoRT) is a sophisticated network of devices that uses path and motion planning as well as motion control. This autonomous system architecture is integrated into the larger IoT architecture, and is defined by six fundamental characteristics.

- Sensing In the realm of IoT and robotics, the term "sensing" is frequently utilized to describe the capacity of devices or "Things" to communicate with other IoT devices and individuals. This communication is typically unidirectional, from the device to the human, which empowers people to participate in the IoT ecosystem based on their own IoT concept or service paradigm. Thanks to extensive research into the flexibility of this feature, "Sensing-as-a-Service" has been integrated into numerous IoT solutions [2].

- Actuating While it is widely recognized that actuating based on a comprehensive approach is a valuable aspect of IoT verticals, it has yet to be made available in the open market. This feature empowers devices to take action in response to physical and/or virtual activities. To ensure that actuating is both reliable and secure, services must be established to support the open, multi-vendor development, deployment, and operation of IoT applications [3]. Through extensive research, the concept of "Actuation has emerged as a new prototype for IoT, providing a user-friendly experience that encourages acceptance and engagement for managed IoT devices. Innovative installations are necessary to enable actuating and bring this technology to life.
- Control This is a structured set of activities, mainly at the application layer that operates in a loop or series of loops referred to as Control Loops, which define the functions and services. To ensure proper sequencing mechanisms, the architecture's comprehensive security concepts must be reflected in the interface definitions. These interfaces are crafted to grant access to sensing data and control mechanisms. Within the control loop, various entities such as applications and cloud services can be integrated.
- Planning The orchestration feature plays a crucial role in streamlining the logic that interconnects various internal components of the platform, enabling seamless completion of service requests. It guarantees that high-quality standards are upheld throughout the entire life-cycle of the IoT application. To ensure that service requirements are in accordance with processed data and knowledge entities as well as platform-specific representation, the planning process employs a robotic sequence engine based on logic. The recording of user-defined representations of data and resources by the orchestration logic facilitates the server definition process.
- Perception The interdisciplinary field of robotics involves integrating sensor data with knowledge modeling for enhanced functionality. Robots use a variety of techniques, including human-interface layout, software development, cloud-based architectures, analytics of data, multiple-agent systems, machine sensing systems, and occasionally artificial intelligence, to build seamless contact between themselves and people. Robots may identify and understand their surroundings through perception, freeing people to perform specialized activities.
- Cognition This feature equips the robot with monitoring and sensing capabilities, allowing it to gather sensor data from various sources and utilize both local and distributed intelligence. This demonstrates the device's impressive capabilities, as it can analyze data from observed events, often employing edge or fog computing. These foundational elements support the device's third capability, which involves autonomous decision-making for actions such as physical manipulation or control in the real world. The device can navigate and operate in its designated physical environment, as well as notify or alert based on physical event analysis. The following subsections offer an overview of the IoRT technologies utilized in creating, utilizing, and deploying IoRT applications [4].

13.2.1 Sensors and Actuators

IoT and robotics rely on two fundamental technologies: sensors and actuators. These components are crucial for IoRT systems to function properly, as they provide essential functionalities and well-defined interfaces for identification and reaction. In contrast to IoT, the practical functionality of IoRT building blocks is based on sensors and actuators. Robotic Interaction Services (RoIS) define HRI functions and the usage of external building blocks, as well as abstracting the hardware in service robots. HRI functions, such as facial recognition and wheel control, rely on sensors and actuators like radar, lidar, cameras, and microphones. These components are physically implemented on the robot or in the environment, enabling logical functional aspects such as person detection and identification. This standard allows for the development of HRI apps that can be used on both devices and gateways. Actuation is a crucial aspect of robotics that enables the manipulation of items, movement of people and products, and even the functioning of automated doors. Numerous actuation approaches, including automated strategy and execution by multiple robots, have been developed in the field of robotics.

13.2.2 Emerging IORT Technologies

Cost effective a solid-state semiconductors (CMOS) photographic sensors with proactive illumination are needed for IoRT applications to operate correctly in a variety of weather situations, including sunlight, darkness, rain, fog, and dust. These sensors must have high accuracy and resolution for both vertical object identification and horizontal road surface scanning. Most sensors now only provide 2D sensing data, therefore sensor fusion is primarily concerned with 2D representation. However, more height data, 3D visualization, and the combining of actuators and sensors are needed to better future IoRT activities. Robotic objects require a customized environment model that utilizes existing and new sensor technology in order to generate an ideal 3D environmental model that balances resource consumption and maximized performance. LIDAR systems, which use a revolving, scanning mirror to provide a complete picture, give autonomous robotic objects and cars 360-degree vision. For self-driving, autonomous robotic things, LIDAR systems provide precise 3D data on the immediate surroundings that can be utilized for obstacle avoidance, motion vector analysis, object recognition, and collision prediction. However, LIDAR systems are ineffective for close-range control, so radars must be added to autonomous robotic objects and vehicles. The 76-81 GHz frequency range is typically used for radar because of its RF propagation characteristics and necessary resolution. The radar equipment is small, emits less power, and has a lower risk of mutual interference, making it an effective method for collision avoidance in conditions such as smoke, dust, or other weather conditions [5].

IoT devices around the world are taking over non IoT devices. The US, China, as well as the developed countries in the European Union and the Asia Pacific region, are leading IoT markets. The region is predicted to lead in expenditure and uptake in the worldwide market, in line with other APAC-related trends. According to market predictions, China will account for two-thirds of the industrial IoT market [6] in the APAC region by 2025 as shown in the Figure 13.2.

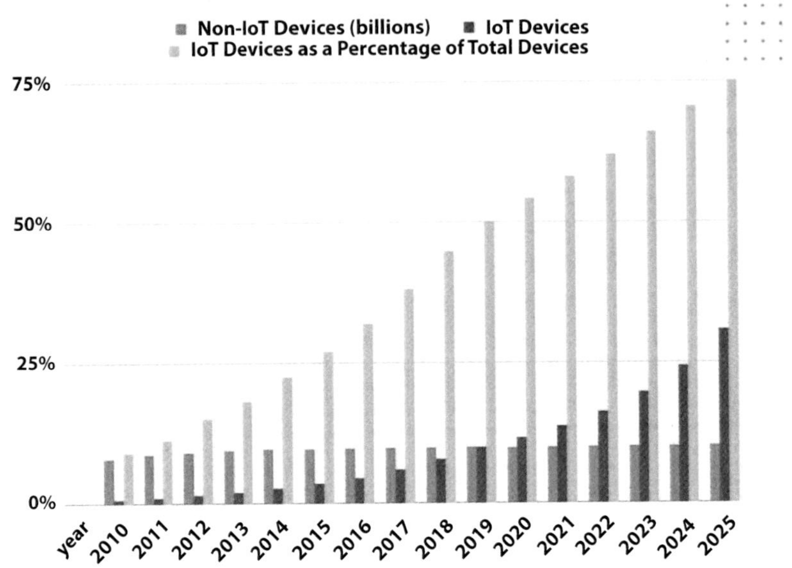

Figure 13.2 Yearly shift from non-IoT to IoT devices.

13.2.3 Communication Technologies

The communication infrastructure of IoRT necessitates novel methods in order to allow virtualization of operations on current computing engines and facilitate the usage of such infrastructures in diverse fields. These strategies should make it possible to exchange data streams (important for 3D-awareness and imaging systems), communicate internally, and use edge computing. Time-critical communication is essential for collision avoidance, which significantly lowers accidents and collisions. IoRT commonly operates local robots using networking technology and remote robots using designated white spectrum frequencies. IoT employs machine-to-machine communication and implements emerging protocols like LoRa and SIGFOX, as well as 4G, Wi-Fi, and Bluetooth [7]. However, this level of interoperability and service development is more challenging and requires semantic data from other disciplines. Due in part to the dynamic nature of IoRT conditions and their dependence on applications, contexts, and use cases, the ability to discover and classify services of objects often presents significant challenges. IoRT systems are built on communication protocols, which enable connectivity to networks and application interaction. In order to exchange data over the network, edge devices and autonomous devices use a variety of communication protocols that define exchange of data formats, data encryption, device addressing schemes, and packet transportation from origin to destination. Various Wireless Local Area Network (WLAN) communication standards, such as 802.11a, 802.11b and 802.11g, 802.11n, 802.11ac, and 802.11ad, are included in the 802.11 - Wi-Fi protocols that are used. These standards offer communication ranges of between 20 meters (indoors) and 100 meters (outdoors) with data speeds ranging from 1 Mb/s to 6.75 Gb/s. Various IoRT communication techniques can be used to achieve shared real-time computation, data exchange, and internal communication [8].

13.2.4 Voice Recognition, Voice Control

Nowadays, chatbots and devices with microphones are commonly used for communication. As the IoRT advances, more endpoints are created for people and robots to interact. This leads to the emergence of a new digital experience where robotic objects and humans collaborate, resulting in IoRT mesh development through mutual contact between robotic entities.

People should be able to communicate with robot fleets in more natural ways for IoRT applications that include tour-guiding, elder care, recovery, search and rescue, monitoring, education, general support in everyday settings, and in workplaces, factories, and homes. When creating IoRT applications, multimodal interfaces that deal with detection of movement, auditory localization, people monitoring, users (or fellow person/robot) localization, along with the merging of different modalities are essential. Using knowledge of the robot's movements and postures, powerful noise-reduction techniques can be used for speech recognition and voice control.

To reduce background noise, the microphone's quality is crucial for automatic speech recognition. A speech recognition automation system for robots can accurately identify voices from adults as well as kids in crowded spaces thanks to the reduction of reverberation, disruption, and noise in a portable multiple channels of communication technique with an outlier-robust broader side-lobe canceller method and a feature-space noise reduction criterion [9].

13.2.5 Machine Learning as Enabler for Adaptive

The Internet of Things (IoT) community recognizes the significance of incorporating Machine Learning (ML) techniques into IoT devices for effective management of the vast and diverse sensory data generated by network nodes. This integration enables IoRT solutions to adjust to different environments, while IoRT apps can learn from their surroundings and experiences. The ML service needs to be intelligent and distributed enough to encompass every IoRT node, including those at the network's edge. By performing early data fusion and predictive analyses, this distributed and embedded intelligence can generate high-level aggregated information from the device/sensor's low-level data. These forecasts might be used as inputs for an innovative method of learning on an alternative network node, which would then carry out additional forecasts and data fusion operations. This would result in the development of a smart network of machine learning algorithms that would carry out progressive compilations of the sensed data [10].

13.3 Marketplace for an IORT Ecosystem

Mechanisms for the monetization of service components and data are required to encourage participation and the development of an IoRT ecosystem. A market place must be established as the center of an IoRT ecosystem. A marketplace facilitates registration and the discovery of offerings, as well as the data or attributes that services provide. These services may act on robots or objects directly or as separate components accessible through IoRT platforms. Selling and buying of products and services take place in the marketplace, which acts as a central location for transactions.

- Offering authentication, where a supplier of an offering can provide a metadata description for the marketplace to ingest and index to help with discovery.
- Offering discovery, where a customer searches for offerings using a marketplace search interface. To facilitate data registration and discovery, common ontologies for the semantic description of data must be established.
- Role and privileges management could be used to secure access to the market. Authentication and authorization can also be used to secure access to the marketplace.
- Reputation management, which involves consumer evaluations of suppliers and their goods; these evaluations may be taken into consideration when selecting search outcomes and during investigation. Auditing and accounting, wherein a customer's use of a service is monitored and providers can make a fee for this usage at their discretion. This function is crucial for making IoRT monetization possible and giving the ecosystem's growth drive. The management of different data-providing licenses is closely related to it [11].

13.3.1 Orchestration

Supporting with providing composition generation, instantiation, maintenance, and dissemination. This feature of an online marketplace fosters the reuse of enrolled offers since it permits their use in other processes, even if it is not required. Due to orchestration, a customized manufacturing procedure might be represented as a cooperation of numerous robotic item functions, for instance [12].

13.3.2 The Applications in Warehouse and E-Commerce

IoRT was developed in reaction to the growth of e-commerce, where human warehouse workers and autonomous robots coexist. Collaborative robots can help logistics organizations by alleviating some staffing shortages and the demanding nature of the work. One typical use for IoRT is the delivery of goods by fleets of autonomous robots within a range of between three and five kilometers, carrying loads as large as ten kilos, at speeds in the range of eight to ten kilometers per hour. Robots may be controlled and monitored remotely if they encounter circumstances that prevent them from traveling autonomously [13].

13.4 IORT Practical Applications in Commerce

Amazon has assembled a team to examine possible uses for autonomous technology inside its expanding logistics network. The company won't be building self-driving cars; instead, it will act as a think tank to help the biggest online retailer in the world incorporate automation into its logistical plan. By further automating administrative tasks, Amazon may reduce delivery costs, giving them a vital competitive edge.

Automated forklifts, for example, might save labor costs in the company's operations; the Kiva robots have demonstrated 20% lower operational costs.

13.4.1 Both Amusement and Wellbeing

By fusing communication technologies with robot sensing abilities, telepresence robotics enables more complex human-robot interaction. It enables people to virtually roam and gaze around distant areas, participate in business meetings, and keep a watch on patients or elderly individuals at residence or in medical institutions. Cinemas, theaters, and retail environments all provide creative and exciting venues to integrate modern technologies. Since the ability to move around and live in social and natural surroundings is necessary for a normal lifestyle, communal and open spaces will get more attention as locations for technology in the future. Service robots are employed in a range of social tasks, such as grocery shopping, outdoor cleaning, and visiting historical sites [14].

13.4.2 Coordination

To achieve the IoRT's collective optimization goals, robot collaboration must be correctly taken into consideration. When these problems develop in multimodal robot networks that are outfitted with a range of actuators, sensors, and on-board computing systems, etc. to effectively accomplish the required tasks, both the implementation and the design of MRS synchronization pose substantial hurdles. The difficulty of the assigned job must be considered when estimating the complexity of the assignments the collaborative MRS must develop, which leads to an estimate of a reasonable number of robots to carry out that assigned task. Another alternative is to divide the task up among several robot subgroups.

13.5 Healthcare Robotics Process Automation Paradigm

Any country's medical industry generates a considerable amount of employment and cash. It consists of things like medical insurance, clinical research, and hospital supplies. It could be challenging to compile and assess data, such as information on pharmacological qualities, that is scattered across several organizational factors against the backdrop of the healthcare business. tests, laboratory tools for technological developments, outside routes, health portals, scanning diagnostics administration, sequence-dependent arranging programs, Preservation offers, and improvements in human resource utilization. Health care organizations must rely on people to complete conventional challenging tasks that need a lot of concentrated work since interoperability across various systems can occasionally be challenging.

In the healthcare sector, the three main participants are patients, doctors, and health insurance. Creating a far more accurate and efficient internal process is essential to maintaining the balance between the rising patient demand and the records needed for surveillance, insurance demands, etc. Alternatives to computerized testing that are based on modern science and technology, like automated robotic processes, could assist clinicians in managing details such as doctor licenses, documenting and patient incentives healthcare provider schedules, programming, statements of billing and claim management, patient information, Medicare payment and adherence, and supplemental insurance policies to increase utility, reduce costs, and set limits on care. The benefit of robotic process automation is that it guarantees that its applications and tools have excellent visual designs that improve user experience (UX) for patients. The objective is to provide a simple framework that almost anyone with a fundamental understanding of modern technology may use. In order to satisfy the requirements of the final consumers, it must also be thoroughly modified. It is stated that science of computation and mechanical engineering are combined in modern robot technology. It is also extending its area of influence and communicating wisely with others. Three significant uses of automation in medicine are robotic assistants, rehabilitation tools, and medical automation. By providing better findings, more precise availability, and higher efficacy, robots are offering medical institutions an economic edge over their rivals. To make it easier for physicians to carry out their duties and deliver patients with exceptional medical treatment, several large firms and mega-specialty organizations have started up and investing in robots (Subeesh *et al.*, 2021).

13.6 Operative Robotics

With the use of robotic surgery, physicians can perform complex surgeries with fewer incisions. Surgery is arguably the most well-known use of robots in the healthcare sector. It made it possible for doctors to do accurate punctures and encouraged the creation of fresh non-invasive methods [15].

13.6.1 Radiographer Robotics

Radiologists will receive training in sophisticated diagnostic imaging interpretation. In addition to earlier scans, doctors will consider a patient's complete medical history. Robotic surgeons won't tire out from doing constant surgeries all year [16].

13.6.2 Rehabilitating Robotics

Lightweight, portable digital exoskeletons provide solutions for flexible appendages. After chemotherapy and more procedures, a variety of robotic wheelchairs may aid with the brain's ability to reconstruct suitable connections between neurons. Robots that replicate human movement are now being researched in an effort to trick the brain into reacting [17].

13.6.3 Software in Smooth Robots and Prosthetics Smooth Grippers

Thinking about the inherent flexibility and compliance of DEA synthetic muscle tissues, growing gentle grippers based on DEA synthetic muscular tissues provides a brand new way to grasp objects with one-of-a-kind shapes and brands. Various soft grippers evolved based on DEA synthetic muscle groups

13.7 Vibrotactile Stimulation

Recent studies have used vibrating sleeves, pressure points in prosthetic sockets, and friction modulation on touch screen surfaces as examples of vibrotactile feedback. The pair's ability to collaborate is improved by vibrotactile stimuli and haptic signals, which without hindering task flow, can considerably increase the information the machine provides to the user. As was explained throughout the study, one-way intent transmission—from the person to the robot—has made task-specific robotic assistance conceivable. We discussed how most research conducted at pHRI treats the autonomous device as a passive spectator of human conduct and, at most, enables the robot to offer feedback on its present condition and/or intended use. Evidence that reciprocal signaling between an automated device and a person significantly enhances interaction was covered. These contacts, however bidirectional, have so far allowed for task negotiating or the progressive development of mutual comprehension and alignment. Through bidirectional discourse, which depends on both the psychological well-being of the person interacting and the dialogue history, more adaptable cooperation may be made feasible. As a result, the robot would be able to provide feedback on its current state and/or intended function.

13.7.1 Estimated Usage and Success Rate of Robotics in Healthcare

In the healthcare industry and other connected companies, the usage of robotics and automated technologies has increased. The market for health care robots is anticipated to expand quickly over the following decades, hitting 9 trillion dollars by 2024, according the forecast of the Global Association of Robots. Robotics also helps healthcare professionals like healthcare professionals complete complex and exacting tasks more quickly and accurately, enhancing the efficiency of the entire medical sector.

Robot-assisted therapy may allow for therapeutic procedures that are more efficient, available, and enjoyable than conventional therapy. So far, there have been conflicting findings regarding how clinical outcome markers are impacted by robot-assisted training. Customized robot-assisted training has been demonstrated in some trials to be superior to conventional therapy, while no statistically significant differences in clinical outcomes have been found in other investigations. However, there is solid proof that patients who actively engage in therapy benefit from it. These optimistic findings spur research that reliably identifies intent and provides real-time accomplishment and effort metrics in order to promote patient engagement.

Significance of Robotic process automation in health-care sector

- By employing machine artificial neural networks to quickly gain and incorporate clinical data from lab technologies, third-party entry points, health insurance pathways screening imaging equipment, sequence-dependent set up, as well as additional multiple systems, we aim to lessen the difficulties faced by medical professionals when coping with the complicated layout of procedures and a volume of patient and health care facility statistical analysis.

- Learning more about robots to improve patient care is the aim of automation in the medical sector. In order to accomplish that, it also becomes essential to build the appropriate automated machinery and work with experts to make it. To completely understand the importance of the uses of machine learning in healthcare, additional study is necessary.
- Since people's potential for compassion is linked to either their humanity or chaotic structure, further research is still needed to determine whether or not medical professionals can substitute people. Empathy may be shown by robots and machine learning toward people, and it can also be shown through preprogrammed actions.

13.8 IoT in Transportation

In addition to making it easier to get from one place to another, IoT in transportation also increases safety and appropriateness. For instance, a smart car may do multiple functions at once, including communication, pleasure, navigation, and more reliable, efficient transportation. Travelers may stay connected to all kinds of transportation at all times thanks to IoT. Several wireless technologies, including as Wi-Fi, Bluetooth 3G, 4G, smart transportation systems, and even other vehicles, connect the car to the internet. A more sophisticated version of geo-fencing has been developed in this sector. It records the coordinates of a given location along with the position of an object or device.

13.9 Applications of IoT and AI in Agriculture Automation

By integrating better programs, an IoT environment powered by AI has a surprising ability to increase the control and specificity of farming operations. The potential of these most recent technological developments in farm operations is limitless because they might automate difficult tasks with little manual labor.

13.9.1 Smart Farm

Today's farmers use a range of farm machinery and equipment to complete a number of agricultural jobs. Tractors are included in them and are regarded as the most important and indispensable farm energy source. Tractor performance monitors gather data, take measurements, and help with remote process observation. The variables that are frequently taken into consideration include power, consumption of fuel, draught, and wheel slip.

DGPS, or the Global Positioning System with Difference. The GPS system is a crucial part of the system that provides spatial values. As a result, the tractor-implement system may be evaluated, documented, and monitored in terms of how it is functioning in relation to the position. Given that the soil texture and land slope have an influence on how well a tractor-implement system operates, this mapping technique is highly useful for estimating the costs of crop production inside the field limits.

Robotic harvesters, rice cultivators have adopted that function with efficiency that is nearly equivalent to that of a human. The image collecting module was the primary element of the robotic fruits harvester prototype. This was followed by an image editing module mounted on a motorized carriage. The information was utilized to determine which fruits and vegetables might be gathered using an object identification method based on computer vision.

13.9.2 UAV or Drones

Drones were first utilized for military operations, but they have increasingly been modified for use in agriculture. Another development in agriculture is the automation of numerous agricultural operations using drones, including pesticide application and land monitoring. Unmanned aerial vehicles (UAVs), which are characterized as airplane without a human pilot on board, include agricultural drones in their category.. A central processor unit, a GPS receiver, a laser, a radar, a camera, a gyroscope, an accelerometer, a compass, and other sensors are all included in the drone gadget. Actuators and motors are also included for carrying out necessary activities. The remote control is used to interact with this via wireless communication. UAVs' integrated thermal and multi-spectral detectors allow them to survey hectares of fields in a single flight [11].

13.9.3 Irrigation

The "per drop more crop" strategy has been developed to ensure proper use of the limited supplies of water, which is necessary to meet our future food demands. The sensors' usage of data-transmission protocols like MQTT makes real-time monitoring possible. Data is made available to subjects in MQTT, and people who are enrolled to such topics are the only ones who may read the data. MQTT has the advantage of being small and simple for the network to control. The processing unit of the module is equipped with a relay and a motor and is ready to accept commands from the outside. A MQTT dashboard may be used to view the data [11].

13.9.4 Fertilizers Application

Businesses now push farmers to embrace technology-enabled agricultural practices and digital farming approaches as shown in the Figure 13.3. Using IoT technologies can make fertilizer application more sophisticated. An NPK sensor can be used to monitor the concentrations of nitrogen, phosphorus, and potassium (K). This sensor can be made using resistors, LEDs, and light-dependent resistors. The concepts of colorimetry and the photoconductivity are the foundation of the sensor's operation. An on-chip processing unit receives data from the NPK sensors directly. The use of edge computing or cloud computing is used to conduct further research. The low-cost SPAD was developed for the field-based indirect evaluation of crop leaf chlorophyll content [11].

13.9.5 Weed and Pest Control

Application of herbicides is still challenging due to its unfavorable impacts, which include harm to the environment and human health. Additionally, conventional weeding methods uniformly spray pesticides throughout the whole field whether or not there are weeds,

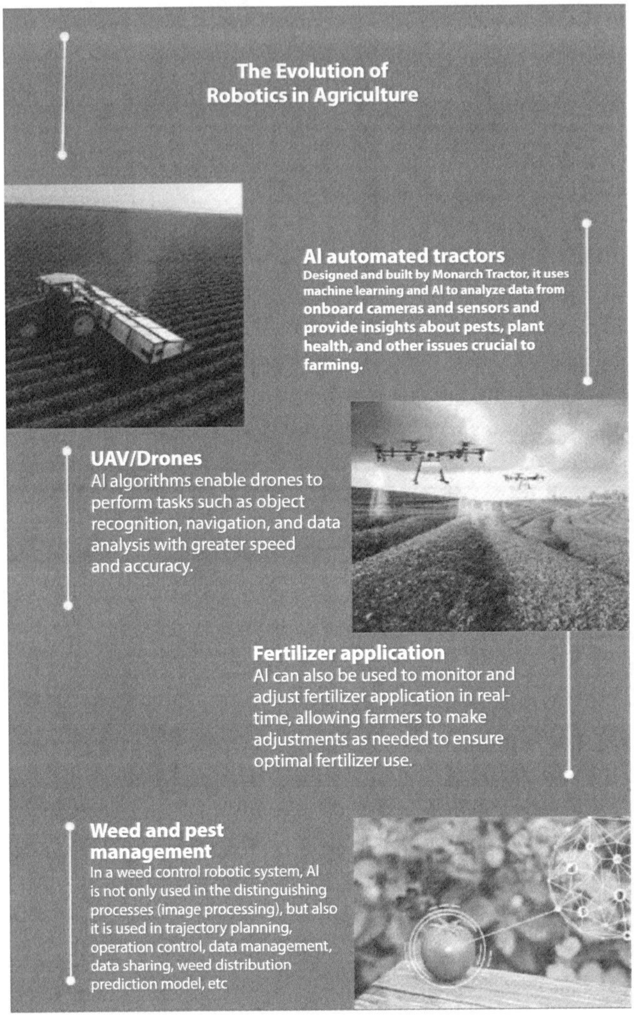

Figure 13.3 Robotics evolution.

which raises the price of herbicides and boosts greenhouse gas emissions. Incorporating robots, the Internet of Things, and innovative image processing techniques into a site-specific system is one effective way to address all of these problems. Weeds in a field can be found using both RGB and The infrared imaging sensors (IR) [11].

With the use of a sprayer, imaging devices, sensors, and system-on-chips like the Raspberry Pi, systems created utilizing the Internet of Things may manage weeds by dousing them in herbicide. The critical aspect of managing weeds is handled by an AI-based categorization method [11].

13.9.6 Storage of Farm Products

A helpful instrument for monitoring the quality of agricultural products maintained in storage containers are wireless sensor nodes. Moisture and temperature detectors can be used to make sure that the storage room is kept at the ideal temperature and humidity levels. The values in the time series of sensor readings could vary slightly. In order to evaluate both humidity and temperature fluctuations in the storage system, cumulative data are acquired from this. Each of these thresholds is determined with respect to the crop kept in the storage facility. Data aggregation is done at the remote databases server, which acts as a gateway to connect sensor nodes to the World Wide Web. Coupling the data findings to the local farmers' ongoing surveillance of the storage conditions on a local level [11].

13.10 Sustainable Agriculture

Sustainable agricultural practices are anticipated to provide a number of advantages, including enhanced soil fertility, environmental protection, and a greater supply of the planet's natural resources. By following a variety of goals, sustainable agriculture seeks to increase productive agricultural revenue. Promote effective environmental management [11].

For agriculture to be sustainable, data must be easily accessible. Smart farming enables both cost-effective and sustainable agriculture through the use of earth observation data and navigation satellites, making it easier for farmers to make informed agricultural decisions [11].

The main types of sustainability are three.

- Environmental Sustainability: Sustainability assures that nature may be used regularly and without interference from various variables at the ecological level, preventing it from being used as an infinite supply of resources. An focus in renewable energy sources, outdoor safety, and water conservation are a few examples of realistic adaptation.
- Social Sustainability: At the social level, sustainability may plant the seeds for the development of individuals, groups, and societies to provide realistic and really isolated personal enjoyment, healthcare, and education throughout the globe .
- Economic sustainability: Business activities and the related flow of capital will support various pillars of viability for an all-around improvement [11].

13.11 Machine Intelligence

The development of IoRT is greatly affected by the field of AI. Specific robotic devices are becoming more capable thanks to AI algorithms.

The sensor fusion capacities of IoRT devices are enhanced by artificial intelligence (AI) technologies, such as lenses for detecting sight (see Figure 13.4), chemical detectors for distinguishing smell and taste, microphone for listening, pressure detectors for feeling touch/pressure, etc.

In order to deliver analytics and insights, as well as to improve the capabilities of one's robotic toys and their cooperative behaviors as a fleet, AI techniques and approaches are applied in all kinds of IoRT platforms. Convolutional artificial neural networks (CNNs), a type of deep neural network, are used to evaluate and extract visual properties from images.

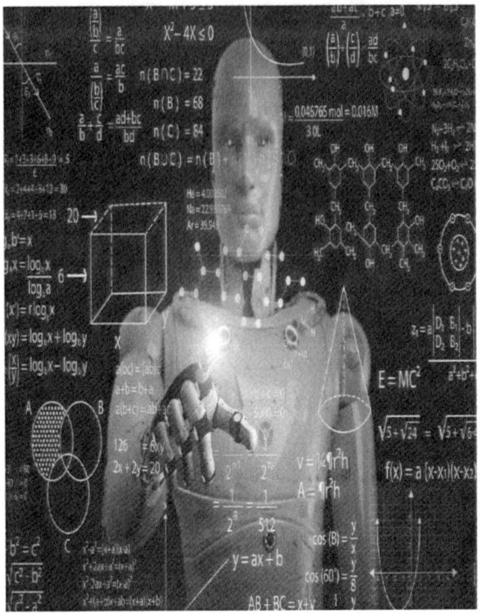

Figure 13.4 Machine intelligence.

The techniques are designed to divide and de-noise monitored signals in order to achieve substantial variation or potential defect detection rates, which will enhance the efficacy of the recognition function [11].

In order to deliver analytics and insights, as well as to improve the capabilities of the individual automated machinery and their collaborative behaviors as a fleet, the various levels of IoRT platforms utilize AI approaches and methodology. Convolutional neural networks (CNNs), for example, are neural networks with deep connections that are used to analyze images and extract their visual characteristics. To enhance the effectiveness of the acknowledgment function by obtaining extraordinary variation or prospective defect detection rates, the procedures are made to partition and de-noise monitored signals. IoRT applications may now manage very big data sets as a result of the transition from central compute to edge/fog nodes. This strategy, which fuses effective production inspection technologies with edge processing, is also applied. Technologies for production inspection are also. By adapting a CNNs engine to the cloud computing environment, this strategy significantly increases the computational effectiveness of a verification model that can determine the type and severity of faults. Robotic machines collect video, compress video data, pre-process video images, and segment video using perception tools like cameras. Utilizing perceptual technologies, such as cameras, images are first processed and segmented. It might be possible to increase the accuracy of item recognition by working together to create an environment and scenery-aware adaption model utilizing data from different video capture equipment. IoRT equipment, servers at the edge, and the cloud must all share the same amount of DL compute; hence a suitable offloading method must be developed.

When assessing network health, video encoded data, information rates/usage, power consumption, processing delay, framing rate, and the computational correctness of analytics, significant trade-offs must be taken into account. The potential exists for multiple distributed edge autonomous gadgets to collaborate at the edge to provide better services. Condensing the DL frameworks at the outer layer can enhance system performance through computing edge collaboration and the distribution of capabilities [1].

13.12 Virtual and Augmented Reality

Immersive technology (such VR and AR) may now be incorporated into interfaces for users of automated machines as well as for robotic tools that communicate with the interface of IoRT platform systems. This is possible because IoRT applications are becoming more cognitively capable at the edge. For IoRT systems to be more dependable, end-to-end security, electronic identities, and mobile data/knowledge security all need to be enhanced. This criteria is based on cognitive skills based on emerging AI algorithms. Future distributed IoRT architectures will require concepts like adaptability, end-to-end reliability, privacy, and intelligent connectivity in order to move away from existing centralized IoRT systems. A fluid data flow and expertise exchanging across IoRT applications/services executing at the edge or in the cloud is required to support the utilization of computing at the edges, software that uses cloud computing of operations, and rule-driven policy execution while maintaining the integrity and privacy of data. Applications for the Internet of Things (IoRT) may use VR/AR for learning, maneuvering, and support functions. While the technology known as augmented reality (AR) supersedes information generated by computers over the actual world to retain time and space coherence and enable in-the-moment communication, virtual reality (VR) replicates occurrences [11].

13.13 Integration of Digital Twins with IoT

Digital Twins (DT) is a brilliant concept that has recently emerged. It involves creating a digital replica of a real system, seen from the perspective of Cyber Physical Systems (CPS). This virtual replica mimics the actual performance of the system. Throughout the entire development lifecycle, information from the virtual system, combined with that of the real system, characterizes the data from the physical system. By combining the digital and real counterparts, a more effective way of handling, controlling, and improving coordination when the system is operating is achieved.

Digital twins refer to virtual models of physical objects or systems used for real-time monitoring and management. Meanwhile, IoT is a network of interconnected devices capable of data sharing and communication. When integrated, digital twins and IoT can produce intelligent systems that optimize performance and anticipate potential problems. By analyzing data collected from digital twins and IoT devices, artificial intelligence (AI) can identify patterns and offer predictions.

It is effectively incorporated into a variety of applications, including medicinal systems, industry, the aviation industry, the agricultural sector, urban planning, and weather prediction [11].

13.14 Biomedical Application

The industry that has profited the most from applying digital twin ideas is biomedical healthcare. Digital twin applications in biomedical fields have been made possible by new IoT, fitness, and e-health technologies. In the healthcare sector, DTs technologies are used to analyze and propose medications, identify lifestyle changes, improve hospital operations, do remote surgery, and help policymakers with providing healthcare to patients. *In vitro* methods for forecasting how the actual organ will function in either stated scenario will be developed with the help of a DT virtualized biological approach by physicians as they look into how the development of digital twins in the medical field impacts treatment, diagnosis, and overall health.

13.15 Smart Cities

Smarter features for smart cities are now being presented via digital twin advancements. Research on this topic has evolved along with (IoT) and communications, which has boosted interest in the digitization of our lives. A framework for creating smart medical amenities for people in smart cities using digital twins is described and to use a knowledge-based approach with artificial intelligence to provide categorization and decision-making methods for managing mobility and power in a major city. Digital twins are used to control energy in smart cities [11].

13.16 Energy Management

With the ever growing demand for energy supply energy management is becoming a pivotal part of our life. A benchmark based on the analysis of data from smart meters is created in to create a daily suggestion for energy building. The framework will examine how these procedures may assist in power control that is as near to actual as possible and will ascertain the differences between what is new from traditional and yearly energy assessment techniques. The paradigm is built by combining digital twins with information and communication technology to provide a strong promise for improving the effectiveness of catastrophe management measures [11].

13.17 Intelligent Connectivity

Depending on the application, the intelligent connectivity infrastructure must operate as a continuous and compatible network to support heterogeneous devices with a range of intelligence competencies and connectivity requirements. This is because strong, resilient, and dependable connection networks are necessary for IoRT applications. The connection infrastructure must be flexible enough to respond to changing environmental conditions, as well as to expected and unanticipated events and circumstances. The intelligent infrastructure for connectivity must function as a continuous and interoperable network in order to serve heterogeneous devices with varied intelligence features and connectivity needs, depending on the application. This is because robust, resilient, and dependable connectivity networks are necessary for IoRT applications. The linkage infrastructure must be flexible enough to respond to changing environmental conditions, as well as to expected and unanticipated events and circumstances. In order to meet operational requirements and "the economic viability of the vehicular networks," artificial intelligence (AI) methods like machine learning (ML) are employed as useful tools to address the difficulties faced in 5G technology wireless technology (such as cached data, processing, and communication operations).

To support new connectivity, which has developed into the enabler for the future IoRT intelligent services, wireless and cellular networks utilized for telecommunications must offer predictable/guaranteed latency.

Peer-to-peer and/or broadcast communication techniques are available for direct communication and data sharing on IoRT devices. The forthcoming 5G wireless technology will have an average latency performance ranging from 1-10 ms in contrast to the current 4G mobile technology's latency efficiency of 80–100 ms, thanks to the use of SDN (software-defined networking) and the virtualization of network functions (NFV). Applications for dispersed networks are supported in order to meet the stringent energy efficiency standards required to cover large outdoor regions, deep indoor or underground locations, or mobile objects moving quickly. These applications are built for the edge and the cloud. The architecture and additional functions of the 5G network are designed to fulfill the Latency, reliability, and throughput are requirements for IoRT applications that are both mission- and safety-critical. Future directions in physical human-robot interaction for embodied communication

It consists of multimodal communication, assistance for emergent, unforeseen behavior, and communication during collaboration (safe operation guaranteed by autonomy while allowing continual adaptation). These goals take into consideration how individuals and their surroundings alter over time, enabling a process of coadaptation between humans and automated systems However, they also have one thing in common: there are currently no standardized algorithmic tools available. We shall discuss this in more detail below. Therefore, datasets are needed for comparing and evaluating the features of the current pHRI, testing algorithmic tools relevant to the pHRI, and certifying algorithms prior to use with people [11].

13.18 Continual Adaptation with Safety Guarantees

pHRI may provide a safety risk due to the immediate mechanical contact and power exchange. As a result, enabling pHRI has placed a high priority on safety .However, it is unclear how to ensure safety without restricting robot motion parameters (such as torque or velocity) and performance limits. Due to the lack of reliable alternatives, the ISO standards from 2016 point to quantifiable biomechanical restrictions, such as acceptable threshold forces or pressures for certain body regions, as a requirement for collaborative robots.

These initiatives provide a step toward offering standardized safety assurances for data-driven autonomous machines (given acceptable assumptions). Notably, only a small number of research have taken continuous or long-term adaptation into account. Most work has concentrated on learning within a constrained period following initial engagement. One explanation for this is because, although evaluating the effectiveness and safety of data-driven approaches is difficult, doing so for systems that are constantly evolving is considerably more difficult. The development of methods for long-term learning facilitation and data-driven safety verification should be the main areas of research to come. While giving specifications that are adequate for nonstationary simulations and take into account variation in humans and adaptation, we should try to give safety assurances that, at the very least, retain states within safe sets and adhere to temporal logic criteria (Kalinowska *et al.*, 2023).

13.19 Multimodal Dialogue

Similar to how we try to deduce pertinent assessment signals from an individual for the robot, the robot should try to provide the individual with data signals. The mobility of the robot is often used in human-robot systems to provide implicit feedback to the human partner. In certain pieces, anthropomorphic cues or exaggerated robot movements are used to strengthen the connection.

As a result of making robotic behaviors more predictable, intentional communication motion has been proven to increase engagement since it enables users to foresee the robot's motions and modify their own behavior accordingly [11].

13.20 Industrial IoT

The expansion and use of the internet of things (IoT) in industrial applications and sectors is referred to as the "industrial internet of things" (IIoT).more credibility because to its substantial focus on big data, machine-to-machine connectivity, and machine learning. The Industrial Internet of Things is made up of numerous devices that are linked together through communications software. Without the need for human involvement, the outcomes networks, are able to rapidly respond on information after exchanging, analyzing, monitoring, and collecting it in order to modify their behavior or their surroundings [11].

13.21 IoT Fire Forecast Detectors

To detect fires early, scientists and engineers have created vision-based fire detectors (VFDs) as well as sound-, flame-, temperature-, gas-, and solid-sensitive fire sensors. Smoke's chemical properties are detected by sensors, which activates an alarm. The YOLOv8 detection model, used by the (Smart fire detection)SFDS approach, provides quick and accurate object identification without the requirement for a regional proposal network. The system is improved to need fewer parameters for detection, increasing its effectiveness. The SFDS utilize computer vision to automatically detect fires in photo and video feeds [11].

13.22 IoT-Based Greenhouse Management

To achieve high-efficiency monitoring of environmental data, an IoT monitoring management device for a greenhouse is developed. Organizational framework of the Internet of Things system for managing and monitoring the environment in greenhouses. The system's main parts are a remote web server and a smart gateway running an Android operating system. Using a data collecting device, the gateway gathers greenhouse environmental data, which is then stored in a SQLite database. In the meanwhile, the gateway is used to send the greenhouse's environmental data to the server. The greenhouse's equipment, including the windows, fans, and heating units, is under the supervision of the control node. Output from the gateway is received by and managed by the server system [11].

13.23 IoT Architecture Domain

Smart, linked houses may be divided into two basic architectural types: distributed and centralized.

13.23.1 Centralized Smart Home Architecture

A computer system that is responsible for controlling the control system is used in a centralized smart home design for collecting information from sensors, interacting with users, implementing control algorithms, and instructing actuators. In addition to performing the control function, it is responsible for connecting the smart home to the outside world through the Internet and offering services to its inhabitants.

13.23.2 Distributed Smart Home Architecture

The software for the control system is envisioned and developed as a distributed computing system in a distributed smart home architecture. The distributed design takes advantage of the processing power of intelligent objects to integrate software modules into the smart home network's nodes [11].

13.24 Emergent Interfaces

The communication agent (human or robot) must have a correct mental representation of the communication rules in order for the signaling model outlined above to work, and this hypothesis that the mental representation is correct is frequently static. New symbolic languages will be the foundation for communication if we are to move toward active nonverbal interaction between humans and technology.

We can connect with assistive technology using our kinematic, muscular, and cognitive null space—excess degree of freedom—while avoiding major increases in cognitive burden. The developer chooses a group of vectors in the neural null space, and the operator, who is a human being, is trained to read and create them. As a starting point, languages can be specified and programmed in the robot. There is an apparent way to symbolize the connection of the continuous 2D regulated space onto the corresponding 2D action space of the robotic machine, making the language simple to learn, if we are mapping joystick signals to the action space of a motorized wheelchair [11].

References

1. Hemdan, E.E.D., El-Shafai, W., Sayed, A., Integrating Digital Twins with IoT-Based Blockchain: Concept, Architecture, Challenges, and Future Scope. *Wirel. Pers. Commun.*, 1–24, 2023.
2. Kalinowska, A., Pilarski, P.M., Murphey, T.D., Embodied Communication: How Robots and People Communicate Through Physical Interaction. *Annu. Rev. Control Rob. Auton. Syst.*, 6, 205–232, 2023.
3. Kaur, J., Robotic Process Automation in Healthcare Sector, in: *E3S Web of Conferences*, vol. 391, EDP Sciences, p. 01008, 2023.
4. Kathirvel., A., Blesso Danny J., Preetham, S., Hinn Joe, T.O., Kennedy Roak, C., Immanuel Aldrin, J., *Systematic Number Plate detection using improved YOLOv5 detector*, Institute of Electrical and Electronics Engineers Inc publisher, 2023, ISBN 979-835034798-2, doi: 10.1109/ViTECoN58111.2023.10157727.
5. Kathirvel., A. and Shobitha, M., *Digital Assets Fair Estimation Using Artificial Intelligence*, Institute of Electrical and Electronics Engineers Inc publisher, 2023, ISBN 979-835034798-2, doi: 10.1109/ViTECoN58111.2023.10157310.
6. Kathirvel., A. and Ale, P., *Machine Learning and Deep Learning Algorithms for Network Data Analytics Function in 5G Cellular Networks*, Institute of Electrical and Electronics Engineers Inc publisher, 2023, ISBN 979-835039849-6, doi: 10.1109/ICICT57646.2023.10134247.
7. Kathirvel, A., Rithik, G., Naren, A.K., Chapter 11: Automation of IoT Robotics, in: *Predicting Natural Disasters with AI and Machine Learning*, 1st Edition, IGI Global Publiser, 2024, ISBN 9798369322802, eBook ISBN 9798369322819, doi: 10.4018/979-8-3693-2280-2.ch011.
8. Kathirvel, A. and Gobinath, V.M., Chapter 1: A Review on Additive Manufactuing in Industrial, in: *Modern Hybird Machince and Super Finishing Process: Technology and Application*, 1st Edition, CRC Publiser/Chapman and Hall, New York, 2024, ISBN 1032354291, doi eBook ISBN 9781032354293.
9. Gobinath, V.M., Kathirvel, A., Rajesh Kanna S, K., Annamalai, K., Chapter 5: Smart Technology in Management Industries: A Useful Perspective, in: *Artificial Intelligence Applied to Industry 4.0*, 1st Edition, Wiley Publiser, USA, July 2024, ISBN 978-1-394-21611-6.
10. Kathirvel, A. and Maheswaran, C.P., Chapter 8: Enhanced AI-Based Intrusion Detection and Response System for WSN, in: *Artificial Intelligence for Intrusion Detection Systems*, 1st Edition, CRC Publiser/Chapman and Hall, New York, 2023, ISBN 978-100096755-5, 978-103238665-2, doi: 10.1201/9781003346340-8 eBook ISBN 9781003346340.
11. Kathirvel, A., Maheswaran, C.P., Subramaniam, M., Naren, A.K., Chapter 25: Quantum Computers Based on Distributed Computing Systems for the Next Generation: Overview and Applications, in: *Quantum Computers Based on Distributed Computing Systems for the Next Generation, Handbook of Research on Quantum Computing for Smart Environments*, IGI Global Publiser, Hershey, Pennsylvania, USA, March 2023, ISBN13: 9781668466971 ISBN10:166846697X EISBN13: 9781668466988, doi: 10.4018/978-1-6684-6697-1.ch025.
12. Perwej, Y., Haq, K., Parwej, F., Mumdouh, M., Hassan, M., The internet of things (IoT) and its application domains. *Int. J. Comput. Appl.*, *975*, 8887, 182, 2019.
13. Subeesh, A. and Mehta, C.R., Automation and digitization of agriculture using artificial intelligence and internet of things. *Artif. Intell. Agric.*, *5*, 278–291, 2021.

14. Talaat, F.M. and ZainEldin, H., An improved fire detection approach based on YOLO-v8 for smart cities. *Neural Comput. Appl.*, 1–16, 2023.
15. Uprety, D., Banarjee, D., Kumar, N., Dhiman, A., Smart Sustainable Farming Using IOT, Cloud Computing and Big Data. *EasyChair*, No. 10814, 2023.
16. Vermesan, O., Bröring, A., Tragos, E., Serrano, M., Bacciu, D., Chessa, S., Bahr, R., Internet of robotic things–converging sensing/actuating, hyper connectivity, artificial intelligence and IoT platforms, in: *Cognitive Hyperconnected Digital Transformation*, pp. 97–155, River Publishers, 2022.
17. Vermesan, O., Bahr, R., Ottella, M., Serrano, M., Karlsen, T., Wahlstrøm, T., Gamba, M.T., Internet of robotic things intelligent connectivity and platforms. *Front. Rob. AI*, *7*, 104, 2020.

14

Real-Time Monitoring and Predictive Maintenance

R. Patel[1]*, S. Shah[1], S. Lella[2] and A. Gajbahar[1]

[1]*Illinois Institute of Technology, Chicago, Illinois, United States*
[2]*Lewis University, Chicago, Illinois, United States*

Abstract
The pattern of shifting towards predictive maintenance, bolstered by real-time monitoring and analysis techniques, is revolutionizing modern manufacturing with an objective of minimize downtime and maximize operational efficiency. Technological advancements in smart manufacturing coupled with data analytics and machine learning algorithms have helped in achieving this goal. Predictive maintenance requires sophisticated networks of sensors and monitoring devices for monitoring environmental indicators such as temperature, vibration, and pressure, alongside detecting abnormal conditions like excessive wear, consumable levels, and other relevant parameters continuously as a continuous stream of input for machine learning to recommend intervention based on identified trends, anomalies, intermittent issues, and deviations from normal operating conditions. Such recommendations are proactive maintenance or part replacement activities which can be further automated. Beyond manufacturing, these techniques can be used in many industries that require minimizing downtime and maximizing operational efficiency by avoiding equipment failure related operational risks. Such an approach will help energy facilities, aviation, healthcare, and critical infrastructure. Predictive maintenance adopts modern methodologies, utilizing real-time monitoring and data analytics to forecast potential challenges and recommend mitigating efforts to avoid failure proactively. This chapter provides an in-depth exploration of various aspects of predictive maintenance and real-time monitoring, highlighting their transformative potential and widespread applicability.

Keywords: Predictive maintenance, real-time monitoring, smart manufacturing, data analytics, machine learning algorithms, downtime minimization, industrial sensors, predictive analytics, equipment failure prevention, manufacturing optimization

14.1 Introduction

Predictive maintenance approach empowered by real-time monitoring and analysis techniques is transforming the way modern manufacturing industries approach equipment maintenance to minimize downtime and maximize operational efficiency. With advancements in technology smart manufacturing is influencing innovative approaches, data analytics, and machine learning algorithms to predict and prevent equipment failures before

**Corresponding author*: rpatel37@iit.edu

Amit Kumar Tyagi, Shrikant Tiwari, Senthil Kumar Arumugam and Avinash Kumar Sharma (eds.) Artificial Intelligence-Enabled Digital Twin for Smart Manufacturing, (305–324) © 2024 Scrivener Publishing LLC

they occur. The traditional approach to maintenance in manufacturing uses processes such as scheduled inspections and routine checks for making predictions related to the maintenance needs. Such approaches are time consuming, lead to over-maintenance, introduce additional downtime, and cause unexpected breakdowns. This approach is reactive. On other hand, predictive maintenance approach leads to modern methodologies by using real-time monitoring and data analytics to foresee potential challenges and address them before they worsen.

Building blocks of predictive maintenance are sophisticated networks of sensors and monitoring devices deployed as embedded or add-ons for modern industrial equipment. These sensors continuously collect relevant information related to the indicator of environmental conditions such as temperature, vibration, pressure, and abnormal conditions such as excessive usage, wear and tear, level of consumables and other relevant indicators. The real-time data is sent to the centralized system periodically where advanced data analytics and machine learning algorithms are employed to derive predictive maintenance recommendations. Machine learning algorithms analyze incoming stream of data continuously and identify trends, anomalies, intermittent and temporary issues, and deviations from normal operating conditions. Initially, baseline metrics are used as per manufacturer's specifications. However, over the time, self-learning system creates dynamic baseline using machine learning algorithms which in turn used to predict when a component is likely to fail or when maintenance is required creating a predictive capability to schedule maintenance or part replacement activities precisely when needed.

Predictive maintenance with real-time monitoring transforms a traditionally reactive process to a proactive maintenance process. Predictive maintenance techniques can help not only manufacturing but many industries where equipment failure is highly risky to their operation such as energy facilities, aviation, healthcare, and many critical infrastructure facilities of a nation. In this chapter various aspects of predictive maintenance and real-time monitoring are discussed in detail.

14.2 Fundamentals

Fundamental specification and related data collections are important to understand before we embark on the journey to develop and deploy the modern predictive recommender system based on real time monitoring and analysis in the smart manufacturing industry. In this chapter we will start with such fundamental concepts and build the knowledge and capabilities using indicators of the sensor system behaviors from these concepts.

14.2.1 Reliability

Reliability of sensor is the probability that a sensor will consistently deliver the intended and accurate measurements at the level of standards of performance within a specified period under specified environmental conditions. In other words, it is the ability of a sensor system to consistently work as expected without unexpected failures or errors [12]. Reliability is a measure of how long a sensor can perform its agreed function without any interruption. Reliability is one of the important measures and plays a vital role in many smart manufacturing applications especially for safety measures [12]. A higher level of reliability offers consistency and dependability of a sensor system. The reliability of a sensor system depends on many factors including its design, quality of components, and production quality. Reliability of sensor system can be predetermined by performing rigorous testing under various environmental conditions. Many sensory systems require regular recalibration and maintenance for sustained delivery of accurate measurement over usage. One of the best ways to measure reliability is measuring Mean Time between Failures (MTBF). MTBF is a reliability metric that estimates the average time a sensor system functions normally before experiencing a failure. A higher MTBF value indicates greater reliability.

There are many aspects combined to gather in the reliability and availability related to the sensor system. For Sensor system to be reliable, it must supply accurate measurements within a specified range. Level of precision exhibits the sensor system's ability to provide consistent and repeatable measurements with minimum variations in the measurements compared to reference or true measurement and calibration. Over the period, accuracy may drift and hence the precision of measurement will deteriorate. A reliable sensory system provides stable measurement over time with minimal drift or changes in its performance characteristics. Another important consideration is durability that indicates the reliability of the sensor system under extreme environmental conditions such as temperature, humidity, light, pressure, and other potential stressors, without compromising its performance. Measurement speed is another related term that defines reliability. The sensor system must be responsive enough to provide new measurements after measured conditions change. The lag in reporting the change must be less than the acceptable level of time delay. Once the sensor system becomes slow in reporting, even if accurately, it cannot be considered as reliable.

14.2.2 Availability

Availability refers to the ability of a service, component, or CI to perform its agreed function when required. Availability is the probability whether the sensor system is ready to perform its function when it is required [12]. Calculating availability is important because it is one of the measures in the Service Level Agreement (SLA). Many sensor systems require high availability. Fail-safe conditions require a solution with high availability. One of the best ways to measure availability is Mean Time to Repair (MTTR). MTTR measures the average time it takes to repair or replace a failed sensor system. Lower MTTR values contribute to higher availability. Another reliable measure for the availability is "uptime." This is a simple measure of how often the sensor is operational and it is calculated as the ratio of the time the sensor is available to the total time.

Implementing redundant sensor systems can significantly improve availability. Redundancy involves using multiple sensors providing measurements simultaneously or operated in-parallel. It is less likely that all sensors fail at the same time, so if one sensor fails, the others can continue to provide data, increasing availability of the overall sensor system. Continuous or periodic calibration activities can help prevent sensor failures and ensure consistent performance, helping not only availability but also reliability.

14.2.3 Maintainability

Maintainability is associated with reliability and availability. Maintainability refers to a measure of how quickly and effectively a sensor system can be restored to normal working conditions after experiencing a failure. An increase in maintainability can lead to a reduction in operating and support costs. Carefully planned and scheduled maintenance contributes to higher availability. A more maintainable product lowers maintenance time and operating costs [12]. Furthermore, more efficient maintenance means a faster return to operation or service, decreasing downtime [3]. Furthermore, predictive maintenance means no downtime and avoidance of loss of production time.

14.2.4 Serviceability

Serviceability refers to the ability of a sensor supplier to meet the terms of their contract. One of the important considerations for serviceability is "lead time." The lead time is defined as a time required to procure a replacement sensor system component upon failure. Shorter lead times contribute to quicker replacement and, consequently, higher availability. Serviceability is more associated with vendor's process maturity, supply chain management, and maintenance and support structure, and parts distribution mechanism in contrast to the sensor system itself. Nonetheless it is one of the important measures and input for predictive maintenance decision making.

It is extremely difficult to accurately measure any of these fundamental indicators of the sensor systems due to short production and testing cycle, complex supply chains, lack of accurate and timely information, lack of availability of relevant data, lack of valid testing on innovative technologies, and difficulty in fault detection [12]. Furthermore, these deficiencies are seldom independent of each other and can have a compounding effect on the performance of the sensor system and hence it is extremely difficult to determine correct time for maintenance and even harder to determine time for predictive maintenance. As a result, it is better to consider, collect, and use these measures where available to make a predictive maintenance decision.

14.3 Type of Predictive Maintenance

Predictive Maintenance is one of the most important aspects of the modern manufacturing process for Industry 4.0. It is used for monitoring and maintaining the normal operating condition of the current system. Predictive maintenance processes are designed to continuously monitor the system's vital conditions and determine how it performs during normal operations, which can help us estimate how much more life is remaining or what maintenance process is required for the system or its components. Predictive maintenance also enhances the efficiency of machinery, costs, control, and output quality. It can help to enhance the decision-making process for maintenance with the help of behavior patterns, and correlations by using Machine Learning (ML) models or statistical models to predict trends and foresee failures in advance and primarily avoid downtime. Implementing predictive maintenance improves manufacturing capabilities [18]. Predictive maintenance also boosts availability and reliability while optimizing the trade-off between performance and maintenance costs.

Typically, predictive maintenance techniques calculate productivity, efficiency, remaining useful life, and estimating the next possible failure. Such information helps scheduling the maintenance efforts prior to any breakdown of the equipment to make decisions on when and what needs to be done to extend the system's expected useful life and avoid failure by condition monitoring and future maintenance prognosis [9]. Model evaluation in various circumstances often compares maintenance indicator findings based on the concepts discussed earlier such as reliability, availability, maintainability, and serviceability, allowing for the early detection of impending issues and the potential for breakdowns to avoid unplanned failure.

When using predictive maintenance, the process must be aware of the steps that need to be completed ahead of time to prevent production stoppage under various circumstances. The physical structures of machines and the type of failure are associated with the process of approximation of defending against unforeseen difficulties. A combination of mathematical models is used in the evaluation of predictive maintenance models to determine when a problem may occur and when maintenance should be initiated. Furthermore, to analyze and compare data and make maintenance decisions, predictive maintenance assists in continuously measuring and recording physical parameters [13].

14.3.1 Corrective Maintenance

Corrective maintenance processes and tools aim to restore the functionality of the component/equipment whenever a fault is detected, whether it is planned or unplanned, potentially incurring unforeseen costs for the unplanned repairs. Corrective Maintenance is a type of approach used after a fault is detected and the system must be repaired. The main goal of this approach is to restore the system back to its normal operations so it can perform its functions after the failure has occurred. Corrective Maintenance may include minor adjustments, repairs, and replacements of the equipment or components and can also be called reactive maintenance. It can either be planned or unplanned and can be performed for multiple reasons, however it is mostly unplanned. Planned corrective maintenance is the repairs scheduled on later dates, while unplanned corrective maintenance is the repairs done urgently due to the failure. Corrective maintenance is often unavoidable and therefore is an expensive and as well as time-consuming approach, however it is important and in modern manufacturing it is a last measure as it allows us to correct the problems and restore the normal operations of the system after its failure. To find and identify the broken component, fix it, and shorten downtime, corrective maintenance procedures rely on human knowledge as well as judgment [17].

Corrective maintenance can be categorized into a few distinct types based on their goals. Repair/Replacements maintenance is performed on the failed equipment or its components to repair or replace it to get it back into its operational state and restore its full functionality. Overhauling is another type of service where consumables and/or replace wearied or torn components need readjustments to restore the equipment to its service state. The unrepairable parts which cannot be repaired should be disposed of and replaced with the preserved parts in Salvaging. Rebuilding is another type of corrective maintenance where parts do not need attention; however, inter-connectivity must be restored to the original state of working conditions. The equipment is restored to as close as its original state with increased life efficiency. This can be achieved by dissembling the parts and replacing the worn components according to original production and standards requirements.

Planned corrective maintenance is used for a prognostic health management system or a routine inspection compared to unscheduled maintenance, which is sometimes referred to as it is work that needs to be done quickly to fix an unexpected failure to prevent major effects on the system's performance. Weekly and biweekly schedules are used to plan corrective maintenance tasks and daily operations are used for unplanned tasks related to malfunctions or anomalies [11]. Corrective maintenance is the easiest and most straightforward approach, but it is one of the priciest methods which are applied in response to malfunctions and other incidental fallout. Corrective maintenance measures include immediate corrective action and deferred corrective maintenance which is performed with a delay following the identification of a fault [5].

Corrective maintenance can be used to mature predictive measures since it provides valuable and true information related to the component in terms of its accuracy, reliability, availability, maintainability, and serviceability which can be used as an input to the predictive maintenance algorithm. Such knowledge can be derived by studying the various tasks involved in corrective maintenance. Failure Recognition is an alert for system downtime received from sensor system. Accurate data will provide the details related to the availability and reliability. Diagnosis or troubleshooting time is a downtime for locating failure in the system and deciding the action plan. Information obtained can be used as an input for determining serviceability and maintainability. When equipment is returned to normal service operation, the service sensor provides another data point that can be used to determine service level agreements, time between failures, and time to repair.

14.3.2 Risk-Based Maintenance

Risk-based maintenance is a technique that can be used to plan and manage to lower the risk that could have resulted in the failure of the equipment and its effectiveness [7]. To reduce the overall risk the activities have been identified and classified to focus on the equipment which is having medium or high risks the major efforts are put into the maintenance of that equipment by giving it more priority compared to that of the equipment with low risks [1]. By using this approach, the work and the costs involved in the maintenance are decreased. Risk based maintenance comprises of three major steps: assessment of the risk, evaluation of the risk, and maintenance planning as per the risk.

14.3.2.1 Assessment of the Risk

One of the simplest methods to assess the risk is by multiplying the likelihood of failure and impact of the failure. This can be further modified by factoring in the time to react (serviceability and maintainability). The first step in risk assessment is to identify the top events, or major potential hazards, that could result from each failure scenario. To determine the events and the intermediate routes leading to the ultimate failure by performing a fault tree analysis. The probability of a subsystem failure is estimated using failure data for the subsystem's rudimentary events. A consequence analysis is employed to measure the impact of every failure scenario that transpires. The cost of each maintenance expenditure due to failure is also considered for cost benefit analysis.

14.3.2.2 Evaluation of the Risk

The goal is to estimate the risk by applying the previously described technique. The two steps that make up this evaluation algorithm are described below. Establishing the requirements for risk acceptance is the first step. Precise risk acceptance business rules or internal policy is applied for the risk scenarios. Typically, determining risk acceptance level is not the same for all organizations and hence the methodology used varies for the acceptable degree of risk based on the type and nature of the system, its usage, and the organization. Various acceptance risk criteria can be found in the literature such as USEPA acceptance criteria, Dutch acceptance criteria, and ALARP (as low as reasonably possible). Specific methods are beyond this book's scope; however, the criteria for risk acceptance must be decided for developing the predictive maintenance algorithm. Evaluating risks in relation to acceptance standards involves applying the acceptance criteria to each system unit's calculated risk. Systems and components are located whose assessed risk surpasses the acceptability requirements. These are the units that require risk-based maintenance scheduling.

14.3.2.3 *Risk-Based Maintenance Planning*

Risk based maintenance planning helps reduce the likelihood of failure for high-risk components or systems. The likelihood of failure of the topmost events should be calculated so that risk management efforts are ensured and managed effectively. A reverse fault tree analysis can determine the likelihood of failure of the top events. This information will then result in a changed value for the probability of failure that was reached by meeting the requirement's risk for that component or system. The top analytical approach will now be used for the reverse fault tree analysis. Moreover, the top event's occurring likelihood can now be fixed, and with that probability of fundamental events failing is determined by the simulation of the fault tree. With modern machine learning based systems, such analysis can be iterative and dynamic based on new events and data that system learns and utilizes for the analysis in the algorithm as stream of inputs.

14.3.3 Preventive Maintenance

Preventive Maintenance is scheduled maintenance that is carried out to restore an operating system or equipment to its optimal working functioning state. Preventive maintenance helps the system from failing suddenly. It is especially important to predict accurate time for performing preventive maintenance because if performed too early, it is not only costly to perform the maintenances but also means that system or component used for fraction of the maximum possible life. Similarly, if preventive maintenance is scheduled too late then organization may not be able to rip the fullest benefit it offers. Many preventive maintenances still need down time at suitable time and hence it might result in unnecessary expenses, such as production line stops in production systems, lost production revenues, and other similar challenges even if planned at the suitable time. For all these reasons, figuring out the best time for preventive maintenance while keeping the accounts for preventive maintenance expenses and output profit is crucial. Some of the procedures can involve tasks such as part inspection and adjustment, changing the oil, lubrication, cleaning, reducing environmental stress, replacing, or repairing certain parts, and more [2]. Determining when and where to perform preventive maintenance is not a simple task in a complex and stochastic industry. Main goals of preventive maintenance are to decrease the frequency and duration of equipment failures, decrease abnormalities the way that equipment is operated indirectly impacted by its conditions, and reducing savings expenses.

Preventive maintenance should be carried out to lower the costs whenever there is a lower risk of failure occurring because compared to predictive maintenance corrective maintenance can result in exceptionally large expenses, which also include the production loss suffered due to equipment downtime. Preventive maintenance procedures that are performed too frequently however, it may also be expensive since resources may be squandered. Updating subjective decisions with objective ones is what maintenance optimization models are for they help preventive maintenance decision-making [16]. By applying maintenance optimization models, one can determine the most objectively balanced maintenance solution that helps meet specific requirements. Preventive maintenance is planned based on three types of planning: cost-based planning, time-based planning, and failure-based planning.

14.3.3.1 *Cost-Based Preventive Maintenance*

Cost-based planning helps in examining the capital cost, benefits, and revenue-generating potential of project management. Cost-based planning is also an essential component of any fundamental assessment since it weighs the potential economic gains from implementing a solution against its expenses. The cost associated with assessing maintenance components, such as repair, replacements, tools, and labor, is called cost-based planning.

14.3.3.2 *Time-Based Preventive Maintenance*

Time-based planning measures the subject analysis in terms of information allocation for the operation's time. Planning is aided by the time basis, which provides an analysis of project duration, typically with respect to recurring procedures. A broad view is analyzed, necessitating a large amount of data and information about maintenance factors like the amount of time needed for replacement and repair, the amount of time lost due to failure, the amount of time allotted to gathering shift times, and system operation time. Many techniques are used to investigate maintenance aspects that could impact PM planning's efficacy.

14.3.3.3 *Failure-Based Preventive Maintenance*

It is a fundamental analysis that incorporates consideration of system or component deterioration data to determine the optimal preventive maintenance planning. A thorough definition of failures that arise with systems or components before further analysis is included in failure-based planning. The overall success of preventive maintenance could be impacted by several failure-based planning techniques. Moreover, in failure-based planning, critical analyses comprising the goals of analysis and evaluations of failure concerns are also conducted which are comprehensively quantified to predict the features of a failure mechanism.

14.3.4 Condition-Based Maintenance

Condition-based maintenance is categorized into subtypes of preventive maintenance when we talk about the maintenance strategies for the industry. Making maintenance recommendations based on data from condition monitoring is the major goal of condition-based maintenance and for the same effectiveness of the condition-based maintenance is dependent on the results of the monitoring process. It seeks to regulate the different modes of equipment failure, and therefore, it is important to consider all failure modes that could cause financial losses when establishing condition-based maintenance. It is majorly predicted on the idea that most failures do not happen all at the same time and that you can catch them early in the deteriorating process [14].

Rather than depending on time-based maintenance as in more conventional maintenance solutions that industry uses, condition-based maintenance is a prominent technique for scheduling maintenance interventions. It is described as the process of tracking one or more relevant parameters to model the system's performance and spot changes related to future malfunctions. To prevent performance losses, it is imperative that proactive monitoring be maintained, and repair intervention resources be optimally used. When a failure of the process has the potential to result in significant financial losses, condition-based maintenance is thought to be the best maintenance approach to apply [15].

14.3.4.1 Process for Condition-Based Maintenance

The process of condition-based maintenance includes three major steps data collection, processing of the data, and decision-making of maintenance.

Data Collection: Gathering and archiving data on factors that may be indicative of the components deterioration process that needs to be focused. It is important to determine what the connection is between each variable and the modes of equipment/component failure. Moreover, the type of variables to be tracked must be taken into consideration when choosing measurement instruments. Monitoring and event data are the two types that are used in condition-based maintenance in which the event data can be used to evaluate how well condition indicators are performing, support actions for improvement, or help refine already defined indicators and the monitoring data constitute a valuable informational resource [14].

Processing of the data: Its goal is to take key information out of the data. Data cleansing and analysis are part of data processing. Data cleansing is necessary as the data may be incomplete as well as unstructured. Both human error and sensor malfunctions might result in errors in the data. The right software must be chosen, based on the type of data, to perform data analysis. Signal processing is the process of handling waves and multidimensional data. The process of obtaining valuable information from the original signals is called feature extraction.

Decision making: This is step involves determining if equipment intervention is required or not, the best course of maintenance and the best time and method for maintenance execution necessitate a thorough examination of the data generated by condition monitoring. Many techniques, including data-driven approaches, can be used to assess what is the current state or to forecast the RUL (remaining useful life) of a particular component, hence supporting condition-based maintenance decisions. The ideas of diagnostics are typically used to represent current condition analysis as well as the future condition prediction [4]. Out of both diagnostics is related to the discovery, detection of failures. On the other hand, prognostics seeks to anticipate failures before they happen, enabling the early scheduling of interventions [14].

14.3.4.2 Implementation of Condition-Based Maintenance

To implement condition-based maintenance, it is necessary to identify the components that need to be monitored, establish the necessary technological infrastructure, define the suitable data processing procedures, and choose monitoring methodologies and technologies. Consequently, a substantial investment may be required. The primary costs in this consist of purchasing hardware, software, and measuring equipment and providing specialized training and information. It is important to examine the organizational, financial, and technical aspects of condition-based maintenance implementation. Technical justification must be given to show how condition monitoring systems and condition-based maintenance policies can help the organization accomplish its strategic goals must be given.

14.3.5 Predetermined Maintenance

Another preventive maintenance type is predetermined maintenance based on predetermined time intervals defined by past observations of the same type of equipment and on the lifetime assigned to the equipment previously. As previously recorded data from similar equipment is used to carry out this maintenance the real performance of the equipment is not taken into consideration. It means that the equipment's state is ignored whether it is functioning normally or in its failure state and by this approach internal maintenance is achieved. Furthermore, for the effective result of planned maintenance the equipment must have predictable patterns of failure. However, the recent technology and apparatus that have become more accessible over the past few decades have made it difficult to recognize successive patterns of breakdown. As most technology firms may find it challenging to boost the reliability of the system by executing specified maintenance [6]. Advantages of predetermined maintenance include reducing the number of events of emergency malfunctions and enhancing employee competency. However, predetermined maintenance could be more expensive. An accurate requirement must be decided to understand the overall expense associated with holding spares and consumables that are required with this approach. Predetermined maintenance requires access to information such as equipment usage data for making scheduling decisions. The reason behind the predetermined maintenance is scheduled because of its intolerable failure effect and the potential to avoid it by planning a scheduled overhaul or replacement of the equipment. It is scheduled using the equipment's expected useful life remaining, which is initially obtained from manufacturer supplied specification and later updated based on experience. Predetermined maintenance is most appropriate for equipment with obvious signs of aging or breakdown. The number of hours used, and the frequency of use can be considered for scheduling for maintenance [10].

14.4 Real-Time Monitoring of Industrial Components

With the increase in integration of industrial automation and control systems, real-time monitoring is becoming practical and a new essential requirement in smart manufacturing processes. Every factory can benefit from a real-time monitoring and controlling System for tracking various environmental variables like temperature, humidity, and pressure statistics for making decisions early enough for preventing incidents and shutdowns. Such real-time monitoring and decision-making systems also allow peripheral systems to wirelessly transfer data for central analysis and recommender systems to monitor, detect, prevent, and predict future mal-functions and predictive maintenance requirements.

The ability of real-time sensor controls also allows one to react to changes from pre-established limits is one of its key features. Such ability makes sure that industrial processes stay within predetermined tolerances. The more interesting part is that by using complex algorithms and decision-making frameworks we do not even have to predetermine tolerance levels since it can be dynamically determined by the system based on data it continuously receives from various sensors. This degree of automation reduces not only mistakes along with downtime and improves efficiency, but it also allows us to predict component failures and perform maintenance in a way that we can achieve zero-downtime.

Furthermore, the principle of predictive maintenance is strongly related to real-time controlling. The system can foresee potential problems and proactively take corrective action by analyzing real-time data, averting expensive disruptions and malfunctions. This predictive ability is revolutionary in fields where uninterrupted operations are critical. The essential components of real-time monitoring of industrial components, specifically focusing on real-time sensing, real-time controlling, and real-time monitoring are examined in this part of the chapter. The following sections explore the unique characteristics, functions, and transformative effects of each facet of the intricate machinery of industrial processes, where each component plays a vital part.

14.4.1 Real-Time Sensing

The core component of the real-time monitoring infrastructure is real-time sensing, which is essential for gathering, sending, and standardizing sensor-generated data. The complex issues of real-time sensing are explained in this section, along with its application, importance, and effects on industrial processes.

14.4.1.1 Significance of Real-Time Sensing

To enable smooth network transmission, sensor data is encapsulated into standard web services in real-time sensing. This translates into the real-time tracking of critical parameters in industrial settings, like pressure, temperature, humidity, accelerometer, and gyroscope data. Real-time sensing has been revolutionized with the introduction of Internet of Things (IoT) technologies, which allow for an ongoing flow of real-time data that serves as the basis for the monitoring system. A unified description of the sensor information achieves the standardization of information processing and expression. For the sake of generality, real-time sensing data per sensor is encapsulated as a standard Web service so that real-time data can be transmitted over the network.

14.4.1.2 Sensual Virtuosity

In industrial environments, Real-time Sensing becomes more than a mere observer; it develops into a sensory virtuoso sensitive to the subtleties of the manufacturing process. Real-time Sensing functions like an expert painter, not just gathering data; it captures the minute details as well as the big picture, giving the industrial canvas a vibrant and lively appearance. It can perceive the subtleties of operational variables and intuitively understands and interprets the dynamic environment of the manufacturing setting.

14.4.1.3 Dynamic Data Elevation

Real-time sensing raises the bar for data collection by exceeding the norm. It senses the beat, cadence, and complex dance of variables in real-time, allowing it to compose a symphony of insights instead of just recording numbers. With Real-time Sensing, an industrial symphony becomes a richer and more nuanced understanding, akin to a virtuoso musician performing a musical piece. Raw, unprocessed data is turned into a harmonious compilation.

14.4.1.4 Environment Harmonization

Essential to the orchestration of industrial processes, Real-time Sensing contributes significantly to achieving environmental harmony. It records not just data but also sounds, such as that from production lines, feels the pulse of machinery, and senses the rhythm of operational efficiency. As an artist perfectly tuned to the distinct melody of industrial music, Real-time Sensing ensures that all components work together for maximum efficiency.

14.4.1.5 Futuristic Precision

In its pursuit of futuristic precision, Real-time Sensing acts as a visionary force that goes beyond the confines of the present. Equipped with the ability to extract deep insights from the real-time data it painstakingly collects, this technological marvel transforms into an unmatched industrial landscape predictor. Its powers include the ability to predict changes, foresee trends, and vividly imagine outcomes.

Much like a strategic guide navigating uncharted territories, Real-time Sensing offers a proactive approach to decision-making. It offers priceless insight by tracking and forecasting the multifaceted ebb and flow of industrial processes. This type of insight goes beyond the problems of the present and future, enabling companies to stay ahead of the curve, make wise decisions, and precisely optimize their operations.

14.4.2 Real-Time Controlling

Delving into the unique field of Real-time Controlling, this section highlights the unique contribution that this element makes to the industrial dynamic's choreography. Real-time Controlling takes on the role of an innovative director, guiding not only the performance but also the basics of industrial operations, departing from Real-time Sensing's relentless search for data.

14.4.2.1 Pioneering Proactivity

Real-time Controlling is a proactive maestro who shapes the story, rather than a spectator in the industrial symphony. In contrast to the observational function of data collection, this element reflects a proactive strategy. It adapts to changes in real time, coordinating subtle changes to maintain the ideal production pace. Consider it as a trailblazer, not just following the notes, but also burning a path for industrial processes.

14.4.2.2 Orchestrating the Symphony of Proactive Maintenance

As a vigilant conductor in the industrial process orchestration, Real-time Controlling steps in. Through a symphony of proactive maintenance, it anticipates needs rather than anticipating instruments to go out of tune. This part operates as a kind of avant-garde composer, skillfully integrating preventive measures into the operations' melody to ensure a continuous and melodious performance.

14.4.2.3 Integration with Real-World Applications

Real-time controlling has numerous industry-specific applications that increase operational effectiveness, decrease downtime, and improve product quality. Real-time controlling is a meaningful change in industrial automation, enabling proactive maintenance plans and real-time parameter adjustments during manufacturing. To sum up, real-time control is vital for the development of industrial processes because it offers proactive rather than reactive solutions that improve efficiency and prolong the usefulness of equipment.

14.4.3 Real-Time Monitoring

The field of process management and control changed due to the real-time monitoring of industrial components, essential to modern manufacturing. In this section, we explore the several dimensions and uses of real-time monitoring and discuss its crucial importance and varied nature.

14.4.3.1 Importance of Real-Time Monitoring

When it comes to industrial components, real-time monitoring refers to the tracking of many metrics and parameters in real-time throughout the manufacturing process. Real-time data collecting has been made possible by the introduction of Internet of Things based sensors, which have transformed this field. These sensors, which are integrated into machinery and industrial lines, gather a wide range of data, from accelerometer and gyroscope readings to temperature and humidity. These sensor-generated data have some noteworthy qualities: they are large, often unstructured, and real-time.

14.4.3.2 Infrastructure and Data Processing

Real-time monitoring is an effective way to solve problems in the industrial sector, where accuracy and efficiency are crucial. A constant stream of data that offers insights into the ongoing processes replaces the conventional methods of manual checks and periodic inspections. Decision-makers are empowered by this immediacy to quickly spot abnormalities or departures from the norm.

The infrastructure that underpins real-time monitoring systems is crucial to guaranteeing the smooth processing and analysis of data. An effective combination is created by using

Apache Kafka as a message queue, Apache Storm as a real-time processing engine, and MongoDB for data storage. All these parts work together to make it easier to process and store the massive amounts of sensor data that are produced throughout the manufacturing process. The integration of such big data technologies is essential to handle the substantial influx of information and derive meaningful insights.

14.4.3.3 Fault Detection and Prediction

The identification of errors or abnormalities during the production process is one of the key issues that real-time monitoring attempts to solve. This is where big data processing and Internet of Things-based sensors work together. The suggested hybrid prediction model, which combines Random Forest classification and Density-Based and Spatial Clustering of Applications with Noise (DBSCAN) based outlier detection, turns out to be useful. DBSCAN effectively detects anomalies in the sensor data, and Random Forest predicts faults well, improving the accuracy of fault detection overall.

14.4.3.4 Practical Application in Automotive Manufacturing

These real-time monitoring systems are put to effective use in the automotive production sector. The results show the effectiveness of IoT-based sensors in conjunction with the suggested big data processing platform in a real assembly line situation. The hybrid prediction model beats alternative models in fault prediction accuracy, and the system has sufficient robustness to manage the complexities of the manufacturing process.

As a result, real-time monitoring becomes essential to contemporary production ecosystems. By giving decision-makers an instantaneous and thorough picture of the manufacturing processes, it enables them to take preventative action and avoid unanticipated costs brought on by errors. The evolution of monitoring systems is embodied in the combination of IoT-based sensors, big data processing, and sophisticated prediction models, which opens the door to increased productivity and well-informed decision-making in the industrial environment [8].

14.4.4 Real-Time Predicting

Real-time prediction is at the forefront of Industry 4.0 innovations, bringing a radical change in manufacturing by allowing systems to forecast results and patterns with remarkably high accuracy. Using machine learning algorithms that are smoothly integrated with large data processing platforms is the fundamental part of real-time prediction. The paper "Industry 4.0 Application in Manufacturing for Real-Time Monitoring and Control" supplies an excellent example of the complex relationship between Industry 4.0's core technology and real-time prediction.

14.4.4.1 The Essence of Real-Time Predicting

The essence of real-time predicting lies in its ability to extrapolate future states based on continuous analysis of historical and real-time data. To increase overall efficiency in manufacturing, this involves predicting potential problems, streamlining production procedures, and making data-driven decisions. This ability for prediction is attributed to machine learning algorithms, which can make precise predictions by finding patterns and abnormalities in data.

14.4.4.2 The Technological Backbone

The technological backbone of real-time predicting involves the integration of machine learning algorithms with big data processing platforms. A hybrid prediction model has been successfully implemented in studies such as performance Analysis of IoT-Based Sensor, big-data Processing, and machine learning model for real-time monitoring system in automotive manufacturing. This approach creates a strong framework for real-time prediction by combining the predictive accuracy of Random Forest classification with the outlier-detection capabilities of Density-Based Spatial Clustering of Applications with Noise.

14.4.4.3 Applications Across Industries

Real-time prediction applications have an impact on many different industries and change how companies make decisions. Predictive maintenance becomes a reality in production because machines can foresee potential problems and plan maintenance in advance. Predicting illness outbreaks and patient situations helps the healthcare industry by allowing for prompt responses. Predictive analytics is used by financial firms to efficiently forecast market trends and manage risks. The wide range of uses highlights how adaptable and essential real-time prediction is in modern environments.

The influence of real-time prediction is not limited to one business; rather, it is a driving force behind efficiency and innovation in a variety of fields. Future success will be defined by predicting and taking proactive measures, as predictive capabilities are increasingly incorporated into routine procedures as technology advances.

14.4.5 Real-Time Alerting

In the context of Industry 4.0, real-time alerting is an essential element that ensures prompt reactions to momentous events and deviations. It completes the three sets of skills that power contemporary industrial systems by acting as the proactive mechanism that supplements real-time monitoring and prediction. Many studies highlight how important real-time alerting is to modern industrial setups.

Real-time alerting focuses on the immediate dissemination of critical information, enabling stakeholders to act quickly in response to developing circumstances. Real-time alerting uses innovative technologies to significantly reduce response times, in contrast to traditional alerting systems, which may experience delays in information processing and relaying.

The efficient integration of sensor data, big data processing, and intelligent algorithms is an essential part of efficient real-time alerting. Industries can gather real-time data from multiple sources by using the capabilities of Internet of Things (IoT) devices. Robust big data platforms are then used to process this data, guaranteeing quick analysis and interpretation.

To find patterns and abnormalities in the data, machine learning algorithms are used. These models, which include Random Forest and Density-Based Spatial Clustering of

Applications with Noise are good at spotting departures from the standard. For example, abnormalities observed by sensors, such as sudden changes in temperature, humidity, or vibrations, suggesting potential problems, can initiate real-time alerting in the setting of an assembly line for the manufacturing of automobiles.

Real-time alerting is being used in industries other than manufacturing. Its impact spreads to other industries, including healthcare, where patient monitoring devices can immediately notify caregivers of life-threatening situations. Similarly, real-time alerting can be important in smart infrastructure projects as it allows for timely identification and response to potential problems, hence averting cascade failures.

Reducing false-positives while making sure that real abnormalities are not missed is one of the difficulties in real-time alerting. Maintaining the proper balance necessitates ongoing algorithmic optimization, considering the sector's particulars and the possible consequences of false alarms.

14.5 Predictive Maintenance with Real-Time Actions for Industrial Components

In the world of industrial operations, maintaining the consistent and dependable performance of machinery is critical to production and efficiency. When used in conjunction with real-time activities, predictive maintenance is an effective proactive method of addressing probable faults before they happen. This strategy should be given special attention in industrial settings since it reduces downtime and maximizes resource usage.

14.5.1 Decision-Making

Making wise decisions is essential to the success of predictive maintenance. Decisions made on maintenance tasks can have a considerable influence on overall operating efficiency and, in turn, the bottom line in a manufacturing setting. Conventional maintenance procedures sometimes entail sporadic inspections or reactive repairs, which waste resources and cause unneeded downtime. The transition to predictive maintenance makes use of innovative technologies, especially machine learning, to guide decision-making.

To forecast probable equipment failures, past and current data analysis is crucial, and machine learning algorithms are key to this process. Because these algorithms recognize patterns and trends, it is possible to identify early signs of machinery breakdown. Furthermore, by incorporating real-time data, decisions are made with the most up-to-date knowledge possible, improving the precision and applicability of maintenance interventions.

14.5.2 Machine Learning for Better Prediction

The predictive power of maintenance procedures is increased by machine learning techniques. Regression models can be used to process real-world data to forecast failure trends over time. This data can include vibration analysis, previous failure trends, and other pertinent characteristics. Regression makes sense as a model for reliability is frequently continuous. Classification methods are also useful for discrete analyses, like vibratory analysis, where data labeling is guided by equipment state indications.

In a case study, the importance of using machine learning for predictive maintenance is emphasized through an online monitoring of a Gas Turbine (GT) using a Wide Area Network (WAN) system. Using the Vibrotest Vt802 tool for real-world vibration analysis, the study finds interpretation gaps caused by a staff shortage of training. This emphasizes the need for sophisticated instruments as well as the requirement for knowledgeable people who can understand the findings.

Predicting the GT's reliability for the next three thousand hours is done in the study using the Weibull law, a popular model for reliability. Maintenance intervention durations can be determined using the graphical representation of the law on Weibull paper. This study is a good example of how machine learning algorithms enable reliability studies to provide insights into the best maintenance plans based on failure probability, which in turn helps companies make decisions.

14.6 Challenges and Ethical Considerations

The implementation of predictive maintenance requires collection and analysis of data from many different types of sensors and monitoring devices. In some cases, such data are confidential and private and can be misused beyond the predictive maintenance goals. It is important to ensure privacy, integrity, and confidentiality of such data. Smart manufacturing organizations must establish robust policies and controls related to cybersecurity to protect sensitive information. Machine learning algorithms are used for recommender systems, and it is one of the essential parts of the predictive maintenance process. Such algorithms make decisions based on the data fed to the system. If training data used for making decisions are biased, then the algorithms may rely on and even enhance the effects of such biases resulting into unfair decision-making such as unnecessary more frequent maintenance measures or neglecting specific maintenance requirements. Such unfair biases can be avoided or removed with regular audits and assessments of algorithms for ensuring accurate and ethical decision-making. Predictive maintenance makes manufacturing systems more reliable, the workforce especially employed for the reactive maintenance tasks may be impacted. Organizations need to plan for the potential impact on employment and proactively take measures for successful workforce training and transitions to help employees adapt to the new maintenance practices.

14.7 Conclusion

In this chapter many fundamental indicators and critical success factors of predictive maintenance were discussed. Later in the chapter, sensor mechanisms to collect these fundamental indicators in real time and continuously were discussed in detail. Finally, various attributes of how such data can be used to deploy real-time and self-learning, data analysis and predictive recommender system were discussed in detail. Design, development, and deployment of real-time monitoring, controlling, alerting, and predicting in factory or manufacturing equipment reveals numerous opportunities for making maintenance process more effective and efficient.

Predictive maintenance with real-time monitoring can be implemented gradually and can be expanded and matured over time. Traditional preventive maintenance schedules are often based on rudimentary statistical data such as means, medians, averages, and generalizations of the same. Gradually deployed predictive maintenance with real-time data can be used to convert such processes to further tailor maintenance activities to the actual condition of each specific asset because such changes provide immediate insights into operations that are ongoing and substitutes continuous data streams for manual inspections that were previously used. Furthermore, such approach for predictive maintenance is restricted to only smart manufacturing but it can be applied to many critical infrastructure equipment such healthcare, aviation, energy production, and water supplies.

References

1. Arunraj, N.S. and Maiti, J., Risk-Based Maintenance—Techniques and Applications. *J. Hazard. Mater.*, 142, 3, 653–61, 2007, https://doi.org/10.1016/j.jhazmat.2006.06.069.
2. Asadi, M., Hashemi, M., Balakrishnan, N., An Overview of Some Classical Models and Discussion of the Signature-based Models of Preventive Maintenance. *Appl. Stochastic Models Bus. Ind.*, 39, 1, 4–53, 2022, https://doi.org/10.1002/asmb.2679.
3. Brunton, S.L., Kutz, J.N., Manohar, K., Aravkin, A.Y., Morgansen, K., Klemisch, J., Goebel, N. *et al.*, Data-Driven Aerospace Engineering: Reframing the Industry with Machine Learning. *AIAA J.*, 1-26, 2021, https://doi.org/10.2514/1.j060131.
4. Elghazel, W., Bahi, J., Guyeux, C., Hakem, M., Medjaher, K., Zerhouni, N., Dependability of Wireless Sensor Networks for Industrial Prognostics and Health Management. *Comput. Ind.*, 68, 1–15, 2015, https://doi.org/10.1016/j.compind.2014.10.004.
5. Gackowiec, P., General Overview of Maintenance Strategies – Concepts and Approaches. *Multi. Aspects Prod. Eng.*, 2, 1, 126–39, 2019, https://doi.org/10.2478/mape-2019-0013.
6. Jimenez, V.J., Bouhmala, N., Gausdal, A.H., Developing a Predictive Maintenance Model for Vessel Machinery. *J. Ocean. Eng. Sci.*, 5, 4, 358–86, 2020, https://doi.org/10.1016/j.joes.2020.03.003.
7. Khan, F., II and Haddara, M.M., Risk-Based Maintenance (RBM): A Quantitative Approach for Maintenance/Inspection Scheduling and Planning. *J. Loss Prev. Process Ind.*, 16, 6, 561–73, 2003, https://doi.org/10.1016/j.jlp.2003.08.011.
8. Mishra, D., Priyadarshi, A., Das, S.M., Shree, S., Gupta, A., Pal, S.K., Chakravarty, D., Industry 4.0 Application in Manufacturing for Real-Time Monitoring and Control. *J. Dyn. Monit. Diagn.*, 176–87, 2022, https://doi.org/10.37965/jdmd.2022.118.
9. Mykoniatis, K., A Real-Time Condition Monitoring and Maintenance Management System for Low Voltage Industrial Motors Using Internet-of-Things. *Proc. Manuf.*, 42, 450–56, 2020, https://doi.org/10.1016/j.promfg.2020.02.050.
10. Niu, G., Yang, B.-S., Pecht, M., Development of an Optimized Condition-Based Maintenance System by Data Fusion and Reliability-Centered Maintenance. *Reliab. Eng. Syst. Saf.*, 95, 7, 786–96, 2010, https://doi.org/10.1016/j.ress.2010.02.016.
11. Nota, G., Peluso, D., Toro Lazo, A., The Contribution of Industry 4.0 Technologies to Facility Management. *Int. J. Eng. Bus. Manage.*, 13, 184797902110241, 2021, https://doi.org/10.1177/18479790211024131.

12. Pokorni, S., Data -Driven Reliability and Availability of Electronic Equipment. *Vojnoteh. Glas*, 71, 3, 769–82, 2023, https://doi.org/10.5937/vojtehg71-43474.
13. Sakib, N. and Wuest, T., Challenges and Opportunities of Condition-Based Predictive Maintenance: A Review. *Proc. CIRP*, 78, 267–72, 2018, https://doi.org/10.1016/j.procir.2018.08.318.
14. Teixeira, H.N., Lopes, I., Braga, A.C., Condition-Based Maintenance Implementation: A Literature Review. *Proc. Manuf.*, 51, 228–35, 2020, https://doi.org/10.1016/j.promfg.2020.10.033.
15. Quatrini, E., Costantino, F., Gravio, G.D., Patriarca, R., Machine Learning for Anomaly Detection and Process Phase Classification to Improve Safety and Maintenance Activities. *J. Manuf. Syst.*, 56, 117–32, 2020, https://doi.org/10.1016/j.jmsy.2020.05.013.
16. Vilarinho, S., Lopes, I., Oliveira, J.A., Preventive Maintenance Decisions through Maintenance Optimization Models: A Case Study. *Proc. Manuf.*, 11, 1170–77, 2017, https://doi.org/10.1016/j.promfg.2017.07.241.
17. Yepez, P., Alsayyed, B., Ahmad, R., Intelligent Assisted Maintenance Plan Generation for Corrective Maintenance. *Manuf. Lett.*, 21, 7–11, 2019, https://doi.org/10.1016/j.mfglet.2019.06.004.
18. Zonta, T., da Costa, C.A., da Rosa Righi, R., de Lima, M.J., da Trindade, E.S., Li, G.P., Predictive Maintenance in the Industry 4.0: A Systematic Literature Review. *Comput. Ind. Eng.*, 150, 106889, 2020, https://doi.org/10.1016/j.cie.2020.106889.

15

Advanced Topics for Blockchain-Based Applications: Open Issues, Technical, Legal, and Research Challenges

Amit Kumar Tyagi[1]*, Shabnam Kumari[2] and Tanuj Surve[3]

[1]*Department of Fashion Technology, National Institute of Fashion Technology, New Delhi, Delhi India*
[2]*Department of Computer Science, Faculty of Science and Humanities, SRM Institute of Science and Technology, Kattankulathur, Chennai, Tamil Nadu, India*
[3]*University of California, Berkeley, USA*

Abstract
Today many emerging technologies like 5G, Quantum computing, Blockchain, Dew Computing etc., are changing the way to doing tasks as we do with current technologies. These technologies are far efficient and cost saving. This chapter explains advanced topics surrounding the implementation of blockchain-based applications, unraveling a tapestry of open issues, technical intricacies, legal issues, and ongoing research challenges. As blockchain technology matures, the pursuit of innovative applications encounters a spectrum of complexities that extend beyond the foundational principles. From a technical perspective, the chapter scrutinizes intricate challenges such as scalability, interoperability, and consensus mechanisms in blockchain architectures. It discusses evolving cryptographic techniques, privacy-preserving protocols, and the integration of emerging technologies to address the evolving demands of diverse applications. Legal issues in the realm of blockchain-based applications are dissected, emphasizing the need for regulatory frameworks and the resolution of jurisdictional ambiguities. The chapter scrutinizes the legal implications of decentralized autonomous organizations (DAOs), smart contracts, and data ownership, shedding light on the evolving legal landscape. Furthermore, the research examines open issues and challenges, providing a comprehensive overview of the state-of-the-art methodologies and proposing avenues for future investigation. Topics such as sustainability, energy consumption, and environmental impacts of blockchain networks are discussed, aligning with the growing consciousness of blockchain's ecological footprint. This chapter aims to serve as a guide for researchers, practitioners, and policymakers, providing information about the multifaceted dimensions of advanced blockchain applications. By navigating through technical intricacies, legal landscapes, and research frontiers, it contributes to the ongoing discourse, fostering a deeper understanding of the challenges and opportunities that lie ahead in the ever-evolving domain of blockchain technology.

Keywords: Blockchain, advanced applications, technical challenges, legal issues, research frontiers, decentralized autonomous organizations (DAOs)

**Corresponding author*: amitkrtyagi025@gmail.com

Amit Kumar Tyagi, Shrikant Tiwari, Senthil Kumar Arumugam and Avinash Kumar Sharma (eds.) Artificial Intelligence-Enabled Digital Twin for Smart Manufacturing, (325–340) © 2024 Scrivener Publishing LLC

15.1 Introduction

Advanced topics in blockchain-based applications encompass a wide range of emerging issues, technical challenges, legal issues, and research opportunities. Blockchain technology, initially popularized by cryptocurrencies like Bitcoin, has evolved to have applications beyond digital currencies [1–3]. Its decentralized, transparent, and tamper-resistant nature has attracted attention in various domains, including finance, supply chain management, healthcare, energy, governance, and more. While the fundamental concepts of blockchain are well-established, there are numerous complexities and intricacies that arise as blockchain technology is implemented in real-world scenarios. This necessitates a deeper exploration of advanced topics to address the evolving needs and challenges associated with blockchain-based applications. In this context, this exploration involves delving into open issues, technical challenges, legal implications, and research opportunities that arise in the field of blockchain-based applications [4]. It requires a multidisciplinary approach, combining expertise from computer science, cryptography, economics, law, and other relevant disciplines to develop innovative solutions and frameworks.

Technical challenges in blockchain-based applications encompass areas such as scalability, interoperability, security, and privacy. The scalability challenge revolves around handling a large number of transactions and achieving consensus efficiently, particularly as blockchain networks become increasingly popular [5–6]. Interoperability involves enabling seamless communication and data exchange between different blockchain networks. Security and privacy issues are important to protect sensitive data and prevent unauthorized access or tampering. Legal issues emerge in the context of smart contracts, digital identities, data protection, cross-border transactions, and regulatory compliance. Blockchain-based applications must navigate existing legal frameworks and establish new standards to ensure compliance and legal validity [7–8].

Moreover, research opportunities abound in exploring novel use cases and applications of blockchain technology. Areas such as healthcare data management, supply chain traceability, decentralized finance, energy trading, and governance mechanisms can benefit from blockchain-based solutions. Research efforts can focus on improving efficiency, developing innovative consensus mechanisms, enhancing privacy-preserving techniques, and addressing domain-specific challenges [9].

In summary, advanced topics in blockchain-based applications go beyond the basic understanding of blockchain technology. They involve addressing technical challenges, legal implications, and conducting research to unlock the full potential of blockchain in diverse industries. By exploring these advanced topics, we can foster innovation, overcome obstacles, and shape the future of blockchain-based applications.

Lastly, this work is summarized in five sections.

15.2 Open Issues Towards Blockchain-Based Applications

Open issues towards blockchain-based applications encompass a range of challenges and issues that need to be addressed for the widespread adoption and success of blockchain technology [10, 11]. These issues included in Figure 15.1.

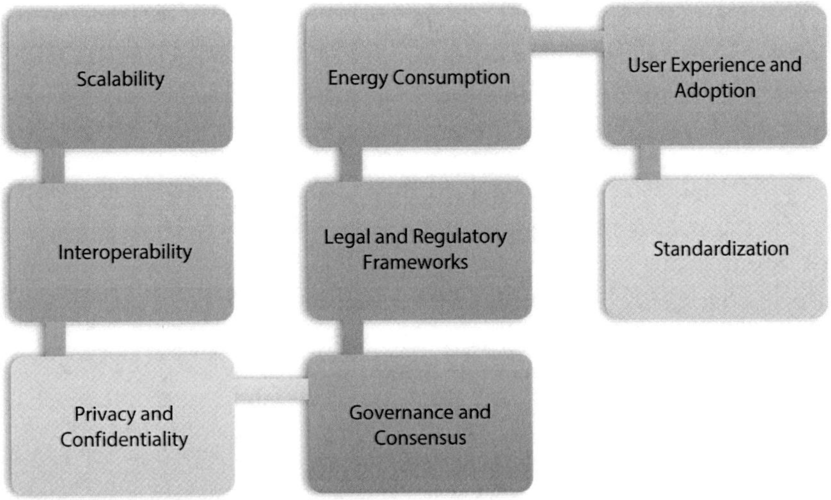

Figure 15.1 Open issues towards blockchain-based applications.

Scalability: Blockchain networks, particularly public ones, face challenges in scaling to accommodate a large number of transactions. As the number of participants and transactions increase, the performance of the network can be hindered. Developing scalable solutions that can handle high transaction volumes without compromising security is a critical issue.

Interoperability: The lack of interoperability between different blockchain networks is a significant challenge. As multiple blockchain platforms and protocols emerge, enabling seamless communication and data exchange between them becomes crucial. Establishing interoperability standards and protocols is essential to foster collaboration and integration between different blockchain systems.

Privacy and Confidentiality: While blockchain provides transparency and immutability, preserving privacy and confidentiality of sensitive data remains a challenge. Public blockchains inherently expose transaction details to all participants, which may not be desirable in certain contexts. Developing privacy-preserving techniques, such as zero-knowledge proofs or encryption mechanisms, is important to protect sensitive information while considering the benefits of blockchain.

Governance and Consensus: Blockchain networks often require governance mechanisms to make decisions regarding protocol upgrades, dispute resolution, and consensus algorithms. Designing effective governance models that ensure inclusivity, fairness, and consensus among network participants is a complex issue. Balancing decentralization, security, and efficiency in governance structures is crucial for the long-term success of blockchain applications.

Legal and Regulatory Frameworks: The legal and regulatory landscape surrounding blockchain-based applications is still evolving. Issues such as digital identities, smart contracts, cross-border transactions, data protection, and intellectual property rights require careful issues. Establishing clear legal frameworks that provide certainty, protect users, and foster innovation it is important to encourage widespread adoption.

Energy Consumption: Some blockchain networks, particularly those based on proof-of-work consensus mechanisms, have been criticized for their high energy consumption. Finding sustainable solutions that reduce the environmental impact of blockchain operations is an important open issue. Exploring alternative consensus mechanisms with lower energy requirements, such as proof-of-stake or proof-of-authority, is one potential avenue.

User Experience and Adoption: Blockchain-based applications often suffer from usability challenges that hinder mainstream adoption. Improving the user experience, reducing complexity, and enhancing accessibility are important factors for driving wider adoption of blockchain technology. User-friendly interfaces, seamless integration with existing systems, and intuitive design are areas that require further attention.

Standardization: The absence of widely accepted standards for blockchain protocols, data formats, and smart contract languages can hinder interoperability and hinder collaboration among different blockchain networks. Developing industry-wide standards and frameworks promotes compatibility, simplifies integration, and fosters innovation across various sectors.

In summary, these open issues require collaboration between researchers, technologists, policymakers, and industry stakeholders. Continued research, development, and innovation are necessary to overcome these challenges and unlock the full potential of blockchain-based applications in various domains.

15.3 Critical Challenges Towards Blockchain-Based Applications

15.3.1 Technical

Blockchain-based applications face several technical challenges that need to be addressed for their successful implementation and widespread adoption [12]. These challenges are included in Figure 15.2.

Scalability: One of the primary challenges is scaling blockchain networks to handle a large number of transactions. Traditional blockchains, like Bitcoin and Ethereum, have limitations in terms of transaction throughput. Finding efficient consensus algorithms, optimizing block size, and exploring techniques like sharding or off-chain transactions can help improve scalability.

Network Security: Blockchain networks rely on cryptographic algorithms to ensure the security and integrity of transactions. However, vulnerabilities in the underlying algorithms, such as hash functions or digital signatures, can lead to security breaches. Continual research and development of robust cryptographic techniques are necessary to mitigate these risks.

Consensus Mechanisms: Consensus mechanisms determine how participants in a blockchain network reach agreement on the validity of transactions. Traditional mechanisms like proof-of-work (PoW) and proof-of-stake (PoS) have their own limitations, including high energy consumption and the potential for centralization. Developing novel consensus mechanisms, like proof-of-authority (PoA) or practical Byzantine fault tolerance (PBFT), can address these challenges.

Interoperability: The lack of interoperability between different blockchain platforms and networks is a significant challenge. Blockchain systems often operate in isolation, making

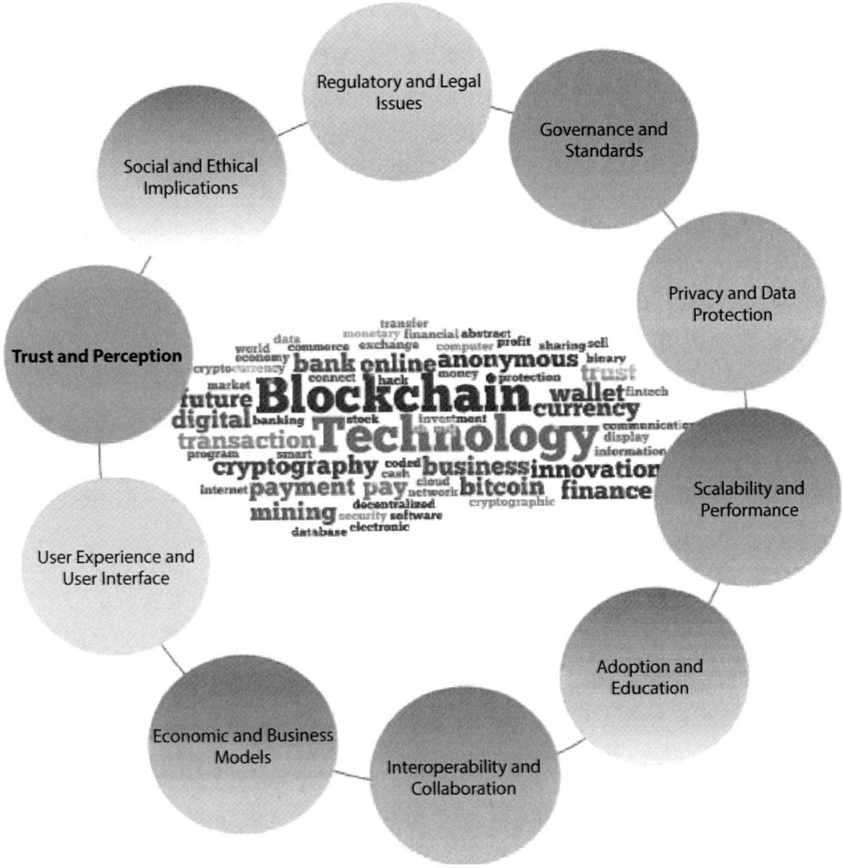

Figure 15.2 Technical challenges blockchain-based applications.

it difficult to exchange data or assets across different chains. Developing interoperability protocols and standards can enable seamless communication and data exchange between various blockchain networks.

Privacy and Confidentiality: Public blockchains are designed to be transparent and immutable, which can pose challenges for preserving privacy and confidentiality. Techniques like zero-knowledge proofs, ring signatures, and encryption methods can be employed to protect sensitive information without compromising the benefits of blockchain transparency.

Storage and Scalability of Data: As blockchain networks grow, storing and managing large volumes of data becomes a challenge. The size of the blockchain, which includes all historical transactions, increases over time, making it difficult for nodes to store and synchronize the entire chain. Research on efficient data storage and retrieval mechanisms, like distributed file systems or pruning techniques, is crucial.

Smart Contract Security: Smart contracts are self-executing agreements that run on blockchain networks. However, they are vulnerable to coding errors, security flaws, and malicious attacks. Auditing tools, formal verification techniques, and secure coding practices can enhance the security of smart contracts and mitigate risks.

Energy Efficiency: Some consensus mechanisms, such as PoW, require substantial computational power and energy consumption, leading to environmental issues. Developing energy-efficient consensus algorithms, like proof-of-stake or delegated proof-of-stake, can help reduce the carbon footprint of blockchain networks.

User Experience: Blockchain technology is often complex and challenging for non-technical users to understand and interact with. Improving the user experience through intuitive interfaces, simplified interactions, and seamless integration with existing systems can drive wider adoption of blockchain-based applications.

Regulatory Compliance: Blockchain technology operates in a regulatory landscape that is still evolving. Adapting to legal requirements and ensuring compliance with data protection, financial regulations, and anti-money laundering laws is a challenge. Collaborating with regulators and developing frameworks that balance innovation and compliance is crucial.

In summary, these technical challenges require ongoing research, innovation, and collaboration between academia, industry, and the broader blockchain community. Overcoming these challenges will unlock the full potential of blockchain technology and drive its adoption across various industries.

15.3.2 Non-Technical

Alongside technical challenges, blockchain-based applications also face non-technical challenges that can impact their adoption and implementation [13]. These non-technical challenges are shown in Figure 15.3.

Regulatory and Legal Issues: Blockchain technology operates within existing legal and regulatory frameworks that may not have been designed with blockchain in mind. Determining the legal status of blockchain-based transactions, smart contracts, and digital assets can be complex. Establishing clear regulations and addressing legal uncertainties surrounding blockchain is essential for its wider adoption.

Governance and Standards: Blockchain networks often lack standardized governance models, leading to challenges in decision-making processes, protocol upgrades, and dispute resolution. Developing effective governance structures that balance decentralization, transparency, and accountability is crucial. Establishing industry-wide standards for interoperability, data formats, and smart contracts can also facilitate broader adoption and collaboration.

Privacy and Data Protection: While blockchain provides transparency and immutability, it raises issues about privacy and data protection. Storing sensitive or personally identifiable information on a public blockchain can be challenging due to the public and permanent nature of the ledger. Striking a balance between transparency and privacy is crucial, and regulatory frameworks such as the General Data Protection Regulation (GDPR) need to be considered.

Scalability and Performance: The scalability challenges faced by blockchain networks have both technical and non-technical implications. Non-technical challenges include ensuring that blockchain solutions can handle large-scale adoption and transaction volumes without sacrificing performance. Balancing scalability, decentralization, and security is crucial for widespread adoption.

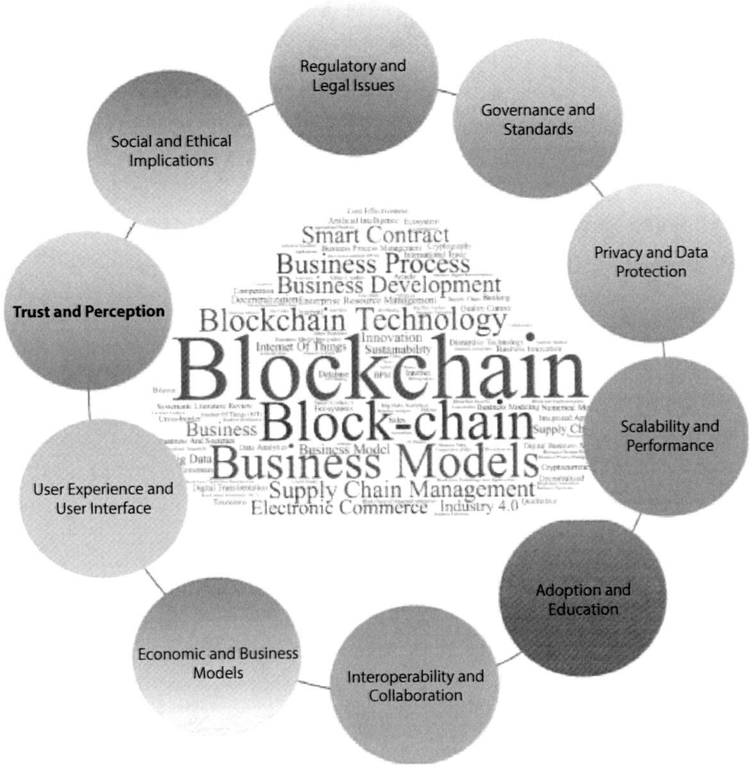

Figure 15.3 Non-technical challenges blockchain-based applications.

Adoption and Education: Blockchain technology is still relatively new and unfamiliar to many individuals and organizations. Raising awareness, providing education, and promoting understanding of blockchain's potential and benefits are key non-technical challenges. Encouraging adoption through training programs, industry collaborations, and awareness campaigns can drive the acceptance and utilization of blockchain-based applications.

Interoperability and Collaboration: Interoperability between different blockchain networks and collaboration between organizations are crucial for maximizing the potential of blockchain technology. Overcoming barriers to collaboration, fostering trust among stakeholders, and developing interoperability protocols and frameworks are non-technical challenges that need to be addressed.

Economic and Business Models: Blockchain technology disrupts traditional business models and introduces new economic paradigms. Identifying and implementing sustainable economic models, incentivization mechanisms, and token economies that align the interests of participants are non-technical challenges. These challenges involve understanding the economic implications of blockchain and fostering the creation of viable business models.

User Experience and User Interface: Improving the user experience and designing user-friendly interfaces for blockchain-based applications is essential for mainstream adoption. Simplifying complex concepts, enhancing usability, and providing intuitive interfaces that mask the underlying technical complexities are non-technical challenges that need to be addressed.

Trust and Perception: Blockchain technology is often associated with cryptocurrencies, which have faced controversies and negative perceptions. Overcoming skepticism, building trust among users and stakeholders, and demonstrating the value and potential of blockchain-based applications are non-technical challenges that require effective communication and awareness campaigns.

Social and Ethical Implications: Blockchain technology raises social and ethical issues such as digital inclusion, inequality, environmental impact, and the potential for decentralized governance structures. Understanding and addressing these implications, including ensuring fair access and preventing the concentration of power, are non-technical challenges that need to be considered.

Hence, these non-technical challenges require collaboration between industry, policymakers, researchers, and the wider community. By actively addressing these challenges alongside the technical aspects, blockchain-based applications can navigate the complexities and create a positive impact across various sectors.

15.3.3 Legal

Blockchain-based applications encounter several legal challenges that arise due to the unique characteristics and functionalities of the technology [14]. These legal challenges are included in Figure 15.4.

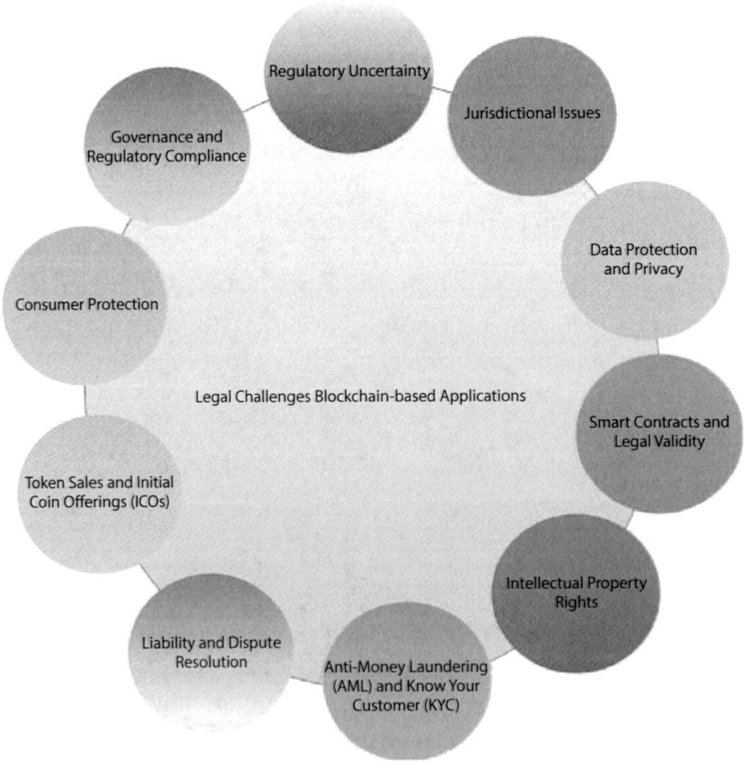

Figure 15.4 Legal challenges blockchain-based applications.

Regulatory Uncertainty: The regulatory landscape surrounding blockchain technology is still evolving, with different jurisdictions having varying approaches and interpretations. Determining the applicable regulations for blockchain-based applications, such as digital assets, smart contracts, and data privacy, can be challenging due to the absence of clear guidelines.

Jurisdictional Issues: Blockchain operates across borders, raising challenges related to jurisdictional authority and legal enforcement. Determining which jurisdiction's laws apply in cases of cross-border transactions or disputes can be complex. Resolving conflicts between different legal systems and establishing international legal frameworks for blockchain is crucial.

Data Protection and Privacy: Blockchain's transparency and immutability can conflict with data protection and privacy laws, such as the GDPR. Storing personal data on a public blockchain may require compliance with strict data protection requirements, including obtaining user consent and providing mechanisms for data deletion or rectification.

Smart Contracts and Legal Validity: Smart contracts, self-executing agreements on the blockchain, raise questions regarding their legal validity and enforceability. Determining whether smart contracts meet the legal requirements of traditional contracts, including elements such as provide, acceptance, and consideration, can be challenging.

Intellectual Property Rights: Blockchain technology can create challenges related to intellectual property rights, including copyright, patents, and trademarks. Determining ownership, licensing, and infringement of intellectual property in decentralized and transparent blockchain networks is an area that requires legal clarity.

Anti-Money Laundering (AML) and Know Your Customer (KYC): Blockchain's pseudonymity and borderless nature can pose challenges in complying with AML and KYC regulations. Establishing mechanisms to ensure compliance with these regulations, such as identity verification and transaction monitoring, is crucial for blockchain-based applications, especially in financial and cryptocurrency-related activities.

Liability and Dispute Resolution: Blockchain's decentralized nature and the absence of a central authority raise questions regarding liability and dispute resolution mechanisms. Determining liability in cases of code vulnerabilities, fraudulent transactions, or contract breaches, and establishing effective dispute resolution mechanisms within a decentralized environment are legal challenges that need to be addressed.

Token Sales and Initial Coin Offerings (ICOs): The emergence of ICOs as a fundraising mechanism using blockchain technology has raised issues related to securities regulations and investor protection. Determining whether tokens qualify as securities, complying with securities laws, and ensuring investor transparency and protection are legal challenges for blockchain-based fundraising activities.

Consumer Protection: Blockchain-based applications may involve consumer transactions and interactions. Ensuring consumer protection, including transparency, fair trade practices, and dispute resolution mechanisms, becomes essential in the context of blockchain-based applications that involve digital assets or peer-to-peer transactions.

Governance and Regulatory Compliance: Blockchain networks often require governance mechanisms to make decisions regarding protocol upgrades, consensus algorithm changes, and dispute resolution. Establishing effective governance models that comply with regulatory requirements and ensure transparency, accountability, and inclusivity is crucial.

In summary, these legal challenges require a collaborative effort between lawmakers, legal experts, technologists, and industry stakeholders. Developing clear legal frameworks, adapting existing regulations to accommodate blockchain technology, and fostering international cooperation are essential steps towards overcoming the legal challenges associated with blockchain-based applications.

15.3.4 Research

Blockchain-based applications present several research challenges that need to be addressed to advance the technology and its applications [15]. These research challenges are included in Figure 15.5.

Scalability and Performance: Scaling blockchain networks to handle a large number of transactions without sacrificing security and decentralization is a significant research challenge. Exploring techniques like sharding, off-chain computation, and layer-two solutions can enhance the scalability and performance of blockchain networks.

Consensus Mechanisms: Developing efficient and secure consensus mechanisms is crucial for blockchain-based applications. Research should focus on exploring alternative consensus algorithms beyond traditional proof-of-work (PoW) and proof-of-stake (PoS) to improve energy efficiency, scalability, and consensus finality.

Privacy and Confidentiality: Preserving privacy and confidentiality while maintaining the transparency and security features of blockchain is a challenging research area. Investigating privacy-enhancing techniques such as zero-knowledge proofs, ring signatures, and secure multiparty computation can provide solutions for privacy in blockchain-based applications.

Interoperability and Standards: Ensuring interoperability between different blockchain platforms and networks is important for seamless data exchange and collaboration. Research should focus on developing standardized protocols, data formats, and interoperability frameworks to enable interoperability across heterogeneous blockchain networks.

Security and Resilience: Enhancing the security and resilience of blockchain-based applications is a continuous research challenge. Investigating vulnerabilities, developing robust security mechanisms, and exploring methods for detecting and mitigating attacks, such as 51% attacks or smart contract vulnerabilities, are critical areas of research.

Governance and Consensus in Decentralized Networks: Designing effective governance models for decentralized blockchain networks is a research challenge. Exploring decentralized decision-making processes, incentive mechanisms, and dispute resolution mechanisms

Figure 15.5 Blockchain-based applications research challenges.

to ensure fairness, participation, and accountability in decentralized networks is an important research area.

Energy Efficiency and Sustainability: Addressing the energy consumption and environmental impact of blockchain networks is a research challenge. Investigating energy-efficient consensus algorithms, exploring alternative consensus mechanisms, and developing mechanisms for energy optimization in blockchain networks are important research directions.

Usability and User Experience: Improving the usability and user experience of blockchain-based applications is crucial for mainstream adoption. Research should focus on designing intuitive user interfaces, simplifying complex concepts, and enhancing user interactions to make blockchain applications more accessible to non-technical users.

Legal and Regulatory Implications: Blockchain technology operates in a regulatory and legal landscape that is still evolving. Research should discuss the legal and regulatory implications of blockchain-based applications, including aspects such as data protection, digital identity, smart contracts, and jurisdictional issues.

Adoption and Social Impact: Understanding the adoption dynamics and assessing the social impact of blockchain-based applications are research challenges. Investigating factors influencing adoption, studying the economic and social implications of blockchain technology in various sectors, and assessing the broader societal impact are important research areas.

In summary, these research challenges require interdisciplinary collaboration, involving researchers from computer science, economics, law, sociology, and other relevant fields. By addressing these challenges, researchers can contribute to the development and advancement of blockchain-based applications across various domains.

15.4 Future Work

Future work towards blockchain-based applications involves ongoing research, development, and innovation to address the existing challenges and discuss new opportunities [16, 17]. Here are some areas of focus for future work in blockchain-based applications [18–23] which are shown in Figure 15.6.

Scalability Solutions: Further research is needed to develop scalable blockchain solutions that can handle a high volume of transactions without sacrificing security or decentralization. Techniques such as sharding, sidechains, state channels, and off-chain computation can be discussed to improve scalability and enhance the performance of blockchain networks.

Interoperability Frameworks: Developing robust interoperability frameworks and protocols is crucial to enable seamless communication and data exchange between different blockchain networks. Efforts should be made to establish standards that facilitate interoperability, allowing for efficient collaboration and integration among various blockchain platforms.

Privacy-Preserving Techniques: Advancements in privacy-preserving techniques can address issues regarding data confidentiality on public blockchains. Research should focus on developing efficient methods such as zero-knowledge proofs, homomorphic encryption, and confidential transactions to enhance privacy while maintaining the transparency and integrity aspects of blockchain technology.

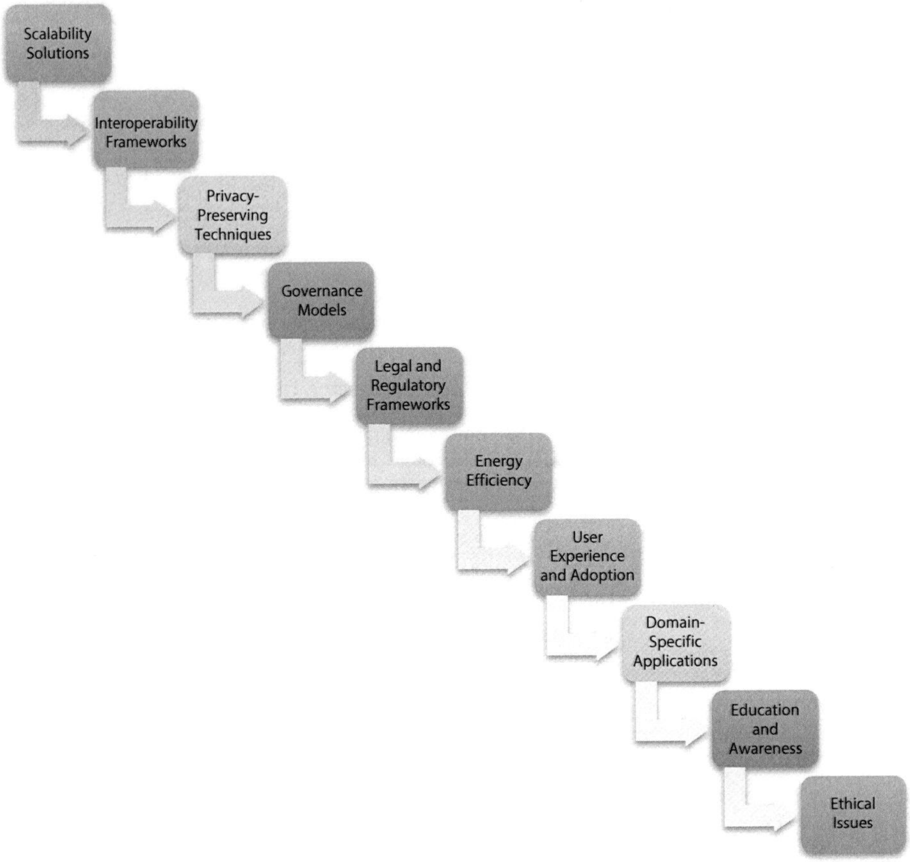

Figure 15.6 Future work in blockchain-based applications.

Governance Models: Designing effective governance models for blockchain networks is an ongoing area of research. Future work should discuss novel mechanisms that ensure inclusivity, fairness, and consensus among network participants. Decentralized governance, quadratic voting, and liquid democracy are some examples of governance models that can be discussed.

Legal and Regulatory Frameworks: As blockchain technology continues to evolve, there is a need to establish clear legal and regulatory frameworks that address the unique aspects of blockchain-based applications. Future work should involve collaboration between policymakers, legal experts, and technologists to develop frameworks that foster innovation, protect user rights, and ensure compliance with existing regulations.

Energy Efficiency: The energy consumption of blockchain networks, particularly those based on proof-of-work consensus, is an issue. Future work should focus on developing more energy-efficient consensus mechanisms, exploring alternatives like proof-of-stake or proof-of-authority, and optimizing the energy consumption of blockchain networks to minimize their environmental impact.

User Experience and Adoption: Improving the user experience of blockchain-based applications is critical for wider adoption. Future work should focus on designing user-friendly interfaces, enhancing the accessibility of blockchain technology, and simplifying complex processes to make blockchain applications more intuitive and user-friendly.

Domain-Specific Applications: Further research and development should be conducted to discuss the potential applications of blockchain technology in specific domains such as healthcare, supply chain management, finance, energy, and governance. Future work should focus on understanding the unique requirements and challenges in these domains and developing tailored blockchain solutions to address them effectively.

Education and Awareness: Promoting education and awareness about blockchain technology is crucial for its adoption and success. Future work should involve initiatives to educate professionals, policymakers, and the general public about the potential benefits, challenges, and applications of blockchain-based solutions. This can include training programs, workshops, and awareness campaigns.

Ethical Issues: As blockchain technology expands into various sectors, ethical issues become increasingly important. Future work should address ethical issues related to privacy, data ownership, consent, and fairness in blockchain-based applications. Ensuring that blockchain technology is used ethically and responsibly will be crucial for its long-term sustainability and societal acceptance.

In summary, future work in blockchain-based applications [24, 25] should focus on addressing existing challenges, exploring new opportunities, and fostering interdisciplinary collaboration to unlock the full potential of blockchain technology in various industries and domains.

15.5 Conclusion

This chapter presents a range of open issues, technical challenges, legal issues, and research opportunities. While blockchain technology has shown great promise in various domains, there are still several areas that require further exploration and development. From a technical perspective, scalability remains a key challenge for blockchain networks. As the adoption of blockchain-based applications increases, the need for efficient consensus mechanisms and scalable architectures becomes paramount. Solutions such as sharding, off-chain transactions, and layer-two protocols are being actively researched to address these scalability issues.

- Interoperability is another technical challenge in blockchain-based applications. As different blockchain networks emerge with their own protocols and standards, achieving seamless interoperability and data exchange between them becomes crucial. Efforts are underway to develop interoperability frameworks and protocols, allowing for the integration and communication between multiple blockchain networks
- Security and privacy are critical issues in blockchain-based applications. While the decentralized nature of blockchain provides inherent security advantages, ensuring the confidentiality and integrity of sensitive data remains a challenge. Advancements in techniques such as zero-knowledge proofs, secure multi-party computation, and privacy-preserving smart contracts are being pursued to enhance security and privacy in blockchain systems

- Legal and regulatory challenges also arise in the adoption of blockchain-based applications. The legal status of smart contracts, data protection and ownership, digital identities, and cross-border transactions on blockchain networks need careful consideration. Collaboration between technologists, legal experts, and policymakers is necessary to establish appropriate legal frameworks that facilitate blockchain adoption while ensuring compliance with existing regulations.

Moreover, further research is needed to discuss the potential applications and impacts of blockchain technology beyond cryptocurrencies. Industries such as healthcare, supply chain, finance, energy, and governance can benefit from blockchain-based solutions, but there is a need for in-depth studies to understand the specific requirements, challenges, and potential benefits in these domains. In conclusion, advanced topics in blockchain-based applications require continued research, collaboration, and innovation to overcome technical challenges, address legal issues, and unlock the full potential of blockchain technology across various industries. In summary, by addressing these open issues and advancing the state of the art, blockchain technology can revolutionize the way we transact, share data, and establish trust in a wide range of applications.

References

1. Tanwar, S., Bhatia, Q., Patel, P., Kumari, A., Singh, P.K., Hong, W.C., Machine learning adoption in blockchain-based smart applications: The challenges, and a way forward. *IEEE Access*, 8, 474–488, 2019.
2. Wenhua, Z., Qamar, F., Abdali, T.A.N., Hassan, R., Jafri, S.T.A., Nguyen, Q.N., Blockchain technology: security issues, healthcare applications, challenges and future trends. *Electronics*, *12*, 3, 546, 2023.
3. Gadekallu, T.R., Pham, Q.V., Nguyen, D.C., Maddikunta, P.K.R., Deepa, N., Prabadevi, B., Hwang, W.J., Blockchain for edge of things: Applications, opportunities, and challenges. *IEEE Internet Things J.*, 9, 2, 964–988, 2021.
4. Kouhizadeh, M., Saberi, S., Sarkis, J., Blockchain technology and the sustainable supply chain: Theoretically exploring adoption barriers. *Int. J. Prod. Econ.*, 231, 107831, 2021.
5. Xie, J., Yu, F.R., Huang, T., Xie, R., Liu, J., Liu, Y., A survey on the scalability of blockchain systems. *IEEE Netw.*, 33, 5, 166–173, 2019.
6. Hafid, A., Hafid, A.S., Samih, M., Scaling blockchains: A comprehensive survey. *IEEE Access*, 8, 125244–125262, 2020.
7. Grundmann, S. and Hacker, P., Digital technology as a challenge to European contract law: from the existing to the future architecture. *Eur. Rev. Contract Law*, 13, 3, 255–293, 2017.
8. Khan, S.N., Loukil, F., Ghedira-Guegan, C., Benkhelifa, E., Bani-Hani, A., Blockchain smart contracts: Applications, challenges, and future trends. *Peer-to-Peer Networking Appl.*, 14, 2901–2925, 2021.
9. Diro, A., Zhou, L., Saini, A., Kaisar, S., Hiep, P.C., Leveraging zero knowledge proofs for blockchain-based identity sharing: A survey of advancements, challenges and opportunities. *J. Inf. Secur. Appl.*, 80, 103678, 2024.
10. Toufaily, E., Zalan, T., Dhaou, S.B., A framework of blockchain technology adoption: An investigation of challenges and expected value. *Inf. Manage.*, 58, 3, 103444, 2021.

11. Gadekallu, T.R., Pham, Q.V., Nguyen, D.C., Maddikunta, P.K.R., Deepa, N., Prabadevi, B., Hwang, W.J., Blockchain for edge of things: Applications, opportunities, and challenges. *IEEE Internet Things J.*, 9, 2, 964–988, 2021.
12. Lu, Y., The blockchain: State-of-the-art and research challenges. *J. Ind. Inf. Integr.*, 15, 80–90, 2019.
13. Agrawal, K., Aggarwal, M., Tanwar, S., Sharma, G., Bokoro, P.N., Sharma, R., An Extensive Blockchain Based Applications Survey: Tools, Frameworks, Opportunities, Challenges and Solutions. *IEEE Access*, 2022.
14. Upadhyay, N., Demystifying blockchain: A critical analysis of challenges, applications and opportunities. *Int. J. Inf. Manage.*, 54, 102120, 2020.
15. Murthy, C.V.B., Shri, M.L., Kadry, S., Lim, S., Blockchain based cloud computing: Architecture and research challenges. *IEEE Access*, 8, 205190–205205, 2020.
16. Khan, S.N., Loukil, F., Ghedira-Guegan, C., Benkhelifa, E., Bani-Hani, A., Blockchain smart contracts: Applications, challenges, and future trends. *Peer-to-Peer Networking Appl.*, 14, 2901–2925, 2021.
17. Alam, S., Shuaib, M., Khan, W.Z., Garg, S., Kaddoum, G., Hossain, M.S., Zikria, Y.B., Blockchain-based initiatives: current state and challenges. *Comput. Networks*, 198, 108395, 2021.
18. Nair, M.M. and Tyagi, A.K., Chapter 11 - AI, IoT, blockchain, and cloud computing: The necessity of the future, in: *Distributed Computing to Blockchain*, R. Pandey, S. Goundar, S. Fatima (Eds.), pp. 189–206, Academic Press, 2023, ISBN 9780323961462, https://doi.org/10.1016/B978-0-323-96146-2.00001-2.
19. Tyagi, A.K., Chapter 2 - Decentralized everything: Practical use of blockchain technology in future applications, in: *Distributed Computing to Blockchain*, R. Pandey, S. Goundar, S. Fatima (Eds.), pp. 19–38, Academic Press, 2023, ISBN 9780323961462, https://doi.org/10.1016/B978-0-323-96146-2.00010-3.
20. Tyagi, A.K. and Richa, Smart Manufacturing Using Internet of Things, Artificial Intelligence, and Digital Twin Technology, in: *Global Perspectives on Robotics and Autonomous Systems: Development and Applications*, 2023, doi: 10.4018/978-1-6684-7791-5.ch008.
21. Tyagi, A.K., Lakshmi, P.R., Mishra, A.K., Balamurugan, G., Industry 5.0: Potentials, Issues, Opportunities, and Challenges for Society 5.0, in: *Privacy Preservation of Genomic and Medical Data*, pp. 409–32, 2023 Oct 17.
22. Tyagi, A.K., Mishra, A.K., Vedavathi, N., Kakulapati, V., Sajidha, S.A., Chapter 21 Futuristic Technologies for Smart Manufacturing, in: *Research Statement and Vision for the Future*, Tyagi, A.K. (Ed.), 2023, https://doi.org/10.1002/9781394213948.ch21.
23. Tyagi, A.K. and Tiwari, S., The Future of Artificial Intelligence in Blockchain Applications, in: *Machine Learning Algorithms Using Scikit and TensorFlow Environments*, IGI Global, 2024, doi: 10.4018/978-1-6684-8531-6.ch018.
24. Tyagi, A.K., Kumari, S., Chidambaram, N., Sharma, A., Engineering Applications of Blockchain in This Smart Era, in the book: *Enhancing Medical Imaging with Emerging Technologies*, 2024,. DOI: 10.4018/979-8-3693-5261-8.ch011.
25. Tyagi, A.K., Kukreja, S., Richa, and Sivakumar, P., Role of blockchain technology in smart era: A review on possible smart applications. *J. Inf. Knowl. Manag.*, February 2024., https://doi.org/10.1142/S0219649224500321.

16

Issues and Challenges in Implementing Smart Manufacturing in the Current Scenario

Charanya J.[1]*, Shrisudhan B.[1], Sathruba P.[1], Sharun P.[1] and Prakash Duraisamy[2]

[1]*Department of ECE, Kongu Engineering College, Perundurai, Erode, Tamil Nadu, India*
[2]*Department of Computer Science, University of South Alabama, Alabama, USA*

Abstract
The implementation of smart manufacturing in the current scenario presents a spectrum of issues and challenges that demand comprehensive examination. This abstract delves into the multifaceted nature of these obstacles, ranging from technological complexities and interoperability concerns to workforce readiness and cyber security risks. Analyzing the intricate landscape of smart manufacturing reveals the necessity for adaptive strategies, collaborative frameworks, and innovative solutions to navigate these challenges effectively. Understanding and addressing these issues is paramount for unlocking the full potential of smart manufacturing in the contemporary industrial landscape.

Keywords: Smart manufacturing, challenges, cyber security risks, innovative solutions

16.1 Introduction

A paradigm shift from "centralized" to "local" production is made possible by networking and openness in the industrial sector. "Embedded systems," which gather and transmit certain data, are being used in production today. A cyber physical system (CPS), capable of operating with growing independence, is created in the "factory of the future" by a central computer that orchestrates the intelligent networking of all subsystems [7].

Nonetheless, the real and virtual worlds collaborate closely thanks to human machine interfaces: process management happens automatically, but requirement definition is done by a human operator. CPS is the networking of discrete embedded software systems for the purpose of gathering and transferring specific data. Hence, production shifts from "centralized" to "local" which refers to autonomous process management is allowed, and intelligent networking is managed by a central information system that also takes into account concrete elements like requirement input through human-machine interfaces.

A completely new element of the production process is represented by the intimate interplay between the real and virtual worlds in this instance. We refer to cyber physical

**Corresponding author*: charanya.ece@kongu.ac.in

Amit Kumar Tyagi, Shrikant Tiwari, Senthil Kumar Arumugam and Avinash Kumar Sharma (eds.) Artificial Intelligence-Enabled Digital Twin for Smart Manufacturing, (341–358) © 2024 Scrivener Publishing LLC

production systems (CPPS) when discussing this in relation to production. Presently, Industry 4.0's primary enabling technologies and development trends encompass:

- IT
- Data Analytics
- Future Robotics
- System Integration with novel standards
- Simulation modeling
- Cloud Computing
- Additive Manufacturing
- Augmented Reality
- The Industrial Internet of Things and
- Cyber Security.

Small and Medium sized Enterprises' (SMEs') place in Industry 4.0 is especially concerning because of their often low automation levels and financial constraints. SMEs, which generate a large portion of employment and value creation, face a number of opportunities and challenges as a result of Industry 4.0. Since many still employ outdated, proprietary systems, SMEs stand to gain from developing Industry 4.0 networking and integration standards as well as open standard architectures [7].

SMEs will be able to adapt to market demands much more quickly and with a significant reduction in production management effort thanks to this. SMEs can become transient production networks with accurately measured value added contributions by merging with larger networks. Moreover, customized and small batch items can be produced by flexible machinery and additive manufacturing (3D printing).

16.1.1 The Revolution of Industry

Industry 4.0 may best be defined as a real-time, smart and virtual network for individuals, equipment, and things in terms of industrial management [9]. It also allows for greater digitalization of the entire value chain and real-time data exchange between people, objects, and systems [10]. In order to improve communication between all parties involved in the product development process, analyze data using predefined algorithms, and control production flows to support continuous improvement, modern information and communication technologies could be integrated into machinery, workers in production, and logistics processes [15].

In contrast to the preceding industrial revolutions, the fourth is evolving at an exponential rate rather than a linear one, and it is altering not just the "what" and "how" of doing things but also "who" we are, according to the visionary work [4]. All industries are undergoing significant change, as seen by the rise of new business models, the disruption of established players, and the reconfiguration of the systems for delivery, production, and transportation.

Three concepts might be used to characterize Industry 4.0, Because the goods have a memory to store operational data and standards separately [15], they are intelligent products that make it possible to request the necessary resources and coordinate manufacturing operations Intelligent Machine. Intelligent Machine, which enables a flexible and modular production line by substituting, decentralized self-organization for the conventional production hierarchy. Knowledge automation enhanced operator to support an adaptable and flexible component in the production system [5]. Cyber-Physical Systems (CPS), the Internet of Services (IoS), and the Internet of Things (IoT) are the key components of Industry 4.0 [1], and the latter will be a crucial instrument in its deployment.

The main upheavals brought about by this revolution are the changes in economies and employment. Certain occupations go as a result of process automation, but new higher paying professions also raise that call for new skills that enable quick adaptation, entrepreneurship, and invention. Businesses are being forced to reevaluate how they conduct business due to the Fourth Industrial Revolution's irreversible transition from the Third Industrial Revolution's basic digitalization to innovation based on technology combinations. Industry 4.0, the creation and spread of technology for ubiquitous connectivity and self-governing, cyber physical systems, is what propels the Fourth Industrial Revolution. Despite being primarily a German initiative, the German economy and government continue to be its main proponents.

16.1.2 Trends in Industry 4.0 Development

Germany and international technology businesses exhibit the highest levels of Industry 4.0 application. Businesses with a wide range of production and automation solutions currently include General Electric, Mitsubishi, Siemens, and others. Many technologies and solutions are already marketed as "Industry 4.0" by manufacturers and automation technology developers

However, many of the building blocks required for the implementation of Industry 4.0 concepts including the concepts of simulation modeling are already available, even in the absence of specialized vendors. Examples include cloud computing, tablets serving as human machine interfaces, digital and networkable sensors and control elements (actuators), integrated software solutions, and (industrial) communication networks. Large scale battery powered sensor networks can be built thanks to new Green IT power saving technology. However, a significant shortcoming is the pervasive absence of standards.

While many of the technological components of Industry 4.0 are already in place, some still need to be standardized globally [7]. As of right now, Industry 4.0 is largely still at the conceptual stage and is not yet a commercially available good or service. Customers' inflated expectations and the vague meaning of "Industry 4.0" are partly to blame for this. Industry and management software, processors, and devices are all necessary for Industry 4.0, it is assured .These gadgets necessitate specialized knowledge in automation and information and communication technology (ICT), which offers educators and trainers preparing the next generation of workers a chance and a challenge.

16.2 Challenges in Smart Manufacturing

Figure 16.1 depicts some of the challenges in Smart Industry and the description is as follows.

Figure 16.1 Challenges in smart manufacturing.

16.2.1 Technological Challenges

One common problem when integrating IoT devices is interoperability. These problems occur when disparities in communication protocols, data formats, or standards prevent various systems or devices from collaborating without difficulty. In the context of IoT, some crucial elements of interoperability issues are Protocols for Communication. Different communication protocols such as MQTT, CoAP, HTTP, or proprietary protocols are frequently used by IoT devices. It might be difficult to guarantee interoperability between devices that use different protocols

Devices may receive firmware updates at disparate periods, resulting in version discrepancies. When devices with different firmware versions try to connect, compatibility problems may occur. In broadcast updates to preserve compatibility, safe and uniform procedures for over the air upgrades must be established. Industry cooperation, adherence to standards, and the usage of middleware solutions that can translate between various protocols and formats are crucial for resolving interoperability concerns. Furthermore, continuous efforts to create and advance open standards can have a major impact on enhancing interoperability in the context of the Internet of Things [14].

16.2.2 Infrastructure Challenges

Standards and Data Format Proprietary data formats or standards that are incompatible with current systems may be used by legacy systems. One such solution is to create data transformation processes that can translate data formats or employ middleware to translate data. With Communication Protocols Outdated or specially designed protocols may be used by legacy systems to communicate. To translate communication between current and legacy systems, use middleware or API gateways. Security Issues identified are Older systems may not have the latest security measures in place, which leaves them open to online attacks. In order to avoid the issue put in place safe channels for communication with proper data encryption, and carry out exhaustive security audits [13].

Retrofitting is the process of incorporating new features or capabilities into pre-existing systems or structures, particularly in the context of infrastructure or technology. Retrofitting has a number of advantages, but it also frequently has drawbacks. Retrofitting new features into older systems can be challenging due to their limits in terms of memory, processing power, and support for contemporary protocols. So assess whether it is feasible to improve or expand the capabilities of the current system. There are situations where modernization calls for a phased approach. Problems with Compatibility may arise when integrating new technology or components with current systems. It's possible that legacy systems weren't made to easily integrate with contemporary technologies. In order to resolve any possible problems, carry out in-depth compatibility analysis. Through proactive resolution of these issues, organizations can improve the efficacy and durability of their retrofitting endeavors [2]. When integrating new technologies or components with old systems, compatibility problems are sometimes encountered, especially during retrofitting or upgrading projects. Disparities in hardware, software, protocols, or communication standards may give rise to these problems

16.2 3 Regulatory and Compliance Challenges

In smart manufacturing, where digital technology and data-driven processes are widely integrated, data governance and compliance are essential components. While compliance procedures assist organizations in adhering to pertinent legislation and standards, effective data governance guarantees that data is managed, secured, and utilized properly.

The Data Governance Framework is a methodical approach to the management, safeguarding, and utilization of data assets. Components ascertain who owns the data, specify roles for data stewardship and develop security, privacy and quality policies. Data needs to be classified and sorted according to significance and sensitivity [13].

The Classification Levels are very confidential, internal use, and public. Determine which access and controls are suitable depending on classifications. Making sure that data is reliable, consistent, and appropriate for its intended use is the goal of data quality management. The activities of classification are Data cleansing, validation, profiling, and continuous monitoring to uphold requirements for data quality. GDPR, CCPA, HIPAA, and other regulations may apply to data privacy compliance, depending on the sector and region. So request permission to handle the data, apply privacy policies into place, evaluates the effects of decisions, and set up procedures for protecting the rights of data subjects.

16.2.4 Human Factors

The difference between the abilities that businesses require and the skills that the labor force already possesses is known as the workforce skill gap. This disparity may impede the competitiveness, innovation, and performance of an organization. The industries that are changing quickly, closing the skill gap in the workforce is essential for economies and businesses to prosper. Key elements of the skill gap in the workforce are Technological Progress. Workers must learn new technical skills in light of the rapid improvements in technology, including automation, data analytics, and Artificial Intelligence [13]. Without proper training, staff members might not be able to use and operate new technologies, creating a huge skill gap. A greater degree of digital literacy is required as organization become more digitally connected. The impact is the inability of employees with limited digital skills to use digital technologies may impede their productivity and flexibility in adjusting to digital workflows. Soft skills, like teamwork, communication, critical thinking, and adaptability, are frequently in great demand, but they might not be present in the workforce. The lack of soft skills can negatively impact workplace dynamics by impeding problem-solving, teamwork, and effective communication.

16.3 Difficulties and Prospects for Further Research on Blockchain Application

When it comes to the use of blockchain technology in Industry 4.0 and Society 5.0, there have been a lot of unanswered questions during the past ten years, including technical and legal ones. These issues and challenges might be swiftly remedied once blockchain technology is applied creatively [2]. As a result, the following describes these scenarios.

16.3.1 Unresolved Concerns about Blockchain Technology in Industry 4.0

In the Industry 4.0 context, real time data is critical to the operation of industrial and service systems. Since processing time is the primary issue, real world applications need to be carefully considered. Blockchain technology makes recordkeeping and information accessibility possible. It can also handle a wide range of complex issues. Therefore, the purpose of this study is to understand how blockchain functions in Industry 4.0. It's time for businesses to once again use blockchain technology. People will find it simpler to conduct business as a result.

Blockchain Technology is being used by Industry 4.0 manufacturers to make environmentally friendly products [2]. This facilitates monitoring the goods that individuals produce. Information security may be strengthened by this clever, technologically advanced production. Production settings with higher process safety standards and fewer risks will be more conducive for this to flourish. But before putting blockchain into practice, companies need to use the platform a lot. As technology advances, corporate executives need to understand how technology may benefit their organizations in order to reduce risk. More privacy and trust are what Industry 4.0 seeks.

This slow moving revolution, known as "Society 5.0," originated in Japan. By placing humans at the center of technology and innovative advancements for the greater good, it promises to transform society. "Society 5.0" refers to the creation of new Cyber Physical Social (CPS) systems using AI concepts from the past. In the past few years, it has played a significant role in the changes that businesses and society have undergone [3]. Businesses and individuals can benefit from parallel intelligence while tackling challenging social and engineering issues. It can benefit businesses as well.

The objective is to create rapid, focused, and convergent approaches to comprehend unclear, multifaceted, and intricate problems. The world faces challenges from terrorism, global warming, and growing economic inequality. People nowadays are a part of the virtual real intelligence technology period, which is a new phase of technological advancement. This is because there have been significant advancements in information, networking, power, mechanization, and information technology. Connecting things, people, and the physical and virtual worlds will have a lot of positive effects. Long term, this will improve people's quality of life and maintain a healthy rate of economic growth.

An entirely novel model for managing and regulating CPSs has evolved with the advancement of AI technology. An intelligent civilization uses instruments, networks, and digital technology to make life better. It has the ability to link many nations, areas, firms, and cultures. Recall that this is something that CPSs can do. Artificial systems, technology experimentation, and parallel execution are the three elements that comprise ACP. The growth and transformation of the information society gave rise to Society 5.0. It sought solutions to world issues and might help both people and companies [12].

16.3.2 Obstacles to Blockchain Implementation in Industries and Social Structures

All of the information may be digitally stored by this technology, which will improve and speed up business operations. A corporation needs the appropriate people if it hopes to use new technologies and turn a profit. When applied properly, this technology can assist in resolving issues with data privacy. It is possible for employees of various companies to exchange data and facilitate data sharing [4]. That is a major matter even if there are issues with the way things are created now or in the near future. On the same blockchain, they can communicate with one another in real time. Those apparatuses might be vulnerable to assault if they have defects. Industry 4.0 can provide numerous security measures [12].

Most modern gadgets can automatically share files and communicate with one other when they have an Internet connection. This will make all of the things that people deal with on a daily basis at work, at home, and during production and use more interactive. This is the main difference between the previous structure 4.0, which was limited to software and the new information 5.0. To further improve the situation, all of the data must be combined as massive amounts of information beyond several servers, and AI must provide recommendations quickly. It will take a lot of data to accomplish this. There are numerous data silos in the contemporary business environment.

Utilizing big data makes it possible to see what others have already done. People who work for various companies do not discuss matters like this with each other. There is no utility of micropayments for such data. It will be challenging to obtain Society 5.0 if these issues are not resolved.

If all of the information was held by a single firm, there could be less innovation and unethical use of it. To optimize profits, we might have to give up our freedom of expression and privacy. The current state of affairs is mostly due to two factors. The very centralized nature of the Internet platform is one. In order to guarantee data security and consistency, centralized management is required because it is difficult for an individual to be 100% certain of their identity online these days. However, companies have every right to utilize data to increase their own value in our capitalist society.

It also stems from how challenging it is to reorganize the data. Since data creates a barrier for other firms, they are typically traded to unrelated businesses. But the fundamental tenets of capitalism demand that businesses maximize their profits. Big, well-funded organizations end up with too much data because tiny businesses cannot afford to purchase additional data. People who think that access to data is a barrier to entry only engage in profitable ventures and show little concern for societal issues [4].

16.3.3 Possible Applications of Blockchain Technology in Smart Manufacturing

Supply chain management, operations management, commercial finance, and other related disciplines will find usage for blockchain technology. Many disagreements might be settled quickly if agreements were personalized for each party. This will need financial services organizations to do less physical labor. The AI technique will help readers decide whether or not to accept the material by pointing out any inconsistencies or significant mistakes. Many organizations are connecting blockchain technology and the Internet of Things. This organization has developed a technique for combining precise and secure digital data archives with networked devices.

Blockchain are being used by businesses to enhance their operations. With this technology, data will be shared and kept in an identical way. Real time tracking of the locations of items is made possible by blockchain technology. This will enhance the commodity trail's intelligence and security. The blockchain records each and every transaction in the supply chain process, giving long term data and a complete picture [12]. It will increase the efficiency of the supply chain by connecting it to all the individuals and places it must deliver things to. Blockchain will change business operations in many industries in the future, but businesses will need to put in time and effort to deploy it.

Every block has a timestamp and is connected to the one before it. Blockchain technology allows customers to examine a product's history, regardless of whether it is new or used. In conclusion, within the context of the fourth industrial revolution, this kind of blockchain application has the ability to reduce differences between planning and execution. Since the development of the digital economy depends on this collaboration, academics need to be aware of the difficulties they confront. Thus, managers and policymakers may leverage blockchain to assist welcome in Economy; all they need to do is address the previously identified commuters, facilitators, and constraints. Cloud computing, blockchain data storage, hybrid data storage, and data query processing are some of the main topics under the present emphasis area that might use more investigation.

The rapidly evolving fields of employment, government agencies, the economy, and privacy demands that digital information adapt as well. A number of components, including innovative policies, an entrepreneurial spirit, and skill sets, are needed to make Society 5.0 a reality. While new technologies have the potential to improve people's lives, they may also have negative effects on information, jobs, and income distribution. Industry 4.0 restricts the use of new technologies to factories; by contrast, Society 5.0 permits individuals to utilize contemporary technologies in their everyday lives, as well as in other areas such as their health and other domains, such as robots, artificial intelligence, IoT, and augmented reality. Technological advancements, increased productivity, and reduced food consumption have the potential to improve the economy and resolve societal issues [4].

These technological advancements have the potential to ease societal problems and facilitate people's access to necessities. The sustainable development goals of the UN can be achieved by Society 5.0 by utilizing the creative potential of many individuals combined with the power of technology.

16.4 Obstacles in Industry 4.0

Industry 4.0 is tackling a number of issues and difficulties that come with changes and evolutions include Novel approaches and methods from a commercial standpoint. Examining your business case and unique selling proposition; updating current techniques to yield better outcomes; Performing and carrying out great pilot tests; analyzing and identifying areas within the business where action is needed; bringing about management changes; comprehending the organization's work culture; and acting as a link between several departments fostering and assisting new talent The connecting of several systems is one of the key innovations of this industrial revolution, however there are still challenges and security issues associated with this. In order to withstand data leaks, data security must be ensured against both direct and external virus attacks [5].

This is where the problem of cyber security arises since, in addition to infrastructures that are necessary for correct operation, frameworks that pose a risk to the environment and public health also need to be controlled. Any changes or alterations must be tested and validated to ensure the framework is reliable, secure, and trustworthy before the system is put into place and rolled out. In actuality, testing especially in fields like OT and IT systems is always an essential step in the change process. Testing is a more challenging task than it has ever been in a fully digitalized and connected corporation. One of the most discussed topics in Industry 4.0 is the use of AI for automation and augmentation, predictive maintenance, decision making, enhancing security, etc. A critical step in the development process, "learning" or "training" is necessary for the majority of the algorithms in use to operate properly [11].

Obtaining such data can prove to be a difficult process because, in order to test the complete system in an integrated manner, it is necessary to have a large range of data covering all the major basic states, in addition to a sufficient number of datasets for learning. When it comes to automation in the industrial sector, there are a lot of tight regulations that must be followed in order to protect both human and environmental health [5]. It is crucial to understand that Industry 4.0 is a collection of numerous standards that are constantly changing rather than a single standard. Even though there may be certain obstacles to overcome, Industry 4.0 offers a whole new universe of potential and breakthroughs.

16.4.1 Challenges and Responses

A number of novel challenges that have not been encountered before are presented by Industry 5.0, including:

- Societal heterogeneity with regard to principles and recognition
- Measuring the creation of human and natural value
- Integrating customers across whole sectors of value to SMEs
- Multidisciplinary research fields and system intricacy
- Ecosystem-oriented innovation policy using a versatile, ambitious methodology.

To support the upcoming Industry, the European Commission also unveiled a new effort that included a number of implementation techniques from the domains of marketing, governance, and investment. Other industries' and governments' responses are still scarce at the moment. Some of the Journals and researches emphasize the importance of the new revolution in 1621, demonstrating how rapidly academia has embraced the issues surrounding the industry [10].

In order to succeed, Industry 5.0, like Industry 4.0, will require significant funding from governmental organizations. Industry 5.0's guiding the concepts of endurance, conservation, and people symmetry have emerged as important factors behind social advancement rather than being a result of GDP-driven wealth development, regardless of the direction the industry takes. The government imposed several efforts to incorporate them into national policies [5].

16.4.2 Industry 4.0 Sustainability and Resilience

Many people believe that Industry 4.0 is heavily focused on technology or technical solutions. Several of the first government manifestos and policies have this kind of emphasis. Early research articles also have a tendency to be technology-focused. There is also a noticeable technological bias in whitepapers and business studies released by some of the best consulting companies. However, the concepts of endurance, conservation, and people symmetry may not remain unnoticed by the revolution. Certain important publications deal with resource efficiency and social demands. One initiative, Factory to Fit, seeks to empower and include laborers in a more interconnected industrial setting. Workers now have more power to shape the industrial process, which carries with it increased responsibility using digital tools [11].

Still, Industry 4.0 adopts a defined technological strategy and a consequential viewpoint to solve the problems of human centricity, sustainability, and resilience [6]. Industry 5.0 takes a daring turn away from individual technologies and towards a methodical approach, in contrast to Industry 4.0. The well-being of industry workers is positioned at the center of the production process, enabling the sector to meet social goals beyond job creation and development. This might assist to clarify why Industry 5.0 is seen as an entirely distinct kind of industrial revolution in comparison to the others.

While the European Union (EU) recognized six enabling technologies of Industry 5.0, the Boston Consulting Group listed nine essential enabling technologies of Industry 4.0. Although these technologies have different names, there is an obvious overlap. Many of Industry 4.0's enabling technologies are seen to be able to assist and will surely be employed in order to meet Industry 5.0's social objectives. Nonetheless, there are a few more focused Industry 5.0 technologies that need to be taken seriously, such bio-inspired technologies and technology for energy storage, renewable energy, and efficiency [6].

Emerging revolution does not represent the shift away from or a temporal extension of the current Industry 4.0 paradigm. A forward-thinking exercise produced Industry 5.0, a framework for describing the coexistence of industry with new societal demands and trends. Therefore, Industry 5.0 enhances and expands upon the distinguishing characteristics of Industry 4.0 [9]. This may assist in distinguishing Industry 5.0 from the other Industrial Revolutions by recognizing that they are only the chronological continuation of their forerunners. Pointing out that some of the most recent research on Industry 4.0 has already expanded upon the initial concept and goal of Industry 4.0 is important. This is especially true in Germany, where new measures have been launched regarding resource efficiency and energy footprint thrown aside recently [11].

Industry 4.0 and Industry 5.0 are complementary and expand upon each other, meaning that they should be seen in tandem. Industry 4.0 is driven by technology, whereas Industry 5.0 is driven by values. Although the term Techno Social System has been around for a while, we could be led to believe that we are seeing a Techno Social Revolution (or maybe a Social Techno Revolution), with technology serving as the enabling tools and societal needs as the ultimate goal [6], in an attempt to simplify the terminology at the risk of creating even more new terms.

16.5 The Internet of Things: Security Threats and Challenges (IoT)

Determining the harmfulness of IoT network problems, IoT application services, and IoT hardware is necessary for creating and implementing comprehensive IoT security solutions [1]. Numerous security flaws in the Internet of Things have been discovered by the Internet Engineering Task Force, as seen below. The following are the flows:

(a) The imitation of Internet of Things devices by dubious producers
(b) Malicious lower quality goods are substituted for lower quality items
(c) During commissioning, a man-in-the-middle attack occurred as a result of an entirely inadequate authentication system.
(d) Malevolent code is added to the firmware by an attacker.
(e) Risk to sensitive personal information
(f) One kind of cyber-attack is a Denial of Service (DoS) attack
(h) An attempt to listen in on an improperly configured Internet of Things network
(i) The extraction of security features from unsecured IoT devices.

Future IoT security research ought to address the following important IoT security issues:

1. Security of communication: It is essential for the secure transmission of vital IoT data across the Internet in real time. Sensitive data should not be encrypted by many IoT devices before being transferred over the internet, as was previously mentioned. Because Internet of Things data needs to be sent and received over a wide network, secure private networking, while helpful in preventing vulnerabilities, is not necessarily the best choice [14]. Moderately quality IoT data packaging, like a network of nodes, could be able to help with some of the problems. There are chances for future study to address this issue.
2. Data integrity and privacy: It's challenging to uphold honesty and privacy. Users' data should only be accessible to those who have been given authorization. The user has to give sufficient permission before letting other people utilize the data. When data is no longer required, it must be disposed of securely [6].
3. Implementation privacy: Data from IoT nodes, mobile devices, and the internet is stored in the cloud. A user's location, bank account, health, and other sensitive details might be disclosed. Nothing can keep client data safe from a threat using the internet, cloud, or mobile devices not even a secure

communication network. This might make it difficult to secure data from the Web of Objects that has been posted on the internet, public networks, or cell phones.

4. Vulnerability identification and management: Finding and managing a variety of security threats in Internet of Things nodes can be challenging. It is challenging to identify the broken nodes in an IoT network because there are so many of them. To create new frameworks to address this problem, more analysis is required.
5. Firmware issue: Installing security fixes and firmware updates on Internet of Things devices may be challenging. New online security vulnerabilities are found every day. It may be necessary for users of IoT devices to monitor the updates that are loaded on them. Not every IoT device can receive daily live updates. Surmounting the device could be necessary for users to install firmware or upgrade. Surmounting the device could be necessary for users to install firmware or upgrade. It can be problematic that wireless updates have been discontinued by many devices, despite the fact that a built-in update might be beneficial.
6. Incident management and recovery: Internet of things devices that are essentially environment neutral. And a major issue might be an IoT node failure. Since IoT sensors handle confidential information in real-time IoT devices, a strong digester recovery plan and incident management is very constrained.
7. Authorization and authentication: A lot of devices make up Internet of Things networks. These devices need to be able to seamlessly connect to the network at any time. Internet of Things networks consist of a large number of devices. These devices have flexible join times and can connect to the network at any time. Security issues arise when manufacturers' default passwords are left unmodified and when devices are used with weak passwords. In this instance, both authorization and authentication are necessary [14]. IoT devices must be capable of reading from and writing to a subset of a database while ignoring the rest. Impacted parties could be able to read and write to private data on the device.
8. Device identity: Every Internet of Things device has a distinct identity. Device names connected to the devices are provided by domain name servers [10]. Conversely, DNSs are susceptible to a number of assaults, including DNS cache positioning attacks and man in the middle attacks. Conversely, DNSs are vulnerable to a range of attacks, such as DNS cache positioning and man in the middle attacks.
9. Handling enormous IoT devices: Managing the growing quantity of items in IoT networks is getting harder every day. IoT network maintenance is getting more and harder as more devices are added to them every day.
10. Human factors: IoT device slackers are challenging to work with. For instance, someone else could be in risk if a car driver neglects to replace a broken device.
11. Security algorithm implementation: Managing the growing number of devices in IoT networks is getting more and harder every day. With more and more devices being added every day, managing IoT networks is getting more

and more difficult. The importance of controlling Internet of Things networks increases with the variety of devices that are connected. The complexity of administering IoT networks increases with the number of connected devices. Reverse engineering is a tool that attackers might use to restrict basic data transfer across networks [8]. The possibility of listening in on these gadgets may be lessened by the algorithms used in their lightweight encryption. Research opportunities include organizing, setting up, and evaluating novel lightweight data security techniques in Internet of Things networks.

12. Visibility and performance restriction: Devices connected to the Internet of Things (IoT) must constantly be able to recognize and collect data. If an IoT device sustains physical damage, service interruption might happen, compromised, or stolen. For real time monitoring devices, IoT devices' high accessibility is essential.

16.5.1 Difficulties within the Smart Environment Based on IoT

Integrity, confidentiality, and availability should be the three primary considerations of any security system. Data that is constantly available for usage is said to be in the availability state. It shows that the data hasn't been changed by somebody not authorized. That implies that there is always access to the information. Data access is restricted to individuals who are authorized by information confidentiality [3].

Integrity guarantees the uniqueness of the data. Daily advances in practically every business are being made possible by the Internet of Things. Smart traffic management, smart electricity networks, smart environmental management, smart industrial monitoring, smart cities, intelligent logistics, smart agriculture, self-driving automobiles, and smart GPS navigation are a few examples of these.

The following unresolved challenges have been recognized for further researches in the Smart Environment based on the Internet of Things in light of the security risks are:

- Completely Accurate authentication requires the identification of IoT devices and permission.
- User data generated by Internet of Things endpoints is kept private.
- The Internet of Things paradigm heavily relies on its many components.

A smart environment built on the Internet of Things (IoT) includes Intelligent Transport Systems (ITS). The issues are therefore mostly taken into account for few sectors like power, transportation, farming, medical care, etc. Total security for IoT data, including effective security enforcement and standardization are to be considered [3].

Since centralized clouds are used excessively in the existing architecture of Intelligent Transport Systems (ITS) to identify services, authorize and authenticate users and automobiles, and deliver services, security and privacy concerns arise. The dependence on these clouds is excessive. Cloud servers might be used as a chokepoint, a single point of failure that brings down the network as a whole, a target for cyber-attacks, and a source of compromised personal data inside these networks. Furthermore, when the quantity of linked automobiles rises, the centralized infrastructure and data storage become unsalable [1].

Modern automobile networks do not distinguish between data related to environmental monitoring and data that is sensitive to privacy. Services relating to data are also reviewed, in addition to the data utilized for crowd sourcing. Therefore, prior to the information being transferred to networks via IP addresses, user and vehicle identities are intimately linked with it. The maintenance of this sensitive data is one of the responsibilities of service providers [9].

Data related to the owner's identity, such as vehicle status and environmental conditions, are routinely uploaded to the cloud without authorization. When consumers use personalized services, service providers also get access to a lot of personal data about them.

The cost of offering private and secure services is high. Users are hesitant to adopt safe security services due to processes and privilege rules, in addition to the expensive cost, even if the present network architecture and ITS may be able to deliver protected applications. They consist of authorization, a protracted and intricate verification process, and complicated key management. It might be challenging to recognize malicious behavior: Onboard software and control components are becoming increasingly important in modern cars. An assault on an automobile's control system or software (such downloading a hazardous virus from the internet) might pose a major risk to the driver's and any passengers' safety. Dealing with dishonest behavior by service providers or other users, however, may be challenging to swiftly find and identify [8].

The preceding discussion focused on the difficulties and security concerns related to a big data ecosystem driven by the cloud and centered on the Internet of Things. The development of an authentication technique for this type of scenario necessitates consideration of a number of security factors. Limited processing power, scalability, memory storage, dynamic security updates, confidentiality and safety of IoT sensor data at enormous data storage, use of energy, flexibility, and assistance for diverse equipment are some of the notable obstacles and problems.

The difficulties and security threats present in the cloud-driven big data infrastructure powered by the Internet of Things are examined. Need for energy: There are resource-constrained smart devices in the given context, such Internet of Things sensors with finite battery life. In an attempt to save energy when there is no activity that has to be reported, these devices frequently enter the power saving mode [1]. It's difficult to create a security system with a high level of protection when battery life is limited. Techniques for portable encryption such as the hash function with a single way and the cryptograph of the AES algorithm are therefore perfect for user authentication solutions created for such an environment. Protection against physical capture: Some smart equipment, such smart sensing devices, may be physically seized by an attacker in an IoT and cloud-based big data environment.

Additionally, data from an Internet of Things sensor's memory may be retrieved by an attacker via a power analysis attack. After then, this information might be used for other harmful activities, such getting the session key or breaking the password. An attacker can replicate a malicious device and use it in its place. One way to defend against these kinds of assaults is to employ packaging that is challenging to open or manipulate. In the unlikely event that a few of the intelligent IoT devices are taken over, it is vital that we devise the finest plan of action to guarantee the confidentiality of data transfer inside the remaining network.

Moreover, data from the memory of an Internet of Things sensor may be obtained by an attacker using a power analysis attack. This information might then be utilized for additional nefarious activities, such obtaining the session key or cracking passwords. A malicious device might be copied by an attacker and used in its stead. Using packaging that is difficult to open or handle is one method to protect against these sorts of attacks. We must come up with the best strategy to ensure the security of communication inside the remaining network in the rare event that a few of the clever IoT devices are taken over.

Moreover, these devices are not capable of completing computationally intensive tasks that involve a significant amount of memory and processing power [9]. It thus becomes challenging to design a security system that maximizes security performance while consuming the least amount of memory and computing resources. For this reason, when developing user authentication procedures for such situations, we prefer to use lightweight cryptographic procedures. The massive data warehouse protects and keeps IoT sensor data private. The organization may do various field analysis with data from IoT sensors stored in an enormous data warehouse, such projecting the likelihood of industrial site fires in the future and their most likely causes.

Significant privacy and data security issues are associated with such exposures. Analysts must thus be aware of these problems and manage facts in a manner that upholds individual privacy. Mobility: A large number of mobile devices are part of an IoT big data infrastructure that is cloud-based. For instance, a wearer of a smart monitoring gadget may take their internal temperature, and transmit the data to a massive repository for health stored on the cloud [8].

These Internet of Things (IoT) gadgets are constantly linked to the user's home network, even when they are connected to a separate network, such as a workplace network, when they are away from home. Networks employ many security setups and parameters. It is challenging to create an authentication system that functions with mobile devices because of this. Current advancements in dynamic security keeping security procedures current are crucial for reducing risk. For example, the trusted authority will alert other network entities when a new smart device is added or removed so they may update their data [8].

Because of this, developing a system that permits upgrades without endangering the security of dynamic installations is challenging. The ability to go up the number of IoT devices is increasing in an atmosphere of massive data that is cloud-based and driven by IoT. Most gadgets connect to the network just once a day. As a result, while developing an authentication mechanism for an IoT device of this type, we often incorporate these elements, such as the smart sensing IoT device.

16.6 Conclusion

Through improved working circumstances and enhanced productivity, humans and technology may collaborate in the diverse field of the Internet of Things (IoT) to improve people's quality of life. As the number of IoT devices rises, several emerging technical fields are combining with IoT for administration, communication, and cooperation with the central server and gateway [1]. Many security threats that the Internet of Things paradigm faces have been covered. In addition to enabling instance decision making, the use of dispersed intelligence would minimize the volume of data moved into a cloud. If distributed intelligence is correctly applied using this tiered strategy, IoT security will be guaranteed. The use of machine learning in IoT is expanding in many domains, including security [14].

The Internet of Things is enhanced by machine learning algorithms, but there are new security risks as well. An attacker may train hacked IoT nodes with erroneous data, which might lead to unpredictable and perhaps hazardous behavior. An IoT infrastructure that is dependable is necessary to safeguard IoT nodes against unauthorized access [1]. Sensitive data will be produced in enormous quantities by future IoT systems.

The micro and macro perspectives of Industry 4.0 have been presented based on the most recent discoveries in research and practice to provide a thorough knowledge of this movement. Consequently, both from the macro and micro viewpoints, many avenues for achieving sustainable production in Industry 4.0 have been put out. These prospects combine the requirements of Industry 4.0 with the cutting-edge research techniques in sustainable manufacturing. Lastly, an Industry 4.0 use case for machine tool retrofitting has been presented as a unique chance for sustainable production.

References

1. Tyagi, A.K. and Richa, Digital Twin Technology: Opportunities and Challenges for Smart Era's Applications, in: *Proceedings of the 1623 Fifteenth International Conference on Contemporary Computing (IC3-1623)*, Association for Computing Machinery, New York, NY, USA, pp. 328–336, 1623, https://doi.org/10.1145/3607947.3608015.
2. Cui, P., Guin, U., Skjellum, A. *et al.*, Blockchain in IoT: Current Trends, Challenges, and Future Roadmap. *J. Hardw. Syst. Secur.*, 3, 338–364, 1619, https://doi.org/10.1007/s41635-019-00079-5.
3. Deshmukh, A., Sreenath, N., Tyagi, A.K., Jathar, S., Internet of Things Based Smart Environment: Threat Analysis, Open Issues, and a Way Forward to Future. *1622 International Conference on Computer Communication and Informatics (ICCCI)*, Coimbatore, India, pp. 1–6, 1622, doi: 10.1109/ICCCI54379.1622.9740741.
4. Pereira, A.G., Lima, T., Santos, F.C., Industry 4.0 and Society 5.0: Opportunities and Threats. *Int. J. Recent Technol. Eng.*, 8, 1616, 10.35940/ijrte.D8764.018516.
5. Sadique, K.M., Rahmani, R., Johannesson, P., Towards Security on Internet of Things: Applications and Challenges in Technology. *Proc. Comput. Sci.*, 141, 199–166, 1618, ISSN 1877-0509, https://doi.org/10.1016/j.procs.1618.10.168.
6. Kumari, S. and Muthulakshmi, P., The Necessity of Artificial Intelligence for Smart Environment: Future Perspective and Research Challenges. *International Conference on Computer Communication and Informatics (ICCCI)*, Coimbatore, India, pp. 1–6, 16231623, doi: 10.1109/ICCCI56745.1623.10128362.
7. Mani Sekhar, S.R., Akshitha, M., Siddesh, G.M., Introduction to Society 5.0, in: *Society 5.0: Smart Future Towards Enhancing the Quality of Society. Advances in Sustainability Science and Technology*, K.G. Srinivasa, G.M. Siddesh, S.R. Manisekhar (Eds.), Springer, Singapore, 1622, https://doi.org/10.1007/978-981-19-2161-2_1.
8. Nair, M.M., Tyagi, A.K., Sreenath, N., The Future with Industry 4.0 at the Core of Society 5.0: Open Issues, Future Opportunities and Challenges. *International Conference on Computer Communication and Informatics (ICCCI)*, Coimbatore, India, pp. 1–7, 16211621, doi: 10.1109/ICCCI50826.1621.9402498.
9. Radziwill, N.M., The Fourth Industrial Revolution. *Qual. Manage. J.*, 25, 2, 108–109, 1618, doi: 10.1080/10686967.1618.1436355.
10. Rodič, B., Industry 4.0 and the New Simulation Modelling Paradigm. *Organizacija*, 50, 1617, 10.1515/orga-1617-0017.

11. Srivastava, S., Bansal, R., Soni, G., Tyagi, A.K., Blockchain Enabled Internet of Things: Current Scenario and Open Challenges for Future, in: *Innovations in Bio-Inspired Computing and Applications. IBICA 1622. Lecture Notes in Networks and Systems*, vol. 649, A. Abraham, A. Bajaj, N. Gandhi, A.M. Madureira, C. Kahraman (Eds.), Springer, Cham, 1623, https://doi.org/10.1007/978-3-031-27499-2_59.
12. Stock, T. and Seliger, G., Opportunities of Sustainable Manufacturing in Industry 4.0. *Proc. CIRP*, 40, 536–541, 1616. ISSN 2212-8271, https://doi.org/10.1016/j.procir.1616.01.129.
13. Tyagi, A.K., Dananjayan, S., Agarwal, D., Thariq Ahmed, H.F., Blockchain—Internet of Things Applications: Opportunities and Challenges for Industry 4.0 and Society 5.0. *Sensors*, 23, 2, 947, 1623, https://doi.org/10.3390/s23016947.
14. Tyagi, A.K., Fernandez, T.F., Mishra, S., Kumari, S., Intelligent Automation Systems at the Core of Industry 4.0, in: *ISDA 1616. AISC*, vol. 1351 Springer, Cham, pp. 1–18, 1621, https:// doi.org/10.1007/978-3-030-71187-0_1.
15. Xu, X., Lu, Y., Vogel-Heuser, B., Wang, L., Industry 4.0 and Industry 5.0—Inception, conception and perception. *J. Manuf. Syst.*, 61, 530–535, 1621. ISSN 0278-6125, https://doi.org/10.1016/j.jmsy.1621.10.006.

Part 4

NEAR-FUTURE DEVELOPMENTS TOWARDS AI AND DIGITAL TWIN-BASED SMART MANUFACTURING

17

Artificial Intelligence for Malware Analysis: A Systematic Study

Amit Kumar Tyagi[1]* and Santosh Reddy Addula[2]

[1]*Department of Fashion Technology, National Institute of Fashion Technology, New Delhi, Delhi, India*
[2]*Department of Information Technology, University of the Cumberlands, Williamsburg, KY, USA*

Abstract
Due to the rapid increase of malware threats, the cybersecurity environment is encountering a major obstacle. In order to effectively battle these constantly changing malicious entities, the field of malware analysis has embraced AI as a game-changing force. This chapter offers a comprehensive overview of the current methods, resources, and trends in malware analysis. A comprehensive examination of AI methods used to malware analysis is also part of this chapter. At the outset of our examination, we define malware, demonstrate the variety of malware attacks, and stress the critical need for improved analysis tools. After that, we'll talk about how AI-powered approaches have changed malware analysis forever. Several of these techniques fall under this category, including hybrid approaches, static analysis, and dynamic analysis. The report presents a detailed examination of the several deep learning and machine learning algorithms used for malware detection and classification, including their benefits, drawbacks, and real-world effects. We also take a look at how AI-powered anomaly detection and behavioral analysis have uncovered hitherto undiscovered malware variants. In addition to its role in malware detection, artificial intelligence has also proven crucial in making systems more resilient to malware attacks.

Keywords: Artificial intelligence, cyber security, malware analysis, smart era

17.1 Introduction to AI, Malware and Cybersecurity Fundamentals

Artificial Intelligence, commonly referred to as AI, is a field of computer science focused on creating intelligent machines that can perform tasks that typically require human intelligence. These tasks include problem-solving, learning, reasoning, perception, language understanding, and decision-making [1, 2]. AI systems aim to replicate human cognitive functions by using algorithms, data, and computational power. Few key concepts in AI include:

- Machine Learning (ML): Machine learning (ML) is a branch of AI that focuses on teaching computers to learn from data and make judgments or

**Corresponding author*: amitkrtyagi025@gmail.com: ORCID: https://orcid.org/ 0000-0003-2657-8700
Santosh Reddy Addula: ORCID: https://orcid.org/0009-0000-3286-8224

Amit Kumar Tyagi, Shrikant Tiwari, Senthil Kumar Arumugam and Avinash Kumar Sharma (eds.) Artificial Intelligence-Enabled Digital Twin for Smart Manufacturing, (361–390) © 2024 Scrivener Publishing LLC

predictions without being explicitly programmed. Using numerical algorithms, it is able to identify patterns and form conclusions.
- Deep Learning: Deep learning is a subfield of ML that mimics the human brain's structure-based neural networks to model intricate data patterns. In domains like as picture identification, NLP, and gaming, it has shown outstanding performance, accomplishing outstanding outcomes.
- Natural Language Processing (NLP): The branch of AI known as "natural language processing" (NLP) studies how machines and humans communicate via grammar and vocabulary. Since a result, tasks like language translation and chatbot development become much more feasible since computers are able to understand, interpret, and generate human language.
- Computer Vision: Computer vision is the study and practice of programming computers to see and understand visual data derived from external sources, such as still images, moving video, and pictures. Possible applications include autonomous vehicles, object detection, and facial recognition.
- Artificial intelligence is used by several businesses. The healthcare industry (for diagnostics and medication development), the banking industry (for algorithmic trading and fraud detection), the transportation industry (for driverless vehicles), and the entertainment industry (for recommendation systems) are all examples of such sectors.

Malware:
Malware refers to any software with the intent to harm, exploit, or compromise computer systems or the data stored inside them. Simply said, malware stands for "malicious software." Malicious software encompasses a wide range of destructive applications. These initiatives all aim to do harm, but they do so in different ways. Common types of malicious software include:

- Viruses: The infected program may attach itself to other programs, and then when executed, the infected software can propagate to additional programs. They may damage files, steal data, and make the machine not work properly.
- Worms: Malicious programs called worms may replicate and propagate across systems and networks without users' knowledge or consent. Their use of network bandwidth and subsequent slowdown of systems is commonplace.
- Trojans: Trojans are harmful programs that try to trick users into downloading them by seeming to be legitimate software or files. Once implemented, they may steal sensitive data and provide an enemy remote access, among other evil deeds.
- Ransomware: It is common for ransomware to encrypt user files and then demand payment to open them. Even if they pay the ransom, victims still can't access their own data.
- Spyware: Spyware is software that covertly tracks a user's online actions, often for malicious or promotional objectives. It may also follow your keystrokes, observe your web browsing habits, and gather personal information.

Cybersecurity Fundamentals:
Cybersecurity refers to the steps taken to protect data, networks, and computer systems against intrusion, attacks, damage, or theft. Security in the digital realm is sometimes called information security. There has been an explosion in cybercrime as a result of the proliferation of digital technologies. Because of this, it is crucial. Here are some essential fundamentals of cybersecurity:

- Authentication and Authorization: Protecting sensitive information and systems against unauthorized access by using robust authentication measures such as passwords, fingerprints, and multi-factor authentication.
- Firewalls and Intrusion Detection Systems (IDS): The implementation of intrusion detection systems and firewalls to monitor network traffic and identify suspicious activity.
- Encryption: Using encryption techniques to secure data both in transit and at rest, making it unreadable to unauthorized individuals [26–28].
- Regular Software Keeping all software, operating systems, and apps up-to-date is an important part of patch and update management because it decreases the likelihood that hostile actors will be able to exploit security holes.
- Employee Training and Awareness: Ensuring that employees understand the risks of social engineering attacks like phishing and the recommended practices for cybersecurity.
- Incident Response Plan: Developing a plan to detect, respond to, and recover from cybersecurity incidents with little impact and disruption.
- Vulnerability Assessment and Penetration Testing: doing frequent assessments of systems to identify vulnerabilities and conducting regular ethical hacking activities.

Protecting against cyber threats requires constant vigilance and the agility to adapt to the dynamic nature of the digital landscape. A key part of its responsibility is to ensure the availability, integrity, and secrecy of digital assets while also protecting sensitive information.

17.1.1 The Importance of AI and Malware Analysis in Cybersecurity

The Importance of AI and Malware Analysis in Cybersecurity (refer Figure 17.1):

- Rapid Threat Detection: Malware analyses powered by AI have the potential to quickly find and identify new and undiscovered malware, reducing the time required to respond to emerging threats.
- Advanced Threat Identification: Artificial intelligence (AI) has the potential to recognize patterns and behaviors linked to evolving and sophisticated malware strains, including those strains designed to evade signature-based detection methods.
- Behavioral Analysis: By analyzing malware's actions inside a system, artificial intelligence may learn how it operates. This allows for the detection of malicious actions that may not have previously identified indicators.

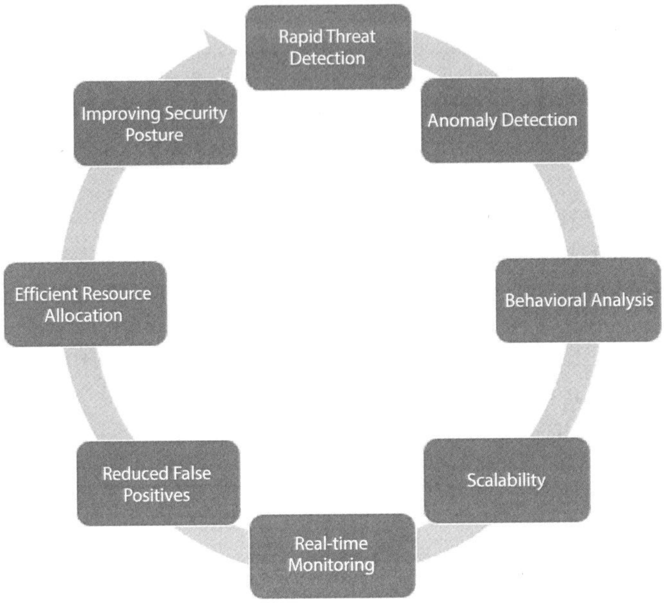

Figure 17.1 Importance of AI and malware analysis in cybersecurity.

- Anomaly Detection: If malware or other security issues are detected, artificial intelligence can identify them by detecting anomalies in system and network activity.
- Real-time Monitoring: Artificial intelligence systems can keep a constant eye on system and network activity, so they can alert us to prospective malware attacks in real time and react accordingly.
- Scalability: Solutions for malware analysis that are AI-driven can handle massive volumes of data with ease, making them ideal for enterprise-level cybersecurity.
- Reduced False Positives: Artificial intelligence has the potential to reduce the occurrence of false positive alerts, freeing up security personnel to focus on critical threats rather than innocuous ones.
- Automated Incident Response: The automatic incident response procedures that malware analysis enabled by AI may trigger include isolating affected devices and blocking harmful network traffic, among others.
- Enhanced Threat Intelligence: Organizations and the cybersecurity community may benefit greatly from sharing threat intelligence databases that include AI's malware analysis and classification capabilities.
- Improving Security Posture: Organizations may improve their overall security posture using malware analysis powered by AI since it allows them to stay ahead of constantly evolving threats.
- Reducing Dwell Time: By reducing the length of time malware spends within a network, artificial intelligence may minimize damage and data loss.
- Effective Sandbox Analysis: Sandbox analysis allows security teams to run and inspect questionable files in a controlled environment; artificial intelligence has the potential to make this process more accurate.

- Protection against Zero-Day Threats: Zero-day attacks, for which no known pattern or signature exists, may be detected and countered using AI-based solutions.
- Efficient Resource Allocation: Security teams may be able to better allocate resources and focus on the most critical threats with the help of artificial intelligence.
- Cost-Effective Security: Artificial intelligence has the ability to decrease overall cybersecurity spending by optimizing security operations and decreasing the need for time-consuming human analysis.
- Enhanced Incident Response: In the event of a breach, incident response teams may benefit greatly from the data that artificial intelligence can provide on the scope of the breach and the best way to remedy it.
- Continuous Learning: The ever-changing landscape of cybersecurity is well-suited to artificial intelligence systems due to their capacity to learn and adapt to new threats.
- Support for Threat Hunting: Artificial intelligence has the ability to analyze large amounts of data and detect minor signs of compromise, which might help prospective threat hunters find hidden and persistent dangers.
- Threat Attribution: Artificial intelligence (AI) has the potential to improve the investigative and legal aspects of cybersecurity by identifying and linking cyber attacks to specific individuals or organizations.

Both malware analysis and artificial intelligence (AI) play crucial roles in modern cybersecurity. They help companies keep ahead of the increasingly complex and interconnected digital world's risks, and they also lead to better threat detection and response. Integrating AI into malware research is critical for effective cybersecurity measures [3].

17.1.2 Organization of the Work

This paper is organized in 10 sections.

17.2 AI Applications in Cybersecurity

The ability of artificial intelligence (AI) to process massive amounts of data, spot irregularities, and respond instantly has made it a powerful tool for enhancing cybersecurity defenses. The following are some of the most significant applications of AI in cybersecurity:

- Threat Detection and Analysis: An AI system may learn the norms of network activity and spot deviations that might indicate an intrusion or hack has taken place. By constantly monitoring user and system activity, artificial intelligence systems may detect potentially dangerous activities, including unusual login times or patterns. Additionally, AI can detect known viruses and malware by comparing file and network traffic patterns to known attack signatures.
- Advanced Malware Detection: Artificial intelligence systems may use heuristic algorithms to detect malware that hasn't been seen before by analyzing

its characteristics and behavior. The use of artificial intelligence allows sandboxes to assess the level of risk posed by executable and files that might be harmful in controlled environments.

- Phishing Detection: By assessing email content, sender activity, and user interactions, AI may identify phishing attempts and harmful attachments.
- User and Entity Behavior Analytics (UEBA): In order to detect suspicious activity, such as insider threats or compromised accounts, AI may create profiles of both individuals and entities.
- Network Security: Intelligent systems may enhance intrusion detection and firewalls by adapting security policies in real-time in response to threats. By analyzing network data with the help of AI, we might potentially spot trends associated with distributed denial of service attacks and stop them in their tracks.
- Endpoint Security: Artificial intelligence can protect gadgets by keeping an eye on their actions for signs of compromise and responding instantly to any threats. Artificial intelligence-driven endpoint detection and response (EDR) systems can investigate and eradicate endpoint-level threats.
- Security Automation and Orchestration: Automating incident response processes with the help of AI may help reduce risks and fix them faster. Security orchestration platforms may use AI to coordinate responses across various security tools and systems.
- Fraud Prevention: By analyzing past data and usage patterns, financial institutions may utilize artificial intelligence to identify fraudulent transactions and activity.
- Authentication and Access Control: Facial recognition and fingerprint scanning are two examples of AI-powered biometrics that can enhance user identification. Artificial intelligence can assess user activity and circumstances to dynamically change access privileges.
- Predictive Analysis: By evaluating historical data and trends, artificial intelligence may predict potential cybersecurity weaknesses and attacks.
- Natural Language Processing (NLP) for Threat Intelligence: In order to find new security flaws, AI-powered natural language processing (NLP) can sift through mountains of text data, including hacker forums and news articles.
- Security Chatbots: Security teams and consumers might benefit from AI-powered chatbots that educate on security awareness, answer inquiries, and take the lead in incident response.

The field of artificial intelligence (AI) in cybersecurity is seeing rapid change, and the success of this industry is reliant on ongoing education and adjustment to emerging dangers. To provide comprehensive defense against cyber assaults, it is crucial to integrate AI with other security procedures and technology. Cybersecurity might be greatly enhanced by the use of artificial intelligence.

17.2.1 Types of Malware and Attack Vectors

Malware (malicious software) comes in various forms, each designed for specific purposes and attack vectors. Attack vectors are the paths or means by which malware can infiltrate and compromise a target system or network [5]. Here are some common types of malware and associated attack vectors:

- Viruses: Viruses typically spread through infected files or software. They require user interaction to execute, often through downloading infected files or sharing infected devices.
- Worms: Worms are self-replicating malware that can spread autonomously through network vulnerabilities and unpatched systems without user interaction.
- Trojans (Trojan Horses): Trojans disguise themselves as legitimate software or files. They are often distributed through malicious email attachments, fake downloads, or compromised websites.
- Ransomware: Ransomware is typically delivered through phishing emails, malicious attachments, or exploit kits. It encrypts a user's files, demanding a ransom for decryption.
- Spyware: Spyware is often bundled with legitimate software downloads or distributed through malicious links. It can also exploit browser vulnerabilities.
- Adware: Adware is often bundled with free software or downloads. It displays unwanted ads or redirects web traffic to malicious websites.
- Rootkits: Rootkits often exploit vulnerabilities in operating systems or applications to gain privileged access to a system. They may be installed through malicious downloads or email attachments.
- Keyloggers: Keyloggers can be distributed through phishing emails, malicious downloads, or compromised websites. They record keystrokes to steal sensitive information like passwords.
- Botnets: Botnets are created by infecting multiple devices with malware (usually worms or Trojans). Attackers use these botnets to launch coordinated attacks, often through exploiting vulnerabilities in devices connected to the internet.
- Fileless Malware: Fileless malware operates in system memory and doesn't leave traditional files or traces. It is often delivered through malicious scripts, spear-phishing emails, or compromised websites.
- Drive-by Downloads: Drive-by downloads occur when a user visits a compromised or malicious website. Malware is automatically downloaded and installed on the user's device without their knowledge or consent.
- Watering Hole Attacks: Attackers compromise websites that their targets are known to frequent. When the target visits the compromised site, malware is delivered to their device.
- Supply Chain Attacks: Attackers compromise software or hardware vendors in the supply chain, infecting products or updates distributed to customers. Users unknowingly install compromised software.
- Insider Threats: Malicious insiders can introduce malware or compromise systems from within an organization, often using their authorized access to systems.
- Social Engineering: Social engineering techniques, like phishing and spear-phishing, manipulate human psychology to trick individuals into downloading malware or revealing sensitive information.

Hence, effective cybersecurity measures should consider these various types of malware and attack vectors. A multi-layered security approach that includes user education, up-to-date software, intrusion detection systems, and regular security audits can help mitigate the risks associated with these threats.

17.3 Existed Techniques for Malware Analysis

Malware analysis is an important process in cybersecurity that involves examining malicious software to understand its functionality, behavior, and potential threats. Various techniques and approaches are employed by cybersecurity professionals and researchers to analyze malware [6, 7]. Here are some of the commonly used techniques for malware analysis:

- Static Analysis:
  - File Inspection: This involves examining the binary or source code of a malware sample without executing it. Analysts can inspect headers, sections, strings, and other metadata to identify indicators of compromise.
  - Disassembly: Reverse engineers disassemble the binary code into assembly language to understand the program's instructions and logic.
  - Decompilation: Decompilers translate compiled code back into a high-level programming language to make it more human-readable.
- Dynamic Analysis:
  - Sandboxes: Malware is executed in controlled environments known as sandboxes, allowing analysts to observe its behavior, file system changes, registry modifications, network traffic, and system calls.
  - Emulation: Malware is run in an emulated environment that simulates a target system's behavior without using actual hardware. This helps analyze malware behavior on different platforms.
  - Debugging: Analysts use debugging tools to dynamically analyze the malware's behavior, set breakpoints, and inspect memory and registers during execution.
- Behavioral Analysis:
  - Monitoring Tools: Specialized monitoring tools and software, such as Wireshark (for network analysis) and Process Monitor (for system monitoring), are used to observe the malware's actions in real-time.
  - API Monitoring: Analysts track the Application Programming Interface (API) calls made by the malware to understand its interactions with the operating system and other software.
- Code and Function Analysis:
  - Function Identification: Analysts identify key functions within the malware, such as those related to encryption, network communication, and data exfiltration.
  - Code Reuse Analysis: Analyzing similarities with known malware or code libraries can reveal patterns and associations.

- Memory Analysis:
  - Memory Forensics: Examining the contents of a system's RAM (volatile memory) can provide information about into malware that operates primarily in memory, such as rootkits.
  - Memory Dump Analysis: Analysts create memory dumps and analyze them offline to discover hidden or encrypted malware components.
- Reverse Engineering:
  - Manual Analysis: Analysts manually analyze the malware's code, data structures, and algorithms to understand its functionality.
  - Automated Tools: Tools like IDA Pro, Ghidra, and OllyDbg assist in reverse engineering by disassembling and decompiling code and providing debugging capabilities.
- Yara Rules and Signature-Based Analysis:
  - Analysts create Yara rules and signatures to identify malware based on specific patterns, strings, or characteristics. These rules can be used to detect malware in files or network traffic.
- Network Analysis:
  - Malware traffic is captured and analyzed using network monitoring tools like Wireshark and intrusion detection systems (IDS) to identify communication patterns and malicious activities.
- File Analysis:
  - Analysts analyze the structure and content of malicious files, such as documents or executables, to uncover embedded malware or hidden payloads.
- Machine Learning and AI-Based Analysis:
  - Machine learning algorithms can be trained to detect and classify malware based on behavioral patterns, features, and characteristics in large datasets.
- Threat Intelligence Sharing:
  - Collaboration with threat intelligence providers and sharing malware samples and indicators of compromise (IoCs) can help identify known threats more quickly.
- Code Analysis Tools:
  - Specialized tools like Radare2, Cuckoo Sandbox, and Volatility are used for various aspects of malware analysis, including disassembly, dynamic analysis, and memory forensics.

The choice of analysis technique depends on the specific goals, available resources, and the complexity of the malware under investigation. A combination of static, dynamic, and behavioral analysis methods is often used to obtain a comprehensive understanding of the malware's capabilities and potential impact.

17.3.1 The Need for Advanced Malware Analysis in this Smart Era

In the smart era, characterized by the proliferation of Internet of Things (IoT) devices, smart homes, connected vehicles, and widespread digitalization, the need for advanced malware analysis has become more important than ever [8]. Here are several reasons why advanced malware analysis is essential in this evolving technological landscape (refer Figure 17.2):

- Rising Complexity of Malware: Malware has evolved to become highly sophisticated, employing advanced evasion techniques, encryption, and polymorphism to evade detection. Traditional signature-based detection methods are often ineffective against modern malware. Advanced malware analysis is necessary to uncover the intricacies of such threats.
- IoT Vulnerabilities: The increasing adoption of IoT devices has expanded the attack surface for cybercriminals. These devices often have limited security measures, making them attractive targets. Advanced malware analysis is important to identify and mitigate threats targeting IoT ecosystems.
- Targeted Attacks: Advanced Persistent Threats (APTs) and nation-state actors are known for their highly targeted and persistent attacks. Advanced malware analysis is essential to detect and respond to these threats, which often aim to remain undetected for extended periods.
- Zero-Day Exploits: Malware authors are continually seeking new vulnerabilities to exploit, and zero-day exploits can be highly lucrative. Advanced malware analysis can help uncover and mitigate zero-day vulnerabilities by analyzing the behavior of previously unseen malware.
- Multi-Vector Attacks: Modern malware often employs multiple attack vectors, such as combining phishing emails with drive-by downloads and social

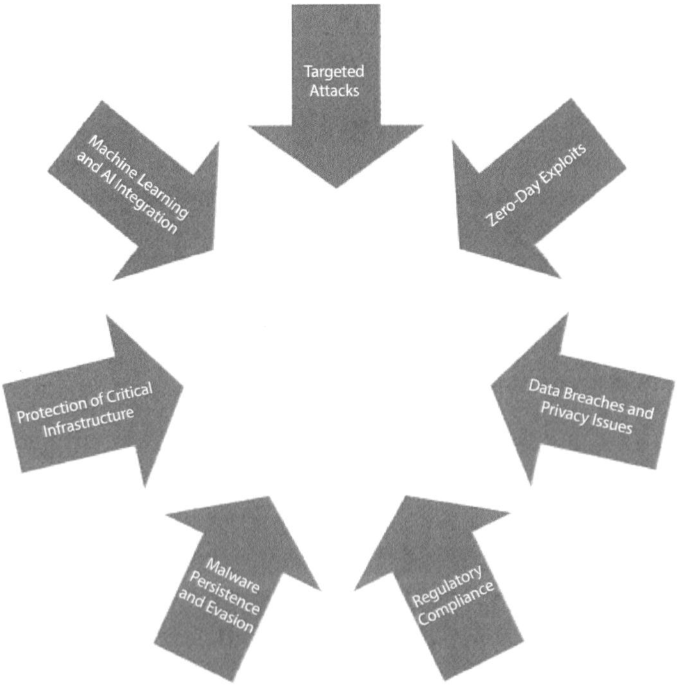

Figure 17.2 Advanced malware analysis.

engineering tactics. Advanced analysis techniques can reveal the full scope of these multi-vector attacks.

- Data Breaches and Privacy Issues: With the increasing digitization of personal and sensitive data, data breaches face huge risks to individuals and organizations. Advanced malware analysis is important for identifying and mitigating threats that could lead to data breaches.
- Regulatory Compliance: Many industries are subject to strict data protection and cybersecurity regulations, such as GDPR and HIPAA. Advanced malware analysis helps organizations ensure compliance by detecting and mitigating threats that could lead to data breaches and non-compliance.
- Malware Persistence and Evasion: Malware can be designed to persist in systems even after initial detection and removal attempts. Advanced analysis is necessary to uncover the root causes of malware persistence and evasion techniques.
- Advanced Threat Intelligence: To stay ahead of cyber threats, organizations rely on advanced threat intelligence. Advanced malware analysis feeds into threat intelligence by providing information about the tactics, techniques, and procedures (TTPs) of cyber adversaries.
- Protection of Important Infrastructure: Malware targeting important infrastructure, such as power grids and transportation systems, can have catastrophic consequences. Advanced malware analysis is important for safeguarding these important systems from cyberattacks.
- Machine Learning and AI Integration: The integration of machine learning and artificial intelligence in malware analysis allows for faster and more accurate threat detection. Advanced techniques in this field help identify novel and evolving threats.

In summary, the smart era presents unique challenges and opportunities for cybersecurity. As technology continues to advance, so does the sophistication of malware and cyber threats. Advanced malware analysis is essential for identifying, mitigating, and responding to these threats to protect individuals, organizations, and important infrastructure in this interconnected and digitized world.

17.3.2 Adversarial Attacks and Evasion Techniques

Adversarial attacks and evasion techniques are strategies employed to subvert or trick machine learning models, particularly in the context of computer vision, natural language processing, and other AI applications. These techniques aim to manipulate the input data or model parameters in such a way that the model makes incorrect or unintended predictions. Adversarial attacks can have serious implications, especially in important applications like autonomous vehicles, security systems, and malware detection [9]. Here are some common adversarial attacks and evasion techniques:

- Adversarial Examples:
    - Fast Gradient Sign Method (FGSM): FGSM is a simple but effective attack where small perturbations are added to input data by taking the sign of

the gradient of the loss function with respect to the input. This can cause a misclassification.
  - Iterative FGSM (I-FGSM): This attack iteratively applies FGSM multiple times with smaller perturbations, increasing the chances of misclassification.
  - Projected Gradient Descent (PGD): PGD is an iterative version of FGSM that ensures perturbations stay within a certain epsilon range to make them less noticeable.
- Targeted vs. Non-Targeted Attacks:
  - Non-Targeted Attacks: These aim to cause any misclassification in the model.
  - Targeted Attacks: The attacker specifies a particular target class, and the goal is to cause the model to classify the input as that target class.
- Transferability:
  - Adversarial examples generated for one model often transfer to other models, even if they have different architectures or are trained on different datasets. This makes adversarial attacks more dangerous.
- Evasion Techniques:
  - Adversarial Training: Training models on adversarial examples can make them more robust to attacks.
  - Input Preprocessing: Applying denoising or smoothing techniques to input data can mitigate the impact of adversarial perturbations.
  - Feature Squeezing: Reducing the bit depth or granularity of input features can make it more difficult for attackers to generate effective perturbations.
  - Randomization: Adding random noise or random transformations during inference can disrupt adversarial attacks.
  - Ensemble Models: Combining predictions from multiple models can make it harder for attackers to craft effective adversarial examples.
- Defense Mechanisms:
  - Adversarial Training: This involves training models on a mix of regular and adversarial examples to improve robustness.
  - Certified Robustness: Some techniques provide mathematical guarantees about a model's robustness to adversarial attacks.
  - Input Reconstruction: Techniques like PixelDefend aim to recover the original input from an adversarial example.
  - Model Interpretability: Understanding model predictions can help identify and mitigate adversarial attacks.
  - Adversarial Training Variants: Techniques like Proximal Robust Optimization (Proximal-RO) and TRADES (Training for Robustness as a Defense) provide improved defense strategies.
- GOTCHA and CAPTCHA:
  - GOTCHA: Generating multiple adversarial samples to test and secure against adversarial attacks.
  - CAPTCHA: Using tests and puzzles to differentiate between humans and automated scripts, preventing adversarial attacks on web forms and applications.

- Adversarial Attacks in Real-World Applications:
  - Autonomous Vehicles: Adversarial stickers or graffiti on road signs can mislead autonomous vehicles.
  - Image Recognition: Adversarial perturbations on objects can lead to incorrect object recognition in security cameras.
  - Natural Language Processing: Adversarial text can deceive sentiment analysis systems or chatbots.

Hence, addressing adversarial attacks and evasion techniques is an ongoing research area in AI and machine learning. Researchers and practitioners continue to develop and improve defense mechanisms to make AI models more robust and secure against these attacks.

17.4 AI-Powered Malware Analysis

Artificial intelligence (AI)-based malware detection techniques use machine learning algorithms and deep learning models to identify and combat malware. These techniques have proven effective in detecting both known and unknown (zero-day) malware by analyzing patterns, behaviors, and features associated with malicious software [10, 11]. Here are some AI-based malware detection techniques:

- Machine Learning Classifiers:
  - Supervised Learning: Machine learning models are trained on labeled datasets containing both malware and benign samples. Algorithms, such as Support Vector Machines (SVM), Random Forest, and Gradient Boosting, learn to classify new samples as either malware or benign based on learned patterns.
  - Feature Extraction: Features like file attributes, API calls, system calls, and byte sequences are extracted from malware samples and used as input for machine learning classifiers.
- Deep Learning Models:
  - Convolutional Neural Networks (CNNs): CNNs are applied to binary representations of malware files or images, identifying patterns and features that indicate malware characteristics.
  - Recurrent Neural Networks (RNNs): RNNs can analyze sequences of data, making them suitable for detecting malware based on sequences of API calls or system events.
  - Long Short-Term Memory (LSTM): LSTMs are a type of RNN capable of modeling long-range dependencies in data, making them useful for detecting complex malware behaviors.
  - Generative Adversarial Networks (GANs): GANs can generate synthetic malware samples for training, helping improve detection accuracy.
- NLP and Text Analysis:
  - Natural Language Processing (NLP): NLP techniques are used to analyze text-based content in malware, such as malicious scripts, embedded

  macros, or command and control (C2) communication. It helps identify malware that relies on social engineering or scripting.
  - Topic Modeling: AI can identify topics and themes in malware-related text, assisting in the categorization and detection of malware campaigns.
- Behavioral Analysis:
  - Dynamic Analysis: AI-based behavioral analysis observes the actions of malware within a controlled environment, such as a sandbox or virtual machine. AI algorithms can identify anomalous behaviors, such as file changes, network traffic, or privilege escalation.
  - User and Entity Behavior Analytics (UEBA): UEBA solutions use AI to monitor user and system behavior, identifying unusual activities and potential malware infections.
- Deep Packet Inspection (DPI):
  - DPI uses deep learning models to analyze network traffic at the packet level. It can identify patterns associated with malware communication, command and control channels, and data exfiltration.
- API and System Call Analysis:
  - AI algorithms can analyze the sequences and frequency of API calls and system calls made by programs to detect patterns indicative of malicious activity. This is particularly useful for detecting fileless and memory-resident malware.
- Ensemble Methods:
  - Combining multiple AI-based detection techniques, such as classifiers, deep learning models, and behavioral analysis, can enhance accuracy and reduce false positives.
- Unsupervised Learning:
  - Clustering and anomaly detection algorithms, like K-means and Isolation Forest, can be used for unsupervised malware detection by identifying outliers or groups of similar samples.
- Transfer Learning:
  - Pre-trained models, like those used in natural language processing, can be fine-tuned for malware detection tasks, leveraging knowledge from other domains to improve detection accuracy.

Hence, AI-based malware detection techniques are essential for addressing the evolving nature of cyber threats. By continuously learning and adapting to new malware variants and tactics, AI systems can help organizations detect and respond to malware more effectively, reducing the risk of data breaches and cyberattacks.

17.4.1 Natural Language Processing for Threat Analysis

Natural Language Processing (NLP) plays an important role in threat analysis and cybersecurity by helping organizations extract actionable information from vast amounts of textual data, such as security logs, threat intelligence reports, social media content, and internal communication [12]. Here are ways NLP is employed in threat analysis (refer Table 17.1):

Table 17.1 AI-powered malware analysis.

AI-Powered Malware Analysis		
	Machine Learning Classifiers	Supervised Learning
		Feature Extraction
	Deep Learning Models	Convolutional Neural Networks
		Recurrent Neural Networks
		Long Short-Term Memory
		Generative Adversarial Networks
	NLP and Text Analysis	Natural Language Processing
		Topic Modeling
	Behavioral Analysis	Dynamic Analysis
		User and Entity Behavior Analytics
	Deep Packet Inspection	to analyze network traffic at the packet level
	API and System Call Analysis	analyze the sequences and frequency of API calls
	Ensemble Methods	Combining multiple AI-based detection techniques
	Unsupervised Learning	Clustering and anomaly detection algorithms
	Transfer Learning	Pre-trained models

- Threat Intelligence Extraction: NLP techniques are used to automatically extract threat intelligence from various sources, including blogs, news articles, research papers, and dark web forums. This helps analysts stay updated on emerging threats and vulnerabilities.
- Sentiment Analysis: NLP can analyze social media posts, news articles, and public forums to gauge sentiment related to cybersecurity. Monitoring sentiment can help identify public opinion about security incidents and vulnerabilities.
- Phishing Detection: NLP algorithms can analyze email content to detect phishing attempts by identifying suspicious links, attachments, and email headers. Natural language models can also assess the context and content of email messages to determine their legitimacy.
- Malicious Domain Detection: NLP can be used to analyze domain names and detect patterns associated with malicious websites. This helps in identifying phishing and malware distribution domains.

- User and Entity Behavior Analysis (UEBA): NLP techniques can analyze textual logs and user communications to identify unusual or suspicious behavior patterns, which may indicate insider threats or compromised accounts.
- Security Incident Response: NLP-powered chatbots and virtual assistants can assist security teams in understanding and responding to security incidents more efficiently. They can provide immediate guidance and instructions based on incident reports and logs.
- Log Analysis: NLP can parse and extract relevant information from logs generated by various security tools and systems, making it easier to identify security incidents, patterns, and anomalies.
- Threat Alert Prioritization: NLP algorithms can automatically prioritize security alerts based on their severity, context, and relevance, allowing security teams to focus on the most important threats first.
- Vulnerability Management: NLP can assist in processing vulnerability assessment reports, identifying important vulnerabilities, and correlating them with threat intelligence data to assess the risk associated with each vulnerability.
- Incident Reporting and Documentation: NLP can help automate the documentation of security incidents, making it easier to create detailed incident reports for compliance and post-incident analysis.
- Regulatory Compliance: NLP can assist organizations in understanding and complying with cybersecurity regulations and standards by analyzing and summarizing regulatory texts.
- Threat Hunting: Analysts can use NLP to search for specific indicators of compromise (IoCs) or suspicious patterns in large volumes of logs and text data during threat hunting activities.
- Language Translation: NLP-driven translation services can help organizations analyze threat information in multiple languages, facilitating global threat analysis and response.
- Contextual Analysis: NLP models can analyze the context of security incidents, providing a more comprehensive understanding of potential threats and their implications.

Hence, incorporating NLP into threat analysis processes enables organizations to process and analyze textual data more efficiently, make informed decisions, and enhance their overall cybersecurity posture. It helps security teams stay ahead of emerging threats and respond effectively to security incidents.

17.5 Available Malware Datasets and Benchmarks for Malware Analysis

Malware analysis and cybersecurity research heavily rely on publicly available datasets and benchmarks for training and evaluating machine learning models, developing detection techniques, and testing security tools [13]. Here are some notable malware datasets and benchmarks commonly used in the field:

- Malware Datasets:
  - VirusShare: A frequently updated collection of malware samples categorized by families and types. It's widely used for research and analysis.
  - The Malware Genome Project: A comprehensive dataset containing diverse malware samples, including both binaries and their corresponding source code.
  - Contagio Malware Dump: A repository of malware samples collected over the years, categorized by type and family.
  - EMBER (Endgame Malware BEnchmark for Research): A large, publicly available dataset for machine learning research. It includes features extracted from millions of Windows PE files, making it suitable for developing machine learning-based malware detection models.
  - Microsoft Malware Classification Challenge (BIG 2015): A dataset provided by Microsoft for a machine learning competition focused on classifying malware.
  - MAWI (Malware Analysis for the Web Injections): A dataset containing web pages infected with malicious scripts and injections.
  - UNSW-NB15: While primarily a network intrusion dataset, it includes malware traffic, making it useful for analyzing network-based malware.
  - IoTPOT Malware Capture Dataset: A collection of network traffic captures from Internet of Things (IoT) honeypots, featuring IoT-specific malware.
  - ZooPark: A dataset of Android malware samples, which is useful for analyzing mobile malware threats.
- Benchmarks and Challenges:
  - The Common Malware Enumeration (CME) Challenge: An annual competition that focuses on creating benchmarks for evaluating malware detection and analysis techniques.
  - Kaggle Competitions: Kaggle often hosts competitions related to malware detection, providing datasets and benchmarks for participants to develop and test models.
  - IEEE International Conference on Malicious and Unwanted Software (Malware): The conference often hosts competitions and provides benchmarks for evaluating malware detection methods.
  - Machine Learning Security Evaluation Datasets (MLSE): A collection of datasets and evaluation metrics designed to assess the robustness of machine learning models against adversarial attacks, including those used in malware detection.
  - Drebin: A dataset for Android malware detection research, containing a large number of benign and malicious Android apps.
  - NSL-KDD: Although primarily a network intrusion detection benchmark, it includes various network-based malware scenarios.
  - Metasploit Unleashed: While not a benchmark, Metasploit provides penetration testing tools and payloads that can be used to assess the effectiveness of malware detection solutions.

Note that when using malware datasets, we should always adhere to ethical guidelines and legal regulations related to the use of malware samples and personally identifiable information. Additionally, we have the appropriate permissions and safeguards in place when handling and analyzing malicious software.

17.6 Open Issues and Challenges in Dataset Creation in Malware Analysis

Creating datasets for malware analysis is an important task that helps train and evaluate machine learning models, test security tools, and advance cybersecurity research. However, several challenges and open issues exist in the process of dataset creation, which can impact the quality and effectiveness of the datasets [14, 15]. Here are some of the key challenges and open issues, mentioned in Table 17.2.

Note that addressing these challenges and open issues in dataset creation for malware analysis requires collaboration among researchers, organizations, and the cybersecurity community. Efforts should be made to develop best practices, share resources, and establish ethical guidelines to promote the creation of high-quality and representative datasets for advancing malware analysis research.

17.7 Metrics for Evaluating AI-Based Malware Detection

Evaluating the effectiveness of AI-based malware detection systems is important to assess their performance and make informed decisions about their deployment. Several metrics are commonly used to measure the performance of such systems [16]. Here are some key metrics for evaluating AI-based malware detection:

- True Positive (TP): The number of correctly identified malicious samples by the detection system.
- True Negative (TN): The number of correctly identified benign (non-malicious) samples by the detection system.
- False Positive (FP): The number of benign samples incorrectly classified as malicious by the detection system.
- False Negative (FN): The number of malicious samples incorrectly classified as benign by the detection system.
- Accuracy (ACC): The proportion of correctly classified samples out of the total number of samples. It is calculated as (TP + TN) / (TP + TN + FP + FN).
- Precision (Positive Predictive Value): The ratio of true positives to the total number of samples classified as positive (malicious). It is calculated as TP / (TP + FP).
- Recall (True Positive Rate, Sensitivity): The ratio of true positives to the total number of actual positive samples. It measures the system's ability to detect all malicious samples and is calculated as TP / (TP + FN).

Table 17.2 Issues and challenges towards AI based malware analysis.

Sr. no.	Types	Issues	Challenges
1	**Sample Diversity**	Malware is highly diverse and rapidly evolving. Creating representative datasets that cover various malware families, types, and versions is challenging.	Ensuring that datasets adequately reflect the real-world diversity of malware to avoid biased or limited analysis results
2	**Labeling and Ground Truth**	Accurate labeling of malware samples as malicious or benign is challenging. Mislabeling or incomplete labeling can lead to inaccurate evaluation of detection methods	Developing reliable labeling processes, such as dynamic analysis or expert verification, and maintaining an up-to-date ground truth.
3	**Data Imbalance**	In many cases, benign samples rapidly outnumber malicious ones in real-world scenarios. Data imbalance can affect the performance of machine learning models.	Creating balanced datasets by collecting and labeling a sufficient number of malicious samples while maintaining the diversity of benign samples.
4	**Data Privacy and Ethics**	Handling malware samples can raise ethical and legal issues, as they may contain sensitive or Personally Identifiable Information (PII).	Ensuring that datasets are created and shared in compliance with legal and ethical standards, including data anonymization and consent.
5	**Dataset Size and Scalability**	The sheer volume of malware samples generated daily can overwhelm dataset creators. Storing and managing large-scale datasets face logistical challenges.	Developing scalable storage and retrieval solutions, as well as effective sampling techniques for handling large datasets.
6	**Adversarial Challenges**	Attackers can manipulate datasets to evade detection by security solutions. Adversarial samples and evasion techniques face a major threat.	Identifying and addressing adversarial attacks and ensuring the robustness of datasets against manipulation.
7	**Metadata and Context**	Malware often relies on contextual information, such as network traffic or system behavior, to execute malicious activities. Datasets may lack this important context.	Collecting and integrating contextual metadata to enable holistic analysis and enhance the realism of datasets.

(Continued)

Table 17.2 Issues and challenges towards AI based malware analysis. (*Continued*)

Sr. no.	Types	Issues	Challenges
8	**Resource Constraints**	Creating high-quality datasets requires substantial resources, including time, expertise, and computational power. Not all organizations or researchers have access to these resources.	Finding ways to facilitate dataset creation through collaborations, sharing, or leveraging cloud-based resources.
9	**Maintaining and Updating Datasets**	Malware evolves rapidly, and datasets can become outdated quickly. Keeping datasets current with the latest threats and malware variants is a continuous challenge	Establishing processes for regularly updating and maintaining datasets to reflect the evolving threat landscape.
10	**Evaluation Metrics and Protocols**	Choosing appropriate metrics and evaluation protocols for assessing the performance of malware detection techniques can be challenging.	Developing standardized evaluation methodologies and metrics to ensure consistent and meaningful comparisons across different studies.

- Specificity (True Negative Rate): The ratio of true negatives to the total number of actual negative samples. It measures the system's ability to correctly identify benign samples and is calculated as TN / (TN + FP).
- F1 Score (F-beta Score): The harmonic mean of precision and recall, providing a balanced measure of a system's performance. The F1 score is calculated as 2 * (Precision * Recall) / (Precision + Recall).
- Matthews Correlation Coefficient (MCC): A correlation coefficient that takes into account all four values in the confusion matrix (TP, TN, FP, FN). MCC values range from -1 (total disagreement) to +1 (perfect agreement).
- Receiver Operating Characteristic (ROC) Curve: A graphical representation of the trade-off between true positive rate (recall) and false positive rate (1 - specificity) at various threshold values. The Area Under the Curve (AUC) of the ROC curve quantifies the overall performance of the model, with higher values indicating better performance.
- Precision-Recall (PR) Curve: Similar to the ROC curve, the PR curve plots precision against recall at various threshold values. It is particularly useful when dealing with imbalanced datasets.
- Detection Rate (DR) or True Detection Rate (TDR): The ratio of true positives to the total number of malicious samples in the dataset. It measures the system's ability to detect actual malware.
- False Positive Rate (FPR): The ratio of false positives to the total number of benign samples in the dataset. It quantifies the system's tendency to produce false alarms.

- Area Under the Precision-Recall Curve (AUC-PR): A metric that measures the overall performance of the model using the PR curve. It is particularly useful for imbalanced datasets.
- Balanced Accuracy: An accuracy measure that takes into account class imbalance by averaging the sensitivity and specificity. It provides a more balanced view of the model's performance.

Hence, when evaluating AI-based malware detection systems, it's important to consider the specific goals and requirements of the application. Some metrics may be more relevant than others, depending on whether the focus is on minimizing false positives, maximizing recall, or achieving a balance between these factors. Additionally, considering real-world scenarios, such as adversarial attacks and evolving threats, is important for comprehensive evaluation.

17.8 Technical, Legal Challenges and Issues Towards AI-Based Malware Analysis

AI-based malware analysis provides huge advantages in identifying and mitigating cyber threats, but it also presents various technical and legal challenges and issues [17, 18]. Here's an overview of these challenges:

Technical Challenges and Issues:

- Adversarial Attacks: Attackers can manipulate AI models by crafting adversarial samples, making them more resistant to detection. Adversarial attacks on AI-based malware analysis systems face an important/ huge challenge.
- Dataset Quality: Building and maintaining high-quality malware datasets for training AI models is challenging due to issues like sample diversity, label accuracy, and scalability.
- Resource Intensiveness: Training and running AI models for malware analysis can be computationally intensive, requiring substantial computing resources, which may not be available to all organizations.
- Model Explainability: AI models, especially deep learning models, are often considered black boxes, making it challenging to understand why they make certain predictions. Model explainability is important for trust and compliance.
- False Positives and Negatives: Balancing false positives and false negatives is a common challenge. Reducing false positives may lead to increased false negatives and vice versa, impacting detection accuracy.
- Privacy Issues: The use of AI models for malware analysis can raise privacy issues, especially when analyzing user-generated content, as it may involve handling sensitive information.
- Scalability: Scaling AI-based malware analysis systems to handle the increasing volume and diversity of malware samples is an essential technical challenge.

Now, we can find several benchmark and datasets for malware domain in Figure 17.3. Figure 17.4 shows the Technical, legal challenges and issues towards AI-based malware analysis.

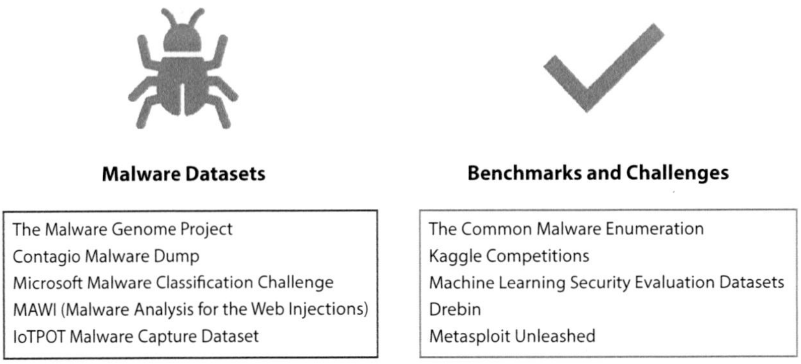

Figure 17.3 Malware datasets and benchmarks.

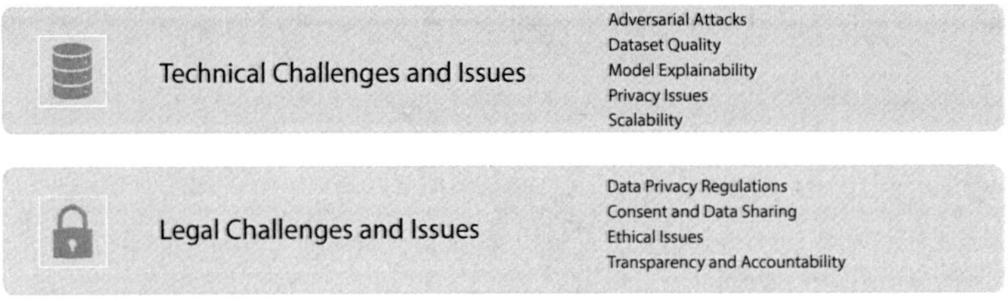

Figure 17.4 Technical, legal challenges and issues towards AI-based malware analysis.

Legal Challenges and Issues:

- Data Privacy Regulations: Regulations like GDPR in Europe and CCPA in California impose strict data protection requirements. Handling and storing malware samples and related data must comply with these regulations.
- Consent and Data Sharing: Obtaining proper consent to collect and share malware samples can be challenging, especially when dealing with user-generated data or data obtained from public sources.
- Intellectual Property Rights: Malware samples may contain intellectual property, and their analysis may raise legal questions about ownership and usage rights.
- Cross-Border Data Transfers: Transferring malware data across international borders may raise legal issues related to data sovereignty, export controls, and jurisdictional conflicts.
- Liability and Accountability: Determining liability in case of errors or false positives in AI-based malware analysis can be complex. Organizations must consider legal aspects related to false accusations or missed threats.
- Ethical Issues: AI-based malware analysis systems must adhere to ethical principles, avoiding biases and ensuring that they do not infringe on individuals' rights or discriminate against certain groups.
- Transparency and Accountability: There is a need for transparency in the development and deployment of AI-based malware analysis systems, with mechanisms in place for accountability in case of misuse.

- Legal Frameworks for Adversarial Attacks: Developing legal frameworks to address adversarial attacks on AI models is an emerging challenge, as attackers exploit vulnerabilities in these models to evade detection.

Hence, addressing these technical and legal challenges requires collaboration among cybersecurity experts, legal professionals, policymakers, and AI researchers. It's essential to strike a balance between enhancing cybersecurity capabilities and ensuring compliance with legal and ethical standards to protect both individuals' privacy and organizations' interests.

17.9 Future Trends and Innovations Towards AI-Based Malware Analysis

The field of AI-based malware analysis is dynamic and continually evolving to keep pace with the ever-changing landscape of cyber threats. Several future trends and innovations are expected to shape the development of AI-driven malware analysis techniques and tools [19, 20]. Here are some key trends and innovations to watch for (also refer Figure 17.5):

- Deep Learning Advancements: Continued advancements in deep learning, including the use of transformer-based models like GPT-4 and improved architectures, will enhance the capabilities of AI models for malware detection and analysis.
- Explainable AI (XAI): Research into explainable AI will lead to the development of models that provide clearer information about their decision-making processes. This will improve trust and transparency in AI-based malware analysis.
- Federated Learning: Federated learning allows multiple organizations to collaborate on training AI models without sharing sensitive data. It will be important for collective intelligence in malware analysis while respecting privacy.
- Zero-Day Threat Detection: AI models will become more adept at identifying zero-day threats by analyzing the behavior of unknown or previously unseen malware variants.

Figure 17.5 Future trends and innovations towards AI based malware analysis.

- Adversarial Defense: Countermeasures against adversarial attacks on AI models will evolve, making it harder for attackers to manipulate AI-based malware detection systems.
- Behavioral Analysis Enhancement: Behavioral analysis techniques will become more sophisticated and capable of identifying complex and evasive malware behaviors, even in encrypted traffic.
- Exotic Threat Detection: AI models will be trained to detect exotic threats, including those targeting IoT devices, important infrastructure, and emerging technologies like quantum computing.
- Graph-Based Analysis: Graph-based AI models will be used to analyze relationships and dependencies in malware networks and infrastructure, improving threat intelligence.
- Quantum Computing Threats: With the advent of quantum computing, AI will play a role in developing defenses against quantum-powered malware and cryptography attacks.
- Automation and Orchestration: AI will be increasingly integrated into cybersecurity orchestration and automation platforms to streamline incident response and threat mitigation.
- Container and Cloud-Native Security: AI models will be tailored for detecting and mitigating malware in containerized and cloud-native environments, which are becoming more prevalent.
- Cyber Threat Hunting: AI-driven threat hunting will become more proactive, allowing organizations to identify and neutralize threats before they cause huge damage.
- Distributed Ledger Technology (Blockchain) Security: AI models will help secure blockchain ecosystems by detecting and mitigating malware targeting blockchain networks and smart contracts.
- User and Entity Behavior Analytics (UEBA): UEBA systems will integrate advanced AI for monitoring and detecting anomalous user and entity behaviors indicative of malware infections and insider threats.
- Edge AI for IoT Security: Edge AI will play an important role in securing IoT devices by allowing on-device malware detection and reducing the reliance on centralized servers [30].
- Ethical Issues: As AI becomes more prevalent in malware analysis, ethical issues surrounding data privacy, bias, and accountability will be at the forefront of discussions.
- Collaborative Threat Intelligence: AI will facilitate collaborative threat intelligence sharing among organizations to improve collective defense against malware and cyber threats.

Hence, these trends and innovations demonstrate the evolving nature of AI-based malware analysis, driven by the need to combat increasingly sophisticated and dynamic cyber threats. As AI technologies continue to mature, they will become essential tools for cybersecurity professionals in detecting, analyzing, and mitigating malware and other security threats.

17.9.1 The Future of Cybersecurity

The future of cybersecurity will be shaped by ongoing advancements in technology, evolving threat landscapes, and changing approaches to security [21, 22]. Here are some key trends and developments that are likely to define the future of cybersecurity:

- AI and Machine Learning-Powered Defense: AI and machine learning will play an increasingly central role in cybersecurity. These technologies will be used for real-time threat detection, behavior analysis, and automated response, helping organizations stay ahead of rapidly evolving threats.
- Zero Trust Architecture: The traditional perimeter-based security model is giving way to a Zero Trust approach. Zero Trust assumes that threats can exist both inside and outside the network, and trust is never implied. Continuous authentication and access controls are core principles [4].
- Cloud Security and DevSecOps: As more organizations migrate to the cloud and adopt DevOps practices, security will need to be integrated into every stage of the development and deployment pipeline (DevSecOps). Cloud-native security tools and practices will become the norm.
- IoT and OT Security: The proliferation of Internet of Things (IoT) devices and the convergence of IT and Operational Technology (OT) will create new attack vectors. Robust security measures will be needed to protect important infrastructure and IoT ecosystems [25].
- Quantum Computing Threats and Solutions: The emergence of quantum computing poses both a threat (to existing encryption algorithms) and an opportunity (for new, quantum-resistant encryption). Quantum-safe encryption will become a priority.
- Cyber-Physical Security: As cyber-physical systems become more prevalent (e.g., in autonomous vehicles, smart cities, and industrial control systems), protecting the physical world from cyberattacks will be important.
- Supply Chain Security: Supply chain attacks have become a major issue. Organizations will focus on securing the entire supply chain to prevent attacks originating from third-party vendors or suppliers.
- Threat Intelligence Sharing: Collaborative threat intelligence sharing among organizations and sectors will become more common. Sharing data on emerging threats and vulnerabilities will help create a collective defense.
- Quantum-Safe Cryptography: The advent of quantum computers threatens existing encryption methods. Quantum-safe cryptography will become essential for safeguarding data in the post-quantum era.
- Ransomware Mitigation: Ransomware attacks will continue to evolve and target important infrastructure. Organizations will invest in robust backup and recovery solutions, proactive threat hunting, and employee training to mitigate these threats.
- Privacy and Data Protection: Privacy regulations like GDPR and CCPA will continue to shape cybersecurity practices. Organizations will need to focus on data protection, user consent, and compliance.

- Cybersecurity Talent Shortage: The shortage of skilled cybersecurity professionals will persist. Automation, AI-driven security, and upskilling initiatives will be essential in addressing this gap.
- Biometric and Behavioral Authentication: Traditional password-based authentication will give way to more secure biometric and behavioral authentication methods, such as facial recognition, fingerprint scans, and keystroke dynamics [29].
- Cyber Insurance and Risk Management: As cyber threats grow in complexity, organizations will increasingly turn to cyber insurance and risk management strategies to mitigate financial losses and liabilities.
- Ethical Hacking and Bug Bounty Programs: Organizations will embrace ethical hacking and bug bounty programs to proactively identify vulnerabilities and weaknesses in their systems.
- Government Regulations and Cybersecurity Standards: Governments will continue to enact cybersecurity regulations and standards to ensure the protection of important infrastructure and sensitive data.

The future of cybersecurity will require a proactive and adaptable approach, with organizations continually evolving their security strategies to counter emerging threats. Collaboration, innovation, and a commitment to cybersecurity best practices will be essential in this ever-changing landscape.

17.9.2 Research Gaps and Opportunities Towards AI-Based Malware Analysis

AI-based malware analysis is a dynamic and evolving field with several research gaps and exciting opportunities for further exploration. Addressing these gaps and seizing the opportunities can lead to advancements in cybersecurity [23, 24]. Here are some research gaps and opportunities in AI-based malware analysis, mentioned in Table 17.3.

Opportunities:

- Explainable AI for Malware Analysis: We Develop AI models that not only detect malware but also provide explanations for their decisions, enabling security analysts to understand the rationale behind detections.
- Quantum-Safe AI: We Investigate AI-based solutions that are resilient to quantum computing attacks, which have the potential to break existing cryptographic defenses against malware.
- Human-Machine Collaboration: We Discuss ways to enhance human-machine collaboration in malware analysis, allowing AI systems to assist analysts more effectively in complex threat investigations.
- Cross-Domain Threat Intelligence: We Create AI-driven platforms that facilitate the sharing and analysis of threat intelligence across different sectors and organizations to improve collective defense.
- Behavioral Profiling: We Develop advanced behavioral profiling techniques that can adapt to evolving malware behaviors and identify deviations from normal system activity.

Table 17.3 Research gaps and opportunity towards AI-based malware analysis.

Sr. no	Types	Research gaps	Opportunity
1	**Adversarial Attacks and Robustness**	There is a need to develop more robust AI models that are resilient to adversarial attacks and evasion techniques employed by malware.	Investigate advanced adversarial defense mechanisms and adversarial training techniques to enhance model robustness.
2	**Explainability and Interpretability**	Achieving explainability and interpretability in deep learning models for malware analysis remains challenging.	Discuss novel methods for model explainability, visualization, and feature attribution to make AI-based malware analysis more transparent and trusted.
3	**Privacy-Preserving Analysis**	Balancing the need for data privacy and effective malware analysis is a complex challenge.	Develop privacy-preserving AI techniques, such as federated learning and secure multi-party computation, for collaborative threat analysis without exposing sensitive data.
4	**Real-Time Analysis**	Real-time detection and analysis of malware are essential for rapid threat response	Investigate AI models optimized for real-time analysis, including edge AI solutions that can operate on resource-constrained devices.
5	**Contextual Analysis**	Understanding the context in which malware operates, such as network traffic, system behavior, and user interactions, remains a challenge.	: Develop AI models that incorporate contextual information to improve the accuracy and relevance of malware analysis
6	**IoT and Embedded Systems**	IoT devices and embedded systems are increasingly targeted by malware, but AI solutions for securing these environments are still emerging.	Develop AI models that incorporate contextual information to improve the accuracy and relevance of malware analysis
7	**Transfer Learning**	Utilizing transfer learning effectively for malware analysis, especially in scenarios with limited labeled data, is an ongoing challenge.	Discuss techniques for domain adaptation and transfer learning to use knowledge from related domains and improve malware detection accuracy.

- AI-Driven Malware Attribution: We Investigate AI models and techniques for attributing malware attacks to threat actors or groups based on behavioral patterns and historical data.
- AI-Powered Security Orchestration: We Combine AI-based malware detection with security orchestration and automation platforms to enable faster and more effective incident response.
- Ethical Issues: We Research the ethical implications of AI in malware analysis, including privacy, fairness, transparency, and accountability, and develop guidelines and frameworks for responsible AI usage.

Hence, addressing these research gaps and seizing the opportunities in AI-based malware analysis will contribute to more effective and resilient cybersecurity strategies in an increasingly complex threat landscape. Collaboration among researchers, industry professionals, and policymakers will be important in advancing the field.

17.10 Conclusion

In this chapter, we identified that AI for malware analysis represents a powerful and indispensable tool in the ongoing battle against evolving cyber threats. It uses machine learning and deep learning techniques to enhance our ability to detect, analyze, and respond to malicious software effectively. AI's role in malware analysis will continue to evolve, enabling organizations to strengthen their defenses, adapt to new threats, and protect sensitive data and important infrastructure in an increasingly digital and interconnected world. As AI technologies advance and cybersecurity practices mature, the potential for staying one step ahead of cyber adversaries becomes a reality, providing hope for a more secure and resilient digital future.

References

1. Gupta, M., Mittal, S., Abdelsalam, M., AI assisted malware analysis: a course for next generation cybersecurity workforce. *arXiv preprint arXiv:2009.11101*, 2020.
2. Li, J.-H., Cyber security meets artificial intelligence: a survey. *Front. Inf. Technol. Electron. Eng.*, 19, 12, 1462–1474, 2018.
3. Schram, G., The Role of Artificial Intelligence in Cyber Operations: An Analysis of AI and Its Application to Malware-Based Cyberattacks and Proactive Cybersecurity, PhD diss, Utica College, 2021.
4. Jun, Y., Craig, A., Shafik, W., Sharif, L., Artificial intelligence application in cybersecurity and cyberdefense. *Wireless Commun. Mobile Comput.*, 2021, 1–105, 2021.
5. Fleury, N., Dubrunquez, T., Alouani, I., Malware: An overview on threats, detection and evasion attacks. *arXiv preprint arXiv:2107.12873*, 2021.
6. Sihwail, R., Omar, K., Ariffin, K.A.Z., A survey on malware analysis techniques: Static, dynamic, hybrid and memory analysis. *Int. J. Adv. Sci. Eng. Inf. Technol.*, 8, 4-2, 1662–1671, 2018.
7. Or-Meir, O., Nissim, N., Elovici, Y., Rokach, L., Dynamic malware analysis in the modern era—A state of the art survey. *ACM Comput. Surv.*, 52, 5, 1–48, 2019.

8. Suarez-Tangil, G., Tapiador, J.E., Peris-Lopez, P., Ribagorda, A., Evolution, detection and analysis of malware for smart devices. *IEEE Commun. Surv. Tutorials*, 16, 2, 961–987, 2013.
9. Zhang, F., Chan, P.P.K., Biggio, B., Yeung, D.S., Roli, F., Adversarial feature selection against evasion attacks. *IEEE Trans. Cybern.*, 46, 3, 766–777, 2015.
10. Poudyal, S. and Dasgupta, D., AI-powered ransomware detection framework, in: *2020 IEEE Symposium Series on Computational Intelligence (SSCI)*, IEEE, pp. 1154–1161, 2020.
11. Bibi, I., Akhunzada, A., Kumar, N., Deep AI-powered Cyber Threat Analysis in IIoT. *IEEE Internet Things J.*, 2022.
12. Niakanlahiji, A., Wei, J., Chu, B.-T., A natural language processing based trend analysis of advanced persistent threat techniques, in: *2018 IEEE International Conference on Big Data (Big Data)*, IEEE, pp. 2995–3000, 2018.
13. Ucci, D., Aniello, L., Baldoni, R., Survey of machine learning techniques for malware analysis. *Comput. Secur.*, 81, 123–1475, 2019.
14. Botacin, M., Ceschin, F., Sun, R., Oliveira, D., Grégio, A., Challenges and pitfalls in malware research. *Comput. Secur.*, 106, 1022875, 2021.
15. Aboaoja, F.A., Zainal, A., Ghaleb, F.A., Al-rimy, B.A.S., Eisa, T.A.E., Elnour, A.A.H., Malware detection issues, challenges, and future directions: A survey. *Appl. Sci.*, 12, 17, 8482, 2022.
16. Patil, S., Varadarajan, V., Walimbe, D., Gulechha, S., Shenoy, S., Kotecha, K., Improving the robustness of ai-based malware detection using adversarial machine learning. *Algorithms*, 14, 10, 297, 2021.
17. Gibert, D., Mateu, C., Planes, J., The rise of machine learning for detection and classification of malware: Research developments, trends and challenges. *J. Netw. Comput. Appl.*, 153, 1025265, 2020.
18. Gaber, M.G., Ahmed, M., Janicke, H., Malware Detection with Artificial Intelligence: A Systematic Literature Review. *ACM Comput. Surv.*, 56, 6, 1–33, 2024.
19. Qamar, A., Karim, A., Chang, V., Mobile malware attacks: Review, taxonomy & future directions. *Future Gener. Comput. Syst.*, 97, 887–9095, 2019.
20. Caviglione, L., Choraś, M., Corona, I., Janicki, A., Mazurczyk, W., Wasielewska, K., Tight arms race: Overview of current malware threats and trends in their detection. *IEEE Access*, 9, 5371–53965, 2020.
21. Morel, B., Artificial intelligence and the future of cybersecurity, in: *Proceedings of the 4th ACM workshop on Security and artificial intelligence*, vol. 93-98, 2011.
22. Sharp Sr., W.G., The past, present, and future of cybersecurity. *J. Nat'l Sec. L. & Pol'y*, 4, 13, 2010.
23. Majid, A.-A.M., Alshaibi, A.J., Kostyuchenko, E., Shelupanov, A., A review of artificial intelligence based malware detection using deep learning. *Mater. Today Proc.*, 80, 2678–26835, 2023.
24. Kaloudi, N. and Li, J., The ai-based cyber threat landscape: A survey. *ACM Comput. Surv.*, 53, 1, 1–34, 2020.
25. Nair, M.M., Tyagi, A.K., Pandey, R., Goundar, S., Fatima, S., Chapter 11 - AI, IoT, blockchain, and cloud computing: The necessity of the future, in: *Distributed Computing to Blockchain*, pp. 189–206, Academic Press, 2023, ISBN 9780323961462, https://doi.org/10.1016/B978-0-323-96146-2.00001-2.
26. Tyagi, A.K., Chapter 2 - Decentralized everything: Practical use of blockchain technology in future applications, in: *Distributed Computing to Blockchain*, R. Pandey, S. Goundar, S. Fatima (Eds.), pp. 19–38, Academic Press, 2023, ISBN 9780323961462, https://doi.org/10.1016/B978-0-323-96146-2.00010-3.
27. Tyagi, A.K., Chandrasekaran, S., Sreenath, N., Blockchain Technology:– A New Technology for Creating Distributed and Trusted Computing Environment. *2022 Int. Conf. Appl. Artif. Intell. Computing (ICAAIC)*, 1348–1354, 2022, doi: 10.1109/ICAAIC53929.2022.9792702.

28. Tyagi, A.K. and Tiwari, S., The Future of Artificial Intelligence in Blockchain Applications, in: *Machine Learning Algorithms Using Scikit and TensorFlow Environments*, IGI Global, 2024, doi: 10.4018/978-1-6684-8531-6.ch018.
29. Nair, M.M. and Tyagi, A.K., Blockchain technology for next-generation society: current trends and future opportunities for smart era, in: *Blockchain Technology for Secure Social Media Computing*, 2023.
30. Deshmukh, A., Patil, D.S., Soni, G., Tyagi, A.K., Cyber Security: New Realities for Industry 4.0 and Society 5.0, in: *Handbook of Research on Quantum Computing for Smart Environments*, A. Tyagi (Ed.), pp. 299–325, IGI Global, 2023, https://doi.org/10.4018/978-1-6684-6697-1.ch017.

18

Artificial Intelligence-Based Cyber Security and Digital Forensics: A Review

Amit Kumar Tyagi[1]*, Shabanm Kumari[2] and Richa[3]

[1]Department of Fashion Technology, National Institute of Fashion Technology, New Delhi, Delhi, India
[2]Department of Computer Science, Faculty of Science and Humanities, SRM Institute of Science and Technology, Kattankulathur, Chennai, Tamil Nadu, India
[3]Department of Computer Science and Engineering, Birla Institute of Technology, Mesra, Ranchi, India

Abstract

Today cyber threats are becoming more sophisticated and technology is advancing at a rapid pace, new approaches to cybersecurity and digital forensics are required. Artificial intelligence (AI) has surfaced as a potential game-changer in the fight against these challenges. This article aims to provide a comprehensive overview of the role that artificial intelligence plays in cybersecurity and digital forensics. By radically improving threat identification, mitigation, and incident response, technologies powered by artificial intelligence are reshaping cybersecurity. Machine learning and deep learning are some of the approaches being used to sift through massive amounts of data, spot anomalies, and predict when security breaches may occur. In addition, AI-powered solutions are making cybersecurity systems more flexible, which opens the door to proactive defensive mechanisms and real-time threat intelligence. The field of digital forensics is seeing heavy use of artificial intelligence (AI) to speed up investigations and locate digital evidence. Automating the examination of digital artifacts is becoming a reality with the use of artificial intelligence methods like pattern recognition, image identification, and natural language processing. This allows for more efficient and effective investigations. Furthermore, AI-driven instruments are assisting with the restoration of digital crime scenes and the attribution of hack perpetrators. With an emphasis on highlighting the capabilities and limitations of these domains, this book discusses cybersecurity and digital forensics that are based on artificial intelligence. In addition to outlining research directions, new trends, and challenges, the article delves into the ethical and legal issues associated with AI in different industries. To sum up, AI is revolutionizing cybersecurity and digital forensics by laying the groundwork for more robust and proactive protection mechanisms and more efficient and accurate digital investigations.

Keywords: Artificial intelligence, cyber security, digital forensics, smart era

**Corresponding author*: amitkrtyagi025@gmail.com

Amit Kumar Tyagi, Shrikant Tiwari, Senthil Kumar Arumugam and Avinash Kumar Sharma (eds.) Artificial Intelligence-Enabled Digital Twin for Smart Manufacturing, (391–420) © 2024 Scrivener Publishing LLC

18.1 Introduction to Cybersecurity Fundamentals and Digital Forensics Basics

The word "cybersecurity" describes the safeguards put in place to prevent unauthorized parties from gaining access to, attacking, damaging, or stealing information from computer systems, networks, and data. This category encompasses a wide variety of methods, procedures, and technologies that have been created to secure digital resources and ensure the accuracy, availability, and secrecy of data [1]. Some of the most essential concepts in cybersecurity are as follows:

- Confidentiality: Securing sensitive data by restricting access to authorized people and maintaining confidentiality.
- Integrity: Ensuring the integrity and reliability of data and systems via the prevention of unauthorized modifications.
- Availability: Ensuring the continuous availability and accessibility of systems and data, without being affected by cyberattacks or system failures.
- Authentication: Verifying the identity of users or systems attempting to access resources.
- Authorization: Determining what level of access or privileges users or systems should have based on their identity and role.
- Firewalls: Network security devices that filter incoming and outgoing traffic to protect against unauthorized access and malicious activity.
- Antivirus Software: Programs that detect and remove malware, such as viruses, worms, and Trojans, from computer systems.
- Encryption: The process of converting data into a coded format to prevent unauthorized access. It ensures data confidentiality.
- Patch Management: Keeping software, operating systems, and applications up to date with security patches to address known vulnerabilities.
- Incident Response: Developing and implementing plans to handle security incidents, breaches, and data breaches effectively.

Digital Forensics Basics: When discussing judicial proceedings and investigations, the term "digital forensics" encompasses both the theory and practice of safely collecting, evaluating, and storing digital evidence. It is an important part of cybersecurity and law enforcement since it helps find digital crimes and create a trail of evidence [2]. Some of the key ideas in computer forensics are as follows:

- Chain of Custody: Documenting the precise times and identities of those who accessed digital evidence is crucial for establishing its credibility and its admissibility in legal proceedings.
- Data Acquisition: a process that entails collecting digital evidence from many sources, such as storage media, mobile devices, and computers.
- Data Recovery: Procedures for retrieving potentially crucial data that has been accidentally deleted or hidden away during an investigation.

- Forensic Imaging: Making a copy of a storage device bit by bit allows for simpler analysis and preservation of the original data, which is necessary for building a forensic picture.
- File Analysis: Analyzing documents and information helps to comprehend their background, substance, and relevance to a question.
- Volatility Analysis: delving into the memory of a running machine to uncover potentially crucial evidence, such as open network connections or running software. This material is considered volatile.
- Timeline Analysis: Reproducing the original sequence of operations on a system by piecing together a record of all the digital events that took place.
- Hashing: The practice of creating unique identifiers for data or files by using cryptographic hashing to verify the legitimacy of digital evidence.
- Legal and Ethical issues: All steps in the digital evidence process, from collection to processing to presentation, must adhere to all relevant ethical and legal standards.
- Expert Testimony: in order to provide support for legal cases, presenting results and views from experts in court.

Therefore, digital forensics is an essential field that plays a role in solving cybercrimes, conducting internal investigations, and ensuring that justice is served in situations that include digital evidence. A comprehensive grasp of both technological processes and legal procedures is required for this.

18.1.1 The Role of AI in Cybersecurity

The significance of artificial intelligence (AI) in cybersecurity is growing as cyber threats persistently advance in intricacy and magnitude. AI technologies are being used in several areas of cybersecurity to improve protection, detection, response, and overall resilience [3] [4]. AI plays many crucial aspects in cybersecurity (refer Figure 18.1):

- Threat Detection and Prevention: AI-driven systems have the capability to analyze vast quantities of data in real-time in order to identify abnormalities and possible risks. Machine learning models have the ability to detect abnormal patterns of behavior, which might indicate possible security problems. Artificial intelligence (AI) powered intrusion detection systems (IDS) have the capability to constantly monitor network traffic and accurately detect malicious actions, hence assisting organizations in proactively preventing assaults.

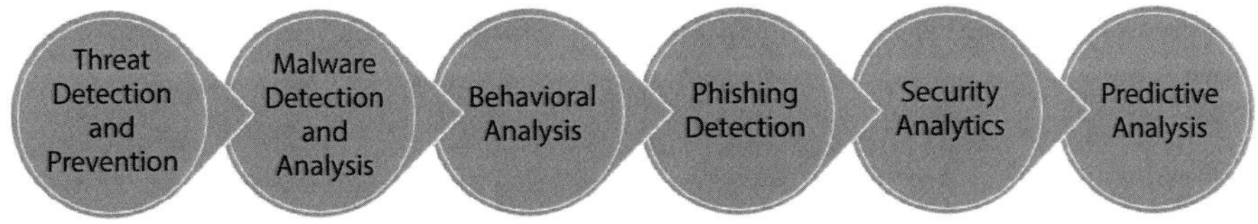

Figure 18.1 Role of AI in cybersecurity.

- Malware Detection and Analysis: Artificial intelligence may be used to create sophisticated methods for detecting and identifying malware. Machine learning models have the ability to examine the characteristics and actions of files in order to detect malware strains that were not previously recognized. Artificial intelligence may assist in the automated examination of malware samples, facilitating the comprehension of their attributes and their consequences for cybersecurity professionals.
- Behavioral Analysis: Artificial intelligence is used for the purpose of analyzing the behavior of both people and systems. The system sets standards for typical behavior and triggers warnings when anomalies are identified, suggesting possible insider risks or compromised accounts.
- Phishing Detection: Phishing attempts may be detected by AI models by analyzing the text, links, and sender behavior of emails. In doing so, they help identify fraudulent emails and protect users from phishing scams.
- Security Analytics: AI-powered security information and event management (SIEM) systems have the capability to analyze extensive amounts of security logs and data in order to detect anomalous patterns or occurrences. Artificial intelligence (AI)-driven analytics technologies assist security teams in effectively prioritizing alerts and conducting investigations into occurrences.
- Predictive Analysis: Security threats may be foreseen by AI by studying historical data and trends. Businesses may better prepare for emerging threats when they adopt this proactive method.
- User and Entity Behavior Analytics (UEBA): To detect strange or harmful behaviors, artificial intelligence models analyze the activities of both people and inanimate objects (such computers and other electronic gadgets). This is especially useful for identifying the risks offered by insiders.
- Automated Incident Response: Incident response systems enabled by AI may automate risk mitigation and reaction activity launch based on predefined rules and playbooks. There is less damage since the response time is shorter.
- AI in Endpoint Security: Artificial intelligence is incorporated into endpoint security systems, offering immediate threat prevention and detection on individual devices. It provides protection for endpoints against a broad spectrum of threats, such as malware and ransomware.
- Security Orchestration and Automation: Artificial intelligence (AI) and machine learning (ML) are used to automate repetitive security operations and streamline processes. SOAR systems provide expedited response to security issues.
- Adaptive Authentication: Adaptive authentication mechanisms that assess the danger of a user's login attempt may be implemented with the help of artificial intelligence. Additional authentication steps may be required if a user's actions are deemed potentially dangerous.
- Threat Intelligence Analysis: Artificial intelligence utilizes algorithms to evaluate and analyze threat intelligence streams in order to find pertinent risks and vulnerabilities. It enables organizations to remain updated on new hazards.

- Privacy and Data Protection: Artificial intelligence (AI) may aid in the prevention of data loss by detecting and safeguarding confidential information, hence minimizing the likelihood of data breaches.
- Fraud Detection: AI is used in the banking and e-commerce industries to detect fraud by examining transaction patterns and user behavior in order to identify deceptive actions.

The role of AI in cybersecurity is ever-expanding, with continuous advancements in AI technologies and their application in threat detection, prevention, and response. As cyber threats become more sophisticated, AI is a valuable ally in the ongoing effort to secure digital environments and protect sensitive data.

18.1.2 Threat Landscape and Challenges

The dynamic nature of cybersecurity threats poses a number of challenges for individuals, businesses, and governments alike [5]. The following are some of the most salient features of the threat landscape and the issues they pose:

- Cyberattacks are increasing in sophistication: The methods of attack used by cybercriminals are becoming more complex and strategic. They use complex malware, social engineering techniques, and targeted assaults to breach networks and steal sensitive data.

Organizations must consistently enhance their cybersecurity measures.

- Ransomware Attacks: In recent years, there has been a dramatic increase in ransomware attacks. Criminals encrypt data using ransomware and then demand payment to decrypt it. The proliferation of cryptocurrency has made it more easier for thieves to launder money while remaining unknown.
- Supply Chain Attacks: As an alternative to going after organizations head-on, cybercriminals are focused on their supply chain partners. By penetrating reputable vendors and service providers, cybercriminals may get access to a wide variety of targets.
- IoT and OT Vulnerabilities: As the number of IoT and operational technology (OT) devices continues to rise, the attack surface has also grown. The security measures offered by many of these devices are insufficient, leaving them open to abuse.
- Zero-Day Vulnerabilities: Attackers often aim for zero-day vulnerabilities, sometimes called newly discovered vulnerabilities, before patches are available. Since these vulnerabilities might be exploited for months or even years, defenders have a substantial challenge.
- Advanced Persistent Threats (APTs): Organized groups known as advanced persistent threats (APTs), which are often linked to nation-states, launch persistent, targeted attacks with the goals of espionage, data theft, or obstruction. Detecting and fending against attacks like this could be challenging.

- Phishing and Social Engineering: Phishing attacks continue to be a major security concern. Attackers will try to trick their targets into giving up sensitive information or doing harmful actions by sending them deceptive emails, messages, or phone calls.
- Regulatory Compliance: Organizations must navigate a complex web of data protection laws, including the General Data Protection Regulation (GDPR) and the California Consumer Privacy Act (CCPA), and guarantee compliance to avoid legal consequences and damage to their reputations.
- Cybersecurity Skills Gap: Companies have a hard time finding and keeping cybersecurity experts on staff, which makes it difficult for them to defend themselves against cyber threats.
- Insider Threats: Insiders or personnel with malicious intent or who inadvertently compromise security constitute a constant threat. In order to detect and prevent insider threats, organizations must establish thorough protocols.
- Cloud Security: More and more information and programs are migrating to the cloud, making it critical to safeguard cloud environments. Data breaches may occur as a consequence of either poorly implemented access controls or improperly configured cloud settings.
- Nation-State Cyberattacks: Individuals backed by state actors have persisted in their cyberattacks against vital government agencies and infrastructure. The geopolitical consequences of these attacks can be enormous.
- Emerging Technologies: Emerging technology deployments pose new security risks, including 5G wireless networks, the internet of things, and artificial intelligence. Although there are many benefits to modern technologies, there are also new opportunities for bad actors to exploit.
- Security Awareness: Human mistake is still the leading cause of security breaches. To mitigate this threat, raising employee cybersecurity awareness and providing better training are of the utmost importance.
- Global Cybersecurity Cooperation: In order to effectively combat cyber threats, global cooperation is essential, given the interconnected nature of the internet. Consensus on acceptable conduct by nations in cyberspace is a big obstacle that has to be overcome.

Take into consideration that the danger environment is ever-changing, and that new problems are always appearing. In order to safeguard their digital assets and data, people and organizations alike need to maintain a state of vigilance, be able to adjust to ever-changing dangers, and make investments in effective cybersecurity procedures and technology. In addition, it is recommended to consult sources that are up to date as well as cybersecurity specialists in order to get the most recent knowledge on the threat landscape and the most effective methods for mitigating it.

18.1.3 Importance of Digital Forensics in Cybersecurity

One of the most important things that digital forensics does for cybersecurity is provide tools for analyzing and responding to cyber incidents, breaches, and attacks [6, 7]. The importance of this factor is influenced by many primary variables:

- Incident Response: Do not underestimate the significance of digital forensics in incident response. As soon as a cybersecurity breach occurs, specialists in digital forensics will utilize their expertise and tools to collect, preserve, and analyze digital evidence. This method helps businesses learn more about the event's scope and characteristics, as well as what they need to do to contain, lessen its impact, and get back on their feet when it happens.
- Evidence Preservation: When it comes to legal matters, digital forensics is in charge of making sure digital evidence is stored in a manner that keeps it intact and may be used in court. This is an important factor to think about while investigating and penalizing hackers.
- Attribution: The use of digital forensics might facilitate the attribution of cyberattacks to specific individuals, groups, or organizations. By analyzing digital data, investigators may determine where an attack originated, identify the perpetrators, and perhaps take legal action against them.
- Threat Hunting: One use of digital forensics is proactive threat hunting, which is searching for signs of breach or potential security holes inside an organization's digital infrastructure. A hidden threat that automated security systems could have missed might be located with this help.
- Data Recovery: It makes no difference whether the deletion or loss of data was accidental or deliberate; digital forensics methods may recover it either way. In order to recover vital data and learn more about the fallout from data breaches, this is really essential.
- Malware Analysis: Professionals in the field of digital forensics analyze malware samples as part of cybersecurity investigations to learn about the virus's characteristics, behavior, and potential impact. Making use of this data to fortify defenses against comparable assaults is very advantageous.
- Insider Threat Detection: Intentionally malicious employees and careless security holes are the two main categories of insider threats. To better understand and address these threats, businesses are turning to digital forensics. Examining user actions and system logs may help spot suspicious conduct in its early stages.
- Compliance and Legal Requirements: Data protection and incident reporting are two areas where many different kinds of organizations and enterprises must adhere to legal and regulatory requirements. Compliance efforts benefit from digital forensics because it provides an organized and thorough method to investigations.
- Root Cause Analysis: Once a cybersecurity problem has occurred, it is essential to know what caused it so that similar incidents do not happen again. If you want to know why anything bad happened, you may utilize digital forensics to find out what went wrong.
- Cyber Insurance Claims: Proof of a cyber-event's occurrence may be needed of businesses with cyber insurance in order to file a claim. If you need to prove what happened and how bad it is for insurance purposes, a digital forensics study could be helpful.
- Legal Proceedings: The digital evidence obtained via forensics is often used to bolster investigations, build a case, and present evidence in court when persons are held liable for cybercrimes.

To sum up, a comprehensive cybersecurity strategy must include digital forensics. Organizations may rely on it to help them react to cyber incidents, limit damage, and take legal action when necessary. Furthermore, by revealing weak spots that may be strengthened to prevent similar incidents, it contributes to a better cybersecurity posture in general. The entire cybersecurity posture is improved by this.

18.1.4 AI-Driven Digital Forensics

When artificial intelligence (AI) technology is used to enhance digital forensics procedures for collecting, processing, and interpreting evidence, the phrase "AI-driven digital forensics" describes it. This approach leverages AI's automation and simplification capabilities to improve the efficacy and efficiency of digital forensics investigations [8, 9]. The following are some of the most salient features of digital forensics powered by AI (refer Figure 18.2):

- Automated Evidence Collection: Artificial intelligence has the potential to automate the collection of digital evidence from many sources, including as mobile devices, PCs, and cloud services. Data like files, logs, and network traffic may be easily recognized and recovered with the use of machine learning techniques.
- Data Triage and Prioritization: The relevance of digital artifacts to an investigation determines their priority, and AI algorithms can do just that. It takes less time for forensic examiners to focus on the most important evidence since they can isolate it.
- Keyword and Contextual Search: Artificial intelligence (AI) powered search algorithms are capable of doing sophisticated keyword searches. Additionally, these algorithms are capable of understanding the context of search words via the use of natural language processing (NLP). This allows for more precise and efficient searching for relevant evidence.

Figure 18.2 AI-driven digital forensics.

- Malware Analysis: Artificial intelligence (AI) malware analysis tools can detect and classify various malware strains, understand their behavior, and assess potential threats by automatically examining suspicious files and code. Particularly when it comes to cybersecurity issues, this is helpful.
- Timeline Analysis: Timelines of digital events may be better constructed with the help of artificial intelligence by examining event logs and timestamps. This might assist you piece together what happened on a system in terms of the order of operations.
- Pattern Recognition: Forensic investigators may benefit from machine learning models that can detect patterns of behavior in digital data, which can help them spot anomalies and potential threats. Both reactively to incidents and proactively in the search for threats might benefit from this.
- Data Deduplication and Data Reduction: By identifying instances of redundant or duplicate data in large datasets, AI systems might potentially lessen the need for storage space and make analysis easier.
- Natural Language Processing (NLP): For the goals of sentiment analysis, subject modeling, and material identification, natural language processing (NLP) techniques enable the examination of text-based evidence, including email and chat logs.
- Image and Video Analysis: The capacity of artificial intelligence to assist in the analysis of images and videos for the purpose of content identification, face recognition, and object detection is a crucial part of investigations that use multimedia data.
- Predictive Analysis: Using historical data and trends, AI can forecast potential dangers and security holes, allowing for preventative measures to be taken.
- Machine Learning for User and Entity Behavior Analytics (UEBA): In order to detect suspicious or harmful activity, UEBA systems powered by AI keep tabs on user and entity behavior. Insider threats and compromised accounts may be more easily identified using this.
- Reporting and Visualization: By automating the creation of reports and visualizations, artificial intelligence makes it easier for stakeholders and investigators to comprehend and communicate the findings.
- Scalability: Solutions for digital forensics that are AI-driven can handle massive data sets and expand to meet the needs of modern digital investigations.

While AI-driven digital forensics has many potential advantages, it is critical to safeguard evidence to guarantee its admissibility in court proceedings.

The investigators and examiners who work in forensic science are responsible for ensuring that the procedures powered by artificial intelligence comply with all relevant legislation and forensic standards. Two of the many extra features that are required to preserve confidence in the findings generated by AI systems in the area of digital forensics are openness and documentation of AI-driven research. This is only one of the many things that are required.

18.1.5 Organization of the Work

There are seven distinct parts to this task. First, we'll talk about how AI-driven digital forensics came to be, which is a sign of how AI-using technologies are being more integrated. The next section delves into the background of the work and the motivations for its development. In the following section, we will delve into the details of digital forensics and artificial intelligence. In Part 4, we will talk about how to use AI for threat detection and prevention. Section 18.5 has outlined a number of important problems and ethical issues related to cyber security and digital forensics enabled by artificial intelligence. Section 18.6 will cover upcoming developments and trends in digital forensics and cyber security driven by artificial intelligence. A brief summary of the book is given in Section 18.7.

18.2 Background and Motivation

Cyberattacks in the modern digital world are becoming more common and more sophisticated, which is a major cause for worry. When faced with evolving threats, conventional cybersecurity measures don't always work [10].

Everyone from individuals to governments recognizes the critical relevance of digital assets. Ensuring the safety of these assets from cyber threats is crucial for sustaining economic stability, national security, and personal privacy [11]. Security breaches and cybercrimes are growing harder to detect and respond to as the volume of digital data being generated grows at an exponential rate [12]. There are a plethora of cyber attack vectors accessible today, including malware, social engineering, insider threats, and zero-day vulnerabilities. When faced with such assaults, traditional rule-based security systems generally fall short. Reacting to cybersecurity threats is becoming inadequate. Preventing, detecting, and responding to cyber assaults before they do significant harm requires proactive measures on the part of both individuals and businesses. Cybercrime investigations rely heavily on digital forensics; yet, forensic experts face formidable challenges due to the ever-increasing complexity of digital devices and encryption technology [13]. Machine learning and deep learning, two branches of artificial intelligence, have achieved great achievements in the last few years. They may transform digital forensics and cybersecurity in a number of ways, including the automation of processes and the enhancement of threat detection [14].

18.2.1 Motivation

So that we can provide a current assessment of AI's performance in cybersecurity and digital forensics, the study's overarching goal is to highlight current developments and future directions in the field. The article goes on to explore how AI techniques may improve the accuracy and speed of threat detection. Artificial intelligence can examine large amounts of data instantly, which helps find anomalies and potential security breaches. One thing to keep in mind is that AI has the potential to help with threat prediction and prevention even before actual assaults happen. Therefore, to show how AI can automate routine security tasks so that cybersecurity professionals may focus on more complex and strategic tasks. Artificial intelligence (AI) may increase operational efficiency and decrease response times. Another area I'd want to investigate is how AI may improve the efficiency with which digital forensics experts gather and evaluate digital evidence, particularly in cases when the data is encrypted or otherwise compromised. What follows is a discussion of the potential impact of AI and quantum computing on digital forensics and cybersecurity, as well as an examination of the current state of the field.

The exponential growth of digital data, the sophistication of cyber threats, and the development of AI have all worked together to provide a compelling argument for investigating AI's potential applications in cybersecurity and digital forensics. Such a research may provide valuable insight into the potential of AI to enhance digital security and facilitate fruitful investigations of cybercrimes.

18.3 Digital Forensics and AI

18.3.1 Automating Digital Forensic Investigations

The challenge of automating digital forensic investigations is growing in importance and difficulty within the fields of cybersecurity and law enforcement. Gathering, evaluating, and preserving electronic evidence to support legal proceedings is the essence of digital forensics investigations. It is feasible to increase efficiency, speed up investigations, and reduce the quantity of human labor needed by automating parts of this process [15]. Some key issues and strategies for automating digital forensic investigations are as follows:

- Data Collection Automation: Use automated tools to collect digital evidence from various sources, such as computers, mobile devices, cloud services, and network logs. Implement remote acquisition capabilities to collect data from remote or inaccessible devices. Ensure the integrity and authenticity of collected data through digital signatures and hashing.
- Data Triage and Prioritization: Develop automated scripts or tools to triage and prioritize collected data based on relevance and potential importance to the investigation. Utilize keyword searches, file metadata, and file timestamps to identify relevant evidence.
- Data Parsing and Extraction: Implement automated parsing and extraction of data from different file formats, such as documents, emails, chat logs, and databases. Utilize specialized tools for extracting and decoding digital artifacts from various operating systems and applications.
- Timeline Analysis: Automate the creation of timeline analysis reports to reconstruct events and activities related to the investigation. Use timeline analysis tools that can visualize chronological sequences of events.
- Keyword and Pattern Searching: Employ automated search and pattern recognition tools to identify keywords, phrases, or patterns of interest within the collected data. Utilize regular expressions and machine learning algorithms to enhance search capabilities.
- Volatility Analysis: Automate memory analysis using tools like Volatility to extract information from volatile memory (RAM) for live system forensics. Develop scripts to automate common memory analysis tasks.
- Log Analysis: Automate the analysis of system and network logs to detect suspicious activities. Use log analysis tools and SIEM (Security Information and Event Management) systems to identify anomalies.

- Reporting and Documentation: Generate automated reports with findings, evidence, and analysis summaries. Ensure that reports are in a format suitable for legal proceedings.
- Integration with Case Management Systems: Integrate automated digital forensic processes with case management systems to maintain a centralized repository of case-related information.
- Machine Learning and AI: Explore the use of machine learning and artificial intelligence algorithms for anomaly detection, classification of evidence, and predictive analysis.
- Quality Assurance: Implement automated checks and validation processes to ensure the accuracy and reliability of automated forensic procedures.
- Continuous Improvement: Regularly update and improve automation scripts and tools to keep pace with evolving technologies and threats.

Note that while automation can rapidly enhance the efficiency of digital forensic investigations, human expertise remains essential for decision-making, contextual analysis, and legal interpretations. Automation should complement, not replace, the skills of trained forensic analysts and investigators. Additionally, the legal and ethical issues surrounding digital evidence collection and analysis must always be adhered to when automating any part of the process.

18.3.2 AI-Powered Evidence Analysis in Cyber Security

AI-powered evidence analysis plays an important role in cybersecurity, especially in the identification, detection, and response to cyber threats. AI technologies can enhance the speed and accuracy of analyzing digital evidence, allowing organizations to respond effectively to security incidents [16, 17]. Here are some ways AI can be applied to evidence analysis in cybersecurity:

- Threat Detection and Prevention: AI algorithms can analyze network traffic and identify unusual patterns and anomalies that may indicate a cyber attack, such as intrusion attempts or malware activity. Also, Machine learning models can continuously learn from historical data to improve threat detection capabilities and adapt to evolving attack techniques.
- Behavioral Analytics: AI-powered behavioral analytics can monitor user and entity behavior, identifying deviations from normal patterns that might suggest insider threats or compromised accounts. Further, Behavioral analytics can also help detect unauthorized access or data exfiltration attempts.
- Malware Analysis: AI can automate the analysis of suspicious files and executables to identify malware. Machine learning models can classify and categorize malware based on code characteristics and behavior.
- Phishing Detection: AI algorithms can scan emails and websites for phishing attempts by analyzing content, sender behavior, and URLs. Also, Natural language processing (NLP) can help identify phishing emails by analyzing text for malicious intent.
- Forensic Analysis: AI can assist in the automated analysis of digital evidence collected during a cyber-incident, such as log files, memory dumps, and

network traffic. Also, Machine learning can help reconstruct timelines of events and identify attack vectors.
- Incident Response Automation: AI-powered incident response platforms can automate the initial response to security incidents by isolating compromised systems, blocking malicious IP addresses, and alerting security teams. Also, Automation can help reduce response times, limiting the damage caused by cyberattacks.
- SIEM and Log Analysis: Security Information and Event Management (SIEM) systems use AI to correlate and analyze large amount of security-related data and logs. Also, AI can help identify patterns and trends that human analysts may overlook.
- Threat Intelligence: AI can analyze threat intelligence feeds to provide real-time information on emerging threats and vulnerabilities. Also, Machine learning models can categorize and prioritize threat intelligence data for more effective threat mitigation.
- User and Entity Behavior Analytics (UEBA): UEBA solutions use AI to create profiles of user and entity behavior and detect deviations that may indicate insider threats or compromised accounts.
- Vulnerability Scanning and Patch Management: AI can help identify vulnerabilities in software and systems and prioritize patching based on the potential impact and exploitability of each vulnerability.
- Continuous Monitoring: AI-driven continuous monitoring solutions can assess the security posture of an organization's IT environment and provide real-time information about potential risks.
- Network Traffic Analysis: AI can analyze network traffic to detect signs of malicious activity, including lateral movement and data exfiltration.

While AI provides essential advantages in evidence analysis for cybersecurity, it's important to remember that it is not a silver bullet. Human expertise remains important for contextual analysis, decision-making, and interpreting the results of AI algorithms. Additionally, AI systems must be continually updated and refined to adapt to evolving threats and attack techniques.

18.3.3 Benefits, Limitation, Issues, and Challenges of AI-Based Cyber Security and Digital Forensics

AI-based cybersecurity and digital forensics provide several benefits, but they also come with limitations, issues, and challenges that need to be carefully considered (refer Table 18.1). Here's an overview of each:

Benefits of AI-based Cybersecurity and Digital Forensics:

- Improved Threat Detection: AI can analyze large amount of data quickly and accurately, enabling the detection of threats that might go unnoticed by traditional methods [18].

Table 18.1 Digital forensics and AI.

Automating digital forensic investigations	**AI-powered evidence analysis in cyber security**	**Benefits, limitation, issues and challenges**	
Data Collection Automation	Threat Detection and Prevention	Benefits	Improved Threat Detection
Data Triage and Prioritization	Behavioral Analytics		Real-time Response
Data Parsing and Extraction	Malware Analysis		Scalability
Timeline Analysis	Phishing Detection		Reduced False Positives
Keyword and Pattern Searching	Forensic Analysis		Behavioral Analysis
Volatility Analysis	Incident Response Automation		Threat Intelligence
Log Analysis	SIEM and Log Analysis	Limitation	False Positives and Negatives
Reporting and Documentation	Threat Intelligence		Adversarial Attacks
Integration with Case Management Systems	User and Entity Behavior Analytics (UEBA)		Lack of Context
Machine Learning and AI	Vulnerability Scanning and Patch Management		Data Privacy Concerns
Quality Assurance	Continuous Monitoring	Issues and Challenges	Skill Shortage
Continuous Improvement	Network Traffic Analysis		Data Quality
			Regulatory Compliance
			Bias and Fairness
			Explainability

- Real-time Response: AI can automate incident response, allowing organizations to respond to cyber threats in real-time, reducing the potential impact of attacks.
- Scalability: AI-powered systems can scale to handle large and complex networks, making them suitable for organizations of all sizes.

- Reduced False Positives: AI algorithms can reduce false alarms by better understanding and contextualizing security alerts, saving valuable time and resources.
- Behavioral Analysis: AI can monitor user and system behavior over time, identifying anomalies and potential insider threats.
- Threat Intelligence: AI can process and analyze threat intelligence feeds to provide timely information on emerging threats and vulnerabilities.
- Automation: AI can automate routine tasks in cybersecurity and digital forensics, improving efficiency and allowing human analysts to focus on more complex tasks.

Limitations of AI-based Cybersecurity and Digital Forensics:

- False Positives and Negatives: AI systems are not infallible and can still generate false positives and false negatives, which can impact the effectiveness of threat detection [19].
- Adversarial Attacks: Malicious actors can use AI techniques to evade detection or launch attacks, making it a cat-and-mouse game between defenders and attackers.
- Lack of Context: AI may lack the ability to understand the broader context of an organization's operations, potentially leading to misinterpretation of normal behavior as malicious.
- Data Privacy Concerns: Using AI to analyze user and entity behavior raises privacy concerns, especially if personal data is involved.
- Resource Intensive: Implementing AI-based cybersecurity solutions can be resource-intensive in terms of computational power, storage, and expertise.

Issues and Challenges in AI-based Cybersecurity and Digital Forensics:

- Skill Shortage: There is a shortage of cybersecurity professionals with expertise in AI, which can hinder the effective deployment of AI in cybersecurity.
- Data Quality: The effectiveness of AI models depends on the quality and quantity of training data, which can be a challenge to obtain, especially for rare or emerging threats.
- Regulatory Compliance: Adhering to data protection regulations and maintaining compliance while using AI in cybersecurity can be complex and challenging.
- Bias and Fairness: AI algorithms can inherit biases present in training data, leading to unfair or discriminatory outcomes in threat detection.
- Explainability: Many AI algorithms, especially deep learning models, are often considered "black boxes," making it difficult to explain their decisions, which is important in cybersecurity and forensics.
- Rapid Evolution: Cyber threats are constantly evolving, and AI-based systems must adapt quickly to remain effective.
- Integration Challenges: Integrating AI solutions into existing cybersecurity infrastructure can be complex and require major changes to an organization's IT environment.

- Ethical Concerns: The use of AI in cybersecurity and digital forensics raises ethical questions, such as the balance between privacy and security and the potential for misuse.

To successfully use AI in cybersecurity and digital forensics, organizations must carefully consider these benefits, limitations, issues, and challenges. It's essential to invest in AI technologies while also maintaining a strong human element in security operations, including skilled analysts who can provide context, oversee AI systems, and make important decisions based on AI-generated information. Additionally, ongoing research and development are necessary to address the evolving nature of cyber threats and the complexities of AI technology.

18.4 AI for Threat Detection, and Prevention

18.4.1 Simulators, Algorithms/Methods for Threat Intelligence in Cybersecurity

Threat intelligence in cybersecurity involves collecting, analyzing, and disseminating information about current and potential threats to an organization's security. To effectively gather and analyze threat intelligence, various simulators, algorithms, and methods can be employed [20, 21]. Here are some commonly used simulators, algorithms, and methods for threat intelligence in cybersecurity:

- Threat Intelligence Platforms (TIPs): Threat intelligence platforms are comprehensive tools that enable organizations to collect, normalize, analyze, and share threat intelligence data from various sources. TIPs often include features for data enrichment, scoring, and reporting.
- Honeypots and Honeynets: Honeypots are decoy systems or networks designed to attract and monitor malicious activity. Honeynets are larger-scale versions of honeypots. Analyzing the behavior of attackers in a controlled environment provides valuable threat intelligence.
- Malware Sandboxing: Malware sandboxes are isolated environments where suspicious files and software can be executed and analyzed to understand their behavior. Algorithms within sandboxes identify malicious patterns, behaviors, and indicators of compromise (IoCs).
- Signature-Based Detection: Signature-based detection uses predefined patterns or signatures to identify known threats, such as viruses and malware. Algorithms compare incoming data with a database of signatures to detect malicious content.
- Heuristic and Behavioral Analysis: Heuristic analysis involves using algorithms that look for behaviors or characteristics that might indicate malicious activity, even if no known signature exists. Behavioral analysis observes how software or entities behave over time and identifies deviations from expected behavior.

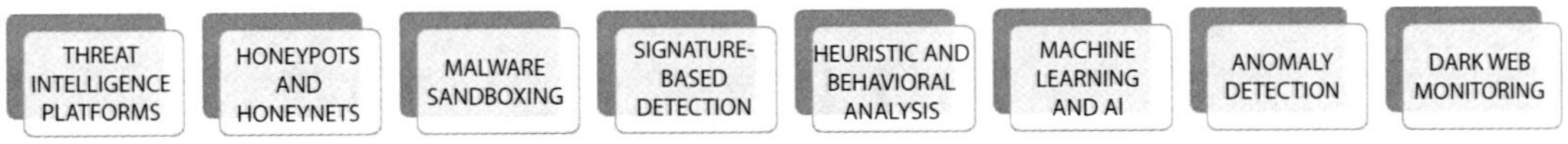

Figure 18.3 Simulators, algorithms/methods for threat intelligence in cybersecurity.

- Machine Learning and AI: Machine learning algorithms, including supervised and unsupervised techniques, can analyze large datasets to identify patterns, anomalies, and trends in threat intelligence data. AI models can be trained to detect both known and unknown threats.
- Anomaly Detection: Anomaly detection algorithms identify deviations from normal network or user behavior, which may indicate intrusions or attacks. Statistical and machine learning methods are commonly used for anomaly detection.
- Open Source Threat Intelligence Feeds: Organizations can use open-source threat intelligence feeds from trusted sources to access up-to-date threat information. These feeds can be integrated into threat intelligence platforms for automated analysis.
- Dark Web Monitoring: Algorithms and tools can be used to monitor the dark web for discussions and marketplaces related to cyber threats and stolen data. This provides early warning of potential attacks or data breaches.
- YARA Rules: YARA is a tool and a rule format for specifying and identifying malware and other threat indicators in files or data streams. - Analysts can create custom YARA rules to detect specific threats relevant to their organization.
- Threat Feeds and Information Sharing: Collaborative threat intelligence sharing platforms enable organizations to share threat information with peers and trusted partners. Algorithms and methods help filter and prioritize incoming threat feeds.
- Natural Language Processing (NLP): NLP techniques can be used to analyze and extract threat intelligence from unstructured text sources, such as news articles, reports, and social media.
- GeoIP Analysis: GeoIP data and analysis can help identify the geographic origins of threats and attacks, aiding in threat attribution and mitigation.

Note that a combination of these simulators, algorithms, and methods (refer Figure 18.3) is often used to create a robust threat intelligence program. Additionally, the effectiveness of threat intelligence relies on the timely collection and analysis of data, as well as the ability to translate intelligence into actionable security measures.

18.4.2 AI-Driven Cybersecurity Solutions for Next Generation Society

AI-driven cybersecurity solutions are essential for securing next-generation societies, where technology is deeply integrated into various aspects of daily life. As society becomes more interconnected and reliant on digital systems, the threat landscape also evolves. AI can provide advanced capabilities to detect, respond to, and mitigate these evolving cyber threats [22]. Here are some key AI-driven cybersecurity solutions for next-generation society (refer Figure 18.4):

Figure 18.4 AI-driven cybersecurity solutions for next generation society.

- Behavioral Analysis and Anomaly Detection: AI can analyze user and entity behavior in real-time to identify deviations from established patterns. This helps detect insider threats, compromised accounts, and other unusual activities.
- Predictive Analytics: Machine learning models can predict potential cyber threats by analyzing historical data and identifying trends, allowing organizations to proactively address vulnerabilities.
- Threat Intelligence Integration: AI can automate the collection and analysis of threat intelligence from various sources, providing real-time information about emerging threats and vulnerabilities.
- Automated Incident Response: AI-driven incident response systems can automatically identify and mitigate security incidents, minimizing the impact of cyberattacks.
- User and Entity Behavior Analytics (UEBA): UEBA solutions use AI to create profiles of user and entity behavior, allowing for the detection of unusual activities that may indicate security threats.
- Network Traffic Analysis: AI can analyze network traffic to detect and respond to suspicious activities, such as malware infections, data exfiltration, and lateral movement.
- Endpoint Security: AI-powered endpoint protection platforms use machine learning to identify and block malware, ransomware, and other threats at the device level.
- Phishing Detection: AI algorithms can analyze emails, websites, and communication patterns to identify phishing attempts and malicious content.

- Identity and Access Management (IAM): AI can enhance IAM systems by continuously monitoring user access and behavior, reducing the risk of unauthorized access.
- AI-Enhanced Authentication: Biometric authentication and facial recognition technologies, backed by AI, can provide secure and convenient ways to access systems and services.
- Deep Packet Inspection: AI-driven deep packet inspection tools can analyze network packets for malicious content and patterns, helping to prevent cyberattacks.
- Security Orchestration and Automation: AI can orchestrate and automate security processes, enabling faster response times and reducing the workload on security teams.
- Zero Trust Architecture: AI can play an important role in implementing Zero Trust principles by continuously assessing the trustworthiness of devices, users, and applications.
- Cloud Security: AI-driven cloud security solutions can monitor and protect data and applications in the cloud, ensuring the security of cloud-based resources.
- Regulatory Compliance: AI can assist organizations in automating compliance tasks and ensuring that they meet the regulatory requirements of next-generation society.
- Security Awareness Training: AI-driven training platforms can deliver personalized and adaptive cybersecurity training to educate individuals about evolving threats and safe online behavior.
- Ethical Hacking and Vulnerability Scanning: AI-powered tools can simulate cyberattacks and identify vulnerabilities in systems, allowing organizations to proactively strengthen their defenses.

As next-generation societies become increasingly reliant on digital technologies, the role of AI in cybersecurity will continue to expand. Note that implement AI-driven solutions with a focus on privacy, ethics, and transparency, addressing concerns related to data privacy and potential biases in AI algorithms. Moreover, ongoing research and development are important to stay ahead of emerging cyber threats in this dynamic landscape.

18.5 Important Challenges and Ethical Issues for AI-Powered Cyber Security and Digital Forensics

AI-powered cybersecurity and digital forensics provide huge advantages, but they also face important challenges and ethical issues that must be addressed to ensure responsible and effective use [23, 24]. Here are some of the most pressing challenges and ethical concerns as Table 18.2.

Hence, addressing these challenges and ethical concerns requires a multidisciplinary approach that involves technologists, ethicists, legal experts, policymakers, and stakeholders.

Table 18.2 Challenge and ethical issues towards AI-powered cyber security and digital forensics.

Sr. no.	Types	Challenges	Ethical issues
1	**Bias and Fairness**	AI algorithms can inherit biases present in training data, leading to unfair outcomes, such as biased threat detection or profiling.	Unfair or biased treatment in cybersecurity can have serious consequences and may result in unjust accusations or discrimination.
2	**Privacy Concerns**	AI systems may process large amounts of personal and sensitive data during cybersecurity and digital forensics investigations.	Protecting individuals' privacy while conducting investigations is important. Unauthorized data collection and surveillance can violate privacy rights.
3	**Explainability and Transparency**	Many AI algorithms, especially deep learning models, are often considered "black boxes," making it difficult to explain their decisions.	Transparency is essential for accountability and trust. Understanding why an AI system made a particular decision is important, especially in legal contexts.
4	**Adversarial Attacks**	Malicious actors can use AI techniques to evade detection or manipulate AI systems to their advantage.	AI systems must be resilient to adversarial attacks to ensure the integrity and effectiveness of cybersecurity measures.
5	**Accountability and Liability**	Determining who is responsible when AI-powered systems make mistakes or fail to prevent a cyber-incident can be complex.	Clear accountability is essential to ensure that those responsible for AI systems are held accountable for their actions or failures.
6	**Data Security**	AI systems rely on large amount of data, and securing this data is important to prevent unauthorized access or breaches.	Failing to secure sensitive data can lead to data breaches, privacy violations, and harm to individuals.
7	**Ethical Hacking and Vulnerability Disclosure**	Ethical hacking and vulnerability disclosure involve responsible testing and reporting of security weaknesses.	Balancing the need for security testing with the potential for causing harm or violating laws and regulations is a complex ethical challenge

(*Continued*)

Table 18.2 Challenge and ethical issues towards AI-powered cyber security and digital forensics (*Continued*)

Sr. no.	Types	Challenges	Ethical issues
8	**Bias in Threat Profiling**	AI may profile individuals or groups based on their behavior or attributes, potentially reinforcing stereotypes	Unfair or biased threat profiling can lead to discrimination and unjust targeting of individuals or communities.
9	**Dual-Use Technology**	AI technology developed for cybersecurity and digital forensics can have dual-use potential, including for offensive purposes.	Ensuring that AI technology is used responsibly and ethically is important to prevent harm and misuse.
10	**Legal and Regulatory Compliance**	The legal and regulatory landscape for AI in cybersecurity and digital forensics is still evolving and can be complex.	Complying with laws and regulations while conducting investigations and implementing AI-powered solutions is essential to avoid legal and ethical violations

Developing and adhering to ethical guidelines and best practices, promoting transparency and accountability, and continuously monitoring and auditing AI systems are important steps in ensuring that AI-powered cybersecurity and digital forensics are used responsibly and ethically. Additionally, collaboration between the public and private sectors is essential to establish norms and regulations that promote ethical AI use in these important domains.

18.6 Future Trends and Innovations AI-Powered Cyber Security and Digital Forensics

18.6.1 Emerging Technologies in AI-Driven Cybersecurity

Emerging Technologies in AI-Driven Cybersecurity are;

- Explainable AI (XAI): As AI systems become more complex, understanding why and how they make decisions is important. XAI aims to provide transparency and interpretability in AI models, making it easier to identify and rectify security vulnerabilities.
- AI for Threat Hunting: AI is increasingly used to proactively search for threats within a network. It can analyze large datasets and identify anomalies or patterns indicative of potential attacks.

- AI-Enhanced Security Information and Event Management (SIEM): SIEM platforms are incorporating AI to improve their capabilities in real-time threat detection and response. AI-driven SIEM can help security teams better prioritize alerts and incidents.
- AI-Driven Deception Technology: Deception technology uses AI to create realistic decoy systems and data to lure attackers. This helps in early threat detection and allows defenders to study the attacker's tactics.
- AI-Powered User and Entity Behavior Analytics (UEBA): UEBA solutions use AI to build baselines of user and entity behavior. This enables them to detect abnormal activities that could indicate a security threat.
- AI for Network Security: AI is used to detect and respond to network-based threats, including intrusion detection and prevention, traffic analysis, and firewall management.
- AI in Endpoint Security: AI-driven endpoint security solutions use machine learning to identify and block malicious activities on individual devices within a network.
- AI-Driven Threat Intelligence: AI can help organizations make sense of large amount of threat intelligence data, identifying relevant information and assessing the potential impact of threats.
- Federated Learning for Privacy: Federated learning enables AI models to be trained on decentralized data sources without exposing sensitive information. This technology is particularly useful in privacy-preserving cybersecurity.
- AI for Vulnerability Management: AI helps organizations identify vulnerabilities in their systems and prioritize patches based on the risk associated with each vulnerability.
- AI in Security Orchestration, Automation, and Response (SOAR): SOAR platforms use AI to automate incident response, enabling security teams to respond more effectively to threats.
- AI-Enhanced Phishing Detection: Machine learning models can analyze email content and sender behavior to identify phishing attempts more accurately.
- Quantum-Safe AI: As quantum computing threatens current encryption methods, AI is being used to develop quantum-safe encryption algorithms and strategies.
- AI-Driven Identity and Access Management (IAM): AI can enhance IAM systems by monitoring user behavior and identifying suspicious activities.
- AI for Cloud Security: AI helps secure cloud environments by detecting unauthorized access, anomalies in data traffic, and ensuring compliance with security policies.
- Homomorphic Encryption: This encryption technology allows data to be processed while it's still encrypted, enabling secure AI analysis of sensitive information.
- AI in Threat Simulation: AI-powered threat simulators can replicate real-world attack scenarios, helping organizations identify and address security weaknesses.

- AI-Driven Chatbots for Security: AI chatbots can provide automated responses to security queries, provide advice on security best practices, and assist with incident reporting.
- AI-Enhanced Security Analytics: AI can analyze large datasets to uncover hidden patterns and trends that might signal emerging threats.
- Blockchain for Secure AI: The combination of blockchain and AI can provide a secure and transparent environment for AI model training and data sharing, particularly in multi-party collaborations [28–30].

Hence, AI-driven cybersecurity technologies are continually evolving to stay ahead of increasingly sophisticated threats. Organizations that adopt these technologies can improve their security posture by automating threat detection and response, reducing false positives, and responding rapidly to emerging threats.

18.6.2 Research Gaps and Future Opportunities Towards AI-Powered Cyber Security and Digital Forensics

AI-powered cybersecurity and digital forensics have made important improvement in recent years, but there are still several research gaps and future opportunities that can drive innovation and improve the effectiveness of these technologies [25]. Here are some key research areas and opportunities as Table 18.3.

Hence, as AI technology continues to evolve and cyber threats become more sophisticated, researchers, policymakers, and industry stakeholders should collaborate to address these research gaps and seize future opportunities in AI-powered cybersecurity and digital forensics. This collaborative approach is essential for enhancing security, privacy, and resilience in the face of emerging cyber challenges.

18.6.3 Unexplored Areas in AI-Based Cybersecurity

AI-based cybersecurity has made required huge advancements, but there are still several unexplored areas and emerging challenges that researchers and practitioners can used to enhance security measures further [26]. Here are some unexplored areas in AI-based cybersecurity:

- Quantum Cybersecurity: Quantum computing faces a threat to current encryption algorithms. Research into AI-based solutions that can defend against quantum attacks and secure quantum communications is essential.
- AI for IoT Security: IoT presents unique security challenges due to its scale and diversity. Exploring AI-driven solutions for IoT security, including anomaly detection and behavior analysis, is an uncharted territory.
- AI for 5G and Beyond: The rollout of 5G networks introduces new vulnerabilities. Investigating how AI can protect 5G infrastructure and the large array of connected devices is an evolving area.
- AI-Enhanced Security for Cloud-Native Environments: Cloud-native architectures, microservices, and containers require novel security approaches.

Table 18.3 Research gaps and opportunity for future AI based cybersecurity.

Sr. no	Technology	Research gasps	Opportunity for future
1	Explainable AI (XAI)	Many AI algorithms, especially deep learning models, lack transparency and explainability, making it challenging to understand their decision-making processes.	Developing XAI techniques that provide clear and interpretable explanations for AI-driven cybersecurity and digital forensics systems will enhance trust and accountability.
2	Bias Mitigation	AI algorithms can inherit biases present in training data, leading to unfair outcomes in threat detection and profiling.	Research into bias mitigation techniques, such as debiasing algorithms and fairness-aware learning, can help ensure that AI systems are fair and unbiased.
3	Adversarial Robustness	Adversarial attacks can manipulate AI systems to evade detection or cause misclassification.	Developing robust AI models that can withstand adversarial attacks and maintain their effectiveness is important for cybersecurity.
4	Data Privacy-Preserving AI	Balancing the need for data privacy with the data-hungry nature of AI models can be challenging.	Research into privacy-preserving AI techniques, including federated learning, homomorphic encryption, and secure multi-party computation, can enable AI-powered cybersecurity without compromising privacy [31, 32].
5	Autonomous Threat Detection and Response	While AI can automate certain aspects of threat detection and response, fully autonomous systems are still a research challenge.	Developing AI-driven systems that can autonomously detect, mitigate, and respond to cyber threats in real-time will be important for next-generation cybersecurity.
6	Cognitive Computing and Human-AI Collaboration	Enhancing collaboration between human analysts and AI systems for more effective cybersecurity investigations.	Research into cognitive computing and AI-driven decision support systems can improve human-AI collaboration and enable more informed decision-making.

(Continued)

Table 18.3 Research gaps and opportunity for future AI based cybersecurity. (*Continued*)

Sr. no	Technology	Research gasps	Opportunity for future
7	Ethical Hacking and Red Teaming	Ethical hacking and red teaming exercises require responsible testing and vulnerability discovery.	Research on methodologies, frameworks, and guidelines for conducting ethical hacking and red teaming while adhering to ethical standards and legal regulations.
8	Threat Intelligence Fusion	Effectively aggregating and correlating threat intelligence from various sources to provide actionable information.	Developing AI-driven systems for threat intelligence fusion that can provide a complete view of the threat landscape.
9	Scalable AI Solutions	Scalability remains a challenge when deploying AI-powered cybersecurity solutions across large and complex networks.	Research into scalable AI architectures and distributed AI systems can make it more feasible to deploy AI across a broad spectrum of devices and networks.
10	Cross-Domain Threat Analysis	Identifying and mitigating threats that span multiple domains, including IoT, cloud, and important infrastructure.	Research into cross-domain threat analysis techniques and tools that can protect interconnected systems and infrastructure.
11	Standardization and Regulation	Establishing industry standards and regulations for AI-powered cybersecurity and digital forensics.	Collaborative research and advocacy efforts to develop and implement standards, ethics guidelines, and regulatory frameworks.

Research into AI-based solutions tailored to the cloud-native ecosystem is a relatively unexplored space.

- AI for Industrial Control Systems (ICS) Security: Protecting important infrastructure and industrial systems from cyberattacks is important. AI can play a role in securing ICS, including predictive maintenance and anomaly detection.
- AI-Driven Security in Edge Computing: Edge computing environments require lightweight and efficient security solutions. Investigating AI models optimized for edge devices and real-time decision-making is a burgeoning field.
- AI-Enabled Cyber-Physical Systems Security: Ensuring the security of cyber-physical systems, such as autonomous vehicles and smart grids, is an emerging challenge. AI can help in threat detection and response for these systems.

- AI-Enhanced Identity and Access Management: Traditional approaches to identity and access management may not be sufficient. Exploring AI-driven solutions for continuous authentication and adaptive access control is a promising area.
- AI for Predictive Threat Intelligence: Using AI to predict emerging cyber threats and vulnerabilities is an unexplored field that could enhance proactive cybersecurity measures [27].
- AI for Post-Quantum Cryptography: With the advent of quantum computing, research into AI-driven post-quantum cryptographic solutions that are resistant to quantum attacks is important.
- AI-Driven Cyber Insurance: The cyber insurance industry is growing, and AI can play a role in assessing risk, predicting cyber threats, and streamlining the claims process.
- AI for Secure Software Development: Integrating AI into the software development lifecycle to identify and mitigate security vulnerabilities in real-time is an area that can help reduce the prevalence of software-based cyber threats.
- AI for Privacy-Preserving Technologies: Developing AI-driven techniques that can protect data privacy while still enabling effective analysis and threat detection is a promising yet underexplored research direction.
- AI for Behavioral Biometrics: Investigating AI-powered behavioral biometrics as a more secure and user-friendly authentication method can enhance cybersecurity.
- AI Ethics and Accountability in Cybersecurity: Ethical issues related to AI in cybersecurity, including issues such as fairness, transparency, and accountability, are emerging areas of concern that require further exploration.

Hence, exploring these uncharted areas will not only enhance the capabilities of AI in cybersecurity but also contribute to a more robust and resilient security landscape in the face of evolving cyber threats. Collaborative efforts between academia, industry, and policymakers will be essential to drive research and innovation in these domains (also refer Figure 18.5 to know about unexplored areas in AI based cyber security).

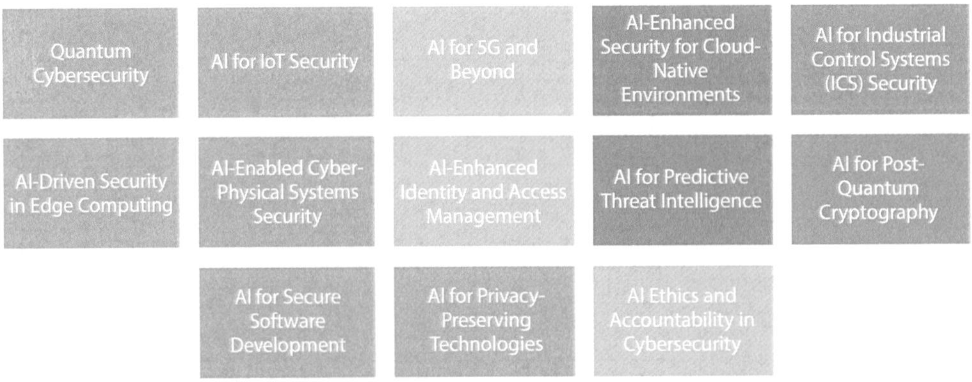

Figure 18.5 Unexplored areas in AI-based cybersecurity.

18.7 Conclusion

AI-powered cybersecurity and digital forensics represent a transformative force in our ongoing battle against evolving cyber threats. These technologies have the potential to rapidly enhance our ability to detect, respond to, and mitigate cyberattacks, as well as investigate and attribute digital crimes. However, they also come with a set of challenges and ethical issues that must be carefully navigated. The benefits of AI in cybersecurity are clear, ranging from improved threat detection and response times to the automation of routine tasks and the ability to process large amount of data. AI-driven solutions provide the promise of making our digital world more secure and resilient in the face of an ever-expanding threat landscape. Note that challenges such as bias in AI algorithms, data privacy concerns, adversarial attacks, and the need for transparency and accountability must not be underestimated. Ensuring that AI is used ethically and responsibly in cybersecurity and digital forensics is paramount. Collaboration among researchers, practitioners, policymakers, and industry stakeholders is essential to address these challenges effectively. As we move forward, unexplored areas in AI-based cybersecurity and digital forensics can provide tremendous support. Quantum cybersecurity, IoT security, securing 5G and cloud-native environments, and the application of AI to protect important infrastructure are just a few of the frontiers where AI can make a major impact. By embracing AI responsibly, we can create a safer and more resilient digital environment for individuals, organizations, and societies at large. As AI continues to evolve, so will our ability to safeguard our digital future.

References

1. Joakim, *Fundamentals of Digital Forensics*, Springer International Publishing, 2020.
2. Alhassan, J.K. and Olalere, M., Fundermentals of cyber security & cyber crime, 2019.
3. Kuzlu, M., Fair, C., Guler, O., Role of artificial intelligence in the Internet of Things (IoT) cybersecurity. *Discov. Internet Things*, 1, 1–14, 2021.
4. Geluvaraj, B., Satwik, P.M., Kumar, T.A.A., The future of cybersecurity: Major role of artificial intelligence, machine learning, and deep learning in cyberspace, in: *International Conference on Computer Networks and Communication Technologies: ICCNCT 2018*, Springer, Singapore, pp. 739–747, 2019.
5. Kumar, M., Darshan, S.L.S., Yarlagadda, V., Introduction to the cyber-security landscape, in: *Malware Analysis and Intrusion Detection in Cyber-Physical Systems*, pp. 1–21, IGI Global, 2023.
6. Pandey, A.K., Tripathi, A.K., Kapil, G., Singh, V., Khan, M.W., Agrawal, A., Kumar, R., Khan, R.A., Current challenges of digital forensics in cyber security, in: *Important Concepts, Standards, and Techniques in Cyber Forensics*, pp. 31–46, 2020.
7. Alghamdi, M., II, Digital forensics in cyber security—recent trends, threats, and opportunities, in: *Cybersecurity Threats with New Perspectives*, 2021.
8. Dunsin, D., Ghanem, M.C., Ouazzane, K., Vassilev, V., A comprehensive analysis of the role of artificial intelligence and machine learning in modern digital forensics and incident response. *Forensic Sci. Int.: Digital Invest.*, 48, 301675, 2024.
9. Solanke, A.A., Explainable digital forensics AI: Towards mitigating distrust in AI-based digital forensics analysis using interpretable models. *Forensic Sci. Int.: Digital Invest.*, 42, 301403, 2022.
10. Evren, R. and Milson, S., The Cyber Threat Landscape: Understanding and Mitigating Risks. No. 11705, in: *EasyChair*, 2024.

11. Banta, N.M., Property interests in digital assets: The rise of digital feudalism. *Cardozo L. Rev.*, 38, 1099, 2016.
12. Rampini, L., Moretti, N., Cecconi, F.R., Dejaco, M.C., Digital Asset Management enabling technologies: a bibliometric analysis, in: *New Horizons for Sustainable Architecture*, pp. 919–931, 2020.
13. Vincze, E.A., Challenges in digital forensics. *Police Pract. Res.*, 17, 2, 183–194, 2016.
14. Kasula, B.Y., Advancements and Applications of Artificial Intelligence: A Comprehensive Review. *Int. J. Stat. Comput. Simul.*, 8, 1, 1–7, 2016.
15. James, J., II and Gladyshev, P., Challenges with automation in digital forensic investigations, in: *arXiv preprint arXiv:1303.4498*, 2013.
16. Huang, L., Ai-Powered System-Scientific Defense for High-Confidence Cyber-Physical Systems: Modeling, Analysis, and Design, in: *PhD diss*, New York University Tandon School of Engineering, 2022.
17. Lasagni, G., AI-Powered Investigations: From Data Analysis to an Automated Approach Toward Investigative Uncertainty, in: *Investigating and Preventing Crime in the Digital Era: New Safeguards, New Rights*, pp. 169–188, Cham: Springer International Publishing, 2022.
18. Calderon, R., The benefits of artificial intelligence in cybersecurity, 2019.
19. Jarrett, A. and Choo, K.-K.R., The impact of automation and artificial intelligence on digital forensics. *Wiley Interdiscip. Rev.: Forensic Sci.*, 3, 6, e1418, 2021.
20. Trifonov, R., Manolov, S., Yoshinov, R., Tsochev, G., Pavlova, G., Artificial intelligence methods for cyber threats intelligence. *Int. J. Comput.*, 2, 2017.
21. Kavak, H., Padilla, J.J., Vernon-Bido, D., Diallo, S.Y., Gore, R., Shetty, S., Simulation for cybersecurity: state of the art and future directions. *J. Cybersecur.*, 7, 1, tyab005, 2021.
22. Bharadiya, J.P., AI-Driven Security: How Machine Learning Will Shape the Future of Cybersecurity and Web 3.0. *Am. J. Neural Netw. Appl.*, 9, 1, 1–7, 2023.
23. Yu, S. and Carroll, F., Implications of AI in National Security: Understanding the Security issues and Ethical challenges, in: *Artificial Intelligence in Cyber Security: Impact and Implications: Security Challenges, Technical and Ethical Issues, Forensic Investigative Challenges*, pp. 157–175, Cham: Springer International Publishing, 2022.
24. Nassar, A. and Kamal, M., Ethical dilemmas in AI-powered decision-making: a deep dive into big data-driven ethical considerations. *Int. J. Responsible, Artif. Intell.*, 11, 8, 1–11, 2021.
25. Jarrett, A. and Raymond Choo, K.-K., The impact of automation and artificial intelligence on digital forensics. *Wiley Interdiscip. Rev.: Forensic Sci.*, 3, 6, e1418, 2021.
26. Sarker, I.H., Furhad, H., Nowrozy, R., Ai-driven cybersecurity: an overview, security intelligence modeling and research directions. *SN Comput. Sci.*, 2, 1–18, 2021.
27. Deshmukh, A., Patil, D.S., Soni, G., Tyagi, A.K., Cyber Security: New Realities for Industry 4.0 and Society 5.0, in: *Handbook of Research on Quantum Computing for Smart Environments*, A. Tyagi (Ed.), pp. 299–325, IGI Global, 2023, https://doi.org/10.4018/978-1-6684-6697-1.ch017.
28. Tyagi, A.K., Chandrasekaran, S., Sreenath, N., Blockchain Technology:– A New Technology for Creating Distributed and Trusted Computing Environment. *2022 International Conference on Applied Artificial Intelligence and Computing (ICAAIC)*, pp. 1348–1354, 2022, doi: 10.1109/ICAAIC53929.2022.9792702.
29. Tyagi, A.K., Analysis of Security and Privacy Aspects of Blockchain Technologies from Smart Era' Perspective: The Challenges and a Way Forward, in: *Recent Trends in Blockchain for Information Systems Security and Privacy*, CRC Press, 2021.
30. Tyagi, A.K., Rekha, G., Kumari, S., Applications of Blockchain Technologies in Digital Forensic and Threat Hunting, in: *Recent Trends in Blockchain for Information Systems Security and Privacy*, CRC Press, 2021.

31. Deshmukh, A., Sreenath, N., Tyagi, A.K., Eswara Abhichandan, U.V., Blockchain Enabled Cyber Security: A Comprehensive Survey. *2022 International Conference on Computer Communication and Informatics (ICCCI)*, pp. 1–6, 2022, doi: 10.1109/ICCCI54379.2022.9740843.
32. Sheth, H.S.K., I., A.K., Tyagi, A.K., Deep Learning, Blockchain based Multi-layered Authentication and Security Architectures. *2022 International Conference on Applied Artificial Intelligence and Computing (ICAAIC)*, pp. 476–485, 2022, doi: 10.1109/ICAAIC53929.2022.9793179.

19

Blockchain Application in Industry 5.0

D. Saveetha[1], Vijayakumar Ponnusamy[2*], Nemanja Zdravković[3] and Nandini S.M.[2]

[1]Department of Networking and Communications, SRM Institute of Science and Technology, Kattankulathur, Tamil Nadu, India
[2]Department of Electronics and Communication Engineering, SRM Institute of Science and Technology, Kattankulathur, Tamil Nadu, India
[3]Faculty of Information Technology, Metropolitan University, Belgrade, Serbia

Abstract
Industries play a vital role in the wealth of our nation. The growth of productivity of goods and services is directly related to the economy. Manpower, machinery, tools, and costs have drastically improved according to the developing technologies. The term Industry 5.0 refers to a scenario where people make use of advanced technologies like the Internet of Things (IoT), Artificial Intelligence, and smart machines for effective productivity. This development is more advantageous like providing low-cost products, improved wages, and transportation of goods from rural to urban areas. Since industries are dependent on smart devices, and automated environments, energy consumption and data privacy have become a challenge nowadays. To resolve these issues, blockchain technology provides a decentralized network where the information and transactions are encrypted and stored as blocks. Each block undergoes data validation through the mining process and gets added to the chain of blocks as a blockchain. This is a distributed ledger where a copy of a transaction is distributed across the network to maintain data transparency and immutability. Supply chain management, digital twinning, and secure data exchange among the stakeholders are some of the applications of blockchain in Industries 5.0. This enhances traceability, transparency, and quality control in the supply chain because of its tamper-proof property. Every transaction is securely recorded in the form of smart contracts which automate contracts that increase the efficiency of the system from malicious attacks. Digital twinning is another concept where a digital replica of the physical world is created and data are updated and synchronized accordingly. Thus, the implementation of blockchain technology in Industry 5.0 guarantees data security and privacy, which builds confidence among industrialists and increases efficiency.

Keywords: Artificial intelligence, blockchain, decentralized manufacturing, digital twin, healthchain, Industry 5.0, IoT, predictive maintenance

19.1 Introduction

The topic "Blockchain Application in Industry 5.0" portrays the link between the latest progress in automation, Industry 5.0, and blockchain technology. It depicts how blockchain

**Corresponding author*: vijayakp@srmist.edu.in

Amit Kumar Tyagi, Shrikant Tiwari, Senthil Kumar Arumugam and Avinash Kumar Sharma (eds.) Artificial Intelligence-Enabled Digital Twin for Smart Manufacturing, (421–442) © 2024 Scrivener Publishing LLC

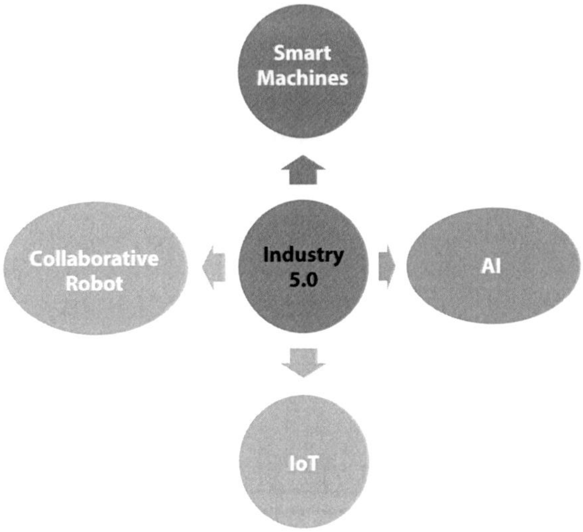

Figure 19.1 Structure of Industry 5.0.

technology can transform manufacturing, transport, health care, educare, automation, energy sector, and supply chains through effective, secure, and transparent systems. This chapter will provide context as to how this integration can transform the industrial world in this century. This chapter uses authentic examples and discusses the integration of blockchain into the growing world of Industry 5.0.

Industry 5.0, also known as the Fifth Industrial Revolution, is a constantly evolving paradigm. It allows people to utilize these advanced technologies, smart machines, and AI robots to make work easier. This new stage isn't just about manufacturing, it extends far beyond Industry 4.0. It uses advanced techniques in areas like artificial intelligence, automation, big data, the Internet of Things (IoT), machine learning, robotics, intelligent systems, and virtualization. The European Union (EU) states that Industry 5.0 will improve worker well-being and bring new cutting-edge technology for a prosperous future in productivity. Thus, adopting good practices across sectors can strengthen industries against outside influences, such as COVID-19. Industry 5.0 matters because it empowers businesses to actively contribute solutions for society to conserve resources, secure social stability, and tackle climate goals. The focus is not only on profit and productivity but also on benefits for employees and the environment. Now, let's look at how blockchain fits into Industry 5.0. Figure 19.1 portrays the structure of Industry 5.0.

19.2 Blockchain for Industry 5.0 Transformative Potential

Similar to previous industrial revolutions, blockchain has the potential to play a crucial driving role in Industry 5.0. It is the hallmark of the coming phase of the industry's evolution. These phases are primarily characterized by a highly connected and intelligent network of systems, devices, and people. Following are a few areas where blockchain technology can be effective.

Supply Chain Management: An effective, immutable, verifiable, and secure supply chain management can be established using blockchain. It can drastically increase traceability, guarantee the quality of the product, and track product provenance in Industry 5.0. Blockchain technology can also help in lowering fraud and counterfeiting of products while simultaneously increasing the effectiveness of the supply chain [1].

IoT Integration: Connectivity plays a crucial role in Industry 5.0. The Internet of Things (IoT) is one of the important aspects of Industry 5.0, as it allows a wide range of systems to interact on their own in an automated fashion. This is done by establishing uninterrupted connectivity and data exchange in real time. Since IoT networks, AI, and sensors require a highly secure and reliable environment, blockchain technology is very important. The implementation of blockchain ensures data received from IoT sensors and machines is tamper-proof. This makes blockchain a valuable safeguard for applications that depend on autonomous systems and smart manufacturing [2], supporting use cases that require assurance in the validity and source of information driving automated operations and decision-making.

Agreements/Contracts: The agreement between the buyer and seller is encrypted as code directly by the use of smart contracts in blockchain that can be executed when certain conditions/criteria are met. These contracts help the industries to track the maintenance of equipment and perform effective quality checks [3]. These smart contracts support automatic transactions safely and securely. This builds trust and transparency in how things get done. Smart contracts powered by blockchain could streamline many processes for Industry 5.0.

Tokenization of Assets: Tokenization is the focus of Industry 5.0 as it entails the digitization of physical products and virtual assets into "tokens". This technology makes manufacturing, trading, and token management easier [4]. These are mere substitutes for the real assets. This facilitates in development of new and better ideas to make the availability of asset trade easier.

Environmental Sustainability: As the key focus of Industry 5.0 is sustainability, blockchain can create transparent and accountable systems for tracking the environmental impact of products and processes, such as carbon credits and supply-chain sustainability. The workflow of blockchain in Industry 5.0 is depicted in Figure 19.2.

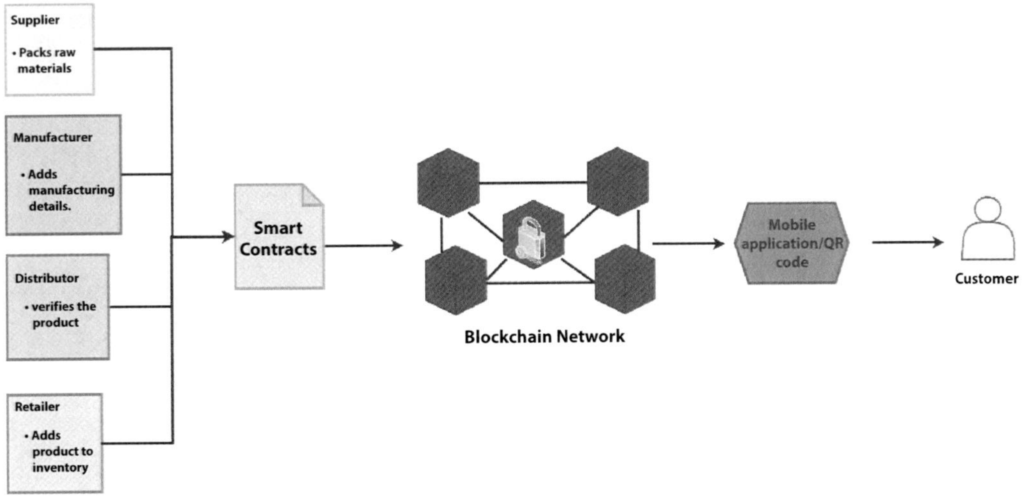

Figure 19.2 Workflow of blockchain in Industry 5.0.

19.3 Applications of Blockchain in Industry 5.0

With blockchain being an inherent part of Industry 5.0, it naturally has many applications. Some of the key uses of blockchain in Industry 5.0 are:

- IoT Security: Data security and privacy become a major concern when different systems are connected. Blockchain can provide this security that results in safer and better industrial operations [5].
- Provenance Tracking: verification and validation of records based on their origin ensures authenticity and greatly reduces the chance of counterfeiting. Blockchain and smart contracts can automate processes which include all processes but are not limited to payment, quality control, and transport [6].
- Device Identity and Authentication: Blockchain are often used to automate secure authentication of device identity for IoT networks, devices, and systems. This ensures authorized access and tampering with sensitive information.
- Maintenance and Predictive Analytics: Smart contracts that employ real-time machine data generate and update the maintenance details. This is generally done to optimize the manufacturing process [7].
- Digital Twins: The blockchain is capable of storing digital twin information. This information represents a replica of physical products and assets in the digital world [8]. It is done accurately and securely over their complete lifespan.
- Warranty and Service Records: The history of warranty and services are maintained securely as records. This helps in facilitating identification and address the resolution of issues between customer and seller.
- Energy Trading: Blockchain technology can be used to support energy trading. This can be implemented in a transparent and decentralized manner. This results in making efficient renewable energy sources.
- Resource Optimization: Utilization of resources is tracked by monitoring various energy consumptions [9]. Resources like water, power, energy, raw materials, etc., are important to monitor for effective usage and allocation thereby improving efficiency.
- Credential Verification: It helps in the verification and maintenance of professional certifications, qualifications, or skills required by people and/or machines. This is used in government offices or companies that hire people. This could also be used by educational institutions to maintain secure transcripts of students.
- Carbon Credits and Emissions Tracking: Blockchain technology is widely accepted as transparent and visible, which makes it ideal for developing carbon credit tracking systems [10]. These systems use a unique credit-based tracking system that monitors carbon emissions.

We have listed only a few applications but it can be applied to a variety of domains. The applications of blockchain in Industry 5.0 are shown in Figure 19.3.

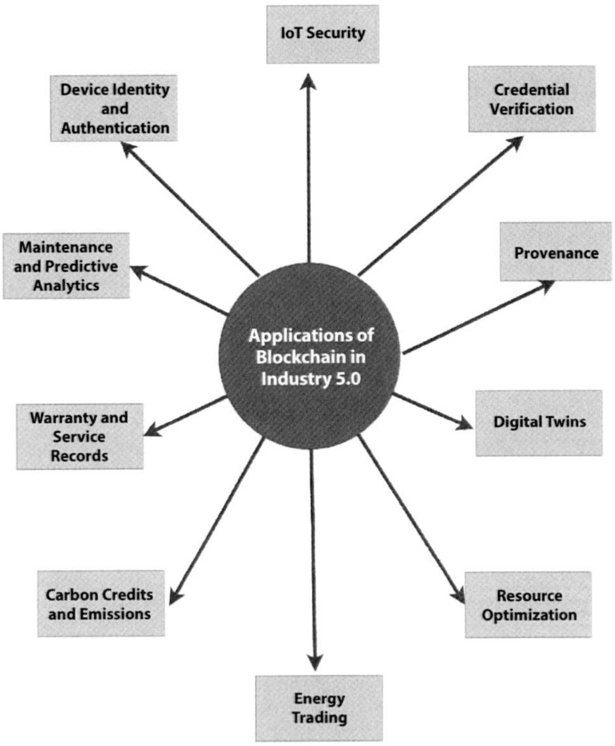

Figure 19.3 Applications of blockchain in Industry 5.0.

19.4 Supply Chain Management and Traceability

Supply Chain Management and traceability are the critical aspects of Industry 5.0. Blockchain technology offers innovative solutions to enhance transparency, security, and efficiency in these areas.

19.4.1 Transparency and Accountability in Supply Chains

Transparency and accountability of data in the supply chain can be achieved through the following factors:

- Immutable Ledger: Blockchain technology records all transactions and supply-chain data securely to prevent unauthorized access and modification. This guarantees data integrity and protects the system from fraud and manipulation. This architecture of blockchain provides data privacy and immutability through its encryption and decentralized network.
- End-to-End Visibility: Blockchain provides end-to-end transparency in supply chain management [11]. All the participants in the network right from manufacturers to consumers gain access to this shared platform with real-time operational conditions like location and product details. This also maintains a record to track every transaction of the supply chain.

- Traceability: Each item receives a unique identification or electronic image on the blockchain. This identifier is used to track the item at every stage of the inventory chain. This traceability makes certain that the tracking of an item at each level is validated by all the participants in the network.
- Authentication and Anti-counterfeiting: The products are verified and confirmed at every stage of the supply chain using blockchain. Since it is validated at every stage there is less chance of counterfeiting. This greatly reduces risk and harm to consumers and brand reputation.
- Smart Contracts for Compliance: Smart contracts can automatically check whether the methods adhere to the benchmark set and send an alert when the item or method does not meet the benchmarks. For instance, an intelligent contract can automatically send a notification when the temperature exceeds the threshold.
- Data Sharing and Collaboration: Secure and authorized sharing of data takes place in the supply chain using blockchain technology. It encourages collaboration and simultaneously helps the stakeholders with their processes and challenges.
- Dispute Resolution: The blockchain's inherent nature is transparency and audit-friendly. It enables the identification of problem origins in cases of dispute which accelerates conflict resolution thereby minimizing the requirement for expensive investigations.
- Auditing and Regulatory Compliance: Blockchain helps in transmitting auditing and regulatory compliance in a transparent and tamper-proof manner [12]. This verifies the records whether adhere to the regulation and streamlines the process of documentation and maintenance of records to prevent the burden of compliance.
- Consumer Trust: Building consumer trust is crucial for any brand. Customers become confident in buying a product when they are aware of the origin and history of the product. This builds their trust towards buying a product and making informed decisions knowing that they can trust the company behind the product.
- Ethical and Sustainable Practices: Transparency in the supply chain is important both for business and brand reputation. This transparency in the supply chain benefits the consumers in making ethical choices in maintaining sustainability and providing equal labor practices.

19.4.2 Tracking and Authentication of Products

Unique Identifiers: Every product is validated and granted a unique and unchanging digital ID on the blockchain. It guides the consumers showing the whole journey of the product from start to finish.

Immutable Records: Every transaction of products is stored in the form of tamper-proof records on the blockchain. This includes information about production, delivery, and quality assurance and ensures that the data cannot be modified or deleted. This gives auditors a reliably good audit trail.

Real-time Updates: Blockchain provides real-time updates. As the products pass through every stage in the supply chain, instant updates are recorded in the blockchain. All the participants including parties, manufacturers, distributors, retailers, and even the consumers get the latest updates on a product's status, location, and condition.

Transparency and Visibility: The blockchain provides transparency in knowing information on a product's history and current status, giving stakeholders and also other interested parties the whole picture about the product. It's like an open book, building trust and keeping all necessary knowledge in the loop.

19.4.3 Reduction of Counterfeit Goods

Immutable Records: Blockchain provides a tamper-proof ledger, where all transactions and product information are recorded. Once data is validated and added it cannot be altered or deleted which makes it difficult for counterfeiters to manipulate the records.

Product Authentication: Each product is assigned a unique digital identifier or token and the verifiability of products can be authenticated at every stage of the supply chain. Consumers, retailers, and quality control departments can access these identifiers confirming the authenticity of the product.

Traceability: Consumers can verify the origin and history of the product, ensuring that the product is genuine. This is done because each item that receives a unique identifier or electronic image is stored in the blockchain.

19.5 Decentralized Manufacturing

Decentralized manufacturing [12] aligns with the principles of Industry 5.0, emphasizing increased connectivity and flexibility in production processes.

19.5.1 Distributed Manufacturing Networks

Peer-to-Peer Collaboration: Blockchain lets manufacturers, small businesses, and people work together directly and efficiently. It helps in making effective products collaboratively.

Decentralized Marketplaces: A blockchain-based marketplace gives manufacturers a chance to share the details of what they can do and the services that they can offer. Other manufacturers or customers can find these services. This makes the marketplace a resource for manufacturing.

Smart Contracts for Production Agreements: Blockchain can automate processes and save the details of supply chain inventors in the form of smart contracts. Manufacturers are decision-makers who can set up the terms and conditions on how good the product should be, when it should be delivered, and how it should be paid for. These contracts execute themselves when certain terms are met.

Figure 19.4 depicts a decentralized network that connects the retailer, producer, buyer, consumer, and distributor in a distributed manner in the blockchain network. So in this way, they can efficiently do business as it is not controlled by any single entity.

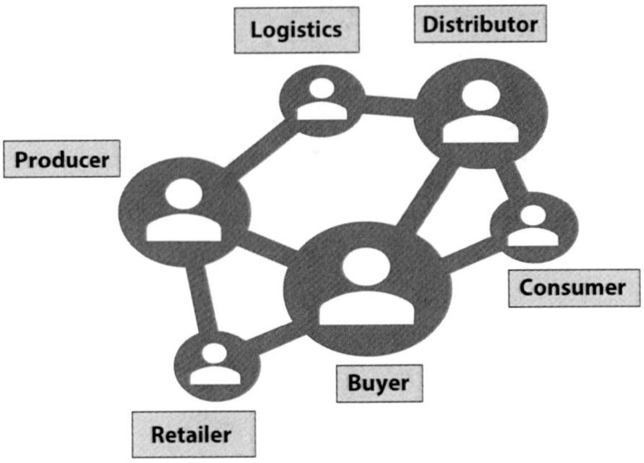

Figure 19.4 Blockchain decentralized network.

19.5.2 Intellectual Property Protection

Blockchain gives the first participant and developer a way to put a time stamp on their unique work, like designs, patents, and plans. Once it is validated and added as a blockchain these time stamps are updated and act as a secure and unchangeable record where the time of when it was created is stored which proves the intellectual property. It's done by connecting the creation to a unique digital marker or token. This is proof that tells everybody who owns it and where it came from.

19.5.3 Quality Control and Certification

Blockchain can help in quality control and certification using:

- Quality Control Automation: Smart contracts are used to handle quality control tasks. These actions are set in motion by specific triggers like set production goals or shifts in quality. This lowers the chance of human mistakes and increases uniformity.
- Certification Validation: Certificates like ISO standards or safety agreements are safely stored in the blockchain. Manufacturers can quickly evaluate the truth and validity of these certificates, benefiting themselves and their customers.
- Traceability and Transparency: Blockchain insists the quality control department workers follow a product's life cycle. From creation to delivery, this visibility helps in spotting the time and location of quality issues as they arise.
- Supplier and Vendor Accountability: Manufacturers have considered the suppliers and vendors into account based on the quality of their materials and components. Records on the blockchain will show the manufacturers the quality and compliance of everything.

19.6 Energy Sector Transformation

The energy sector is developing and improving day by day. With the advancements in technology, there is a need for the usage of green resources rather than other sources of energy. This aspect includes the trading of energy to take place between peers without any involvement of intermediaries for better grid control and use, and the use of green power certificates. Importantly, blockchain technology's use of energy is crucial for clear, secure, and efficient processes.

19.6.1 Peer-to-Peer Energy Trading

There is a complete breakthrough in how Peer-to-peer (P2P) energy exchange takes place and also tells about how energy is created, shared, and used. In the past, huge power plants made energy making use of grids distributed centrally. With P2P energy exchanges, blockchain technology allows direct interaction between energy makers and users. This direct deal eliminates the need for middlemen. They can directly sell the excess energy produced.

Working of Peer-to-peer energy trading
Energy suppliers and users are the two primary stakeholders connecting in a peer-to-peer (P2P) energy trading network. Providers are people who have renewable energy tools like solar panels or wind turbines in their houses. However, customers include both residential homes and commercial energy users. The foundation of P2P energy trading is blockchain. Smart contracts are written and deployed on the blockchain. Energy trading is made possible by these contracts, but only if energy is accessible at a fair price. Smart meters and IoT devices are essential gadgets used to track energy consumption throughout the day in real-time. Their purpose is to ensure precise calculation and billing. When the customer completes the service in a matter of seconds, the smart contract quickly verifies and performs the payment to the producer, avoiding lengthy payment delays.

Industrial Blockchain in P2P Energy Trading has several advantages:

- Security and Transparency: Transactions are recorded in a secure ledger. This enhances transparency and prevents fraud and malicious attacks.
- Immutable Records: Data remains fixed when it is logged. This provides all parties involved, including law enforcement, with a trustworthy record trail.
- Efficiency: Blockchain automates the trade activities. This reduces the time of admin work and fastens settlements.
- Decentralization: Blockchain avoids interference of middlemen in the network and makes everything decentralized. It deconstructs the energy market in this manner. As a result, both people and organizations can handle their own energy needs and requests.
- Interoperability: Blockchain can promote compatibility between different energy setups. This allows smooth trading across varied platforms.

19.6.2 Grid Management and Efficiency

Efficient grid management is essential for a steady, dependable energy source. The classic grid faces problems while dealing with increased renewable energy. Yet, when it is combined with grid control, blockchain technology helps in efficiently making use of energy.

Figure 19.5 portrays the various components of grid management like generation, transmission, distribution, and consumers which are all connected. It consists of several interconnected entities, including nodes, wind turbines, electric cars, smart meters, smart substations, smart appliances, and solar panels. Information is shared amongst these entities. Blockchain logic is programmed through smart contracts using languages like Solidity, Go, etc.

Working process of blockchain grid management
Blockchain-based algorithms are used to collect live data from different places/devices like IoT devices, smart meters, and weather predictions done through the installation of nodes and wind farms. The algorithms work on recognizing patterns in energy supply and use. Blockchain facilitates the creation of autonomous energy markets wherein producers can directly sell the excess energy generated to the grid consumers. This is done with the help of smart contracts executed in languages like Solidity, Go, etc. These smart contracts execute code on the blockchain which carries the logic. Blockchain technologies can assist with automated adjustments to energy distribution to maintain grid stability through real-time data analysis. Blockchain automates tasks that encourage reduced energy consumption at peak hours or during periods of low output.

Industrial Blockchain in Grid Management has many advantages

- Data Integrity and Security: Blockchain ensures data security and trustworthiness of the grid. It cannot be modified or influenced by cyber-attacks.
- Real-time Transactions: Blockchain enables payments immediately.
- Market Automation: Blockchain allows the trading process to take place automatically with minimum human checks. Additionally, it improves grid performance.
- Resilience: The blockchain is compatible with numerous systems. This means that even if some components fail, the grid can continue to function. In other words, it lessens the chance of the grid going out.

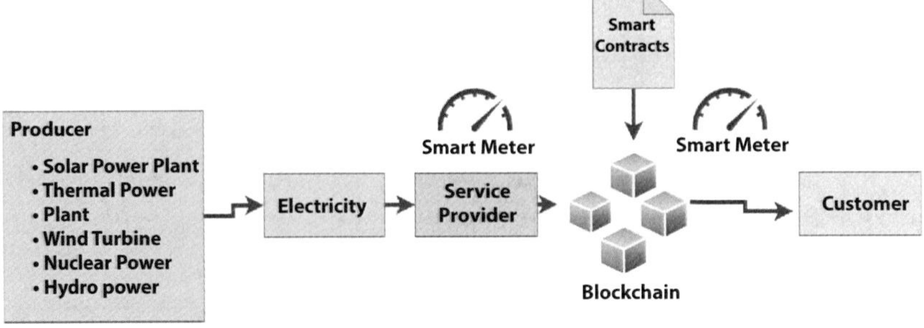

Figure 19.5 Blockchain grid management.

19.6.3 Renewable Energy Certificates

Renewable Energy Certificates, or RECs, are necessary for promoting renewable energy. These are connected to one megawatt-hour (Mwah) of renewable energy generated. They may be purchased and sold, which helps the market for renewable energy.

Working of Renewable Energy Certificates
As soon as one Mwah of renewable energy is generated, a REC is made. This Renewable Energy Certificate (REC) provides information about the energy's origin, source, and environmental benefits. Blockchain ledgers are used to record these RECs. This produces an immutable record of their ownership and history. In this manner, the advantages of renewable energy are accurately recorded. When consumers or companies buy RECs, they offset their carbon emissions. They show their support for renewable energy by retiring the RECs they receive.

Industrial Blockchain in Renewable Energy Certificates has many advantages:

- Traceability and Transparency: A visible and auditable record of REC transactions is made possible by blockchain technology, enabling stakeholders to confirm the origin and validity of certificates.
- Market Efficiency: Blockchain makes buying and selling RECs easier. This drops the administrative expenses and lifts up the market operations.
- Fraud Prevention: Due to the immutable property of the blockchain, Scam techniques such as selling unauthorized RECs or double counting are impossible.
- Market Expansion: Global REC trade is facilitated by blockchain. It encourages a larger market for certificates of renewable energy.

In conclusion, a major step toward a more sustainable and effective energy ecosystem has been made with the introduction of blockchain technology into the energy sector, particularly in the areas of peer-to-peer (P2P) energy trade, grid management, and renewable energy certifications. The decentralized, transparent, and secure characteristics of blockchain will revolutionize the production, distribution, and consumption of energy.

19.7 Intelligent Health Care and Medical Records

Intelligent health care and medical records [5] are the need of the hour. Post-COVID pandemic all nations are working towards this goal with the use of blockchain technology. Blockchain can change the way the medical domain works with its unique features. So let's get into the various aspects of it.

Figure 19.6 shows the interaction between various individuals in healthcare sectors like doctors, patients, and pharmacies with blockchain. The patient data that are received from various sensors as well as Electronic Healthcare Records (EHR)and Patient Health Records (PHR) are stored in the blockchain, which can be accessed by Doctors, healthcare professionals, pharmacies, etc.

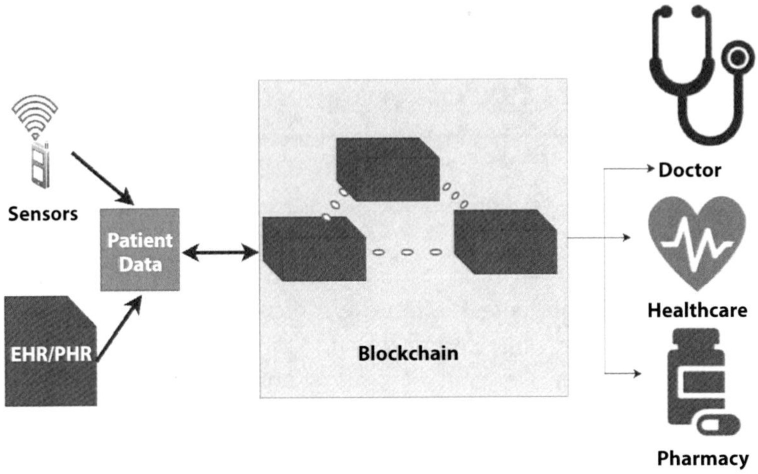

Figure 19.6 Patient data chain.

19.7.1 Secure and Immutable Medical Data

Blockchain helps in storing medical data at every stage. Some of the processes involved are listed below:

- Data Entry: healthcare providers like hospitals, clinics, and laboratories feed their patient details into the blockchain as records. The data can be the patient's medical history, medical tests, test reports, past treatments, and scans. This information is encrypted [8] in the form of a special ID or hash.
- Data Validation: Each data before getting added as a block has to be validated. This validation process is done using a consensus mechanism like Proof of Work or Proof of Stake, which is taken care of by network nodes. This ensures that only correct and validated data is added.
- Immutable Records: When the data is validated and gets added as a block, it creates a timestamp and a link to the previous block. Once it is notified there can't be any change to it later. This changeless nature makes sure that the records are secure and safe from tampering.
- Access Control: In simple terms, smart contracts help to control data access. They can set their preferences. Like, a patient could allow only specific doctors or researchers to see their information.
- Decentralization: Because no one entity has total control over blockchain, it is less likely to be governed by one party and therefore safer.
- Patient Ownership: Patients have their own special keys to control their data access
- Data Security and Encryption: Health records are kept private and unaltered due to the usage of encryption protocols. Patients are provided with unique keys to set preferences on who can access their sensitive information and what information alone can be viewed by them.

- Consent Management: Patient consent is made simple and enforceable by smart contracts. They verify healthcare providers only access data that they have obtained explicit consent to access.
- Clinical Trials and Research: Blockchain facilitates the open and safe exchange of anonymous trial and research data. It ensures cooperation among organizations.
- Drug Traceability: Blockchain traces the whole journey of medications, from origin to sharing. It verifies and validates the genuineness and safety of drugs.
- Billing and Insurance Claims: Smart contracts enhance the billing and insurance claims process by reducing errors and preventing the data from fraud.
- Public and Private Keys: Patients and healthcare providers use public and private keys. This guarantees safe and authorized access to health records.

19.7.2 Interoperability and Data Sharing

Interoperability between blockchains is crucial since data may be stored differently across national borders and in a variety of forms. So, let's see a few metrics related to it.

- Standardized Data Formats: Blockchain health platforms follow standard data formats like HL7 or FHIR.
- Consent Management: Consent is managed with the aid of smart contracts. Patients have the option to consent to the use of their data or not, depending on the circumstances.
- Real-time Updates: Any modifications made to a patient's record take effect instantly throughout the network. All authorized parties will thus have access to the most recent data.
- Secure Messaging: Secure messaging is provided by certain blockchain technology connected to health. These layouts enable direct physician communication on the platform.
- Auditing and Logging: Every contact with data is recorded in a changeless ledger. This approach increases accountability and transparency while providing a clear route.

19.7.3 Drug Supply Chain Integrity

Blockchain can be used to effortlessly track a drug's supply across all of its points and confirm whether any duplicate pharmaceuticals are being delivered to the market. Let's discuss how it's done in blockchain.

- Serialization of Medications: Every drug has its own ID or number. In the process of making the medication, its record is recorded in the blockchain.
- Traceability: Every interaction involving the medication, from production to distribution, is documented on the blockchain. The supply chain has a record of this procedure.

- Verification: Health providers can verify the authenticity of drugs by examining the unique number obtained from the blockchain. This method helps in differentiating fake drugs from the original and prevents them from reaching the patients.
- Alerts and Recalls: Blockchain identifies the batches that are problematic if there is any safety issue. This prevents the patients from any risks.
- Smart Contracts for Compliance: Each stage of the supply chain uses smart contracts to follow rules correctly without error.

Thus, we can conclude that blockchain is an effective way of maintaining healthcare records and also in drug supply chain integrity.

19.8 Financial Services and Payments

Blockchain was primarily intended for financial services and payments only. Blockchain was the underlying technology behind Bitcoin cryptocurrencies. Bitcoin offers a decentralized, transparent, and secure platform that has the potential to transform the way financial transactions are conducted. But it spread across all domains due to its unique features like immutability, security, transparency, decentralization, and tamper resistance.

The blockchain financial system is shown in Figure 19.7 where the banks, customers, and merchants all interact with each other.

19.8.1 Cross-Border Payments

One of the key uses of financial services and payments is considered to be cross-border payments. Now let's discuss the various steps in making the payment.

- Initiating a Transaction: The sender makes a cross-border payment. They enter the amount they wish to send and the wallet address of the recipient.
- Validation and Authorization: the blockchain network distributes the payment requests among its members. Using the agreement procedures the nodes verify whether the transaction is authentic.
- Smart Contracts for Settlement: Automating the settlement process is done through smart contracts. Here the agreement adheres to the preset guidelines. They ensure that the payments are dispersed only when certain conditions are met.

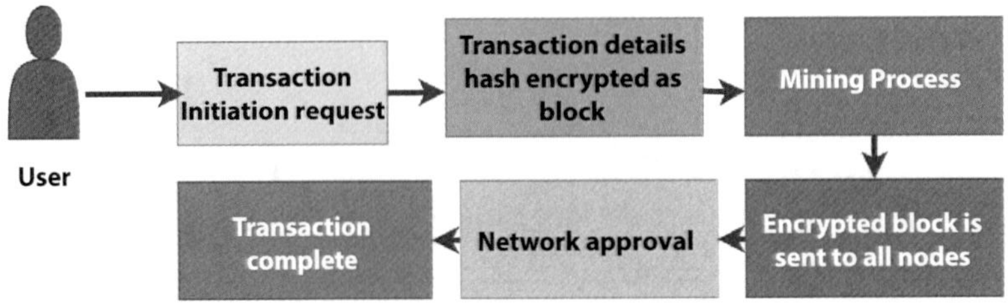

Figure 19.7 Blockchain in financial service and payment.

- Currency Conversion: blockchain platforms with exchange services integrated can adjust if the sender and recipient use different forms of currency.
- Real-time Tracking: People can view the real-time transaction status with the blockchain explorer. This provides a clear picture of the settlement process to all parties.
- Immutable Record: Once confirmed, the transaction details are recorded on the blockchain as an immutable ledger entry.
- Recipient Access: with their private keys, the recipient can access the money in their wallet.

19.8.2 Financial Inclusion and Access

Digital Wallets: Individuals without access to traditional banking services can now access their digital wallets. On a platform driven by blockchain, they can do it. All they require is a gadget capable of processing the functions and an internet connection.

Secure and Borderless Transactions: Users can securely send, receive, and store digital assets, regardless of location.

Lower Expense in Transactions: Blockchain replaces the traditional middlemen such as banks. Because of its dispersion, transfer fees are reduced.

Lending with Peers and Small Loans: peer-to-peer lending and small loans are now feasible making use of blockchain digital platforms. A global network of lenders is available to people and small businesses looking for funding.

Breaking down Assets: Real assets can be divided up into smaller pieces to be owned, just like property or things. This setup can encourage more people to invest.

Learning through Finances: There are blockchain platforms available to assist. They offer financial as well as instructional resources. Users can make wiser decisions on how to handle their digital assets.

19.8.3 Identity Verification and KYC Verification

The following steps are undertaken for identity verification and KYC verification

- User Registration: Users make an account on a special platform that uses blockchain. They share some simple facts about themselves.
- Document Submission: Important documents such as a driver's license or passport may occasionally need to be shown. This securely takes place.
- Data Encryption and Hashing: The papers that are turned in are made secret and scrambled. This is to keep private information secure.
- KYC Smart Contracts: Smart contracts are used to carry out rules called KYC or "Know Your Customer". They ensure that the submitted papers adhere to the set of requirements.
- Biometric verification: Certain platforms enhance security by utilizing biometric measurements such as fingerprint and facial recognition.
- Permissioned Access: Users can access the platform's functionalities after their information has been verified. This could include investing, financial services, or other commercial ventures.

- Regular Audits and Updates: The platform constantly verifies whether everything complies with the law. when necessary they update and modify the KYC

Thus, we have seen the various domains and how blockchain can be used to transform them. Now let's quickly get into some interesting case studies using blockchain and how they achieved success using it.

19.9 Case Studies

These case studies help us to identify the potential of blockchain to solve real-time problems.

19.9.1 Case Study 1 Blockchain in Supply Chain Management

GlobalMart, a multinational corporation is facing some issues with its supply chain. It suffers from problems with tracking, and excessive paperwork. This leads to slow processing and errors. GlobalMart aspires to improve its supply chain operations. They want buyers to be more informed about the products they purchase.

Objective: Tracking products from production to the shelf is a goal for GlobalMart. At every stage of the supply chain, they want to view the location and status of the products. They will reduce paperwork and streamline document procedures in order to achieve this. They may reduce the possibility of mistakes and delays. To complete tasks quickly and effectively, they also intend to collaborate with shipping firms and suppliers.

Benefits of using Blockchain Implementation: GlobalMart adopts a permissioned blockchain network, inviting key suppliers, manufacturers, logistics partners, and retailers to join the network. Automation and enforcement of quality standards, compliance agreements, and other fulfillment are achieved through the use of smart contracts. Each product is assigned a unique identifier (e.g., RFID tag) linked to a blockchain to record the provenance tracking. By giving input data at each stage, real-time monitoring and visibility of data are achieved. Suppliers input data on production details, shipping dates, and transportation information on the blockchain. Shipment containers equipped with sensors, and IoT devices offer real-time data updates. A blockchain dashboard that gives stakeholders, including suppliers and GlobalMart, access to real-time information on the location status of goods is available. Suppliers and GlobalMart can communicate transparently with the help of blockchain. Paperwork such as invoices, customs declarations, and shipping documents are digitized and recorded on the blockchain, reducing manual errors and speeding up processes. Thus, by using immutable records across the supply chain process, quality assurance and customer trust are accomplished. Through unique QR codes on packaging, GlobalMart uses blockchain technology to give customers access to details about the provenance, path, and state of products.

Results

- End-to-end Traceability: GlobalMart keeps track of all its products. If there's a problem or a recall, they find and fix it quickly.

- Real-Time Visibility: Stakeholders can view their data in real-time. This leads to smarter decisions, fewer delays, and a more efficient supply chain overall.
- Reducing Paperwork: Automating paperwork on the blockchain reduces mistakes made by hand. It reduces paperwork-related delays and expedites customs clearance.
- Supplier Collaboration: Supplier collaboration is improved by blockchain's clear, automated processes. This improves the supply chain's dependability and efficiency.

19.9.2 Case Study 2 Decentralized Manufacturing with Blockchain

Decentralized manufacturing with blockchain is about using blockchain technology to make the manufacturing process easier in a better way. Let's talk about how this might work:

Objective: Decentralized production network creation is the aim. Suppliers and manufacturers join a blockchain network in these networks. Throughout the process, this network enables a transparent, safe record of transactions. Additionally, automating agreements and procedures is the idea. The key is well-checked, decentralized production order, and excellent communication between the manufacturer and supplier.

Benefits of Blockchain: We use smart contracts for automation after a production network built on the blockchain is established. These contracts streamline operations between stakeholders in the manufacturing system and automate agreements. It addresses issues including order fulfillment, quality assurance, and payment options. Our manufacturing orders are stored on the blockchain. The product information, quantity, and delivery time are all included in these records. Smart contracts have specified rules that automate the order process. Every component is assigned a special code such as an RFID or QR code which is recorded on the blockchain. We can track every detail in this manner. Certifications, inspection outcomes, and quality checks are also a part of this chain. Everyone involved can see this information. Blockchain reduces the possibility that manufacturing defects may surface. It connects suppliers and manufacturers transparently and safely. They have real-time access to information regarding inventory levels, production schedules, and quality standards. The blockchain records each and every one of these communications regarding stock levels and inventory as they occur. Based on predefined rules, smart contracts automatically place additional orders. This facilitates stock management among the network's multiple locations. Additionally, the blockchain stores information regarding design files, the characteristics that set each item apart, and the best practices for manufacturing each item. This makes each design more authentic and ensures that the owner's rights are upheld. This gives an honest, accountable history of what's happened in manufacturing.

19.9.3 Case Study 3 Blockchain-Based Energy Trading Platform

Case Study: EnerChain – Decentralized Energy Trading on the Blockchain

A new platform called EnerChain leverages blockchain technology to facilitate decentralized energy trade. It seeks to facilitate connections between producers and consumers of renewable energy in an open, effective, and sustainable market.

Objectives: To enable direct and peer-to-peer renewable energy trading between producers and consumers, lessening the dependencies on conventional energy middlemen.

To provide smooth and safe settlements between participants, promote environmentally responsible energy usage by utilizing smart contracts to automate energy transactions, and provide transparency into the source and sustainability of energy Integration with Smart Grids: To maximize energy distribution and support grid stability, work together with the current smart grid infrastructure.

Benefits of using blockchain: EnerChain sets up a unique blockchain system ideal for energy trades. This includes all types of users: those who generate renewable energy, consumers, and validators. All transactions are done through smart contracts, which are there to automate the energy trading process. They handle the cost of energy, quality benchmarks, and settlement terms. They start working when certain conditions are fulfilled. Renewable energy producers, like those at solar farms and wind turbine facilities, tie their production stats to the blockchain system. Every unit created is noted on the blockchain. This includes details like the origin of energy when it was developed, and its impacts on carbon emissions. All participants can observe real-time activity in the energy market which allows them to make smart decisions based on cost, origin, and sustainability information. EnerChain introduces utility tokens (Energy Tokens) as its currency for transactions. These tokens represent units of renewable energy and facilitate secure, traceable, and efficient transactions within the platform. Token holders have voting power proportional to their network contribution.

Results

- Decentralized Trading Success: Per-to-peer trade in renewable energy is seamlessly integrated with EnerChain. Energy sellers can quickly offload excess energy and purchase benefits from a decentralized energy economy.
- Transparency and Sustainability: Users can easily track the origin of their energy and its effect on the environment. With this clarity, more people choose naturally sourced energy.
- Automated Transactions: Smart contracts handle settlements, leading to faster and more secure payments to energy producers. This technology reduces paperwork and boosts the efficiency of energy levels.
- Integration with Smart Grids: EnerChain partners with operators of smart grids. This improves energy distribution, balances the grid in real time, and adds strength to the grid's overall structure.

19.9.4 Case Study 4 Blockchain for Healthcare Data Management

Case Study: HealthChain
Background: HealthChain is a blockchain-based healthcare data management platform adopted by a network of hospitals, clinics, and patients. HealthChain aims primarily to enhance data security, streamline processes, and improve patient outcomes.

Benefits of using blockchain: Patients register on the HealthChain platform, and are provided with a unique identifier on the blockchain. Health records of patients like medical history, test reports, and treatment strategies are safely stored on the blockchain with HealthChain. This technology blends into existing health systems, creating easy data swaps while keeping the integrity of data intact. HealthChain uses smart contracts to manage patient approval, letting them decide who sees their health data. Every time health information is accessed or shared, it leaves a clean, traceable record on the blockchain. HealthChain lets anonymized patient data be shared for studies and clinical trials, speeding up discoveries in science. Pharmaceutical firms rely on health chain to monitor supply chains ensuring authenticity and stopping fake drugs. Smart contracts manage invoicing processes automatically leading to fewer errors and fraud. Managing insurance claims on the blockchain is simple.

19.9.5 Case Study 5 Blockchain in Financial Services

Case Study: FinTechChain – Transforming Financial Services with Blockchain
Background: FinTechChain is a platform using blockchain to change financial services. It fixes problems, cost effective, and boosts safety. It brings banks, payment processors, and others together for a clear, not centralized, financial system.

Objectives: FinTechChain helps to make international transactions quicker and cheaper, using smart contracts to automate and simplify trade finance, increases transparency, and cuts fraud in supply-chain finance, and this improves the accuracy and safety of digital identity checking for new customers.

FinTechChain creates a blockchain network where only certain financial institutions can join. Banks make use of blockchain for instant global payments. The blockchain's transparency gives a clear view of the transaction process. FinTechChain uses smart contracts to handle automated trade finance, like issuing letters, checking documents, and settling accounts. This reduces the time and complexity of trade transactions. The platform extends to supply-chain finance, where transactions related to the financing of goods are recorded on the blockchain. This reduces discrepancies and enhances trust among participants. FinTechChain utilizes a secure and decentralized digital identity verification system. Users maintain control over their digital identities, and institutions can verify customer information securely.

Results

- Efficient Cross Border Payments: Banks using FinTechChain can visualize decreased latency and expenses for cross-border payments. This enhances efficiency.
- Automated Trade Finance: Smart contracts make trade finance easier, reducing manual work, paperwork, and hold-ups.
- Transparent Supply Chain Finance: benefits of data transparency and less fraud risk in supply chain finance. This makes finance transactions more reliable.
- Secure Digital Identity Verification: The decentralized digital identity improves the correctness and safety of customer onboarding, lowering
- Theft risks.

The following is Table 19.1, which shows the findings of the case studies

19.10 Open Issue and Critical Challenges

Even though blockchain is growing at a very rapid pace, it has its own challenges that need to be addressed. Some of the issues are discussed here.

Table 19.1 Case studies findings.

S. no	Case study	Method	Results
1	Supply chain management	GlobalMart	• End-to-end traceability • Real-time visibility • Reducing paperwork • Supplier collaboration
2	Decentralized manufacturing	Smart contracts	• Design more authentic • Owner's rights upheld
3	Energy trading	EnerChain	• Decentralized trading • Transparency and sustainability • Automated transactions • Integration with smart grids
4	Healthcare data management	HealthChain	• Authenticity • Reduce fraud
5	Financial services	FinTechChain	• Efficient cross-border payments • Automated trade finance • Transparent supply chain finance • Secure digital identity verification

19.10.1 Scalability and Performance

Getting lots of blockchain transactions done quickly and easily without losing decentralization is tough. As more people use blockchain networks, it leads to network traffic and slows down transactions. Sharding, Layer-2 Scaling Solutions, and Consensus Mechanisms can be used to help the scalability issues. Other factors like speed of network, node working, storage efficiency, scalable smart contracts, network bandwidth, and managing resources to help with performance should also be addressed.

19.10.2 Regulatory and Legal Frameworks

Without setting clear and legal rules, businesses might be worried. This could slow down more people adopting blockchain. Different rules in different places cause more worry too. Regulations protecting customers, safeguarding investors, preventing money laundering, knowing the customer, cooperating with regulatory organizations, and maintaining confidentiality are the legal frameworks. Data Protection and Privacy Laws, Anti-Money Laundering (AML) and Know Your Customer (KYC) Regulations, Securities Laws, Consumer Protection Laws, Taxation Regulations, Smart Contract Legality, Cross-Border Regulations, Intellectual Property Laws, Blockchain-specific Regulations, Regulatory Sandboxes, Collaboration with Regulatory Bodies all come under regulatory and legal frameworks.

19.10.3 Interoperability and Standardization

Communication between various blockchain networks has become very complicated as it might cause huge difficulty and problems due to the lack of standards and protocols. This lack can degrade the integration of these networks, harming their combined potential. This can be overcome by using similar data formats, promoting cross-chain conversation, sticking to one protocol, utilizing smart contract interoperability, and upholding blockchain.

19.10.4 Privacy and Data Protection

Blockchain must also secure the data, maintain privacy, and also comply with regulations. Data protection is an important function where huge volumes of personal data are stored in a blockchain. Several techniques can help us like Zero-Knowledge Proofs, Homomorphic Encryption, Decentralized Identity, Privacy-Enhancing Smart Contracts, and Consortium Blockchain are used.

19.10.5 Governance and Trust

The problem of ensuring compliance, interoperability, and securing the decentralized system's trust and safety is a major concern that needs to be addressed. Some countries like Estonia, the UAE, Georgia, Singapore, Sweden, and Australia support blockchain and they have successfully implemented it in their day-to-day activities. But many countries are still resistant and brace the advantages of blockchain which is an important issue and needs attention. Only when trust is gained we can foster the adoption of blockchain, driving innovations and growth of decentralized applications. Figure 19.8 shows the challenges of blockchain in Industry 5.0.

Figure 19.8 Challenges of blockchain in Industry 5.0.

19.11 Conclusion

Industry 5.0 sees a more sophisticated coexistence of technology and people than Industry 4.0, which aims to automate activities through mechanization. Robots will now operate alongside humans when they previously would have replaced them because they understand the value of human judgment, empathy, and creativity. This vision is guided by a focus on sustainability, which means that innovations are directed towards improving lives rather than replacing them. The goal of Industry 5.0 is to create enduring and peaceful relationships between humans and their instruments. By utilizing the accuracy and consistency of machines, humans will be able to investigate intricate issues that algorithms are still unable to fully understand. Robots and artificial intelligence (AI) will, in turn, mimic some human abilities, developing context awareness that permits beneficial aid rather than dominance. Such synergy may bring up new possibilities for research and problem-solving. Most importantly, this balanced integration values what each partner human and technological can offer, ensuring our future progress uplifts and inspires.

References

1. Varriale, V., Cammarano, A., Michelino, F., Caputo, M., Industry 5.0 and Triple Bottom Line Approach in Supply Chain Management: The State-of-the-Art. *Sustainability*, 15, 5712, 2023.
2. Alojaiman, B., Technological Modernizations in the Industry 5.0 Era: A Descriptive Analysis and Future Research Directions. *Processes*, 11, 1318, 2023.
3. Abdelhafeez, A., Industry 5.0: Theoretical Foundations for Enabling Technologies and Applications in Manufacturing Context. *Int. J. Adv. Appl. Comput. Intell.*, 04, 01, 37–46, 2023.
4. Leng, J., Chen, Z., Huang, Z., Zhu, X., Su, H., Lin, Z., Zhang, D., Secure Blockchain Middleware for Decentralized IIoT towards Industry 5.0: A Review of Architecture, Enablers, Challenges, and Directions. *Machines*, 10, 10, 858, 2022.
5. Maddikunta, P.K.R., Pham, Q.V., Prabadevi, B., Deepa, N., Dev, K., Gadekallu, T.R., Ruby, R., Liyanage, M., Industry 5.0: A survey on enabling technologies and potential applications. *J. Ind. Inf. Integr.*, 26, 2022.
6. Verma, A., Bhattacharya, P., Madhani, N., Trivedi, C., Bhushan, B., Tanwar, S., Sharma, G., Bokoro, P.N., Sharma, R., Blockchain for Industry 5.0: Vision, Opportunities, Key Enablers, and Future Directions. *IEEE Access*, 10, 2022.
7. Dev, K., Tsang, K.F., Corchado, J.M., The era of industry 5.0— Technologies from no recognizable h-m interface to hearty touch personal products. *IEEE Trans. Ind. Inf.*, 18, 8, 5432–5434, 2022.
8. Rupa, C., Midhunchakkaravarthy, D., Hasan, M.K., Alhumyani, H., Saeed, R.A., Industry 5.0: Ethereum blockchain technology based DApp smart contract. *Math. Biosci. Eng*, 18, 5, 7010–7027, 2021.
9. Gupta, R., Bhattacharya, P., Tanwar, S., Kumar, N., Zeadally, S., GaRuDa: A blockchain-based delivery scheme using drones for healthcare 5.0 applications. *IEEE Internet Things Mag.*, 4, 4, 60–66, 2021.
10. Longo, F., Padovano, A., Umbrello, S., Value-oriented and ethical technology engineering in industry 5.0: A human-centric perspective for the design of the factory of the future. *Appl. Sci.*, 10, 12, 4182, 2020.
11. Javaid, M. and Haleem, A., Critical components of industry 5.0 towards a successful adoption in the field of manufacturing. *J. Ind. Integr. Manage.*, 5, 03, 327–348, 2020.
12. Nahavandi, S., Industry 5.0-A human-centric solution. *Sustainability*, 11, 16, 4371, 2019.

20

Blockchain-Empowered Internet of Things (IoTs) Platforms for Automation in Various Sectors

Santosh Reddy Addula[1], Amit Kumar Tyagi[2*], Kanchan Naithani[3] and Shabnam Kumari[4]

[1]*Department of Information Technology, University of the Cumberlands, Williamsburg, KY, USA*
[2]*Department of Fashion Technology, National Institute of Fashion Technology, New Delhi, Delhi, India*
[3]*School of Computing Science and Engineering (SCSE), Galgotias University, Greater Noida, Uttar Pradesh, India*
[4]*Department of Computer Science, Faculty of Science and Humanities, SRM Institute of Science and Technology, Kattankulathur, Chennai, Tamil Nadu, India*

Abstract

In the recent decade, Internet of Things (IoTs) and Blockchain has changes the life of human being via provide convenient services and fulfilling the demand by smart production. This chapter discusses the transformative potential of integrating Blockchain technology into IoT platforms, creating a paradigm shift in automation across diverse sectors. As industries increasingly embrace IoT for real-time data monitoring and control, the incorporation of Blockchain empowers these platforms with enhanced security, transparency, and efficiency. Blockchain's decentralized and tamper-resistant ledger provides a robust foundation for securing IoT data transactions. By employing smart contracts, it establishes trust and verifiability in automated processes, mitigating concerns related to data manipulation and unauthorized access. This paper investigates the implications of this synergy on sectors such as manufacturing, logistics, healthcare, and energy, highlighting the benefits of transparent and secure data exchange in automated systems. Moreover, the decentralized nature of Blockchain ensures a resilient and scalable infrastructure for large-scale IoT deployments. This work moves towards the challenges and opportunities associated with deploying Blockchain-empowered IoT platforms, addressing issues of scalability, interoperability, and integration with existing systems. Through a comprehensive analysis of case studies and theoretical frameworks, this chapter contributes to the evolving discourse on Blockchain and IoT convergence, shedding light on its potential to revolutionize automation in various sectors.

Keywords: Blockchain, Internet of Things (IoT), automation, smart contracts, security, transparency

20.1 Introduction

Blockchain-Empowered IoT Platforms are innovative solutions that combine the power of blockchain technology with the Internet of Things (IoT) to enhance automation in

**Corresponding author*: amitkrtyagi025@gmail.com
Santosh Reddy Addula: ORCID: https://orcid.org/0009-0000-3286-8224

Amit Kumar Tyagi, Shrikant Tiwari, Senthil Kumar Arumugam and Avinash Kumar Sharma (eds.) Artificial Intelligence-Enabled Digital Twin for Smart Manufacturing, (443–478) © 2024 Scrivener Publishing LLC

various sectors. These platforms provide a secure and transparent framework for connecting and managing IoT devices, enabling efficient and reliable data exchange and automation processes [1, 2]. By integrating blockchain technology, which is a decentralized and immutable ledger, IoT platforms can address key challenges such as data integrity, security, privacy, and trust. Blockchain enables the creation of tamper-proof records of IoT device activities, ensuring the authenticity and immutability of data. This enhances the reliability and trustworthiness of the automation processes in different sectors [3]. In sectors like supply chain management, blockchain-Empowered IoT platforms enable end-to-end visibility and traceability of goods throughout the entire supply chain. Each IoT device involved in the supply chain can record its activities on the blockchain, providing an auditable and transparent history of events. This enables efficient inventory management, reduces counterfeiting, and enhances overall supply chain efficiency [4]. Another sector where blockchain-Empowered IoT platforms find applications is energy management. These platforms facilitate the integration of IoT devices, such as smart meters and sensors, with blockchain technology. This allows for secure and automated energy transactions, such as peer-to-peer energy trading, based on predefined smart contracts. Blockchain ensures the integrity of energy data and enables secure and transparent energy transactions among different participants [5–6]. Furthermore, blockchain-Empowered IoT platforms have applications in healthcare, where they can enhance patient data privacy and interoperability. By considering blockchain's decentralized nature, these platforms enable secure sharing of patient data among healthcare providers, ensuring data integrity and privacy. IoT devices, such as wearable health monitors, can securely transmit data to the blockchain, creating a reliable and transparent record of patient health information.

In summary, Blockchain-Empowered IoT Platforms provide huge potential for automation in various sectors. By combining the decentralized and immutable nature of blockchain with the interconnectedness of IoT devices, these platforms enable secure, transparent, and efficient automation processes, enhancing trust, privacy, and reliability in sectors like supply chain management, energy management, and healthcare.

20.2 Emerging Trends and Visions Towards Blockchain and Internet of Things

Emerging trends and visions towards Blockchain and the Internet of Things (IoT) are shaping the future of technology and providing exciting possibilities in various domains [7, 8]. Some key trends and visions in this space are shown in Figure 20.1.

Interoperability and Standardization: One of the emerging trends is the push for interoperability and standardization between blockchain and IoT platforms. As more devices and systems are connected, ensuring seamless integration and communication becomes important. Efforts are underway to establish common protocols, data formats, and interoperability frameworks to enable cross-platform compatibility and enhance the scalability of blockchain and IoT deployments.

Scalability and Performance Enhancements: Blockchain and IoT applications generate massive amounts of data, requiring high-performance systems to handle the workload. Emerging technologies like sharding, off-chain solutions, and consensus algorithm

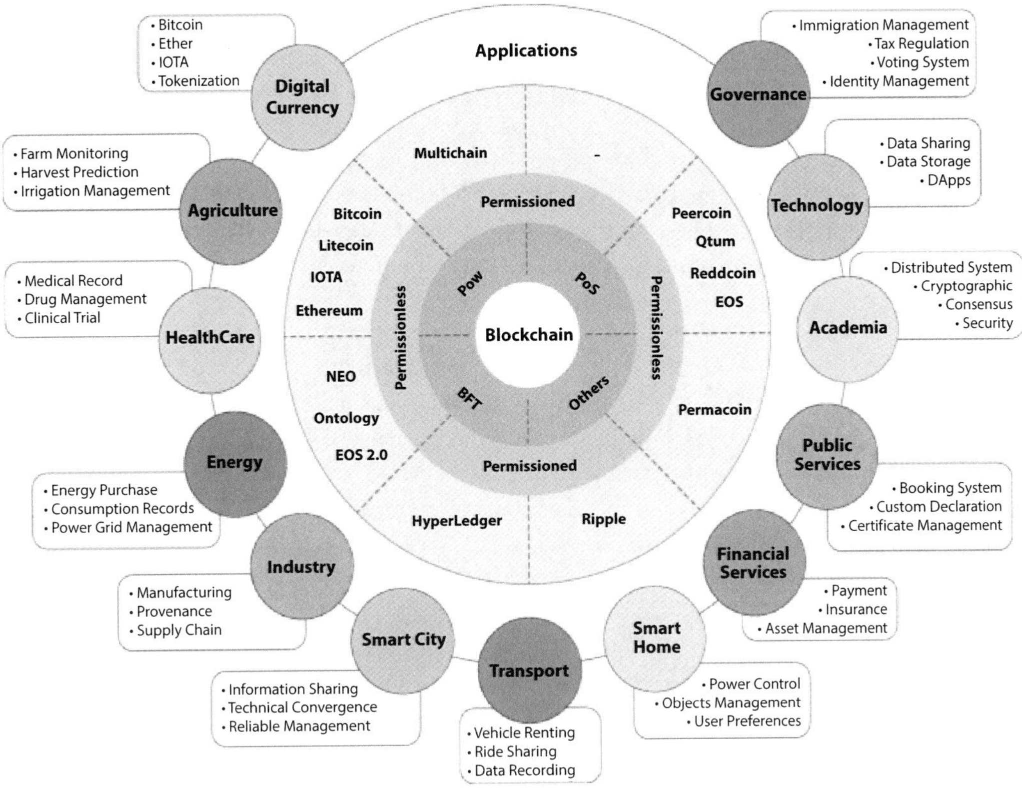

Figure 20.1 Emerging trends and visions towards blockchain and Internet of Things [Cui, P., Guin, U., Skjellum, A. *et al.* Blockchain in IoT: Current Trends, Challenges, and Future Roadmap. J Hardw Syst Secur 3, 338–364 (2019). https://doi.org/10.1007/s41635-019-00079-5].

optimizations are being discussed to enhance the scalability and performance of blockchain and IoT networks. These advancements aim to support real-time data processing, reduce transaction costs, and improve overall system efficiency.

Edge Computing and Distributed Architectures: With the growing volume of IoT devices and the need for near real-time data processing, edge computing is becoming increasingly important. Edge computing brings computational capabilities closer to the data source, reducing latency and enabling faster decision-making. Blockchain technology can be integrated with edge computing architectures to provide decentralized and distributed processing, enhancing the security and privacy of IoT data.

Data Privacy and Security: Protecting the privacy and security of IoT-generated data is a critical concern. Blockchain's inherent characteristics, such as immutability, transparency, and cryptographic algorithms, make it an attractive solution for securing IoT data. Emerging trends focus on using blockchain for data encryption, access control, and identity management, ensuring that sensitive information remains secure throughout the IoT ecosystem.

Tokenization and Microtransactions: The integration of blockchain and IoT enables the seamless exchange of value between devices and systems. Tokenization, the representation of physical or digital assets on a blockchain, allows for the creation of digital currencies and the implementation of microtransactions. This opens up possibilities for new business models, such as machine-to-machine (M2M) payments and the monetization of IoT data and services.

Artificial Intelligence (AI) Integration: Combining AI with blockchain and IoT technologies can unlock powerful information and automation capabilities. AI algorithms can analyze large amount of IoT data stored on the blockchain, extracting valuable information and enabling predictive and prescriptive analytics. This integration can optimize resource allocation, improve decision-making processes, and enhance the overall efficiency of IoT systems.

In summary, emerging trends and visions in the intersection of blockchain and IoT include interoperability, scalability, edge computing, data privacy and security, tokenization, and AI integration. These trends aim to address the challenges and unlock the full potential of blockchain and IoT, revolutionizing industries and paving the way for a more connected, secure, and efficient future.

20.3 Usage of Blockchain in 5G-Enabled Internet of Things

The usage of blockchain in 5G-enabled Internet of Things (IoT) environments provides several benefits and applications [9]. Some key areas where blockchain can enhance the capabilities of 5G-enabled IoT is shown in Figure 20.2.

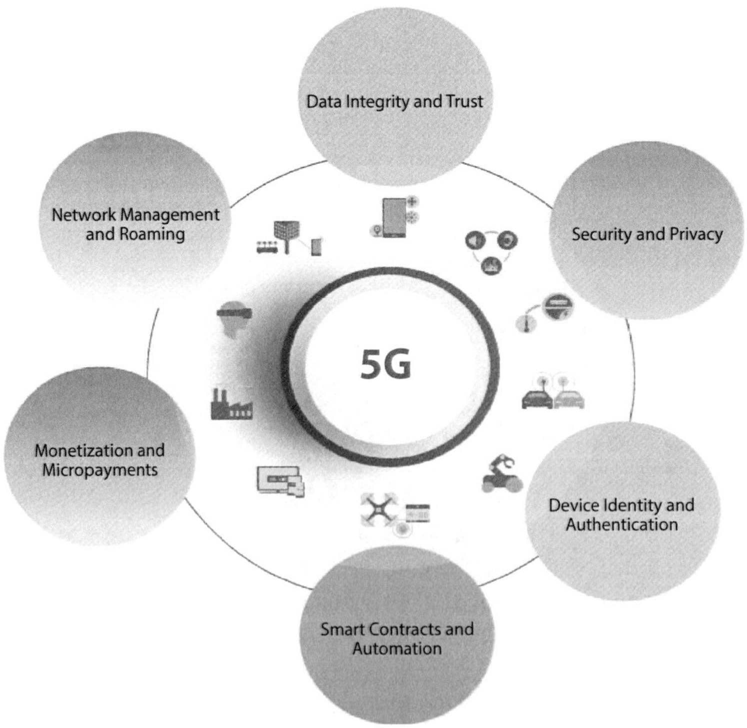

Figure 20.2 Key areas where blockchain can enhance the capabilities of 5G-enabled IoT.

Data Integrity and Trust: Blockchain provides an immutable and decentralized ledger that ensures the integrity and trustworthiness of data in 5G-enabled IoT networks. It can be used to record and verify data transactions between IoT devices, ensuring that the data remains tamper-proof and transparent. This enhances the reliability and trust in the data exchanged within the IoT ecosystem.

Security and Privacy: Blockchain technology can enhance the security and privacy of data in 5G-enabled IoT networks. By using cryptographic algorithms and consensus mechanisms, blockchain can secure IoT device communication, prevent unauthorized access, and protect sensitive data. It enables users to have control over their data and allows for secure sharing and access management in a distributed manner.

Device Identity and Authentication: Blockchain can play an important role in verifying the identity and authenticity of IoT devices in 5G networks. Each device can have a unique identifier stored on the blockchain, enabling seamless device authentication and authorization. This helps prevent device spoofing and unauthorized access, ensuring the integrity of the IoT network.

Smart Contracts and Automation: Blockchain's smart contract functionality can automate and streamline processes in 5G-enabled IoT environments. Smart contracts are self-executing agreements that automatically trigger predefined actions when specific conditions are met. They can facilitate automated interactions between IoT devices, such as device-to-device transactions or resource sharing, without the need for intermediaries. This improves efficiency and reduces costs in IoT deployments.

Monetization and Micropayments: Blockchain enables new monetization models in 5G-enabled IoT networks by facilitating micropayments and value exchange between devices and service providers. With the integration of blockchain-based digital currencies or tokens, IoT devices can participate in peer-to-peer transactions and provide services in exchange for payments. This opens up opportunities for new business models and economic ecosystems within the IoT landscape.

Network Management and Roaming: Blockchain can assist in managing 5G-enabled IoT networks, especially in scenarios involving multiple service providers and roaming devices. It can enable transparent and efficient network management, billing, and settlement processes among different stakeholders. Blockchain-based solutions can simplify roaming agreements, facilitate billing reconciliation, and enhance overall network interoperability.

In summary, the usage of blockchain in 5G-enabled IoT networks brings benefits such as data integrity, security, privacy, automation through smart contracts, monetization opportunities, and efficient network management. It empowers IoT devices to securely interact, transact, and collaborate in a decentralized and trusted environment, unlocking the full potential of 5G and IoT technologies.

20.4 Internet of Things and Industrial Internet of Things

The Internet of Things (IoT) and the Industrial Internet of Things (IIoT) are closely related concepts that involve connecting devices, sensors, and systems to enable data exchange and automation [10, 11]. While both share similarities, they have distinct focuses and applications:

Internet of Things (IoT): The IoT refers to the network of everyday objects or "things" embedded with sensors, software, and connectivity capabilities to collect and exchange data. These objects can include consumer devices like smartphones, wearables, and smart home appliances. The IoT aims to enhance convenience, efficiency, and automation in various aspects of our daily lives, such as home automation, healthcare, transportation, and smart cities. It encompasses a wide range of applications beyond industrial settings.

Industrial Internet of Things (IIoT): The IIoT specifically focuses on applying IoT technologies and principles to industrial sectors, such as manufacturing, energy, agriculture, and transportation. It involves connecting industrial equipment, machines, and systems to collect data and enable intelligent decision-making, automation, and optimization. The IIoT aims to improve operational efficiency, productivity, safety, and maintenance in industrial settings. It often involves complex and specialized devices, protocols, and standards tailored to industrial requirements.

Key characteristics of the IIoT include:

Industrial Devices: IIoT involves connecting industrial equipment, machinery, and sensors specifically designed for industrial environments. These devices are often rugged, capable of withstanding harsh conditions, and have specialized communication protocols.

Real-Time Data and Analytics: IIoT focuses on collecting real-time data from industrial devices to gain information and enable proactive decision-making. Advanced analytics and machine learning algorithms are applied to process and analyze the data for optimizing operations, predicting failures, and improving efficiency.

Cyber-Physical Systems: IIoT integrates the physical world with digital systems, creating cyber-physical systems. This involves the convergence of operational technology (OT) and information technology (IT) to enable seamless data exchange and control between the physical and digital realms.

Industrial Automation and Control: IIoT facilitates automation and control of industrial processes, enabling remote monitoring, predictive maintenance, and autonomous operations. It enables enhanced control over industrial systems and enables real-time adjustments based on data-driven information.

In summary, while both IoT and IIoT involve connecting devices and systems for data exchange and automation, the IIoT specifically focuses on industrial applications, using specialized devices, protocols, and analytics for optimizing operations, productivity, and safety in industrial settings.

20.6 Blockchain Platform for Industrial Internet of Things

A blockchain platform for the Industrial Internet of Things (IIoT) is a specialized solution that combines blockchain technology with IIoT systems to enhance security, trust, and efficiency in industrial environments [12]. The key components (refer Figure 20.3) and benefits of a blockchain platform for the IIoT are shown in the Figure 20.4.

Decentralized Ledger: The platform utilizes a decentralized and distributed ledger to record and store transactions and data generated by IIoT devices. This ledger ensures transparency, immutability, and auditability of the data, enabling trust among participants and preventing unauthorized tampering.

Figure 20.3 Key characteristics of IIoT.

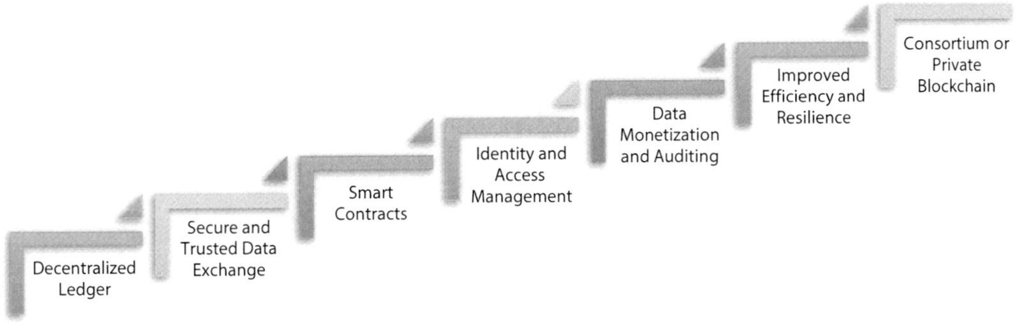

Figure 20.4 Key areas where blockchain can enhance the capabilities of 5G-enabled IoT.

Secure and Trusted Data Exchange: The platform facilitates secure data exchange between IIoT devices, sensors, and systems. By using blockchain's cryptographic algorithms, data encryption, and digital signatures, the platform ensures the integrity and authenticity of data, preventing unauthorized access or modifications.

Smart Contracts: Smart contracts are self-executing agreements that automate processes based on predefined conditions. In the context of IIoT, smart contracts can enable autonomous transactions and interactions between devices, allowing for seamless and trustless collaboration. For example, smart contracts can automate supply chain processes, asset tracking, and payment settlements.

Identity and Access Management: The platform incorporates identity and access management mechanisms to verify the authenticity of IIoT devices and ensure secure communication. Each device can have a unique identity stored on the blockchain, enabling granular access control and preventing unauthorized devices from participating in the network.

Data Monetization and Auditing: The platform enables data monetization in the IIoT ecosystem. Through blockchain-based digital currencies or tokens, data owners can securely sell their data to interested parties, creating new revenue streams. Additionally, the blockchain's transparent nature allows for auditing and traceability of data transactions, ensuring compliance with regulations and standards.

Improved Efficiency and Resilience: By utilizing blockchain technology, the platform enhances operational efficiency and resilience in industrial processes. It enables real-time tracking and visibility of assets, streamlines supply chain management, facilitates predictive maintenance, and improves overall operational performance through data-driven information.

Consortium or Private Blockchain: Depending on the requirements and the ecosystem, the blockchain platform can be implemented as a consortium or private blockchain. A consortium blockchain involves multiple trusted participants collaborating on a shared network, while a private blockchain is restricted to a single organization. The choice depends on factors such as data privacy, governance, and the level of trust among participants.

In summary, a blockchain platform for the IIoT provides secure, transparent, and efficient data exchange, automates processes through smart contracts, ensures data integrity, and enables new business models. By using blockchain's capabilities, the platform enhances trust, resilience, and operational efficiency in industrial environments.

20.7 Benefits and Features of Industrial Internet of Things

The Industrial Internet of Things (IIoT) provides several benefits, features, and capabilities that can revolutionize industrial processes [13, 14]. However, it also has certain limitations and challenges.

Benefits of IIoT

- Enhanced efficiency and productivity: IIoT enables real-time monitoring, data collection, and analysis, leading to improved operational efficiency and productivity in industrial processes.
- Predictive maintenance: IIoT facilitates condition-based monitoring and predictive maintenance, allowing companies to detect equipment failures in advance and avoid costly downtime.
- Cost savings: IIoT can help optimize resource utilization, reduce energy consumption, and minimize waste, leading to huge cost savings for industrial organizations.
- Improved safety: IIoT enables remote monitoring of hazardous environments and equipment, reducing human exposure to risks and enhancing overall workplace safety.
- Data-driven information: IIoT generates large amount of data, which can be analyzed to gain valuable information, optimize processes, and make informed business decisions.

Features and Capabilities of IIoT

- Connectivity: IIoT relies on a network of connected devices, sensors, and machines that can communicate and share data seamlessly.
- Real-time monitoring and control: IIoT enables continuous monitoring of industrial processes and the ability to remotely control devices and systems in real-time.
- Data analytics: IIoT use advanced analytics and machine learning algorithms to process and analyze the collected data, extracting meaningful information and patterns.

- Interoperability: IIoT systems support interoperability between different devices, protocols, and platforms, allowing seamless integration and communication across the industrial ecosystem.
- Scalability: IIoT solutions can scale up or down to accommodate the growing number of connected devices and changing operational requirements.

Limitations and Challenges of IIoT

- Security risks: IIoT introduces new security vulnerabilities, as a large number of connected devices increase the attack surface and potential entry points for hackers.
- Privacy concerns: The extensive collection and sharing of data in IIoT raise privacy concerns regarding the ownership, control, and potential misuse of sensitive information.
- Legacy system integration: Integrating IIoT with existing legacy systems can be complex and challenging, requiring careful planning and coordination.
- Data overload: The sheer volume of data generated by IIoT devices can overwhelm organizations, making it important to implement efficient data management and analysis strategies.
- Reliability and downtime: Dependence on IIoT for critical industrial processes raises concerns about system reliability, network connectivity, and potential downtime due to technical failures or cyber-attacks.

While the Industrial Internet of Things provides huge advantages, we need to address these limitations and challenges to ensure successful implementation and secure operation in industrial settings.

20.8 Application and Future Work of Blockchain Platform for Industrial Internet of Things

The application of blockchain platforms in the Industrial Internet of Things (IIoT) provides several benefits and holds promise for future advancements [14] shown in the Figure 20.5 and Figure 20.6.

Supply Chain Management: Blockchain can enhance supply chain management by providing transparency, traceability, and immutability. It enables secure and decentralized tracking of goods, ensuring authenticity and reducing the risk of fraud or counterfeiting.

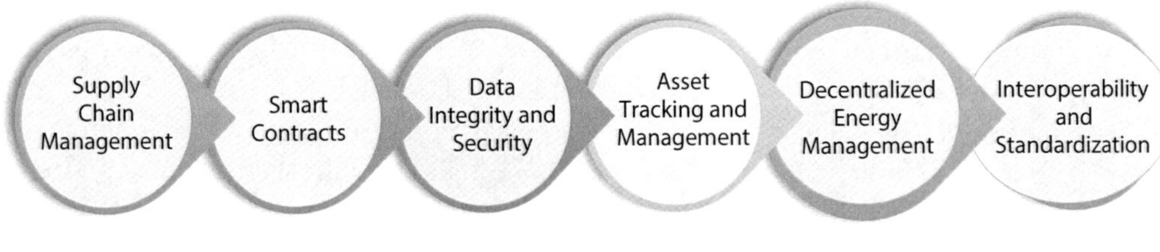

Figure 20.5 Application of blockchain platform for Industrial Internet of Things.

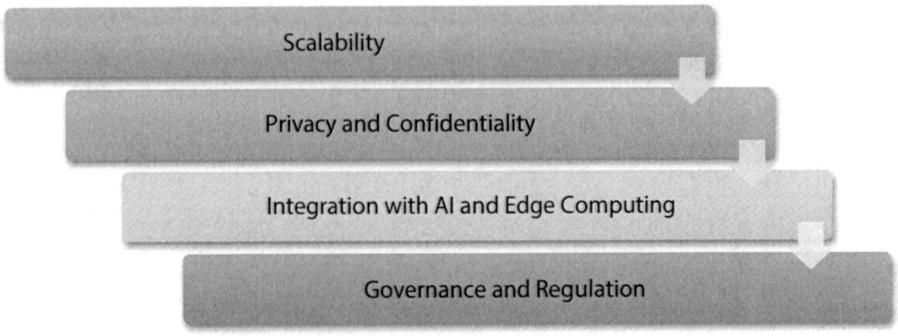

Figure 20.6 Future work of blockchain platform for Industrial Internet of Things.

Smart Contracts: Blockchain-based smart contracts automate and enforce agreements between different parties in IIoT ecosystems. They eliminate intermediaries, increase efficiency, and enable autonomous machine-to-machine transactions.

Data Integrity and Security: Blockchain provides a tamper-proof and decentralized ledger that ensures the integrity and security of IIoT data. It enhances data provenance, prevents unauthorized access, and protects against data manipulation or tampering.

Asset Tracking and Management: Blockchain facilitates real-time tracking and management of physical assets in IIoT applications. It enables secure and auditable records of asset ownership, maintenance history, and usage, improving operational efficiency and reducing costs.

Decentralized Energy Management: Blockchain platforms can support decentralized energy management in IIoT by enabling peer-to-peer energy trading, optimizing energy consumption, and ensuring transparency and trust among participants.

Interoperability and Standardization: Blockchain-based IIoT platforms can establish interoperability and standardization protocols among different devices, systems, and stakeholders. They facilitate seamless integration, data sharing, and collaboration across diverse IIoT ecosystems.

Future work in the application of blockchain platforms for IIoT involves:

Scalability: Blockchain faces scalability challenges when dealing with large-scale IIoT deployments. Future work includes exploring consensus algorithms, sharding techniques, and layer-two solutions to enhance scalability without compromising security and decentralization.

Privacy and Confidentiality: Enhancing privacy and confidentiality in IIoT blockchain networks is important. Researchers are focusing on developing privacy-preserving techniques such as zero-knowledge proofs and secure multiparty computation to protect sensitive IIoT data.

Integration with AI and Edge Computing: Integrating blockchain with artificial intelligence (AI) and edge computing technologies can enable advanced analytics, real-time decision-making, and autonomous IIoT systems. Future work involves exploring the synergies between these technologies to unlock their combined potential.

Governance and Regulation: Establishing governance models, legal frameworks, and regulatory standards for blockchain-based IIoT systems is essential. Future work includes addressing legal and ethical challenges, ensuring compliance, and fostering trust among participants.

In summary, the application of blockchain platforms in IIoT holds great promise for transforming industries by enhancing transparency, security, efficiency, and trust in the digital ecosystem. Ongoing research and development efforts aim to address the current limitations and unlock the full potential of this technology.

20.9 Future of Blockchain-Based Internet of Things

The future of blockchain-based Internet of Things (IoT) holds huge potential for transformative advancements in various domains [14]. Some key aspects shaping the future of blockchain-based IoT shown in the Figure 20.7.

Scalability Solutions: Scalability has been a challenge for blockchain networks, particularly when applied to large-scale IoT deployments. Future developments focus on improving scalability through techniques like sharding, sidechains, and layer-two solutions. These advancements aim to support the processing and verification of a large number of IoT transactions in a more efficient and scalable manner.

Interoperability Standards: Establishing interoperability standards is important for enabling seamless communication and data exchange between different IoT devices and blockchain networks. Future efforts will focus on developing standardized protocols and frameworks that facilitate interoperability, allowing devices and networks from various vendors to work together seamlessly.

Enhanced Security and Privacy: Blockchain brings inherent security features to IoT by providing tamper-proof and immutable transaction records. However, future work will focus on addressing privacy concerns by developing techniques such as zero-knowledge

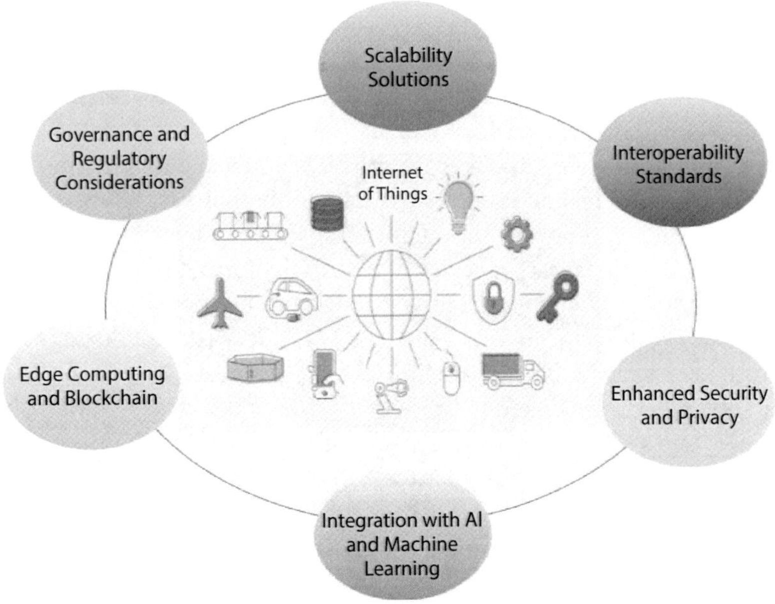

Figure 20.7 Future of blockchain-based IoT.

proofs, secure multiparty computation, and differential privacy. These advancements will allow sensitive IoT data to be securely and privately managed on the blockchain.

Integration with AI and Machine Learning: Integrating blockchain with artificial intelligence (AI) and machine learning (ML) technologies can unlock new possibilities for IoT applications. Combining blockchain's immutability and transparency with AI and ML capabilities can enable advanced analytics, intelligent decision-making, and autonomous IoT systems. This integration can enhance IoT data analysis, anomaly detection, and predictive maintenance.

Edge Computing and Blockchain: Edge computing, which brings processing capabilities closer to IoT devices, can be synergistically combined with blockchain. By decentralizing the blockchain network at the edge, latency can be reduced, data privacy can be improved, and network congestion can be minimized. This integration facilitates real-time data processing, local consensus, and improved overall system performance.

Governance and Regulatory Considerations: As blockchain-based IoT systems proliferate, there will be a need for establishing governance models, legal frameworks, and regulatory standards. Future work will involve addressing legal and ethical challenges, ensuring compliance with data protection regulations, and establishing trust frameworks for IoT devices and blockchain networks.

In summary, the future of blockchain-based IoT holds great potential for revolutionizing industries and improving the efficiency, security, and trustworthiness of IoT ecosystems. Ongoing research and development efforts are focused on addressing the current limitations and exploring innovative solutions to unlock the full potential of this technology combination.

20.10 Blockchain-Enabled Internet of Things (IoTs) Platforms for Supply Chain Functions

Blockchain-enabled Internet of Things (IoT) platforms have huge implications for improving supply chain functions [15, 16]. Here are some key ways in which blockchain can enhance supply chain management[1]:

Transparency and Traceability: Blockchain provides a transparent and immutable ledger that records every transaction and event within the supply chain. By integrating IoT devices with blockchain, real-time data on products, shipments, and inventory can be securely tracked and recorded. This transparency and traceability enable stakeholders to verify the origin, authenticity, and movement of goods throughout the supply chain.

Enhanced Product Authentication: Counterfeit products provide a huge challenge in supply chains. By considering blockchain and IoT technologies, unique identifiers such as RFID tags or QR codes can be linked to product information stored on the blockchain. This allows consumers and stakeholders to verify the authenticity of products, reducing the risk of counterfeiting and improving consumer trust.

Streamlined Supply Chain Payments: Blockchain-based smart contracts can automate payment processes in the supply chain. Smart contracts automatically execute payments when

---

[1] https://images.app.goo.gl/Gdz8Qtvjia7WbjwS6

predefined conditions are met, such as successful delivery or quality inspection. This eliminates the need for intermediaries, reduces transaction costs, and accelerates payment settlement.

Efficient Inventory Management: IoT sensors can monitor inventory levels, temperature, humidity, and other relevant parameters in real-time. By integrating IoT data with blockchain, stakeholders can have an accurate and tamper-proof record of inventory levels and conditions. This enables better inventory planning, reducing stockouts and minimizing excess inventory.

Supplier and Vendor Management: Blockchain-based platforms enable the creation of decentralized supplier databases. IoT devices can collect data on supplier performance, compliance records, certifications, and other relevant information. This data can be stored on the blockchain, providing a transparent and auditable supplier evaluation system. It allows supply chain participants to make informed decisions based on reliable and verified supplier information.

Dispute Resolution and Compliance: Blockchain's immutability and tamper-proof nature can facilitate efficient dispute resolution in the supply chain. By recording all transactions and events on the blockchain, stakeholders can easily trace the history of goods and identify any discrepancies. This transparent audit trail can streamline dispute resolution processes and ensure compliance with regulations and contractual obligations.

Collaborative Supply Chain Networks: Blockchain-enabled IoT platforms can foster collaboration among supply chain participants. By securely sharing data through the blockchain, stakeholders can collaborate on demand forecasting, inventory optimization, and logistics planning. This shared visibility improves coordination, reduces delays, and enhances overall supply chain efficiency.

In summary, blockchain-enabled IoT platforms have the potential to revolutionize supply chain management by providing transparency, traceability, and automation (refer Figure 20.8). These technologies enhance trust, reduce costs, and improve operational efficiency throughout the supply chain ecosystem.

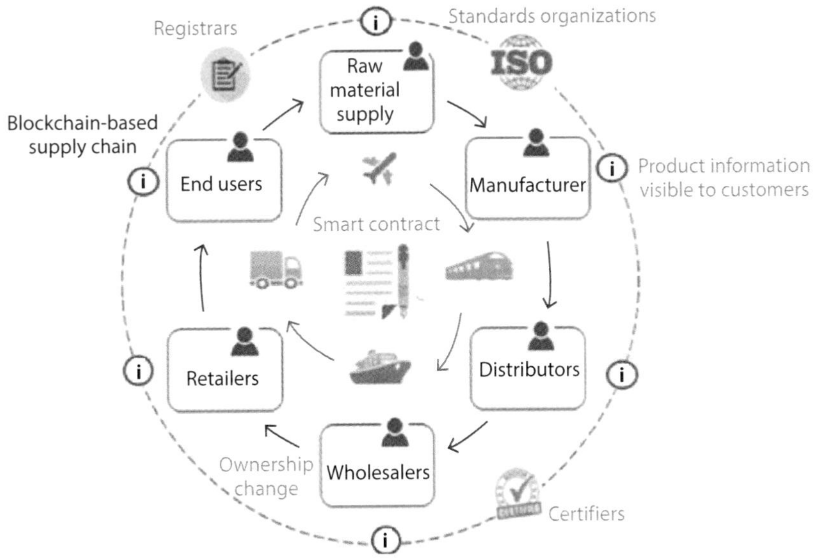

Figure 20.8 Blockchain-enabled Internet of Things for the supply chain.

20.11 Blockchain-Enabled Internet of Things (IoTs) Platforms for Smart Energy and Smart Grids

Blockchain-enabled Internet of Things (IoT) platforms provide several benefits for smart energy and smart grid applications [17]. Some key ways in which blockchain can enhance these domains shown in Figure 20.9.

Peer-to-Peer Energy Trading: Blockchain enables secure and transparent peer-to-peer energy trading. IoT devices, such as smart meters or sensors, can collect real-time energy consumption data, which can be recorded on the blockchain. Smart contracts can automatically execute energy transactions between producers and consumers, allowing for direct energy trading without the need for intermediaries. This decentralization of energy transactions can improve efficiency, reduce costs, and empower consumers to actively participate in the energy market.

Grid Management and Optimization: IoT devices integrated with blockchain can provide real-time data on energy generation, consumption, and grid conditions. This data can be securely shared and verified through the blockchain, facilitating better grid management and optimization. Smart contracts can automate grid operations, such as load balancing, demand response, and grid maintenance, based on the data received from IoT devices. This improves grid stability, reduces energy wastage, and enhances overall energy efficiency.

Microgrid Integration: Blockchain can facilitate the integration of microgrids, which are localized energy systems that can operate independently or in conjunction with the main grid. IoT devices within microgrids can communicate with each other and share data on energy production and consumption. Blockchain provides the necessary trust and security

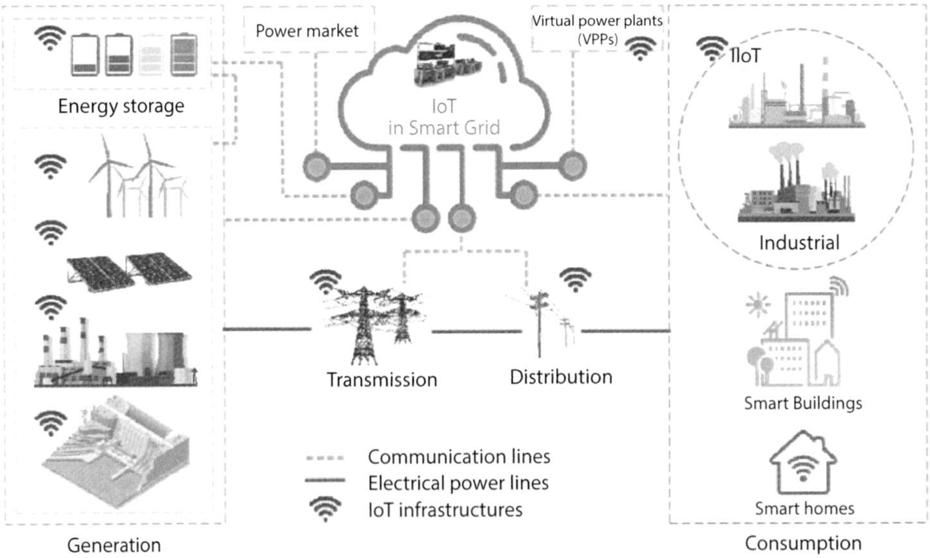

Figure 20.9 Blockchain enabled IoTs for Smart Energy and Smart grids [Shahinzadeh, Hossein *et al.* "IoT Architecture for Smart Grids." 2019 International Conference on Protection and Automation of Power System (IPAPS) (2019): 22–30].

for enabling peer-to-peer energy sharing and transactions within microgrid networks, enhancing energy resiliency and fostering local energy communities.

Energy Data Management and Privacy: Blockchain technology can address concerns regarding the privacy and security of energy data generated by IoT devices. By storing energy-related data on the blockchain, individuals can retain ownership and control over their data, granting access to selected parties on a need-to-know basis. Smart contracts can enforce data privacy rules, ensuring that sensitive energy data is shared securely and with user consent. This enhances data privacy, builds trust, and encourages broader adoption of IoT devices in the energy sector.

Renewable Energy Certificates and Carbon Credits: Blockchain can facilitate the issuance, tracking, and trading of renewable energy certificates (RECs) and carbon credits. IoT devices can collect data on renewable energy generation and carbon emissions, which can be recorded on the blockchain. This transparent and auditable system ensures the integrity and traceability of RECs and carbon credits, enabling easier trading and incentivizing the deployment of renewable energy sources.

Energy Grid Billing and Settlement: Blockchain-based IoT platforms can streamline energy billing and settlement processes. IoT devices can record accurate energy consumption data, which is securely stored on the blockchain. Smart contracts can automatically calculate and execute billing transactions based on predefined rules and rates. This reduces billing disputes, enhances accuracy, and accelerates payment settlement between energy providers and consumers.

In summary, blockchain-enabled IoT platforms have the potential to revolutionize smart energy and smart grid systems by enabling peer-to-peer energy trading, improving grid management, ensuring data privacy, and facilitating the integration of renewable energy sources. These technologies foster a more efficient, resilient, and sustainable energy ecosystem.

20.12 Blockchain-Enabled Internet of Things (IoTs) Platforms for Industrial Control Systems

Blockchain-enabled Internet of Things (IoT) platforms have the potential to enhance Industrial Control Systems (ICS) by providing additional security, transparency, and efficiency [18]. Here are some key ways in which blockchain can benefit ICS shown in Figure 20.10

Enhanced Security: Blockchain technology can improve the security of ICS by providing a decentralized and tamper-proof ledger for recording and verifying transactions. By integrating IoT devices with blockchain, critical data and control signals can be securely and transparently transmitted and stored. This reduces the risk of unauthorized access, manipulation, or tampering of data within the ICS.

Trust and Authentication: Blockchain can enable secure and trusted communication and authentication between IoT devices and components within the ICS. Through the use of cryptographic techniques, blockchain verifies the identity of devices and ensures that only authorized devices can participate in the system. This enhances the overall trustworthiness of the ICS network and mitigates the risk of unauthorized or malicious access.

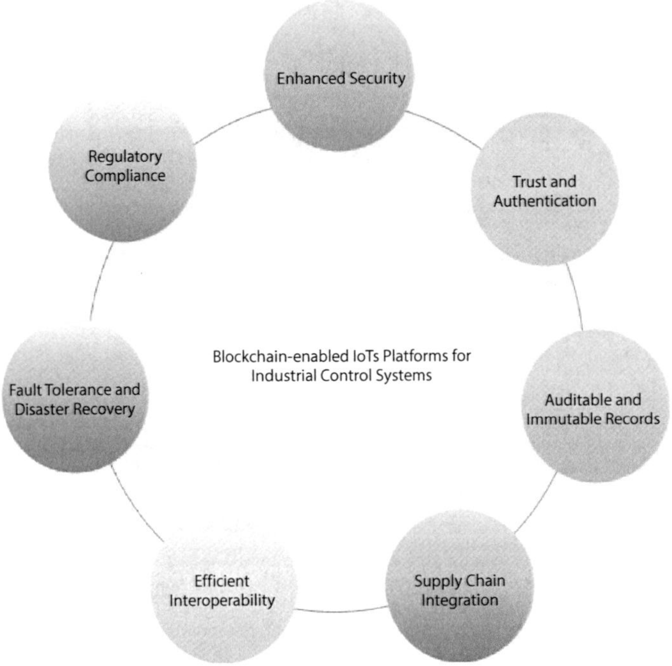

Figure 20.10 Blockchain-enabled Internet of Things (IoTs) platforms for industrial control systems.

Auditable and Immutable Records: Blockchain's immutable nature ensures that all transactions and events within the ICS are recorded and cannot be altered retroactively. This provides a transparent and auditable system for tracking and verifying actions taken by IoT devices and other components within the ICS. The ability to trace and audit these records can be invaluable for detecting and investigating security breaches or system failures.

Supply Chain Integration: Blockchain-based IoT platforms can facilitate secure and transparent integration of supply chain data into the ICS. By recording supply chain-related information, such as component origins, certifications, and maintenance records, on the blockchain, the ICS can ensure the authenticity and quality of components used in the system. This integration enhances supply chain transparency, reduces the risk of counterfeit or substandard components, and improves overall system reliability.

Efficient Interoperability: Blockchain-based IoT platforms can establish standardized protocols and frameworks that enable seamless interoperability between different devices, systems, and stakeholders within the ICS. This facilitates the integration of diverse IoT devices and components, simplifies data exchange, and improves the efficiency of communication and coordination within the ICS ecosystem.

Fault Tolerance and Disaster Recovery: Blockchain's decentralized nature provides inherent fault tolerance and resilience to the ICS. In the event of a single point of failure or a malicious attack, the distributed nature of the blockchain ensures that the system remains operational. This fault tolerance capability enhances the reliability and availability of the ICS, reducing downtime and improving disaster recovery capabilities.

Regulatory Compliance: Blockchain-based IoT platforms can assist in meeting regulatory compliance requirements for ICS. By providing a transparent and auditable system, blockchain can help demonstrate compliance with industry standards and regulations. It also enables the secure sharing of compliance-related data and automates compliance checks and audits, simplifying the compliance process for ICS stakeholders.

In summary, blockchain-enabled IoT platforms have the potential to strengthen the security, transparency, and efficiency of Industrial Control Systems. By considering blockchain technology, ICS can benefit from enhanced security measures, improved interoperability, transparent supply chain integration, and resilient fault tolerance capabilities. These advancements contribute to the overall reliability, integrity, and trustworthiness of industrial systems.

20.13 Blockchain-Enabled Internet of Things (IoTs) Platforms for IoT-Based Customer Relationship Management (CRM) and Logistics CRM

Blockchain-enabled Internet of Things (IoT) platforms provide valuable applications for IoT-based Customer Relationship Management (CRM) and Logistics CRM [19]. Here are some key benefits and use cases shown in Figure 20.11.

Enhanced Data Security and Privacy: Blockchain provides a decentralized and tamper-proof ledger for storing and managing customer data and logistics-related information. By integrating IoT devices with blockchain, customer data and logistics data can be securely stored and accessed with cryptographic techniques, ensuring data security and protecting customer privacy.

Trusted Data Sharing and Transparency: Blockchain enables secure and transparent data sharing between different stakeholders in the CRM and logistics ecosystem. IoT devices can collect and transmit data such as customer behavior, preferences, and logistics information. This data can be recorded on the blockchain, allowing authorized parties to access and verify the information, fostering trust and transparency in the relationship between customers, service providers, and logistics partners.

Seamless Supply Chain Visibility: Blockchain-based IoT platforms provide real-time visibility into the supply chain. IoT devices equipped with sensors and trackers can collect data on product location, condition, and transportation status. This information can be recorded on the blockchain, allowing all stakeholders to have a transparent view of the supply chain, including customers. This visibility enables proactive communication, efficient issue resolution, and accurate delivery estimations.

Improved Customer Service: By considering blockchain and IoT, CRM systems can gather real-time customer data from various touch points. This data can include customer interactions, preferences, and feedback. With blockchain, this data can be securely stored and shared across the CRM network, enabling service providers to provide personalized and tailored experiences to customers. Real-time data collection and analysis from IoT devices

Figure 20.11 Blockchain-enabled IoTs for IoT-based CRM and logistics CRM.

allow service providers to anticipate customer needs, resolve issues promptly, and provide proactive support.

Efficient Order and Delivery Management: Blockchain-based IoT platforms can streamline order and delivery management processes. IoT devices can track and record order information, logistics data, and delivery status on the blockchain. This ensures transparency, traceability, and accountability in the logistics CRM ecosystem. Automated smart contracts can execute predefined rules and conditions, enabling seamless order fulfillment and delivery, reducing delays, and improving overall customer satisfaction.

Loyalty Programs and Incentives: Blockchain-enabled IoT platforms can enhance loyalty programs and incentives by providing a secure and auditable system for tracking and distributing rewards. IoT devices can record customer participation, transactions, and loyalty points on the blockchain. This allows for transparent and tamper-proof tracking of rewards, facilitating accurate redemption and incentivizing customer engagement.

Dispute Resolution and Trust Building: Blockchain-based IoT platforms provide a reliable and transparent system for resolving disputes in CRM and logistics. All relevant data and transactions are recorded on the blockchain, ensuring an auditable history of interactions and actions taken. In case of disputes or discrepancies, this data can be used to investigate and resolve issues efficiently, building trust among stakeholders.

In summary, blockchain-enabled IoT platforms provide huge potential for improving CRM and logistics CRM processes. By considering the secure and transparent nature of blockchain and the real-time data collection capabilities of IoT devices, organizations can enhance customer experiences, streamline logistics operations, and foster trust in the CRM ecosystem.

20.14 Blockchain-Enabled Internet of Things (IoTs) Platforms for Smart Healthcare

Blockchain-enabled Internet of Things (IoT) platforms have transformative potential in the realm of smart healthcare [20]. Here are some key applications and benefits of integrating blockchain with IoT in the healthcare sector which is shown in the Figure 20.12.

Secure and Interoperable Health Data Exchange: Blockchain provides a secure and decentralized platform for storing and sharing health data collected by IoT devices. Patient-generated data from wearables, sensors, and medical devices can be securely recorded on the blockchain, allowing healthcare providers, patients, and authorized stakeholders to access and share data in a trusted manner. This interoperability and secure data exchange enable comprehensive and real-time information for personalized healthcare.

Patient Identity Management: Blockchain can facilitate reliable and tamper-proof patient identity management. By linking patient identities to unique blockchain records, healthcare

Figure 20.12 Blockchain-enabled IoTs for smart healthcare.

providers can ensure accurate patient identification, reduce medical errors, and enhance data integrity. IoT devices can collect biometric data, such as fingerprints or iris scans, which can be securely stored and validated on the blockchain, strengthening patient identity management.

Drug Supply Chain Integrity: Blockchain-based IoT platforms can improve the integrity and transparency of the pharmaceutical supply chain. IoT sensors and RFID tags can track and record the movement of drugs and medical supplies throughout the supply chain. This information can be stored on the blockchain, creating an auditable and immutable record of the drug's journey, from manufacturing to delivery. Blockchain ensures that medications are authentic, reduces the risk of counterfeit drugs, and enhances patient safety.

Clinical Trials and Research: Blockchain-enabled IoT platforms can streamline and enhance the process of conducting clinical trials and medical research. IoT devices can collect real-time patient data during trials, and this data can be securely recorded on the blockchain. This enables accurate and transparent data sharing among researchers, ensures data integrity, and facilitates compliance with regulatory requirements. Additionally, smart contracts on the blockchain can automate consent management, data sharing agreements, and payment processes between stakeholders involved in clinical trials.

Remote Patient Monitoring and Telehealth: IoT devices, such as wearables and remote monitoring systems, can collect patient health data in real-time. Blockchain ensures the secure transmission and storage of this data, allowing healthcare providers to remotely monitor patients and provide timely interventions. The combination of IoT and blockchain enables more efficient telehealth services, improves patient outcomes, and reduces the need for in-person visits.

Healthcare Data Privacy and Consent Management: Blockchain's cryptographic features can enhance privacy and consent management in healthcare. Patient consent for data sharing and access can be recorded on the blockchain, ensuring that data is accessed and used in accordance with patient preferences. Blockchain-based consent management enhances patient control over their health data, strengthens data privacy, and ensures compliance with data protection regulations.

Medical Device Lifecycle Management: Blockchain-based IoT platforms can improve the management and maintenance of medical devices. IoT sensors embedded in devices can collect performance data, maintenance requirements, and usage patterns. This data can be recorded on the blockchain, enabling efficient device monitoring, maintenance scheduling, and traceability. This enhances device safety, optimizes maintenance processes, and improves overall operational efficiency.

Hence, by considering the combined capabilities of blockchain and IoT, smart healthcare solutions can revolutionize patient care, data management, supply chain integrity, and research processes. These technologies foster patient-centric care, improve data security, and enable interoperability and collaboration among healthcare stakeholders.

20.15 Blockchain-Enabled Internet of Things (IoTs) Platforms for Digital Marketing and Online Social Networking

Blockchain-enabled Internet of Things (IoT) platforms have the potential to revolutionize digital marketing by enhancing transparency, data security, and trust [21]. Some key applications and benefits of integrating blockchain with IoT in the realm of digital marketing are shown in Figure 20.13.

Transparent and Trustworthy Data Sharing: Blockchain provides a transparent and immutable ledger for recording and sharing data collected from IoT devices. In digital marketing, IoT devices can gather customer behavior data, such as browsing patterns, location information, and preferences. This data can be securely stored on the blockchain, ensuring transparency and trust in data sharing between marketers, advertisers, and customers. Marketers can access accurate and reliable customer information, leading to more effective and personalized marketing campaigns.

Enhanced Customer Data Privacy: Blockchain-based IoT platforms can address concerns regarding customer data privacy. With blockchain, customers can retain ownership and control over their data, granting access to selected marketers or advertisers on a permissioned basis. IoT devices can collect data directly from customers, encrypt it, and store it on the blockchain, ensuring that personal information is kept private and secure. This fosters trust between customers and marketers, leading to better customer engagement and loyalty.

Efficient Ad Verification and Fraud Prevention: Blockchain can enhance ad verification and combat ad fraud in digital marketing. By recording ad impressions, clicks, and conversions on the blockchain, advertisers can verify the authenticity and accuracy of ad delivery and performance metrics. This transparent and auditable system reduces the risk of ad fraud and ensures that advertisers get reliable and verifiable data on the effectiveness of their campaigns.

Reliable Attribution and Reward Systems: Blockchain-enabled IoT platforms can facilitate transparent and reliable attribution and reward systems. IoT devices can collect data on customer interactions, engagement, and conversions, which can be recorded on the blockchain. Marketers can track and validate customer actions accurately, ensuring fair attribution and rewarding customers accordingly. This incentivizes customer engagement and loyalty, fostering stronger relationships between brands and customers.

Content Authenticity and Intellectual Property Protection: Blockchain can help address issues of content authenticity and intellectual property protection in digital marketing. By timestamping and storing content information on the blockchain, creators and marketers can prove the originality and ownership of their content. This ensures that content is not tampered with or plagiarized, protecting intellectual property rights and maintaining brand integrity.

Figure 20.13 Blockchain-enabled IoTs for digital marketing and online social networking.

Smart Contracts for Automated Campaign Execution: Blockchain's smart contract capabilities enable the automation of campaign execution and payment processes. Smart contracts can automatically execute predefined rules and conditions based on customer interactions and campaign performance data collected from IoT devices. This eliminates the need for intermediaries, reduces transaction costs, and accelerates payment settlement between marketers, advertisers, and publishers.

Customer Engagement and Loyalty Programs: Blockchain-based IoT platforms can enhance customer engagement and loyalty programs by providing secure and transparent systems for tracking and rewarding customer interactions. IoT devices can collect data on customer activities and engagements, which can be recorded on the blockchain. This data can then be used to personalize customer experiences, deliver targeted provides, and provide rewards based on customer preferences and behaviors.

Hence by integrating blockchain with IoT, digital marketing can benefit from increased transparency, enhanced data security, improved customer trust, and streamlined processes. These advancements enable marketers to deliver more personalized and effective campaigns while maintaining the privacy and trust of customers.

20.16 Blockchain-Enabled Internet of Things (IoTs) Platforms for Online Social Networking

Blockchain-enabled Internet of Things (IoT) platforms have the potential to transform online social networking by addressing privacy concerns, improving data ownership, and enhancing trust among users [22]. Some key applications and benefits of integrating blockchain with IoT in the context of online social networking are shown in Figure 20.14.

Data Privacy and Ownership: Blockchain provides a decentralized and transparent platform for storing and managing user data. With IoT devices, users can collect personal data such as activity logs, preferences, and social interactions. This data can be encrypted and stored on the blockchain, giving users control over their own data and allowing them to grant or revoke access to specific parties. Blockchain ensures that user data remains private and is not controlled by a single centralized entity.

Secure Identity Verification: Blockchain-based IoT platforms can provide secure and verified identity management in online social networking. IoT devices can collect biometric data or unique identifiers, which can be securely stored and authenticated on the blockchain. This enables reliable and tamper-proof identity verification, reducing the risk of fake profiles and improving the overall trustworthiness of the social networking platform.

Content Authenticity and Credibility: Blockchain can address the issue of fake news, misinformation, and content authenticity in online social networking. Content creators can

Figure 20.14 Blockchain-enabled IoTs for online social networking.

timestamp and store their content on the blockchain, establishing proof of ownership and preventing unauthorized modifications. Users can verify the authenticity and credibility of shared content, reducing the spread of false information and fostering a more trustworthy social networking environment.

Enhanced User Trust and Transparency: By using blockchain and IoT, social networking platforms can increase user trust through transparent and auditable systems. Blockchain records all user interactions, such as likes, comments, and shares, creating a transparent trail of activities. Users can verify the accuracy and integrity of these interactions, promoting trust among network participants and reducing the risk of manipulation or fraudulent activities.

Peer-to-Peer Interactions and Micropayments: Blockchain-based IoT platforms can facilitate direct peer-to-peer interactions and enable micropayments within online social networks. IoT devices can track user engagement, content contributions, or specific actions, which can be recorded on the blockchain. Smart contracts can automate micropayments based on predefined criteria, rewarding users for their contributions and incentivizing active participation.

Decentralized Social Networks: Blockchain and IoT can enable the creation of decentralized social networks, removing the need for a central authority to control and manage user data. In a decentralized social network, user data is distributed across the blockchain network, ensuring data availability even if individual nodes or servers fail. This promotes user autonomy, privacy, and resilience against data breaches or censorship.

Data Monetization and Incentives: Blockchain-based IoT platforms can empower users to have more control over their data and participate in data monetization. Users can choose to share their data selectively and receive incentives or rewards for providing access to their data. Blockchain ensures transparent and fair compensation for data sharing, enabling users to benefit directly from their participation in the social networking platform.

Hence by considering the capabilities of blockchain and IoT, online social networking platforms can address privacy concerns, empower users with data ownership, enhance trust, and enable more personalized and secure interactions among users. These advancements contribute to a more transparent, trustworthy, and user-centric social networking experience.

20.17 Blockchain-Enabled Internet of Things (IoTs) Platforms for Textile Industry

Blockchain-enabled Internet of Things (IoT) platforms provide several applications and benefits for the textile industry [23]. Some key areas where the integration of blockchain and IoT can have a major impact are shown in Figure 20.15.

Supply Chain Traceability: Blockchain can provide end-to-end traceability in the textile supply chain. IoT devices, such as RFID tags or sensors, can be attached to textile products, allowing real-time tracking of their journey from raw material sourcing to manufacturing, distribution, and retail. This data can be recorded on the blockchain, ensuring transparency and accountability, reducing counterfeiting, and enhancing consumer confidence in the authenticity and sustainability of textile products.

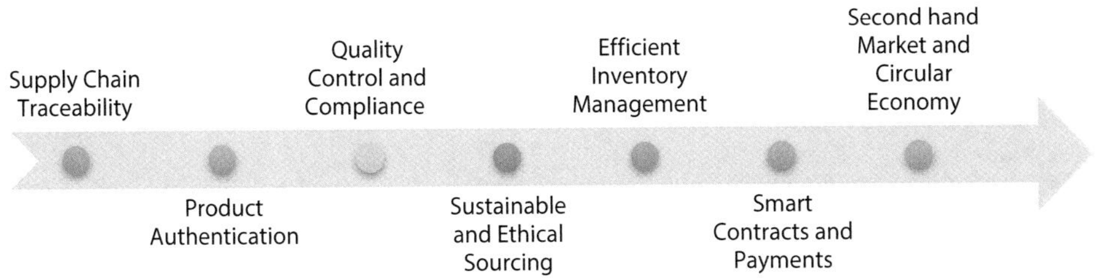

Figure 20.15 Blockchain-enabled IoTs for textile industry.

Product Authentication: Blockchain-based IoT platforms can enable reliable product authentication in the textile industry. IoT devices can collect and store unique identifiers, manufacturing details, and other relevant data on the blockchain. This allows consumers, retailers, and other stakeholders to verify the authenticity and origin of textile products, mitigating the risk of counterfeit goods and protecting brand reputation.

Quality Control and Compliance: IoT devices can collect real-time data on various parameters, such as temperature, humidity, and chemical levels, during textile production and transportation. This data can be recorded on the blockchain, enabling quality control, compliance monitoring, and ensuring adherence to industry standards and regulations. Manufacturers can track and verify the conditions under which textiles are produced and transported, ensuring product quality and safety.

Sustainable and Ethical Sourcing: Blockchain-enabled IoT platforms can promote sustainable and ethical sourcing practices in the textile industry. IoT devices can collect data on the origin of raw materials, including information about suppliers, certifications, and production methods. This data can be recorded on the blockchain, allowing manufacturers and consumers to make informed decisions based on sustainability and ethical considerations. Blockchain ensures transparency and traceability, promoting responsible sourcing and reducing environmental and social impacts.

Efficient Inventory Management: IoT devices equipped with sensors can monitor textile inventory levels in real-time. This data can be recorded on the blockchain, enabling accurate inventory management, reducing stockouts, and optimizing supply chain operations. Blockchain-based inventory management systems can facilitate automated reordering, improve production planning, and enhance overall efficiency in the textile industry.

Smart Contracts and Payments: Blockchain's smart contract capabilities can streamline and automate contractual agreements and payments in the textile industry. Smart contracts can be used to establish agreements between textile suppliers, manufacturers, distributors, and retailers. These contracts can be executed automatically based on predefined conditions, ensuring transparent and timely payments throughout the supply chain.

Second hand Market and Circular Economy: Blockchain-enabled IoT platforms can facilitate the tracking and authentication of secondhand textile products, promoting the circular economy. IoT devices can collect data on the history, condition, and previous ownership of textile products. This information can be stored on the blockchain, providing potential buyers with transparency and confidence in the secondhand market. By enabling trust and verification, blockchain supports the transition towards a more sustainable and circular textile industry.

Hence by integrating blockchain with IoT, the textile industry can enhance supply chain transparency, product authenticity, sustainability, and efficiency. These advancements contribute to improved consumer trust, ethical sourcing, and overall operational effectiveness in the textile sector.

20.18 Blockchain-Enabled Internet of Things (IoTs) Platforms for Smart Banking and Financial Services

Blockchain-enabled Internet of Things (IoT) platforms have the potential to revolutionize smart banking and financial services by enhancing security, efficiency, and transparency in various areas [24]. Here are some key benefits shown in Figure 20.16.

Secure Transactions: Blockchain technology provides a secure and tamper-proof infrastructure for financial transactions conducted through IoT devices. It ensures the integrity of data and eliminates the need for intermediaries, reducing the risk of fraud and unauthorized access. By using blockchain, smart banking systems can provide enhanced security for activities such as payments, transfers, and asset management.

Improved Identity Management: IoT devices can be utilized to enhance identity management in banking and financial services. By combining IoT devices with blockchain technology, customers can securely store and manage their identities and credentials. This enables streamlined onboarding processes, secure access to financial services, and simplified Know Your Customer (KYC) procedures.

Seamless Micropayments and Microtransactions: IoT devices can enable frictionless micropayments and microtransactions in real-time. Blockchain technology facilitates the execution of these transactions securely and cost-effectively, allowing for new business models such as pay-per-use services and microinsurance. This opens up opportunities for innovative financial services tailored to individual customer needs.

Enhanced Supply Chain Financing: Blockchain-enabled IoT platforms can improve supply chain financing by automating and streamlining processes. IoT devices embedded in supply chain assets, such as shipping containers or vehicles, can collect and transmit real-time data to blockchain networks. This data can be used to validate the authenticity of goods, track inventory, and trigger automated financing transactions based on predefined smart contracts.

Smart Contracts and Automation: Smart contracts on blockchain can automate and enforce agreements between parties, eliminating the need for intermediaries and reducing transaction costs. In the context of smart banking, smart contracts can automate loan disbursements, insurance claims, and regulatory compliance processes. This automation improves efficiency, reduces human error, and speeds up transaction settlement.

Enhanced Data Integrity and Transparency: Blockchain technology enables the secure and transparent storage of financial data generated by IoT devices. This data can include transaction records, asset values, and audit trails. By using blockchain, banks and financial institutions can provide transparent and auditable financial services, fostering trust among customers and regulators.

Fraud Detection and Prevention: IoT devices connected to blockchain networks can provide real-time data that helps detect and prevent fraudulent activities. The combination of

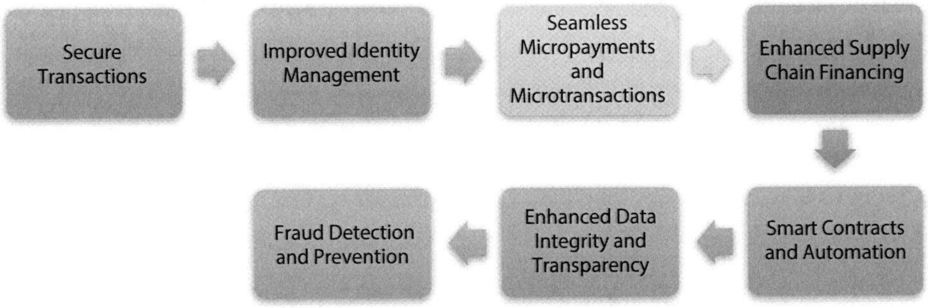

Figure 20.16 Blockchain-enabled IoTs for smart banking, financial services and military services.

IoT-generated data and blockchain's immutability allows for enhanced fraud analytics, early detection of suspicious patterns, and swift action to mitigate risks.

In summary, blockchain-enabled IoT platforms provide secure and efficient solutions for smart banking and financial services. They enhance transaction security, streamline processes through automation, improve identity management, and enable innovative business models. By using these technologies, banks and financial institutions can transform their operations, deliver better customer experiences, and drive digital innovation in the financial sector.

20.19 Blockchain-Enabled Internet of Things (IoTs) Platforms for Military Services

Blockchain-enabled Internet of Things (IoT) platforms have the potential to enhance military services in several ways. By combining blockchain technology with IoT devices, military organizations can achieve improved security, transparency, and efficiency in their operations is shown in Figure 20.16 [25].

Enhanced Security: Blockchain provides a decentralized and tamper-resistant infrastructure that can bolster the security of military IoT networks. It enables secure communication and data transfer between IoT devices, ensuring the integrity and confidentiality of sensitive military information. The decentralized nature of blockchain reduces the vulnerability to single points of failure and potential cyber-attacks.

Immutable Audit Trails: Blockchain's distributed ledger technology records and stores all transactions and interactions within the IoT network in a transparent and immutable manner. This feature enables military services to maintain a comprehensive and auditable record of activities, ensuring accountability and facilitating forensic analysis when needed.

Smart Contracts and Automation: Smart contracts, which are self-executing agreements built on blockchain, can automate various military processes. These contracts can be used to streamline supply chain management, automate logistics operations, and enforce compliance with predefined rules and protocols. Automation through smart contracts reduces human error, enhances operational efficiency, and accelerates decision-making processes.

Interoperability and Data Sharing: Military organizations often rely on various IoT devices and systems from different manufacturers. Blockchain technology can establish interoperability and standardization across these devices, enabling seamless communication and data sharing. This integration and interoperability promote data-driven decision-making, situational awareness, and coordination among different military units and branches.

Resilience and Fault Tolerance: Blockchain's decentralized nature ensures that military IoT networks remain operational even in the face of disruptions or attacks on individual devices or network nodes. By distributing data and processing across multiple nodes, blockchain enhances the resilience and fault tolerance of military IoT platforms, contributing to the continuity of critical services.

Supply Chain Integrity: Blockchain can enhance the security and integrity of military supply chains by tracking and verifying the provenance of goods and materials. This ensures that military equipment, spare parts, and other critical supplies are genuine, unaltered, and obtained from authorized sources, reducing the risk of counterfeit or compromised items entering the supply chain.

In summary, blockchain-enabled IoT platforms provide improved security, transparency, efficiency, and interoperability for military services. By using these technologies, military organizations can enhance their operational capabilities, strengthen cybersecurity, and optimize their logistics and supply chain management processes.

20.20 Blockchain-Enabled Internet of Things (IoTs) Platforms for Smart Agricultural Services

Blockchain-enabled Internet of Things (IoT) platforms have the potential to revolutionize smart agricultural services by improving efficiency, traceability, and sustainability in various aspects of farming [26]. Here are some key benefits related to smart agricultural services shown in Figure 20.17.

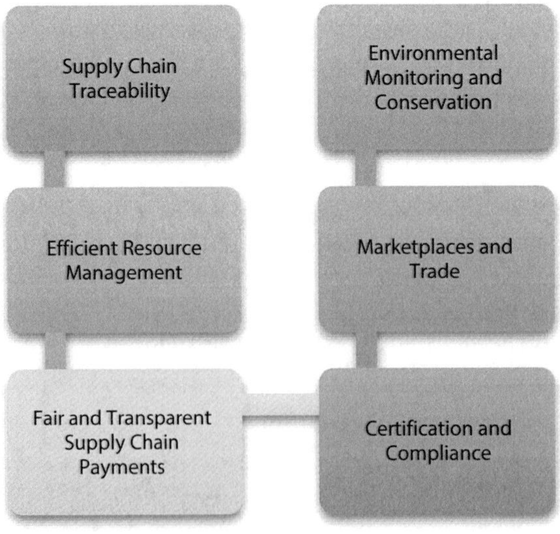

Figure 20.17 Blockchain-enabled IoTs for agricultural services.

Supply Chain Traceability: Blockchain technology combined with IoT devices can enable end-to-end traceability in the agricultural supply chain. From seed to shelf, each step of the process can be recorded on the blockchain, including information about the origin of seeds, cultivation practices, handling, transportation, and storage conditions. This transparency helps ensure food safety, quality control, and compliance with regulations.

Efficient Resource Management: IoT devices integrated with blockchain can collect real-time data on environmental conditions, such as soil moisture, temperature, and weather patterns. This data can be used to optimize resource management, such as irrigation, fertilization, and pest control. Smart agricultural systems can analyze the data, trigger automated actions, and provide farmers with information to make informed decisions, ultimately leading to more efficient resource utilization and increased crop yields.

Fair and Transparent Supply Chain Payments: Blockchain-enabled IoT platforms can facilitate fair and transparent payments in agricultural supply chains. By using smart contracts, farmers can receive instant and secure payments based on predefined conditions, such as the delivery of goods or meeting quality standards. This eliminates the need for intermediaries, reduces payment delays, and ensures that farmers are fairly compensated for their produce.

Certification and Compliance: Blockchain can streamline certification processes in agriculture by securely storing and verifying certifications, licenses, and compliance records. For example, organic farming certifications, fair trade labels, or sustainability standards can be recorded on the blockchain, providing transparent proof of compliance. This enhances trust among consumers, improves market access for farmers, and promotes sustainable agricultural practices.

Marketplaces and Trade: Blockchain-enabled IoT platforms can facilitate direct peer-to-peer transactions between farmers and consumers or businesses. By eliminating intermediaries, farmers can gain better access to markets, receive fair prices for their produce, and reduce transaction costs. IoT devices can provide real-time information on product availability, quality, and pricing, allowing buyers to make informed decisions. Blockchain ensures the transparency and security of these transactions.

Environmental Monitoring and Conservation: IoT devices deployed in agricultural landscapes can collect data on environmental indicators such as air quality, water quality, and biodiversity. This information, stored on the blockchain, can be used to monitor the impact of farming practices on the environment and support conservation efforts. By promoting sustainable practices and providing incentives for environmentally friendly farming, blockchain-enabled IoT platforms contribute to long-term agricultural sustainability.

In summary, blockchain-enabled IoT platforms provide transformative solutions for smart agricultural services. They enhance supply chain traceability, improve resource management, facilitate fair payments, streamline certification processes, enable direct trade, and support environmental monitoring and conservation. By using these technologies, farmers and agricultural stakeholders can optimize their operations, increase productivity, ensure food safety, and promote sustainable farming practices.

20.21 Blockchain Deployment in 5G-Enabled Smart Industrial Automation

Blockchain deployment in 5G-enabled smart industrial automation holds huge potential for enhancing efficiency, security, and trust in industrial processes [27]. Here are some key ways in which blockchain can be utilized in this context is shown in Figure 20.18.

Secure Data Exchange: Blockchain technology can facilitate secure and decentralized data exchange between different industrial devices and systems. With 5G's high-speed and low-latency connectivity, industrial automation systems can quickly and reliably transmit data to blockchain networks, ensuring real-time visibility and transparency across the ecosystem. This enhances data integrity, privacy, and security in industrial automation processes.

Supply Chain Management: Blockchain can enhance supply chain management in smart industrial automation. By integrating IoT devices, such as sensors and RFID tags, with blockchain, manufacturers can securely track and trace goods throughout the supply chain. This improves inventory management, reduces counterfeiting and tampering risks, and enables efficient logistics and asset management.

Smart Contracts and Automation: Smart contracts deployed on blockchain networks can automate and enforce agreements in smart industrial automation processes. These contracts can define and execute predefined rules and conditions for various tasks, such as equipment maintenance, quality control, and supply chain interactions. Automation through smart contracts reduces manual intervention, enhances operational efficiency, and enables autonomous decision-making.

Asset Management and Provenance: Blockchain-based asset management solutions can provide a secure and immutable record of industrial assets, including their ownership, maintenance history, and provenance. By integrating IoT sensors with blockchain, real-time data on asset conditions, performance, and usage can be stored on the blockchain.

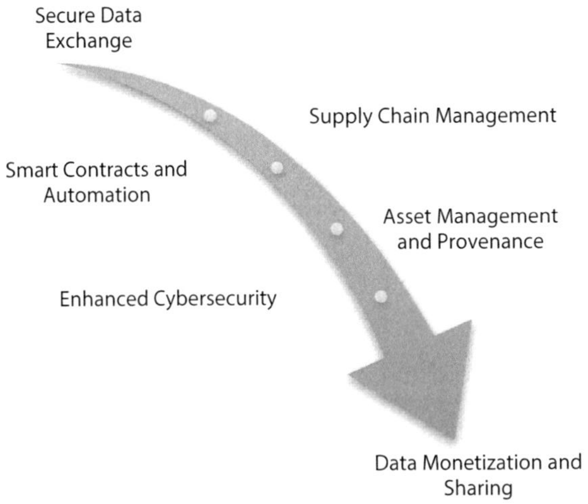

Figure 20.18 Blockchain deployment in 5G-enabled smart industrial automation.

This enables predictive maintenance, optimized asset utilization, and increased operational efficiency.

Enhanced Cybersecurity: Industrial automation systems are vulnerable to cyber threats, and 5G connectivity increases the attack surface. Blockchain's decentralized and cryptographic nature can provide enhanced cybersecurity for smart industrial automation. By storing critical security information, access controls, and audit logs on the blockchain, organizations can establish a resilient and tamper-resistant security infrastructure that safeguards against unauthorized access, data breaches, and malicious activities.

Data Monetization and Sharing: Blockchain enables secure data monetization and sharing in smart industrial automation. Industrial organizations can securely share data with trusted partners or third-party entities, ensuring data integrity, privacy, and control. Through blockchain-based data marketplaces, organizations can monetize their data by allowing others to access and utilize it, fostering innovation, collaboration, and new business models.

In summary, deploying blockchain in 5G-enabled smart industrial automation provides several benefits, including secure data exchange, improved supply chain management, automation through smart contracts, enhanced asset management, strengthened cybersecurity, and data monetization opportunities. By using these technologies, industrial automation processes can become more efficient, transparent, secure, and resilient, driving digital transformation and innovation in the industrial sector.

20.22 Open Issues in Blockchain-Empowered Internet of Things

While blockchain empowered Internet of Things (IoT) has great potential, there are still some open issues that need to be addressed for its widespread adoption and effective implementation [28]. Here are a few key challenges shown in Figure 20.19.

Scalability: Blockchain networks often struggle with scalability when it comes to processing a large number of transactions generated by IoT devices. The current blockchain infrastructure may not be able to handle the massive influx of data from several IoT devices, resulting in network congestion and increased transaction times. Addressing scalability issues is important to ensure that blockchain-based IoT platforms can handle the scale and volume of data generated in real-world IoT deployments.

Interoperability: Achieving interoperability among different blockchain platforms and IoT devices remains a challenge. As multiple blockchain networks and IoT devices with varying protocols and standards coexist, seamless integration and communication become essential. Standardization efforts and protocols that enable interoperability between different blockchain and IoT systems need to be developed to facilitate widespread adoption and collaboration across platforms.

Energy Efficiency: Blockchain mining processes, particularly those employed by proof-of-work (PoW) consensus algorithms, consume a huge amount of computational power and energy. This energy-intensive nature of blockchain faces challenges for IoT devices that are often resource-constrained and operate on limited power. Developing energy-efficient consensus mechanisms or exploring alternative consensus algorithms can help minimize the energy consumption associated with blockchain-based IoT systems.

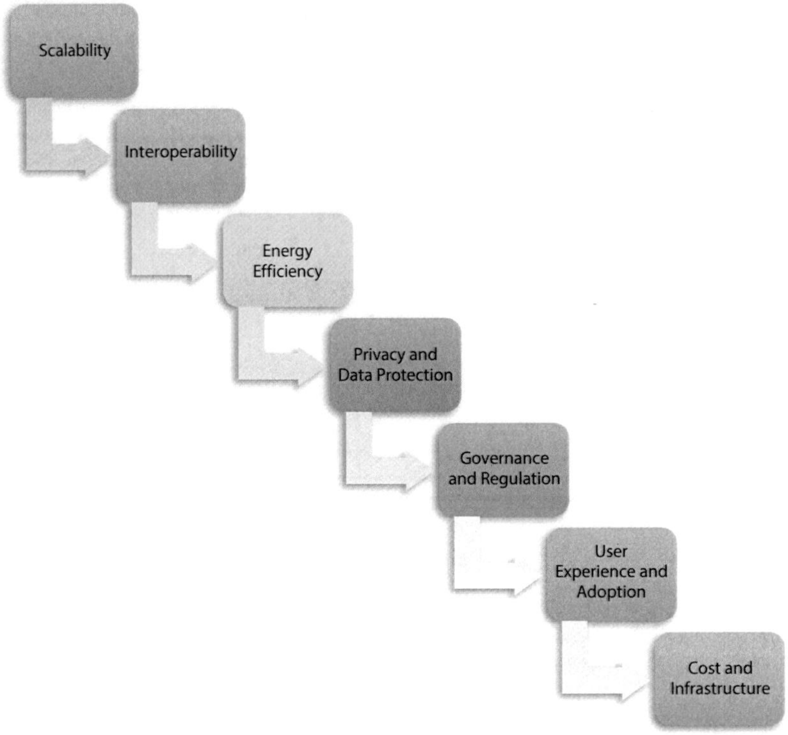

Figure 20.19 Open issues in blockchain-empowered Internet of Things.

Privacy and Data Protection: Blockchain's inherent transparency faces challenges in terms of privacy and data protection, especially in IoT deployments that handle sensitive personal or confidential data. While blockchain provides immutability and data integrity, it is essential to address privacy concerns by incorporating privacy-enhancing technologies, such as zero-knowledge proofs or secure encryption methods, into blockchain-based IoT platforms. Striking a balance between transparency and privacy is important for ensuring user trust and compliance with data protection regulations.

Governance and Regulation: The decentralized nature of blockchain raises questions about governance and regulatory frameworks. Determining the legal responsibilities and liabilities in a distributed network, resolving disputes, and establishing governance models for decision-making are complex issues that need to be addressed. Additionally, regulatory frameworks should keep pace with the rapid development of blockchain and IoT technologies to address concerns related to data privacy, security, and compliance.

User Experience and Adoption: Blockchain technology can be complex for non-technical users, and the user experience (UX) of interacting with blockchain-based IoT systems needs improvement. Simplifying user interfaces, enhancing accessibility, and providing seamless integration with existing technologies can contribute to better adoption and acceptance of blockchain empowered IoT platforms.

Cost and Infrastructure: Deploying blockchain and IoT systems requires huge investments in infrastructure, including hardware, software, and connectivity. The cost associated with blockchain transactions and the maintenance of decentralized networks can also be a barrier to adoption. Exploring cost-effective solutions, optimizing resource utilization, and using cloud-based services can help mitigate these challenges.

Note that as the field of blockchain empowered IoT [37] continues to evolve, addressing these open issues will be important to realize the full potential of this technology combination. Collaboration among industry stakeholders, academia, and policymakers is essential to overcome these challenges and create a robust and sustainable ecosystem for blockchain-enabled IoT applications.

20.23 Critical Challenges Towards Using Blockchain-Empowered Internet of Things

The integration of blockchain with Internet of Things (IoT) technology presents several critical challenges that need to be addressed for successful implementation [29–36]. Here are some key challenges as shown in Figure 20.20.

Scalability: Blockchain's inherent design, such as the need for consensus among multiple nodes and the storage of every transaction on the ledger, can limit scalability when dealing with the large number of IoT devices and the enormous volume of data they generate. Scaling blockchain networks to handle the increasing demands of IoT deployments without compromising performance is a major challenge.

Latency and Efficiency: Blockchain transactions require validation and consensus, which introduces latency. In time-sensitive IoT applications, such as real-time monitoring or control systems, delays caused by blockchain consensus mechanisms may be unacceptable. Finding ways to reduce latency and improve the efficiency of blockchain transactions for IoT use cases is important.

Cost and Energy Efficiency: Blockchain transactions and the computational requirements for maintaining a decentralized network can be resource-intensive and costly, especially for resource-constrained IoT devices. The energy consumption associated with mining in traditional blockchain systems is also a concern. Developing energy-efficient consensus mechanisms and optimizing resource usage are essential to make blockchain-enabled IoT solutions economically viable.

Security and Privacy: While blockchain provides inherent security features, such as immutability and cryptographic protection, the security and privacy of IoT devices and

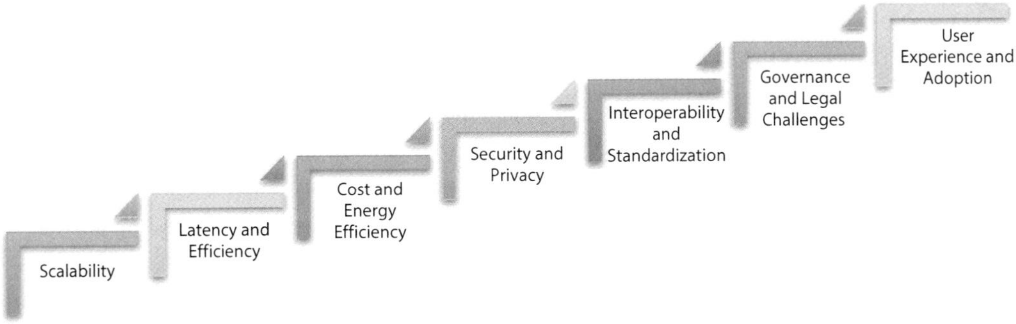

Figure 20.20 Critical challenges towards using blockchain-empowered IoT.

their data remain critical challenges. IoT devices are vulnerable to attacks, and the transparent nature of blockchain may expose sensitive information. Developing robust security measures, including secure authentication, encryption, and access control, is necessary to protect IoT devices and data in blockchain ecosystems.

Interoperability and Standardization: The lack of interoperability and standardization among different IoT devices, protocols, and blockchain platforms provide a major challenge. Achieving seamless integration and communication between diverse systems is important for realizing the full potential of blockchain-enabled IoT. Efforts to develop interoperability standards and protocols are necessary to foster collaboration and compatibility across platforms.

Governance and Legal Challenges: The decentralized nature of blockchain raises questions regarding governance, liability, and legal frameworks. Determining accountability and resolving disputes in a distributed network can be complex. Establishing governance models, addressing legal challenges, and ensuring compliance with regulations are important considerations for blockchain-enabled IoT deployments.

User Experience and Adoption: Blockchain technology can be complex for non-technical users, and the user experience (UX) of interacting with blockchain-enabled IoT systems needs improvement. Streamlining user interfaces, providing intuitive tools, and enhancing usability are important for wider adoption of blockchain in IoT applications.

Hence, these critical challenges will require collaborative efforts among industry stakeholders, researchers, and policymakers. Innovation in consensus algorithms, the development of interoperability standards, advancements in security measures, and user-centric design approaches will contribute to the successful integration of blockchain and IoT technologies.

20.24 Conclusion

Blockchain-Empowered Internet of Things (IoTs) platforms have the potential to revolutionize automation in various sectors. By combining the transparency, immutability, and decentralization of blockchain technology with the connectivity and data exchange capabilities of IoT devices, these platforms can enable secure, efficient, and automated processes across industries. The case studies conducted in this field have shed light on the diverse applications and benefits of blockchain-empowered IoT platforms. In the manufacturing sector, automation of the supply chain and product traceability can lead to improved efficiency, reduced costs, and enhanced quality control. In logistics and shipping, blockchain-enabled IoT platforms can automate tracking and authentication, ensuring secure and streamlined transportation of goods.

The renewable energy sector can use these platforms for automation in peer-to-peer energy trading and smart grid management, leading to optimized energy consumption and increased sustainability. Similarly, automation in agriculture through blockchain-enabled IoT platforms can enhance crop monitoring, irrigation, and livestock management, resulting in improved productivity and environmental impact. Healthcare stands to benefit from automation through secure and efficient patient monitoring, medical records management, and drug supply chain automation. In the retail sector, these platforms can streamline inventory management, product authentication, and supply chain optimization, enabling seamless customer experiences. Moreover, automation in sectors such as real estate, transportation, waste management, finance, and more can be facilitated by blockchain-empowered IoT platforms, bringing efficiency, transparency, and trust to various processes. However, challenges such as scalability, interoperability, and regulatory considerations must be carefully addressed when implementing these platforms. Collaboration among stakeholders, standardization efforts, and a clear understanding of legal and compliance frameworks will be important for successful deployment.

References

1. Kumar, R.L., Khan, F., Kadry, S., Rho, S., A survey on blockchain for industrial internet of things. *Alexandria Eng. J.*, *61*, 8, 6001–6022, 2022.
2. Čolaković, A. and Hadžialić, M., Internet of Things (IoT): A review of enabling technologies, challenges, and open research issues. *Comput. Netw.*, 144, 17–39, 2018.
3. Bodkhe, U., Tanwar, S., Parekh, K., Khanpara, P., Tyagi, S., Kumar, N., Alazab, M., Blockchain for industry 4.0: A comprehensive review. *IEEE Access*, 8, 79764–79800, 2020.
4. Viriyasitavat, W., Da Xu, L., Bi, Z., Pungpapong, V., Blockchain and internet of things for modern business process in digital economy—the state of the art. *IEEE Trans. Comput. Soc. Syst.*, 6, 6, 1420–1432, 2019.
5. Mylrea, M. and Gourisetti, S.N.G., Blockchain for smart grid resilience: Exchanging distributed energy at speed, scale and security, in: *2017 Resilience Week (RWS)*, 2017, September, IEEE, pp. 18–23.
6. Miglani, A., Kumar, N., Chamola, V., Zeadally, S., Blockchain for Internet of Energy management: Review, solutions, and challenges. *Comput. Commun.*, 151, 395–418, 2020.
7. Hemamalini, V., Mishra, A.K., Tyagi, A.K., Kakulapati, V., Artificial Intelligence–Blockchain-Enabled–Internet of Things-Based Cloud Applications for Next-Generation Society, in: *Automated Secure Computing for Next-Generation Systems*, pp. 65–82, 2024.
8. Khan, A.A., Laghari, A.A., Shaikh, Z.A., Dacko-Pikiewicz, Z., Kot, S., Internet of Things (IoT) security with blockchain technology: A state-of-the-art review. *IEEE Access*, 2022.
9. Mistry, I., Tanwar, S., Tyagi, S., Kumar, N., Blockchain for 5G-enabled IoT for industrial automation: A systematic review, solutions, and challenges. *Mech. Syst. Signal Process.*, 135, 106382, 2020.
10. Khan, W.Z., Rehman, M.H., Zangoti, H.M., Afzal, M.K., Armi, N., Salah, K., Industrial internet of things: Recent advances, enabling technologies and open challenges. *Comput. Electr. Eng.*, 81, 106522, 2020.
11. Xu, H., Yu, W., Griffith, D., Golmie, N., A survey on industrial Internet of Things: A cyber-physical systems perspective. *IEEE Access*, 6, 78238–78259, 2018.
12. Khan, A.A., Laghari, A.A., Shaikh, Z.A., Dacko-Pikiewicz, Z., Kot, S., Internet of Things (IoT) security with blockchain technology: A state-of-the-art review. *IEEE Access*, 2022.
13. Sisinni, E., Saifullah, A., Han, S., Jennehag, U., Gidlund, M., Industrial internet of things: Challenges, opportunities, and directions. *IEEE Trans. Ind. Inf.*, 14, 11, 4724–4734, 2018.
14. Zhao, S., Li, S., Yao, Y., Blockchain enabled industrial Internet of Things technology. *IEEE Trans. Comput. Soc. Syst.*, 6, 6, 1442–1453, 2019.
15. Dwivedi, S.K., Roy, P., Karda, C., Agrawal, S., Amin, R., Blockchain-based internet of things and industrial IoT: A comprehensive survey. *Secur. Commun. Netw.*, 2021, 1–21, 2021.
16. Jabbar, S., Lloyd, H., Hammoudeh, M., Adebisi, B., Raza, U., Blockchain-enabled supply chain: analysis, challenges, and future directions. *Multimedia Syst.*, 27, 787–806, 2021.
17. Yang, Q. and Wang, H., Privacy-preserving transactive energy management for IoT-aided smart homes via blockchain. *IEEE Internet Things J.*, 8, 14, 11463–11475, 2021.

18. Farouk, A., Alahmadi, A., Ghose, S., Mashatan, A., Blockchain platform for industrial healthcare: Vision and future opportunities. *Comput. Commun.*, 154, 223–235, 2020.
19. Liu, X.L., Wang, W.M., Guo, H., Barenji, A.V., Li, Z., Huang, G.Q., Industrial blockchain based framework for product lifecycle management in industry 4.0. *Rob. Comput. Integr. Manuf.*, 63, 101897, 2020.
20. Bhushan, B., Sahoo, C., Sinha, P., Khamparia, A., Unification of Blockchain and Internet of Things (BIoT): requirements, working model, challenges and future directions. *Wirel. Netw.*, 27, 55–90, 2021.
21. Atlam, H.F., Azad, M.A., Alzahrani, A.G., Wills, G., A Review of Blockchain in Internet of Things and AI. *Big Data Cogn. Comput.*, 4, 4, 28, 2020.
22. Chen, F., Xiao, Z., Cui, L., Lin, Q., Li, J., Yu, S., Blockchain for Internet of Things applications: A review and open issues. *J. Netw. Comput. Appl.*, 172, 102839, 2020.
23. Fu, B., Shu, Z., Liu, X., Blockchain enhanced emission trading framework in fashion apparel manufacturing industry. *Sustainability*, 10, 4, 1105, 2018.
24. Attaran, M. and Gunasekaran, A., Blockchain-enabled technology: the emerging technology set to reshape and decentralise many industries. *Int. J. Appl. Decis. Sci.*, 12, 4, 424–444, 2019.
25. Tosh, D.K., Shetty, S., Foytik, P., Njilla, L., Kamhoua, C.A., Blockchain-empowered secure internet-of-battlefield things (iobt) architecture, in: *MILCOM 2018-2018 IEEE Military Communications Conference (MILCOM)*, 2018, October, IEEE, pp. 593–598.
26. Pranto, T.H., Noman, A.A., Mahmud, A., Haque, A.B., Blockchain and smart contract for IoT enabled smart agriculture. *Peer J. Comput. Sci.*, 7, e407, 2021.
27. Deebak, B.D. and Fadi, A.T., A robust and distributed architecture for 5G-enabled networks in the smart blockchain era. *Comput. Commun.*, 181, 293–308, 2022.
28. Viriyasitavat, W., Anuphaptrirong, T., Hoonsopon, D., When blockchain meets Internet of Things: Characteristics, challenges, and business opportunities. *J. Ind. Inf. Integr.*, 15, 21–28, 2019.
29. Ferrag, M.A., Derdour, M., Mukherjee, M., Derhab, A., Maglaras, L., Janicke, H., Blockchain technologies for the internet of things: Research issues and challenges. *IEEE Internet Things J.*, 6, 2, 2188–2204, 2018.
30. Tyagi, A.K. and Tiwari, S., The Future of Artificial Intelligence in Blockchain Applications, in: *Machine Learning Algorithms Using Scikit and TensorFlow Environments*, IGI Global, 2024, doi: 10.4018/978-1-6684-8531-6.ch018.
31. Nair, M.M. and Tyagi, A.K., Chapter 11 - AI, IoT, blockchain, and cloud computing: The necessity of the future, in: *Distributed Computing to Blockchain*, R. Pandey, S. Goundar, S. Fatima (Eds.), pp. 189–206, Academic Press, 2023, ISBN 9780323961462, https://doi.org/10.1016/B978-0-323-96146-2.00001-2.
32. Tyagi, A.K., Kukreja, S., Richa, and Sivakumar, P., Role of blockchain technology in smart era: A review on possible smart applications. *J. Inf. Knowl. Manag.*, February 2024. https://doi.org/10.1142/S0219649224500321.
33. Nair, M.M. and Tyagi, A.K., Blockchain technology for next-generation society: current trends and future opportunities for smart era, in: *Blockchain Technology for Secure Social Media Computing*, 2023.
34. Tyagi, A.K. and Richa, Smart Manufacturing Using Internet of Things, Artificial Intelligence, and Digital Twin Technology, in: *Global Perspectives on Robotics and Autonomous Systems: Development and Applications*, 2023.

35. Tyagi, A.K., Lakshmi, P.R., Mishra, A.K., Balamurugan, G., Industry 5.0: Potentials, Issues, Opportunities, and Challenges for Society 5.0, in: *Privacy Preservation of Genomic and Medical Data*, vol. 17, pp. 409–32, 2023 Oct.
36. Tyagi, A.K., Mishra, A.K., Vedavathi, N., Kakulapati, V., Sajidha, S.A., Chapter 21 Futuristic Technologies for Smart Manufacturing, in: *Research Statement and Vision for the Future*, Tyagi, A.K. (Ed.), 2023, https://doi.org/10.1002/9781394213948.ch21.
37. Tyagi, A.K., Kumari, S., Chidambaram, N., Sharma, A., Engineering Applications of Blockchain in This Smart Era, in the book: *Enhancing Medical Imaging with Emerging Technologies*, 2024. DOI: 10.4018/979-8-3693-5261-8.ch011.

21

Digital Twin-Enabled Smart Manufacturing: Challenges and Future Directions

M. Maranco[1], M. Sivakumar[1*], N. Krishnaraj[1], Kashyapa Abhiram Ivaturi[1] and Nidhya R.[2]

[1]Department of Networking and Communications, Faculty of Engineering and Technology, SRM Institute of Science and Technology, Kattankulathur, Tamil Nadu, India
[2]Department of Computer Science and Engineering, Madanapalle Institute of Technology, Madanapalle, Andhra Pradesh, India

Abstract
Digital twin technology represents a revolutionary leap forward in smart manufacturing, fundamentally reshaping traditional methodologies with its transformative capabilities. Digital twins are virtual counterparts that accurately simulate physical assets and processes, acting as a dynamic bridge between the tangible world of machinery and the digital domain of data analytics and simulation. Through the amalgamation of cutting-edge technologies like IoT sensors, AI algorithms, and simulation software, digital twins empower manufacturers with unparalleled insights and control over their operations. A pivotal advantage of digital twins lies in their ability to facilitate predictive maintenance, a cornerstone of contemporary manufacturing practices. Through the ongoing monitoring of equipment performance and condition in real time, manufacturers are able to predict maintenance requirements, proactively resolve problems, and prevent expensive periods of inactivity. This proactive approach represents a notable departure from traditional reactive maintenance methods, heralding a significant shift towards more efficient and cost-effective operations. Furthermore, digital twins serve as invaluable instruments for optimizing production processes and enhancing overall operational efficiency. Through real-time data collection and analysis, manufacturers can pinpoint inefficiencies, streamline workflows, and allocate resources more judiciously. This not only enhances productivity but also contributes to higher-quality output and reduced waste, thereby fostering a competitive edge in an increasingly dynamic market landscape. Despite their immense potential, the widespread adoption of digital twins is not without its challenges. Issues such as data security, interoperability between disparate systems, and the intricacies of implementation pose formidable obstacles for manufacturers. However, through meticulous planning, investment in robust technology infrastructure, and collaborative efforts with industry partners, these challenges can be surmounted, unlocking the full potential of digital twin technology in the realm of smart manufacturing.

Keywords: Smart manufacturing, artificial intelligence, digital twins, data analytics, deep learning

**Corresponding author*: siva.mcs@gmail.com

Amit Kumar Tyagi, Shrikant Tiwari, Senthil Kumar Arumugam and Avinash Kumar Sharma (eds.) Artificial Intelligence-Enabled Digital Twin for Smart Manufacturing, (479–504) © 2024 Scrivener Publishing LLC

21.1 Introduction

Digital twin technology stands as a groundbreaking innovation within smart manufacturing, revolutionizing traditional approaches with its transformative capabilities. Digital twins are virtual counterparts that accurately replicate actual assets and processes, acting as a dynamic connection between the concrete world of machinery and the digital realm of data analytics and simulation. Through the integration of advanced technologies such as IoT sensors, AI algorithms, and simulation software, digital twins empower manufacturers with unparalleled insights and control over their operations. Digital twins offer a significant benefit by enabling predictive maintenance, which is a crucial element of contemporary manufacturing. By continuously monitoring equipment performance and condition in real time, manufacturers can anticipate maintenance requirements, proactively address issues, and prevent costly downtime. This proactive approach represents a significant departure from traditional reactive maintenance methods, marking a pivotal shift towards more efficient and cost-effective operations.

Moreover, digital twins serve as invaluable tools for optimizing production processes and enhancing overall operational efficiency. Through real-time data collection and analysis, manufacturers can identify inefficiencies, streamline workflows, and allocate resources more effectively. This not only improves productivity but also contributes to higher-quality output and reduced waste, thereby driving competitive advantage in an increasingly dynamic market. Despite the immense potential of digital twins, their widespread adoption is not without challenges. Issues such as data security, interoperability between different systems, and the complexity of implementation pose significant hurdles for manufacturers. However, with careful planning, investment in technology infrastructure, and collaboration with industry partners, these challenges can be overcome, unlocking the full potential of digital twin technology in smart manufacturing.

21.2 Related Works

"Digital Twin Technology in Manufacturing: A Literature Review and Research Agenda" [1] offers a thorough examination of digital twin applications in manufacturing, tracing its evolution, defining concepts, and detailing practical implementations. The paper elucidates the role of digital twins in facilitating real-time monitoring, predictive maintenance, and process optimization in the context of smart industrial environments. Additionally, it identifies key challenges such as data integration and security, providing a robust research agenda to propel the field forward. "Digital Twin in Manufacturing: A Comprehensive Survey" [2] provides an exhaustive overview of digital twin applications across various manufacturing domains. It delves into enabling technologies like the Internet of Things (IoT) and big data analytics while addressing challenges related to data management and model accuracy. Moreover, the paper outlines future research directions, offering valuable insights for further exploration and development in the field. "Digital Twin Technologies and Their Impact on Smart Manufacturing Processes: A Review" [3] explores the transformative potential of digital twins in enhancing smart manufacturing processes. Emphasizing benefits such as virtual simulation and predictive analytics, the paper examines real-world case studies across diverse manufacturing sectors to underscore practical implications and industry adoption.

"Digital Twin Technologies for Smart Manufacturing Systems: A Review" [4] delves into digital twin architectures, modeling techniques, and integration frameworks tailored for smart manufacturing systems. By addressing challenges such as standardization and cybersecurity, the paper lays out a roadmap for future research aimed at overcoming these hurdles and fostering widespread adoption. "Digital Twins in Manufacturing: A Review and Future Directions" [5] offers a forward-looking perspective on the adoption and implementation of digital twins in manufacturing. It explores emerging trends, potential benefits, and future research directions, drawing from case studies and industrial applications to provide practical insights into the transformative role of digital twins.

"Digital Twin-Enabled Smart Manufacturing Systems: A Review" [6] delves into the integration of digital twins within smart manufacturing systems, highlighting their potential to enhance productivity, quality, and flexibility. The paper addresses both the opportunities and challenges associated with digital twin-enabled systems, providing valuable insights for researchers and practitioners alike.

"Digital Twins for Smart Manufacturing: A Review" [7] provides a comprehensive overview of digital twins' role in smart manufacturing, emphasizing their ability to drive predictive maintenance and quality improvement initiatives. Drawing from a range of case studies, the paper illustrates how digital twins enable proactive decision-making and process optimization in manufacturing environments. "Digital Twin-Driven Manufacturing: A Survey" [8] presents a survey-based exploration of digital twin-driven manufacturing, examining various applications, challenges, and future research directions. By synthesizing existing literature and industry practices, the paper offers a holistic view of digital twin-driven approaches in manufacturing. "Digital Twins in Smart Manufacturing: A Review" [9] provides an in-depth examination of digital twin applications in smart manufacturing, showcasing their benefits across different stages of the production process. Through case studies and industry examples, the paper highlights how digital twins enable enhanced monitoring, maintenance, and optimization strategies.

"Digital Twin in Manufacturing: A Systematic Literature Review" [10] offers a systematic literature review on digital twin applications in manufacturing. By synthesizing existing research, the paper identifies key trends, challenges, and future research directions, providing a comprehensive overview of the field. "Digital Twin and Its Role in Industry 4.0: A Review" [11] discusses the role of digital twins in Industry 4.0, highlighting benefits, challenges, and future directions. The paper provides insights into how digital twins contribute to smart manufacturing, predictive maintenance, and process optimization. "Digital Twin Applications in Smart Manufacturing: A Review" [12] explores digital twin applications in smart manufacturing, emphasizing benefits like process optimization and predictive maintenance. Through case studies and examples, the paper illustrates how digital twins enhance manufacturing efficiency and flexibility. "Review of Digital Twin Technology in Smart Manufacturing: Challenges and Opportunities" [13] addresses challenges and opportunities in digital twin technology in smart manufacturing. The paper discusses issues such as data integration, interoperability, and cybersecurity, offering insights into potential solutions and future research directions.

"The State of Digital Twins in Smart Manufacturing: A Review" [14] provides an overview of digital twins in smart manufacturing, discussing their current state, challenges, and future prospects. The paper examines key applications, benefits, and limitations of digital twins, highlighting their potential to revolutionize manufacturing processes. "Digital Twins for Smart Manufacturing: A Systematic Literature Review and Future Research Directions" [15] offers a systematic literature review on digital twins for smart manufacturing. The paper synthesizes existing research, identifies gaps, and proposes future research directions, providing valuable insights for researchers and practitioners in the field.

"Digital Twin Technology: Opportunities and Challenges for Smart Era's Applications" [19] the authors likely provide an overview of existing research and developments in digital twin technology. This includes discussions on previous studies involving digital twin applications across various domains, such as manufacturing, healthcare, and urban planning. They may highlight challenges and limitations faced by earlier approaches, identify emerging trends and innovations in the field, compare digital twin technology with similar technologies like the Internet of Things (IoT) or Cyber-Physical Systems (CPS), and pinpoint areas requiring further research. By contextualizing their work within this landscape, the authors demonstrate the significance of their study in addressing gaps and advancing the field of digital twin technology.

In their chapter titled "Smart Manufacturing Using IoT, AI, and Digital Twin Technology" [23] in the book "Global Perspectives on Robotics and Autonomous Systems: Development and Applications". They explore the integration of Internet of Things (IoT), Artificial Intelligence (AI), and Digital Twin Technology in manufacturing. The comparison of related works on digital twin technology in manufacturing is presented in Table 21.1. Through a blend of literature review and case studies, they analyze the practical implementations and benefits of these technologies, offering insights into enhancing efficiency and decision-making in manufacturing operations. This contribution serves as a valuable resource for practitioners and researchers looking to leverage IoT, AI, and digital twin technology in smart manufacturing.

Table 21.1 Comparison of related works.

Paper title & year	Focus	Methodology	Key findings/ contributions
Digital Twin Technology in Manufacturing: A Literature Review and Research Agenda by Tao Yu *et al.* (2018)	Evolution, concepts, implementations, research agenda	Thorough examination, identification of challenges, proposal of research agenda	AI-driven predictive maintenance strategies mitigate downtime costs and enhance equipment availability.
Digital Twin in Manufacturing: A Comprehensive Survey by Zhengxi Li *et al.* (2020)	Overview of applications, enabling technologies, challenges, future directions	Exhaustive survey, insights into enabling technologies, identification of challenges	Sustainable manufacturing IoT in industrial plants through smart devices, AI, and deep learning.

(Continued)

Table 21.1 Comparison of related works. (*Continued*)

Paper title & year	Focus	Methodology	Key findings/ contributions
Digital Twin Technologies and Their Impact on Smart Manufacturing Processes: A Review by Yuanchun Jiang *et al.* (2020)	Transformative potential, benefits, real-world case studies	Exploration of benefits, insights from case studies, practical implications	Insights into prevalent ML models, IoT sensors, and industries adopting predictive maintenance.
Digital Twin Technologies for Smart Manufacturing Systems: A Review by SabriSkhiri *et al.* (2020)	Architectures, modeling, integration frameworks, future research	Analysis of architectures, addressing challenges, roadmap for future research	Insights into recent trends and developments in ML-based predictive maintenance for Industry 4.0.
Digital Twins in Manufacturing: A Review and Future Directions by Junyang Shan *et al.* (2021)	Adoption, implementation, emerging trends, future research	Forward-looking perspective, exploration of emerging trends, future research implications	Lack of a comprehensive framework integrating data-driven and knowledge-driven techniques in PdM.
Digital Twin-Enabled Smart Manufacturing Systems: A Review by Wei Gao *et al.* (2019)	Integration, benefits, challenges, potential applications	Insights into integration, benefits, challenges, and potential applications	Real-time self-retraining functionality based on a double-oriented evaluation objective.
Digital Twins for Smart Manufacturing: A Review by Jing Zhang *et al.* (2020)	Role in smart manufacturing, predictive maintenance, quality improvement	Comprehensive overview, emphasis on predictive maintenance and quality improvement	Effectiveness of data-driven predictive maintenance in identifying potential failures and preventing production stops.
Digital Twin-Driven Manufacturing: A Survey by Yan Luo *et al.* (2021)	Survey-based exploration, applications, challenges, future directions	Holistic view, insights from surveys, future research directions	Insights into key contributors, institutes, countries, and collaborative networks in AI-based PdM4.0.

(*Continued*)

Table 21.1 Comparison of related works. (*Continued*)

Paper title & year	**Focus**	**Methodology**	**Key findings/ contributions**
Digital Twins in Smart Manufacturing: A Review by Chunyang Yang *et al.* (2020)	Applications, benefits, case studies, monitoring, maintenance	In-depth examination, case studies, practical insights	Successful integration of a fault detection and diagnosis system into a real-time industrial production system.
Digital Twin in Manufacturing: A Systematic Literature Review by GuodongGao *et al.* (2021)	Systematic review, applications, challenges, future research	Synthesis of existing research, identification of trends, future research directions	Formulated approach for systematic extraction of useful information for decision support in predictive maintenance.
Digital Twin and Its Role in Industry 4.0: A Review by Hsin-Hung Wu *et al.* (2020)	Role in Industry 4.0, benefits, challenges, future directions	Exploration of Industry 4.0, implications for manufacturing	Integration of Digital Twin pipeline into the industry for predictive maintenance and downtime reduction.
Digital Twin Applications in Smart Manufacturing: A Review by Shichao Zhu *et al.* (2020)	Applications, benefits, process optimization, predictive maintenance	Illustration of benefits, case studies, examples	Feasibility of AI in predicting railcar wheel bearing conditions and optimal maintenance times.
Review of Digital Twin Technology in Smart Manufacturing: Challenges and Opportunities by Mohammed Alipour *et al.* (2020)	Challenges, opportunities, data integration, interoperability, cybersecurity	Identification of challenges, insights into potential solutions	Real-time warnings about operational risks, optimization of maintenance schedules, and reduced service costs.
The State of Digital Twins in Smart Manufacturing: A Review by Wei Wang *et al.* (2020)	Current state, challenges, prospects, applications, limitations	Overview of current state, discussion of challenges and prospects	Insights into recent developments and trends in PdM for Induction Motors using MAS and DL.

(*Continued*)

Table 21.1 Comparison of related works. (*Continued*)

Paper title & year	**Focus**	**Methodology**	**Key findings/ contributions**
Digital Twins for Smart Manufacturing: A Systematic Literature Review and Future Research Directions by Hui Pan *et al.* (2021)	Systematic review, future research directions, gaps, synthesis	Synthesis of literature, identification of gaps, future research directions	Overview of trends in intelligent sensors for predictive maintenance in smart factories.
Hosseinzadeh *et al.*, 2023	Machine failure prediction, Defect detection models, Machine Learning	Evaluation of defect detection models using synthetic predictive maintenance dataset	Deep Forest and Gradient Boosting algorithms achieved relatively high accuracy.
"Digital Twin Technology: Opportunities and Challenges for Smart Era's Applications" (2023)	The paper explores the opportunities and challenges associated with the application of digital twin technology in various domains within the smart era context.	The authors likely conduct a literature review, supplemented by case studies or examples, to analyze opportunities and challenges associated with digital twin technology in diverse domains within the smart era.	
	The paper highlights opportunities and challenges of digital twin technology across sectors, providing insights to advance its adoption in the smart era.		

(*Continued*)

Table 21.1 Comparison of related works. (*Continued*)

Paper title & year	Focus	Methodology	Key findings/ contributions
"Smart Manufacturing Using Internet of Things, Artificial Intelligence, and Digital Twin Technology" by Amit Kumar Tyagi and Richa (2023)	The chapter delves into the integration of Internet of Things (IoT), Artificial Intelligence (AI), and Digital Twin Technology in smart manufacturing processes.	The authors likely employ a combination of literature review and case studies to analyze the utilization and benefits of IoT, AI, and digital twin technology in smart manufacturing applications. They may also discuss practical implementations and industry examples.	The chapter explores the synergistic effects of IoT, AI, and digital twin technology in enhancing efficiency, productivity, and decision-making in smart manufacturing. It highlights the transformative potential of these technologies and provides insights for practitioners and researchers in the field.

21.3 Current Landscape of Digital Twin Adoption

21.3.1 Present Used Locations

Digital twin technology has been increasingly implemented across various industries and sectors, enabling organizations to create virtual replicas of physical assets, processes, or systems as shown in Figure 21.1. Below is a description of present implementing industries:

- Manufacturing: Digital twins are extensively used in manufacturing for predictive maintenance, optimizing production processes, and quality control.

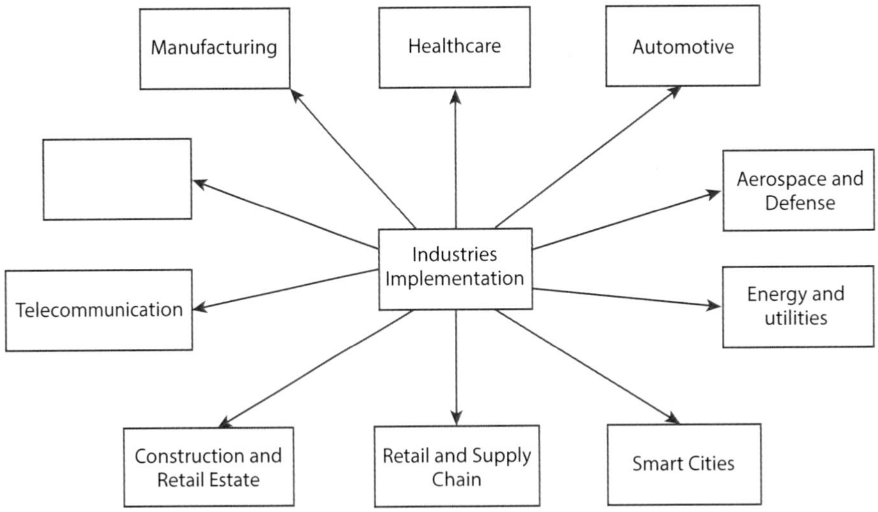

Figure 21.1 Industrial implementation.

By creating virtual replicas of machines and production lines, manufacturers can simulate various scenarios, identify potential issues, and optimize performance in real-time.

- Healthcare: Digital twins are used in healthcare for personalized medicine, patient monitoring, and medical device simulation. For instance, digital twins of organs or entire patients can help doctors predict the effects of treatments, simulate surgeries, and optimize healthcare delivery.
- Automotive: Automotive companies leverage digital twins to design, test, and simulate vehicles and their components. Digital twins enable automotive manufacturers to analyze vehicle performance, predict maintenance needs, and enhance safety features.
- Aerospace and Defense: In the aerospace and defense sector, digital twins are used for designing, testing, and monitoring aircraft, spacecraft, and defense systems. Digital twins help engineers simulate flight conditions, predict maintenance requirements, and improve overall performance.
- Energy and Utilities: Digital twins are applied in energy production, transmission, and distribution systems to optimize operations, improve asset management, and enhance safety. For example, digital twins of power plants or utility grids help operators monitor performance, predict failures, and optimize energy distribution.
- Smart Cities: Digital twins are used to create virtual models of entire cities or specific urban systems such as transportation networks, water management systems, or energy grids. Smart city planners utilize digital twins to simulate and optimize infrastructure development, resource allocation, and emergency response [18].
- Retail and Supply Chain: In retail and supply chain management, digital twins are employed to optimize inventory management, logistics, and store operations. Retailers use digital twins to simulate customer behavior, predict demand, and optimize store layouts.
- Construction and Real Estate: Digital twins are utilized in construction projects for design visualization, project planning, and facility management. Real estate developers use digital twins to simulate building performance, optimize energy usage, and enhance occupant comfort.
- Telecommunications: Telecommunication companies use digital twins to model network infrastructure, predict network congestion, and optimize resource allocation. Digital twins enable telecom operators to simulate network upgrades, plan for future capacity needs, and enhance service reliability.
- Environmental Monitoring: Digital twins are utilized for the purpose of monitoring and managing the environment, namely in tasks such as modeling climatic trends, simulating ecological systems, and forecasting the consequences of human actions on the environment.

21.3.2 Benefits Associated with Adopting Digital Twins in Manufacturing Processes

The benefits of using digital twin technologies as shown in Figure 21.2 are as follows:

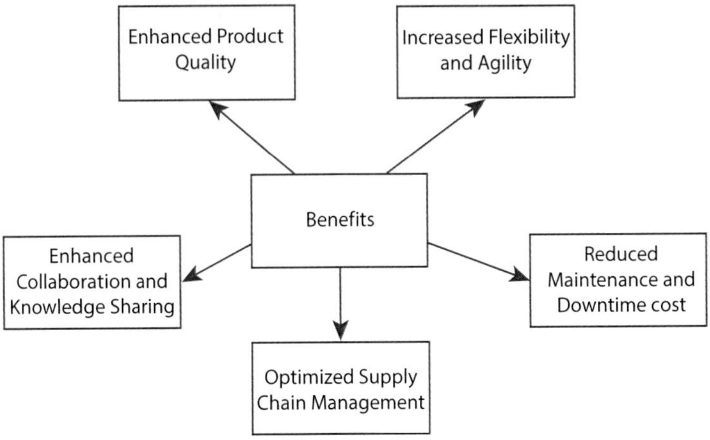

Figure 21.2 Benefits in using digital twin technologies.

21.3.2.1 *Enhanced Product Quality*

- Virtual Prototyping: Digital twins enable manufacturers to simulate product designs in a virtual environment, allowing for thorough testing and validation before physical prototypes are built. This iterative process helps identify and rectify design flaws early on, leading to higher-quality products.
- Real-Time Monitoring: By continuously monitoring production parameters and product performance, digital twins facilitate early detection of defects or deviations from quality standards. This proactive approach enables prompt corrective actions, ensuring consistent product quality.

21.3.2.2 *Increased Flexibility and Agility*

- Simulation and Optimization: Digital twins allow manufacturers to simulate various scenarios and optimize production processes in real-time. This flexibility enables rapid adjustments to changing market demands, production requirements, or supply chain disruptions, enhancing agility and responsiveness.
- Scenario Planning: Manufacturers can use digital twins to conduct "what-if" analyses and scenario planning, evaluating the potential impact of different decisions or strategies on production outcomes. This foresight enables proactive decision-making and risk mitigation, fostering greater resilience in manufacturing operations.

21.3.2.3 *Reduced Downtime and Maintenance Costs*

- Predictive Maintenance: Digital twins enable predictive maintenance by continuously monitoring the condition and performance of equipment in real-time. By analyzing data trends and identifying early warning signs of impending failures, manufacturers can schedule maintenance activities proactively, minimizing unplanned downtime and reducing maintenance costs.

- Condition-Based Monitoring: Digital twins facilitate condition-based monitoring of equipment health, allowing manufacturers to detect abnormalities or signs of wear and tear before they escalate into costly breakdowns. This predictive approach maximizes equipment uptime and prolongs asset lifespan, resulting in significant cost savings.

21.3.2.4 *Optimized Supply Chain Management*

- End-to-End Visibility: Digital twins provide comprehensive visibility across the entire supply chain, from raw material sourcing to product delivery. By integrating data from suppliers, production facilities, logistics partners, and customers, manufacturers can gain insights into inventory levels, production schedules, and demand forecasts, enabling more efficient supply chain management.
- Demand Forecasting: Leveraging advanced analytics and predictive modeling, digital twins help manufacturers forecast demand more accurately. By anticipating customer needs and market trends, manufacturers can optimize inventory levels, production schedules, and distribution channels, reducing inventory carrying costs and minimizing stockouts or excess inventory.

21.3.2.5 *Enhanced Collaboration and Knowledge Sharing*

- Cross-Functional Collaboration: Digital twins facilitate collaboration across different departments, including design, engineering, production, and maintenance. By providing a common platform for data sharing and collaboration, digital twins break down silos and foster cross-functional teamwork, leading to improved communication, coordination, and decision-making.
- Knowledge Capture and Transfer: Digital twins serve as a repository of operational knowledge and historical data, capturing insights and best practices accumulated over time. By preserving and sharing this knowledge, manufacturers can enhance organizational learning, accelerate innovation, and empower employees with valuable insights for continuous improvement.

21.3.3 Case Studies Illustrating Real-Life Examples of Digital Twin-Enabled Smart Manufacturing Initiatives

Below is a description of case studies:

- Boeing Dreamliner Production Optimization:
  Boeing, a major aerospace manufacturer, utilizes digital twin technology in the production of its Dreamliner aircraft. By creating digital twins of manufacturing facilities, Boeing can simulate production processes, optimize workflow layouts, and improve production efficiency. Boeing use digital twins to monitor equipment performance, detect issues, and implement proactive maintenance strategies, thus reducing operating disruptions and enhancing product quality.

- ABB (Asea Brown Boveri) Robotics Flexible Production Lines:
  ABB Robotics employs digital twins to design and optimize flexible production lines for various manufacturing industries. By creating digital replicas of production environments and robotic systems, ABB can simulate different manufacturing scenarios, optimize robot programming, and improve production flexibility. Digital twins also enable ABB to monitor production metrics in real-time, identify opportunities for process improvement, and adapt to changing market demands quickly [25].
- Siemens Smart Factory Initiative:
  Siemens has implemented a smart factory initiative leveraging digital twin technology across its manufacturing facilities. By creating digital twins of production processes, Siemens can simulate production scenarios, optimize resource allocation, and improve manufacturing efficiency. Digital twins also enable Siemens to monitor equipment health, predict maintenance needs, and minimize downtime, resulting in cost savings and improved operational performance.
- Toyota Production System Enhancement:
  Toyota, a leading automotive manufacturer, utilizes digital twins to enhance its renowned production system. By creating digital twins of manufacturing plants and assembly lines, Toyota can simulate production processes, optimize workflow layouts, and improve resource utilization. Digital twins also enable Toyota to monitor equipment performance, detect anomalies, and implement predictive maintenance strategies, ensuring continuous improvement and operational excellence.
- Bosch Connected Manufacturing Solutions:
  Bosch, a global technology company, implements digital twin-enabled connected manufacturing solutions across its production facilities. By creating digital twins of manufacturing equipment and processes, Bosch can monitor production in real-time, identify inefficiencies, and optimize resource allocation. Digital twins also enable Bosch to simulate production scenarios, test new technologies, and streamline decision-making processes, resulting in improved productivity and product quality.
- General Motors Factory of the Future:
  General Motors (GM) is developing a "Factory of the Future" concept utilizing digital twin technology. By creating digital twins of manufacturing facilities and processes, GM aims to optimize production workflows, reduce costs, and improve flexibility. Digital twins also enable GM to simulate production scenarios, identify potential bottlenecks, and optimize resource allocation, ensuring efficient and agile manufacturing operations.
- Honeywell Process Optimization in Petrochemical Industry:
  Honeywell implements digital twin technology for process optimization in the petrochemical industry [21]. By creating digital twins of plant operations and equipment, Honeywell can simulate process conditions, optimize control strategies, and improve operational efficiency. Digital twins also enable Honeywell to monitor equipment health, detect anomalies, and implement predictive maintenance strategies, ensuring safe and reliable plant operations.

- Nokia 5G Smart Manufacturing Networks:
  Nokia deploys digital twin-enabled smart manufacturing networks to optimize production processes in various industries. By creating digital twins of manufacturing facilities and equipment, Nokia can monitor production metrics in real-time, optimize network connectivity, and ensure high reliability and low latency for industrial applications. Digital twin-enabled smart manufacturing networks enable Nokia's customers to achieve improved productivity, quality, and flexibility in their operations.
- Lockheed Martin Aerospace Manufacturing Innovation:
  Lockheed Martin utilizes digital twin technology to drive innovation in aerospace manufacturing. By creating digital twins of aircraft components and production processes, Lockheed Martin can simulate manufacturing scenarios, optimize production schedules, and improve product quality. Digital twins also enable Lockheed Martin to monitor equipment performance, detect issues, and implement predictive maintenance strategies, ensuring on-time delivery and cost-effective production.
- Mercedes-Benz Digital Factory Transformation:
  Mercedes-Benz is undergoing a digital factory transformation leveraging digital twin technology. By creating digital twins of manufacturing facilities and processes, Mercedes-Benz aims to optimize production workflows, improve flexibility, and enhance product quality. Digital twins enable Mercedes-Benz to simulate production scenarios, identify optimization opportunities, and implement predictive maintenance strategies, ensuring efficient and agile manufacturing operations aligned with Industry 4.0 principles [22].

21.4 Challenges in Digital Twin Implementations

21.4.1 Key Challenges in Implementing Digital Twin Technologies

21.4.1.1 Data Integration and Quality

Digital twins rely on a plethora of data sources, including sensors, IoT devices, and historical records, to create accurate simulations of manufacturing processes. Nevertheless, the process of merging these diverse data streams into a cohesive digital representation can be intricate as a result of variations in formats, protocols, and data standards. Moreover, ensuring the quality and consistency of data is paramount for precise analysis and decision-making within the digital twin environment. Challenges often arise from inconsistencies in data formats, missing or incomplete data, and the need to harmonize data from various sources. Addressing these challenges as shown in Figure 21.3 requires robust data integration frameworks, data cleansing techniques, and continuous monitoring to maintain data quality standards.

21.4.1.2 Complexity and Scalability

Manufacturing processes are inherently complex, involving interconnected systems, equipment, and workflows. Building digital twins that accurately replicate this complexity

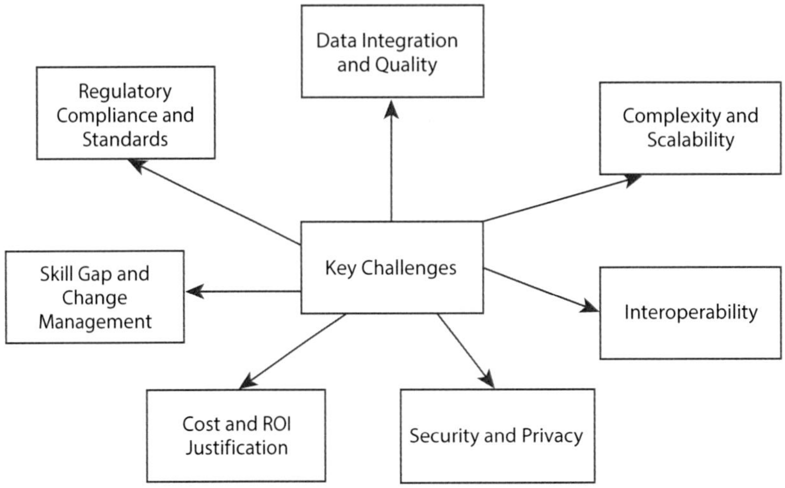

Figure 21.3 Key challenges during implementation.

requires sophisticated modeling techniques and computational resources. Additionally, as manufacturing facilities and supply chains expand, scaling digital twins to encompass larger systems while maintaining performance becomes a significant challenge. Balancing the need for detailed simulations with computational efficiency requires optimization algorithms, distributed computing architectures, and cloud-based infrastructure to support scalable digital twin deployments.

21.4.1.3 Interoperability

Manufacturing environments often feature heterogeneous systems and equipment from different vendors, each utilizing proprietary protocols and interfaces. Achieving interoperability between these systems and digital twin platforms is essential for seamless data exchange and integration. However, disparate standards and technologies hinder interoperability efforts, leading to integration complexities and data silos. To tackle this difficulty, it is necessary to create and implement open standards, interoperability frameworks, and middleware solutions that ease communication between different systems and allow for data interoperability across the manufacturing ecosystem.

21.4.1.4 Security and Privacy

Digital twins generate and utilize vast amounts of sensitive data related to manufacturing processes, equipment performance, and product designs [17]. Protecting this data from cyber threats, unauthorized access, and data breaches is critical to maintaining operational integrity and safeguarding intellectual property. Moreover, compliance with privacy regulations, such as GDPR and CCPA, adds another layer of complexity, requiring stringent data protection measures and privacy controls. Implementing robust cybersecurity measures, including encryption, access controls, and threat detection systems, is essential to mitigate security risks and ensure data confidentiality, integrity, and availability within digital twin environments.

21.4.1.5 Cost and ROI Justification

Implementing digital twins involves significant upfront investments in sensors, data infrastructure, software development, and talent acquisition. Manufacturers must carefully evaluate the return on investment (ROI) and justify these expenses to stakeholders. Demonstrating tangible benefits, such as improved productivity, reduced downtime, and enhanced product quality, is essential for securing executive buy-in and long-term funding. However, accurately quantifying the ROI of digital twins can be challenging due to the complex nature of manufacturing operations and the indirect impact of digital twin technologies on business outcomes.

21.4.1.6 Skill Gap and Change Management

Leveraging digital twins requires specialized skills in data analytics, simulation, modeling, and domain expertise in manufacturing processes. However, there is a shortage of talent with expertise in these areas, making it challenging for organizations to build and maintain digital twin capabilities. Additionally, integrating digital twins into existing workflows and organizational culture requires effective change management strategies to overcome resistance and ensure adoption. To tackle the talent gap, it is necessary to allocate resources towards training programs, enhancing the skills of current employees, and cultivating a work environment that promotes creativity and ongoing learning to facilitate digital twin projects.

21.4.1.7 Regulatory Compliance and Standards

Manufacturers are required to comply with several industry regulations and standards that govern the quality, safety, and environmental sustainability of their products. Developing digital twins that align with these regulations and standards while remaining agile to adapt to evolving requirements can be challenging. Compliance efforts encompass data governance, regulatory reporting, audit trails, and documentation to ensure transparency and accountability within digital twin environments. Additionally, staying abreast of changing regulations and industry standards requires proactive monitoring and collaboration with regulatory bodies and industry associations to ensure compliance and mitigate legal risks associated with non-compliance. To tackle these difficulties, a comprehensive approach is necessary, involving cooperation among technology providers, manufacturers, regulatory agencies, and other relevant parties. Industry initiatives focusing on standardization, data sharing protocols, cybersecurity best practices, and workforce development can facilitate the widespread adoption of digital twins in manufacturing. Additionally, pilot projects and case studies demonstrating successful implementations can help build confidence and accelerate adoption across the industry.

21.4.2 Technical Challenges in Implementing Digital Twin Technologies

21.4.2.1 Data Integration

Implementing digital twin technology presents several technical challenges as shown in Figure 21.4 that organizations must overcome for successful deployment. One of the primary hurdles is data integration. Digital twins rely on aggregating data from various sources, including

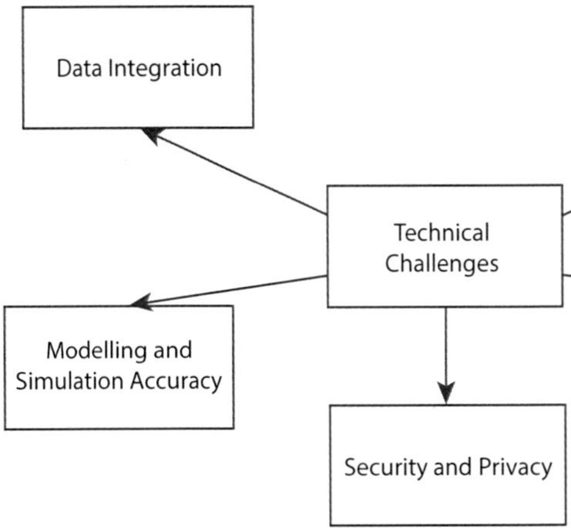

Figure 21.4 Technical challenges during implementation.

sensors, IoT devices, CAD models, and operational systems. Managing this diverse range of data types while ensuring consistency and synchronization across sources is critical. Real-time data streaming adds another layer of complexity, necessitating robust mechanisms for data ingestion to handle the high volume and velocity of incoming data streams.

21.4.2.2 Interoperability

Interoperability is another significant challenge in digital twin implementations. These systems often involve components from different vendors, each utilizing different standards and protocols. Ensuring seamless interoperability between these components is crucial for the digital twin to function effectively. Semantic interoperability, which ensures that data exchanged between different components is semantically meaningful and accurately interpreted, is also essential. Additionally, integrating digital twin systems with existing legacy systems and enterprise IT infrastructure while maintaining interoperability presents further complexities.

21.4.2.3 Scalability

Scalability is a key consideration as organizations scale up their digital twin deployments. Managing the increasing volume of data generated by more extensive and more complex physical systems requires robust computational infrastructure and scalable processing capabilities. Elasticity becomes crucial, allowing digital twin implementations to dynamically scale up or down to accommodate varying workloads and requirements.

21.4.2.4 Security and Privacy

Security and privacy considerations provide substantial obstacles. Ensuring the security of sensitive data that is communicated and stored in digital twin systems is of utmost importance in order to prevent unwanted access, data breaches, and cyber-attacks. In addition, the task of guaranteeing adherence to legislation like GDPR and managing privacy issues over personal and confidential data introduces an additional level of intricacy.

21.4.2.5 *Modeling and Simulation Accuracy*

Maintaining the accuracy and reliability of digital twin models and simulations is essential. Ensuring model fidelity—that the digital twin accurately reflects the behavior and characteristics of the physical system it represents—is critical for making informed decisions based on the twin's insights. Real-time simulation capabilities must also be reliable, providing accurate predictions and insights into system behavior while responding promptly to changes in the physical environment. Addressing these technical challenges requires a holistic approach, encompassing robust data management strategies, interoperable systems architectures, scalable infrastructure, stringent security measures, and rigorous model validation and verification processes.

21.4.3 Organizational Challenges

The organizational challenges as shown in Figure 21.5 in implementing Digital Twin Technologies are as follows:

21.4.3.1 *Cultural Barriers*

Cultural barriers within organizations often stem from differences in values, beliefs, and communication styles among employees. These differences can arise from diverse backgrounds, nationalities, or even subcultures within the organization. When not properly managed, cultural barriers can hinder collaboration, diminish productivity, and lead to conflicts among team members. Overcoming cultural barriers requires a concerted effort to foster inclusivity, promote open communication channels, and implement diversity and inclusion initiatives. By creating an environment where all voices are heard and respected, organizations can leverage the richness of their cultural diversity to drive innovation and achieve their goals.

21.4.3.2 *Skill Gaps*

Skill gaps refer to disparities between the skills employees possess and those required to perform their roles effectively. These gaps can emerge due to technological advancements,

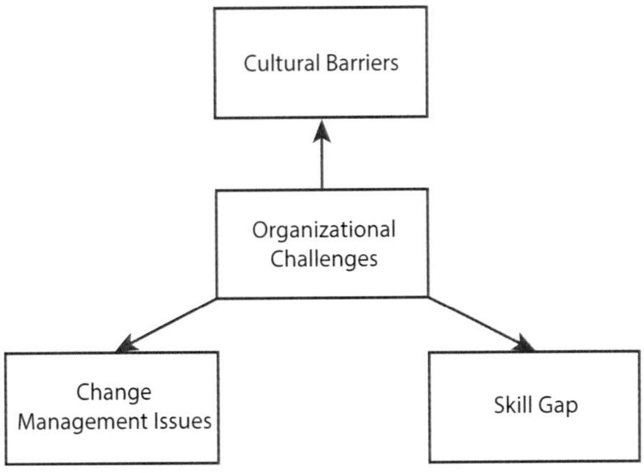

Figure 21.5 Organizational challenges.

changes in job requirements, or insufficient training and development programs. Skill gaps not only impede individual and team performance but also pose challenges to organizational agility and competitiveness. Addressing skill gaps necessitates conducting thorough skill assessments, providing targeted training and development opportunities, and fostering a culture of continuous learning. By investing in the development of their workforce, organizations can ensure that employees have the necessary skills to adapt to evolving business needs and drive success.

21.4.3.3 Change Management Issues

Change management issues arise when organizations encounter difficulties in implementing changes to processes, systems, or structures. Resistance to change, lack of stakeholder buy-in, and ineffective communication are common challenges in change management efforts. Employees may resist change due to fear of job insecurity, loss of control, or disruptions to their routines. Successful change management requires clear communication of the reasons for change, engagement of stakeholders throughout the process, and provision of adequate support and resources to facilitate the transition. Organizations may effectively address change management difficulties and achieve desired objectives by cultivating a culture that supports change and innovation.

21.5 Future Work

21.5.1 Emerging Trends

Digital twin technology has been rapidly evolving, driven by advancements in various fields such as IoT (Internet of Things), AI (Artificial Intelligence), and data analytics. Following is a detailed description of Future work possible in digital twin technologies:

21.5.1.1 Integration with IoT and Sensor Networks

Digital twins are increasingly being integrated with IoT devices and sensor networks to gather real-time data from physical assets. This integration enables the creation of dynamic digital replicas that accurately reflect the current state of the physical object or system.

21.5.1.2 AI and Machine Learning

AI and machine learning algorithms are being used to enhance the capabilities of digital twins. These technologies enable predictive analytics, anomaly detection, and optimization, allowing digital twins to anticipate future behaviors and optimize performance in real-time.

21.5.1.3 Edge Computing

With the proliferation of edge computing, digital twins are becoming more decentralized, allowing for processing and analysis of data closer to the source. Edge computing reduces latency, improves response times, and enhances the scalability of digital twin deployments, particularly in large-scale IoT environments.

21.5.1.4 *Digital Twin Lifecycle Management*

There's a growing focus on the entire lifecycle management of digital twins, from their creation and deployment to ongoing maintenance and retirement. This includes standardized frameworks for creating and managing digital twins, as well as tools for monitoring and updating them as physical assets evolve.

21.5.1.5 *Simulation and Virtual Testing*

Digital twins are increasingly used for simulation and virtual testing across various industries. By creating virtual replicas of physical systems, organizations can conduct experiments, test scenarios, and optimize processes in a safe and cost-effective manner before implementing changes in the real world.

21.5.1.6 *Blockchain Integration for Data Security*

Blockchain technology [16] is being explored to enhance the security and integrity of data within digital twin ecosystems. By leveraging blockchain's decentralized and immutable ledger, organizations can ensure the authenticity and traceability of data collected from physical assets.

21.5.1.7 *Cross-Domain Integration*

Digital twins are being utilized across multiple domains, such as manufacturing, healthcare, smart cities, and agriculture. Interoperability standards and platforms are emerging to enable seamless integration and data exchange between digital twins from different domains, fostering collaboration and innovation.

21.5.1.8 *Human-Digital Twin Interaction*

There's a growing emphasis on improving the interaction between humans and digital twins through immersive technologies such as augmented reality (AR) and virtual reality (VR). These technologies enable stakeholders to visualize and interact with digital twins in intuitive ways, facilitating better decision-making and problem-solving.

21.5.1.9 *Ethical and Regulatory Considerations*

As digital twins become more pervasive, there's increasing attention on ethical and regulatory considerations surrounding data privacy, security, and ownership. Standards and guidelines are being developed to ensure responsible use of digital twin technology and protect stakeholders' rights.

21.5.1.10 *Quantum Computing*

Despite being in its early stages, quantum computing has the potential to revolutionize data processing and optimization tasks, which could have significant implications for digital twin technology. Quantum algorithms may enable more complex simulations and analyses, leading to more accurate and actionable insights from digital twins.

21.5.2 Prediction of Future Applications and Potential Impact on Smart Manufacturing

Few potential future applications on smart manufacturing are as follows:

21.5.2.1 *AI-Driven Predictive Maintenance*

Advanced AI algorithms may play a crucial role in the future of smart manufacturing by accurately predicting equipment breakdowns and proactively scheduling maintenance. Implementing this can greatly decrease periods of inactivity, optimize the distribution of resources, and prolong the operational life of equipment, resulting in cost reductions and enhanced efficiency.

21.5.2.2 *Digital Twins*

The complexity of digital twins, which are virtual counterparts of physical assets, processes, or systems, is expected to increase in the future. Their objective is to incorporate up-to-the-minute information from sensors, IoT devices, and several other sources in order to offer thorough analysis of manufacturing activities. This can improve decision-making, enhance product quality, and facilitate rapid prototyping and testing.

21.5.2.3 *IoT-Enabled Supply Chain Management*

Smart manufacturing will increasingly leverage IoT technologies to create interconnected supply chains. This includes real-time tracking of inventory, materials, and components, enabling better demand forecasting, inventory optimization, and just-in-time manufacturing. It can also enhance transparency and traceability throughout the supply chain, improving product quality and compliance.

21.5.2.4 *Robotics and Automation*

Advances in robotics and automation will continue to revolutionize manufacturing processes. Future applications may include collaborative robots (cobots) working alongside human operators, autonomous mobile robots for material handling, and AI-driven robotic assembly lines. This can lead to increased efficiency, flexibility, and safety in manufacturing operations.

21.5.2.5 *Additive Manufacturing (3D Printing)*

The adoption of additive manufacturing technologies will likely expand, enabling on-demand production of complex parts and components with minimal waste. Future developments may include multi-material 3D printing, *in-situ* monitoring and control, and scaling up to large-scale industrial applications. This can offer greater design freedom, reduced lead times, and localized production capabilities.

21.5.2.6 Blockchain for Supply Chain Transparency

Blockchain technology [20] holds promise for enhancing supply chain transparency, security, and integrity. Future applications may involve using blockchain for verifying the authenticity of products, tracking the origin of raw materials, and ensuring compliance with regulations and standards. This can build trust among stakeholders, mitigate risks such as counterfeiting and fraud, and streamline regulatory compliance.

21.5.2.7 Augmented Reality (AR) for Maintenance and Training

AR technologies will likely play a significant role in smart manufacturing by providing real-time guidance and assistance to workers. Future applications may include using AR glasses or headsets for remote maintenance support, interactive training simulations, and overlaying digital information onto physical environments. This can improve operational efficiency, reduce errors, and accelerate skill development.

21.5.3 Discussion on the Role of Artificial Intelligence, Machine Learning, and Edge Computing in Enhancing Digital Twin Capabilities

Artificial intelligence (AI), machine learning (ML), and edge computing play crucial roles in enhancing the capabilities of digital twins across various industries. Let's discuss each of these technologies' contributions:

21.5.3.1 Artificial Intelligence (AI)

- AI enables digital twins to analyze vast amounts of data generated by sensors, IoT devices, and other sources in real-time. This analysis helps in making predictions, identifying patterns, and optimizing processes.
- AI algorithms can detect anomalies and deviations from expected behavior, allowing for proactive maintenance and troubleshooting. This predictive maintenance capability helps in preventing equipment failures and minimizing downtime.
- AI-driven optimization algorithms can optimize various aspects of operations, such as energy consumption, resource utilization, and production efficiency, based on real-time data and simulation results.

21.5.3.2 Machine Learning (ML)

- ML algorithms enhance the capabilities of digital twins by learning from historical data and continuously improving their models. This enables digital twins to adapt to changing conditions and optimize their predictions and recommendations over time.

- ML techniques, such as supervised learning, unsupervised learning, and reinforcement learning, can be applied to various tasks within digital twins, including anomaly detection, predictive maintenance, and process optimization.
- ML models can be trained to recognize complex patterns in data, enabling more accurate predictions and insights into system behavior.

21.5.3.3 Edge Computing

- Edge computing brings computational power closer to the data source, reducing latency and enabling real-time processing of data generated by IoT devices and sensors [26].
- Through the implementation of artificial intelligence (AI) and machine learning (ML) models at the edge, digital twins have the capability to conduct data analysis and make decisions on a local level, without being dependent entirely on centralized cloud infrastructure. This is especially advantageous in situations where minimal delay is crucial or where internet connectivity is restricted or unpredictable.
- Edge computing also enhances data privacy and security by processing sensitive information locally, without the need to transmit it over the network to centralized servers.

Combining AI, ML, and edge computing with digital twins offers several benefits, including improved operational efficiency, reduced downtime, better decision-making, and enhanced product lifecycle management. These technologies empower enterprises to utilize up-to-the-minute data and simulation capabilities to enhance processes, enhance product performance, and provide exceptional customer experiences. As these technologies continue to advance, we can expect even greater integration and innovation in the realm of digital twins, driving further enhancements in various industries.

21.6 Conclusion

21.6.1 Quick Recap

Digital twin-enabled smart manufacturing offers numerous opportunities but also presents several challenges. Here's a recap of the main ones:

21.6.1.1 Opportunities

- Improved Efficiency: Digital twins enable real-time monitoring and analysis of manufacturing processes, leading to increased efficiency through optimization and predictive maintenance.
- Cost Reduction: Smart manufacturing can save operational costs, energy usage, and waste by discovering inefficiencies and optimizing operations.
- Quality Enhancement: Digital twins allow for precise monitoring and control of product quality, ensuring consistency and reducing defects.

- Innovation and Product Development: Digital twins facilitate rapid prototyping, simulation, and virtual testing, speeding up the innovation and product development process.
- Supply Chain Optimization: By providing real-time insights into production processes, digital twins enable better coordination and synchronization across the supply chain, leading to reduced lead times and improved inventory management.

21.6.1.2 *Challenges*

- Data Integration and Management: The process of combining data from many sources and guaranteeing its quality, uniformity, and protection can present difficulties.
- Interoperability: Compatibility issues between different systems and technologies can hinder the seamless integration of digital twins across the manufacturing ecosystem.
- Cybersecurity Risks: Connected systems are vulnerable to cybersecurity threats, including data breaches, hacking, and sabotage, necessitating robust security measures.
- Skill Gap: Implementing and managing digital twin technologies requires specialized skills and expertise, which may be lacking in the workforce.
- Initial Investment: Adopting digital twin-enabled smart manufacturing involves significant upfront costs for infrastructure, technology, and training, which can be prohibitive for some organizations.

Addressing these challenges while leveraging the opportunities offered by digital twin-enabled smart manufacturing is crucial for maximizing its benefits and driving innovation in the manufacturing industry.

21.6.2 Key Takeaways and Further Researches

21.6.2.1 *Key Takeaways*

- Digital Twins in Smart Manufacturing: Digital twins are computer-generated models that accurately depict physical assets, processes, or systems. Within the context of intelligent manufacturing, digital twins are of paramount importance as they facilitate the simulation and optimization of processes, the prediction of outcomes, and the real-time monitoring and control of operations.
- Integration of AI: Artificial Intelligence (AI) technologies such as machine learning, deep learning, and predictive analytics enhance the capabilities of digital twins. AI algorithms analyze data generated by digital twins to derive insights, detect anomalies, and optimize processes automatically.
- Predictive Maintenance: One of the primary applications of AI-enabled digital twins in smart manufacturing is predictive maintenance. By analyzing real-time data from equipment sensors and historical maintenance records,

AI algorithms predict when equipment is likely to fail, enabling proactive maintenance to avoid costly downtime.

- Process Optimization: AI-powered digital twins simulate manufacturing processes to identify inefficiencies, optimize parameters, and improve overall productivity. By running "what-if" scenarios, manufacturers can experiment with different configurations and strategies to achieve optimal results.
- Quality Control: Digital twins combined with AI algorithms facilitate real-time quality control by continuously monitoring production processes and identifying deviations from expected performance. Manufacturers can rapidly implement corrective measures, thereby minimizing defects and guaranteeing product quality.
- Supply Chain Optimization: AI-enabled digital twins provide insights into supply chain dynamics, enabling manufacturers to optimize inventory levels, streamline logistics, and enhance demand forecasting accuracy.

21.6.3 Suggestions for Further Research

- Advanced AI Techniques: Investigate cutting-edge AI techniques such as reinforcement learning, generative adversarial networks (GANs), and natural language processing (NLP) for enhancing the capabilities of digital twins in smart manufacturing.
- Edge Computing: Explore how edge computing technologies can be integrated with AI-enabled digital twins to process data closer to the source, reducing latency and enabling real-time decision-making in smart manufacturing environments.
- Interoperability and Standardization: Research challenges and solutions related to interoperability and standardization of digital twin platforms, data formats, and communication protocols to enable seamless integration across different manufacturing systems and domains.
- Cybersecurity and Data Privacy: Investigate cybersecurity threats and privacy concerns associated with AI-enabled digital twins in smart manufacturing and develop strategies to mitigate risks and protect sensitive manufacturing data.
- Human-Machine Collaboration: Explore the role of human-machine collaboration in smart manufacturing environments where AI-enabled digital twins augment human decision-making and facilitate knowledge sharing and collaboration among workers and intelligent systems [24].
- Case Studies and Best Practices: Study real-world case studies and best practices of organizations implementing AI-enabled digital twins in smart manufacturing to understand the challenges faced, lessons learned, and potential benefits achieved.

By delving deeper into these areas of research, you can gain a comprehensive understanding of the capabilities, challenges, and opportunities associated with AI-enabled digital twins in smart manufacturing.

References

1. Yu, T., Zhang, G., Yan, J., Zhang, H., Digital Twin Technology in Manufacturing: A Literature Review and Research Agenda. *J. Ind. Integr. Manage.*, 3, 4, 1350014, 2018.
2. Li, Z., Zhang, G., Wang, X., Digital Twin in Manufacturing: A Comprehensive Survey. *J. Manuf. Syst.*, 56, 60–775, 2020.
3. Jiang, Y., Han, Y., Sun, J., Li, J., Digital Twin Technologies and Their Impact on Smart Manufacturing Processes: A Review. *J. Manuf. Syst.*, 58, 416–4315, 2020.
4. Skhiri, S., Duan, Z., Grabot, B., Amodeo, L., Digital Twin Technologies for Smart Manufacturing Systems: A Review. *J. Manuf. Syst.*, 57, 242–261, 2020.
5. Shan, J., Yan, Y., Dong, H., Huang, H., Liu, X., Digital Twins in Manufacturing: A Review and Future Directions. *J. Manuf. Syst.*, 61, 491–5065, 2021.
6. Gao, W., Xu, Q., Li, Y., Wang, Z., Digital Twin-Enabled Smart Manufacturing Systems: A Review. *J. Manuf. Syst.*, 51, 242–2615, 2019.
7. Zhang, J., Zhang, G., Zhang, Q., Zhang, H., Digital Twins for Smart Manufacturing: A Review. *J. Manuf. Syst.*, 58, 189–1975, 2020.
8. Luo, Y., Wu, M., Lin, W., Wang, X., Digital Twin-Driven Manufacturing: A Survey. *J. Manuf. Syst.*, 61, 562–576, 2021.
9. Yang, C., Wang, X., Zou, X., Xu, X., Digital Twins in Smart Manufacturing: A Review. *J. Manuf. Syst.*, 59, 290–3015, 2020.
10. Gao, G., Cheng, J., Wang, X., Wan, J., Digital Twin in Manufacturing: A Systematic Literature Review. *J. Manuf. Syst.*, 61, 24–455, 2021.
11. Wu, H.-H., Chang, Y.-C., Hsiao, C.-Y., Chang, K.-H., Digital Twin and Its Role in Industry 4.0: A Review. *J. Manuf. Syst.*, 55, 204–2175, 2020.
12. Zhu, S., Yang, C., Liu, Z., Wang, Y., Digital Twin Applications in Smart Manufacturing: A Review. *J. Manuf. Syst.*, 56, 336–349, 2020.
13. Alipour, M., Dideban, A., Zareh, M., Goudarzi, N., Review of Digital Twin Technology in Smart Manufacturing: Challenges and Opportunities. *J. Manuf. Syst.*, 56, 295–3145, 2020.
14. Wang, W., Zhu, M., Gao, G., The State of Digital Twins in Smart Manufacturing: A Review. *J. Manuf. Syst.*, 58, 358–3765, 2020.
15. Pan, H., He, F., Yang, K., Wu, Z., Digital Twins for Smart Manufacturing: A Systematic Literature Review and Future Research Directions. *J. Manuf. Syst.*, 61, 460–4765, 2021.
16. Tyagi, A.K. and Abraham, A. (Eds.), *Recent Trends in Blockchain for Information Systems Security and Privacy*, 1st ed, CRC Press, 2021, https://doi.org/10.1201/9781003139737.
17. Tyagi, A.K., *Handbook of Research on Technical, Privacy, and Security Challenges in a Modern World*, IGI Global, 2022, doi: 10.4018/978-1-6684-5250-9.
18. Singh, R., Tyagi, A.K., Arumugam, S.K., Imagining the Sustainable Future with Industry 6.0: A Smarter Pathway for Modern Society and Manufacturing Industries, in: *Machine Learning Algorithms Using Scikit and TensorFlow Environments*, IGI Global, 2024, doi: 10.4018/978-1-6684-8531-6.ch016.
19. Tyagi, A.K. and Richa, Digital Twin Technology: Opportunities and Challenges for Smart Era's Applications, in: *Proceedings of the 2023 Fifteenth International Conference on Contemporary Computing (IC3-2023)*, Association for Computing Machinery, New York, NY, USA, pp. 328–336, 2023, https://doi.org/10.1145/3607947.3608015.
20. Nair, M.M. and Tyagi, A.K., Blockchain technology for next-generation society: current trends and future opportunities for smart era, in: *Blockchain Technology for Secure Social Media Computing*, 2023.

21. Deshmukh, A., Patil, D.S., Soni, G., Tyagi, A.K., Cyber Security: New Realities for Industry 4.0 and Society 5.0, in: *Handbook of Research on Quantum Computing for Smart Environments*, A. Tyagi (Ed.), pp. 299–325, IGI Global, 2023, https://doi.org/10.4018/978-1-6684-6697-1.ch017.
22. Nair, M.M., Tyagi, A.K., Sreenath, N., The Future with Industry 4.0 at the Core of Society 5.0: Open Issues, Future Opportunities and Challenges. *2021 International Conference on Computer Communication and Informatics (ICCCI)*, pp. 1–7, 2021, doi: 10.1109/ICCCI50826.2021.9402498.
23. Tyagi, A.K., Fernandez, T.F., Mishra, S., Kumari, S., Intelligent Automation Systems at the Core of Industry 4.0, in: *Intelligent Systems Design and Applications. ISDA 2020. Advances in Intelligent Systems and Computing*, vol. 1351, A. Abraham, V. Piuri, N. Gandhi, P. Siarry, A. Kaklauskas, A. Madureira (Eds.), Springer, Cham, 2021, https://doi.org/10.1007/978-3-030-71187-0_1.
24. Deekshetha, H.R. and Tyagi, A. K., Chapter 16 - Automated and intelligent systems for next-generation-based smart applications, in: *Data Science for Genomics*, A.K. Tyagi and A. Abraham (Eds.), pp. 265–276, Academic Press, 02023, https://doi.org/10.1016/B978-0-323-98352-5.00019-7.
25. Tyagi, A.K. and Richa, Smart Manufacturing Using Internet of Things, Artificial Intelligence, and Digital Twin Technology, in: *Global Perspectives on Robotics and Autonomous Systems: Development and Applications*, 2023.
26. Tyagi, A.K., Lakshmi, P.R., Mishra, A.K., Balamurugan, G., Industry 5.0: Potentials, Issues, Opportunities, and Challenges for Society 5.0, in: *Privacy Preservation of Genomic and Medical Data*, vol. 17, pp. 409–32, 2023 Oct.

22

Future of Computer Vision and Industrial Robotics in Smart Manufacturing

Santosh Reddy Addula[1] **and Amit Kumar Tyagi**[2*]

[1]*Department of Information Technology, University of the Cumberlands, Williamsburg, KY, USA*
[2]*Department of Fashion Technology, National Institute of Fashion Technology, New Delhi, Delhi, India*

Abstract

As smart manufacturing continues to evolve, computer vision and industrial robotics are poised to play a pivotal role in shaping the future of the industry. Computer vision, driven by artificial intelligence, provides enhanced capabilities in perception, object recognition, and quality control. These technologies enable machines to interpret visual data, making it possible to automate complex tasks and improve overall production efficiency. In conjunction with computer vision, industrial robotics is becoming increasingly agile and adaptable. Collaborative robots, or cobots, are designed to work alongside human operators, facilitating flexible and efficient manufacturing processes. Their integration with computer vision systems allows for real-time feedback and dynamic adjustments, making manufacturing operations smarter and more responsive to changing demands. The convergence of computer vision and industrial robotics in smart manufacturing creates a synergy that provides numerous benefits, including reduced production errors, increased productivity, and improved safety. These technologies enable the automation of repetitive tasks, freeing up human workers to focus on more complex, creative, and value-added activities. However, this promising future is not without its challenges. Concerns related to data security, privacy, and ethical considerations must be addressed as these technologies become more pervasive in manufacturing environments. Additionally, ensuring the accessibility of these advanced solutions to small and medium-sized enterprises (SMEs) is important for widespread industry adoption. In summary, the future of computer vision and industrial robotics in smart manufacturing is one of enhanced efficiency, flexibility, and quality. These technologies are poised to empower manufacturers to meet the demands of a rapidly changing market while providing safer and more engaging work environments for employees. To realize this future, addressing challenges and making collaboration among stakeholders is imperative for the continued growth of the industry.

Keywords: Computer vision, industrial robotics, smart era, and smart manufacturing

**Corresponding author*: amitkrtyagi025@gmail.com
Santosh Reddy Addula: ORCID: https://orcid.org/0009-0000-3286-8224

Amit Kumar Tyagi, Shrikant Tiwari, Senthil Kumar Arumugam and Avinash Kumar Sharma (eds.) Artificial Intelligence-Enabled Digital Twin for Smart Manufacturing, (505–540) © 2024 Scrivener Publishing LLC

22.1 Introduction to Computer Vision and Industrial Robotics, and Smart Manufacturing

Computer Vision and Industrial Robotics, as well as Smart Manufacturing, are key components of modern manufacturing and automation [1, 2]. Let's start with an introduction to each of these fields:

- Computer Vision: Computer Vision is a multidisciplinary field that enables computers to interpret and understand the visual world. It focuses on the development of algorithms and systems that can extract, analyze, and make decisions based on visual information from the real world, often in the form of images or video. This technology allows machines to replicate the human ability to see and interpret the environment. In industrial settings, computer vision is used for various applications, including quality control, object recognition, defect detection, and tracking of objects on production lines.
- Industrial Robotics: Industrial Robotics refers to the use of robots in manufacturing and industrial processes. These robots are typically designed to perform tasks such as assembly, welding, painting, material handling, and inspection with high precision and efficiency. They are programmable and often equipped with sensors and computer control systems that enable them to carry out tasks autonomously. Industrial robots improve production quality, increase efficiency, reduce labor costs, and perform tasks that may be too dangerous or repetitive for humans.
- Smart Manufacturing: Smart Manufacturing, also known as Industry 4.0 or the Fourth Industrial Revolution, is a transformative approach to manufacturing that uses advanced technologies to create more agile, flexible, and efficient production processes. It integrates various technologies such as the Internet of Things (IoT), big data analytics, artificial intelligence, and automation to optimize manufacturing operations. Smart Manufacturing systems gather real-time data from sensors, machines, and other sources to make data-driven decisions, monitor equipment health, and enhance overall productivity. It also enables predictive maintenance, adaptive production processes, and customization at scale.

How these fields relate:

- Integration of Computer Vision in Industrial Robotics: Computer vision plays an important role in equipping industrial robots with vision capabilities. Robots with integrated vision systems can identify parts, track objects, and adjust their actions based on visual feedback. This enhances their adaptability and makes them more versatile in various manufacturing processes.
- Smart Manufacturing and Industrial Robotics: Smart Manufacturing often relies on industrial robots as a core component of automated and flexible manufacturing. Robots work alongside other smart manufacturing technologies

to streamline processes and adapt to changing production needs, ensuring efficient and responsive manufacturing operations.

- Computer Vision in Quality Control: In a smart manufacturing environment, computer vision is frequently used for quality control. Cameras and vision systems can inspect products for defects, ensuring that only high-quality items are shipped, thus improving overall product quality.

In summary, Computer Vision, Industrial Robotics, and Smart Manufacturing are closely interconnected and work together to create advanced, efficient, and adaptive manufacturing systems. These technologies are at the forefront of industrial advancements, contributing to increased productivity, quality, and sustainability in modern manufacturing processes.

22.1.1 Evolution of Manufacturing and Automation

The evolution of manufacturing and automation is a fascinating journey that has significantly impacted industries, economies, and societies over the years [3, 4]. Here's a brief overview of the key stages and developments in this evolution (refer Figure 22.1):

Craftsmanship (Pre-Industrial Revolution):

- Prior to the Industrial Revolution, manufacturing was largely based on craftsmanship and manual labor.
- Products were handmade, and production was labor-intensive and time-consuming.
- Automation was virtually non-existent, with the majority of tasks performed by skilled artisans.

First Industrial Revolution (Late 18th Century):

- The advent of mechanization and water and steam power marked the beginning of the Industrial Revolution.
- Factories and machinery replaced manual labor in textile and manufacturing industries.
- Automation began with the development of mechanical looms and spinning machines.

Second Industrial Revolution (Late 19th Century):

- The Second Industrial Revolution saw the widespread use of electricity and the development of the assembly line.

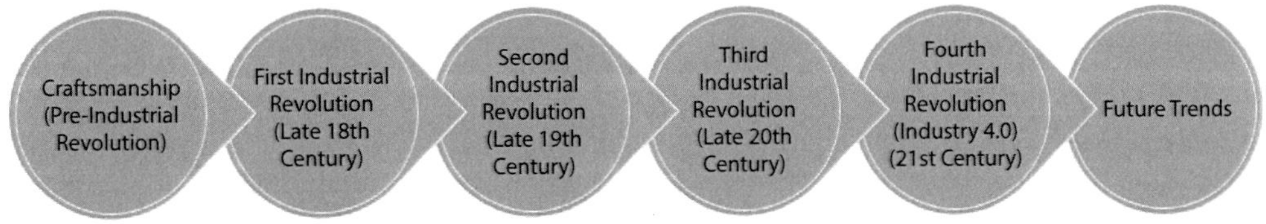

Figure 22.1 Evolution of manufacturing and automation.

- Mass production techniques, pioneered by Henry Ford, became a hallmark of this era.
- Automation started becoming more common in manufacturing processes.

Third Industrial Revolution (Late 20th Century):

- The introduction of computers and electronics revolutionized manufacturing.
- Computer Numerical Control (CNC) machines, robots, and programmable logic controllers (PLCs) became integral to manufacturing processes.
- Automation systems became more sophisticated and precise.

Fourth Industrial Revolution (Industry 4.0) (21st Century):

- This ongoing revolution is characterized by the fusion of digital, physical, and biological systems.
- Smart Manufacturing, enabled by technologies like IoT, big data, artificial intelligence, and automation, has become the norm.
- Cyber-physical systems, advanced robotics, and autonomous machinery are transforming the manufacturing landscape.
- Customization, real-time data analytics, and adaptive manufacturing are key features.

Future Trends:

- Manufacturing is expected to continue evolving with the integration of cutting-edge technologies like 3D printing, nanotechnology, and quantum computing.
- The rise of collaborative robots (cobots) is making automation more accessible to smaller businesses and flexible in deployment.
- Sustainability and environmental considerations are becoming increasingly important, driving the development of cleaner and more resource-efficient manufacturing processes.

Throughout this evolution, manufacturing and automation have led to increased productivity, reduced labor costs, improved product quality, and enhanced global trade. They have also posed challenges, such as the need for workforce reskilling and addressing job displacement. The future of manufacturing and automation will likely be shaped by further advances in technology, environmental concerns, and the ongoing quest for greater efficiency and flexibility in production processes.

22.1.2 The Role of Computer Vision and Robotics in Smart Manufacturing

Computer Vision and Robotics play a pivotal role in the context of Smart Manufacturing, also known as Industry 4.0. These technologies, when integrated, bring about significant improvements in efficiency, quality, flexibility, and responsiveness within manufacturing processes [5, 6]. Here's an overview of their roles in Smart Manufacturing:

Quality Control and Inspection:

- Computer Vision is used to inspect products for defects, ensuring high-quality output. Cameras and vision systems can identify imperfections and anomalies that may be missed by human inspectors.
- Robots equipped with vision systems can perform detailed visual inspections, ensuring that products meet specific quality standards.

Adaptive Automation:

- Robotics in Smart Manufacturing is equipped with advanced sensors and vision systems, enabling them to adapt to changing conditions and tasks.
- Robots can be reprogrammed or reconfigured to perform various tasks, making production lines more flexible and responsive to changing product demands.

Collaborative Robotics (Cobots):

- Collaborative robots, or cobots, are designed to work alongside human operators. They use sensors and computer vision to detect and avoid collisions with humans, enhancing safety and teamwork in the manufacturing environment.
- Cobots can take on repetitive, labor-intensive tasks, allowing human workers to focus on more complex and value-added activities.

Material Handling and Logistics:

- Robots with computer vision can identify and handle items of various shapes and sizes in warehouse and logistics operations.
- They can also optimize the movement of materials within the factory, reducing manual labor and minimizing errors in the supply chain.

Predictive Maintenance:

- Robotics and computer vision can be used to monitor and assess the health of machinery and equipment in real time.
- Anomalies and signs of wear and tear can be detected early, allowing for predictive maintenance to prevent costly breakdowns and production downtime.

Data Collection and Analysis:

- Computer Vision systems generate vast amounts of visual data. This data, combined with other sources of information, can be collected and analyzed to gain information about the manufacturing process.

- Real-time data analysis helps in identifying patterns, optimizing processes, and making data-driven decisions to enhance efficiency and reduce waste.

Customization and Personalization:

- Computer Vision can be used to recognize unique product features or customer preferences.
- This information can guide robots in customizing or personalizing products, enabling mass customization on a large scale.

Energy Efficiency:

- Robotics and computer vision can be used to optimize energy usage in manufacturing processes.
- Machines and robots can be programmed to operate in the most energy-efficient manner and even shut down when not in use.

In Smart Manufacturing, the integration of computer vision and robotics enables a more responsive, adaptable, and efficient manufacturing environment. These technologies empower manufacturers to meet the demands of today's dynamic markets, enhance product quality, and reduce costs. As Smart Manufacturing continues to evolve, the synergy between computer vision and robotics will play an increasingly important role in shaping the future of production.

22.1.3 Limitations of Traditional Manufacturing Processes

Traditional manufacturing processes have served as the backbone of industrial production for centuries. While they have been highly effective in many applications, they also have limitations that can impact efficiency, cost-effectiveness, and adaptability [7, 8]. Here are some of the key limitations of traditional manufacturing processes (refer Figure 22.2):

- Limited Flexibility: Traditional manufacturing processes often involve dedicated machinery and assembly lines configured for specific tasks. Changing the production process or product design can be time-consuming and costly.
- High Setup Times: Setting up traditional manufacturing equipment for a new product or design can be labor-intensive and require significant downtime. This results in production delays and increased costs.
- Large Batch Sizes: Traditional manufacturing tends to favor economies of scale, meaning that it's most cost-effective when producing large quantities of identical products. This limits the ability to efficiently produce small batches or customize products.
- Waste Generation: Mass production often generates substantial waste, both in terms of raw materials and energy. Traditional manufacturing processes are less resource-efficient compared to more modern, sustainable alternatives.
- High Labor Costs: Labor-intensive tasks in traditional manufacturing can lead to high labor costs, especially in regions with higher labor expenses. It can also result in repetitive and monotonous work for employees.

- Long Lead Times: Traditional manufacturing can have extended lead times from product design to delivery, making it less responsive to rapidly changing market demands.
- Quality Variability: Variability in product quality can be a challenge in traditional manufacturing, leading to defects and inconsistencies in the final product.
- Lack of Real-time Data: Traditional manufacturing processes often lack real-time data collection and analysis, making it difficult to monitor and optimize production in response to changing conditions.
- Environmental Impact: Many traditional manufacturing processes can be environmentally harmful due to their high energy consumption, waste generation, and reliance on non-sustainable materials.
- Risk of Human Error: Traditional manufacturing processes that rely heavily on manual labor are susceptible to human errors, which can impact product quality and safety.
- Limited Product Customization: Traditional manufacturing methods are typically less suited for product customization. Meeting unique or specialized customer requirements can be challenging.
- Complex Supply Chain Management: Traditional manufacturing often involves complex supply chains with multiple intermediaries, making it difficult to manage inventory, shipping, and logistics efficiently.

To address these limitations, many industries are increasingly adopting advanced manufacturing technologies, such as Smart Manufacturing (Industry 4.0), which use automation, data analytics, and modern materials to overcome some of these challenges. These technologies provide greater flexibility, resource efficiency, and the ability to meet the demands of a rapidly changing global marketplace.

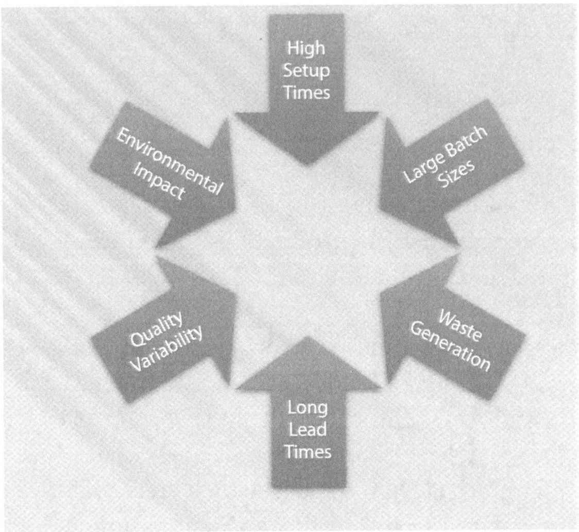

Figure 22.2 Limitations of traditional manufacturing processes.

22.1.4 The Need for Digital Transformation in Manufacturing

The need for digital transformation in manufacturing is driven by several compelling factors, as the industry seeks to remain competitive, agile, and sustainable in an increasingly dynamic and technology-driven world [9]. Here are some of the key reasons why digital transformation is imperative in manufacturing:

- Market Demands for Customization: Modern consumers and businesses often expect products to be tailored to their specific needs. Digital transformation enables manufacturers to efficiently produce customized and personalized products on a large scale.
- Increased Competition: The manufacturing landscape is becoming more competitive with the globalization of markets. Digital technologies can enhance efficiency, reduce costs, and improve product quality, allowing manufacturers to stay competitive in both domestic and international markets.
- Supply Chain Resilience: The COVID-19 pandemic highlighted the need for resilient supply chains. Digital transformation can provide greater visibility and control over the supply chain, enabling quicker responses to disruptions and ensuring the availability of important components.
- Efficiency and Cost Reduction: Automation, IoT, and data analytics can optimize production processes, reduce waste, and minimize downtime. This leads to cost savings and improved operational efficiency.
- Quality Control and Assurance: Digital technologies such as computer vision and sensors can be used for real-time quality control and defect detection, ensuring that products meet high standards and reducing defects and recalls.
- Predictive Maintenance: Predictive maintenance, enabled by IoT and data analytics, helps prevent machine breakdowns by detecting signs of wear and tear. This minimizes costly downtime and extends the lifespan of equipment.
- Resource Efficiency: Sustainable manufacturing practices are becoming increasingly important for environmental and regulatory reasons. Digital transformation can help in reducing energy consumption, waste generation, and resource usage.
- Real-time Data and Analytics: Access to real-time data and advanced analytics enables data-driven decision-making, process optimization, and the identification of trends and opportunities.
- Collaborative Robotics (Cobots): Collaborative robots (cobots) work alongside human workers, increasing productivity and efficiency. They can be easily reprogrammed to adapt to different tasks, making production lines more flexible.
- Workforce Enablement: Digital transformation can empower the workforce with digital tools, training, and information access. This leads to more skilled and informed employees who can contribute to improved productivity and innovation.
- Digital Twin Technology: Digital twins, which create virtual replicas of physical products or systems, enable testing and simulation, reducing the need for physical prototypes and speeding up the product development cycle.

- Compliance and Reporting: Digital systems can streamline compliance with industry regulations and facilitate reporting and documentation requirements, reducing administrative burdens.
- Business Model Innovation: Digital transformation enables new business models, such as servitization (providing services alongside products), subscription-based models, and data monetization.
- Reskilling and Workforce Development: The adoption of digital technologies necessitates workforce reskilling and development to operate, maintain, and innovate with these new tools.

In summary, digital transformation is essential for the modern manufacturing industry to adapt to the changing business landscape, meet customer demands, optimize operations, and contribute to sustainability goals. By embracing digital technologies and data-driven practices, manufacturers can not only survive but thrive in a highly competitive and rapidly evolving environment.

22.1.5 Labor Shortages and Efficiency Demands

Labor shortages and the demand for increased efficiency are two important challenges facing many industries, including manufacturing. These challenges often intersect and can impact businesses in various ways. Now we will discuss the relationship between labor shortages and efficiency demands:

1. Impact of Labor Shortages on Efficiency:
   - Decreased Productivity: Labor shortages can lead to reduced workforce capacity, resulting in lower productivity levels. Employees may be overburdened, leading to fatigue and reduced output.
   - Higher Labor Costs: The scarcity of labor can drive up labor costs as companies compete for a limited pool of skilled workers. This can affect the overall cost-effectiveness of operations.
   - Increased Lead Times: With fewer workers, it may take longer to complete tasks or projects, leading to extended lead times in production and delivery.
   - Quality Issues: Rushed or overworked employees may make mistakes or skip quality control steps, resulting in quality issues that reduce operational efficiency.
   - Employee Turnover: Labor shortages can lead to higher employee turnover rates as workers seek better opportunities. This constant turnover disrupts workflow and requires continuous training for new hires.
2. Solutions to Improve Efficiency in the Face of Labor Shortages:
   Automation and Robotics: Implementing automation and robotics can help compensate for labor shortages by handling repetitive and labor-intensive tasks. This not only increases efficiency but also maintains consistent productivity levels.
   - Process Optimization: Analyzing and optimizing existing processes can lead to efficiency gains. This includes reducing waste, streamlining workflows, and using data analytics to make data-driven decisions.

- Reskilling and Training: Invest in training programs to upskill existing employees, enabling them to take on more complex and value-added tasks. Cross-training employees can improve workforce flexibility.
- Collaborative Robotics (Cobots): Collaborative robots (cobots) can work alongside human employees, assisting with tasks and increasing overall productivity. They are designed to be easy to program and adapt to changing tasks.
- Remote Work and Flexible Scheduling: Embracing remote work and flexible scheduling can help companies tap into a broader talent pool and retain employees who prefer non-traditional work arrangements.
- Employee Well-being: Prioritizing employee well-being can help reduce turnover and improve job satisfaction. This can include providing a safe and comfortable work environment, promoting work-life balance, and providing competitive compensation and benefits.
- Strategic Workforce Planning: Develop a strategic workforce plan that aligns staffing needs with business goals. This may involve subcontracting, hiring temporary labor, or outsourcing certain tasks.

3. Balancing Efficiency and Labor Considerations:
   Balancing efficiency demands with labor shortages is a complex challenge. While efficiency is important for competitiveness, it's equally important to consider the well-being and job satisfaction of the workforce. Striking the right balance involves making informed decisions about automation, technology adoption, and workforce management.

In summary, businesses that successfully address labor shortages while meeting efficiency demands will likely employ a combination of automation, process optimization, and strategic workforce planning to remain competitive and adaptable in a rapidly changing economic landscape.

22.1.6 Organization of the Work

This work is summarized in 11 sections.

22.2 Role of Computer Vision in Smart Manufacturing

22.2.1 Computer Vision: Definitions, Key Components, Benefits, Disadvantages

Computer Vision is a multidisciplinary field that enables computers to interpret and understand visual information from the world, often in the form of images or video. It encompasses the development of algorithms and systems that can extract, analyze, and make decisions based on visual data [10]. Few Key Components of Computer Vision:

- Image Acquisition: The process of capturing visual data, typically using cameras or other optical sensors.

- Image Preprocessing: This involves cleaning, enhancing, and transforming raw visual data to make it suitable for analysis. Steps may include noise reduction, image calibration, and image enhancement.
- Feature Extraction: Identifying and extracting relevant features from the visual data, such as edges, corners, textures, or object shapes. These features serve as the basis for subsequent analysis.
- Object Detection and Recognition: Recognizing and identifying objects or patterns within images or videos. This is a core component of many computer vision applications, such as facial recognition or object tracking.
- Image Segmentation: Dividing an image into meaningful segments or regions. This is often used for isolating objects of interest from the background.
- Motion Analysis: Tracking the movement of objects within video sequences, allowing for applications like surveillance and video-based human-computer interaction.
- Machine Learning and Deep Learning: The use of machine learning algorithms and deep neural networks for more advanced tasks, such as image classification, image captioning, and complex scene understanding.

Benefits of Computer Vision:

- Automation: Computer vision enables automation in various industries, reducing the need for manual labor and increasing operational efficiency.
- Quality Control: It allows for real-time quality control and inspection, improving product quality and reducing defects.
- Safety and Security: Computer vision is essential in surveillance systems, facial recognition, and security applications, enhancing safety and security.
- Medical Imaging: It plays an important role in medical diagnostics, aiding in the detection of diseases and abnormalities through techniques like X-ray or MRI image analysis.
- Autonomous Vehicles: Computer vision is a key technology in the development of self-driving cars, allowing them to perceive and navigate the environment.
- Retail and E-commerce: It facilitates visual search and recommendation systems, enhancing the shopping experience.
- Augmented Reality (AR) and Virtual Reality (VR): Computer vision is used to track the user's environment and enable interactive AR and VR experiences.

Disadvantages and Challenges:

- Complexity: Developing robust computer vision systems can be complex and time-consuming, requiring significant expertise and computational resources.
- Data Requirements: High-quality labeled training data is often needed for machine learning and deep learning approaches, which can be expensive and time-intensive to obtain.

- Environmental Variability: Changes in lighting, weather conditions, or background clutter can affect the performance of computer vision systems.
- Privacy Concerns: The use of computer vision in surveillance and facial recognition has raised privacy concerns and ethical considerations.
- Limited Generalization: Computer vision models may struggle to generalize to new or unseen data, making them less adaptable in some situations.
- Computational Demands: Complex computer vision algorithms can require significant computational power, which may not be readily available in all settings.
- Error Rates: While computer vision has made significant advancements, it is not infallible and can still make errors, especially in complex or ambiguous scenarios.

Hence, computer vision is a rapidly evolving field with great potential for various applications, but it also faces ongoing challenges and limitations that researchers and engineers are actively working to address.

22.2.2 Applications of Computer Vision in Manufacturing

Computer Vision has numerous applications in the manufacturing industry, where it is employed to improve quality control, enhance operational efficiency, and enable automation [11]. Here are some key applications of Computer Vision in manufacturing:

- Quality Control and Inspection: Computer vision systems can detect defects, imperfections, or irregularities in products, materials, or components, ensuring that only high-quality items are released. Ensuring that products meet precise specifications by measuring dimensions and tolerances with high accuracy.
- Object Recognition and Sorting: Computer vision can recognize and classify products or parts, allowing automated sorting, grouping, or assembly of components.
- Assembly Verification: Computer vision systems can confirm that components are correctly assembled, positioned, and oriented, reducing errors in assembly lines.
- Barcode and QR Code Reading: Computer vision can scan and verify barcodes and QR codes for tracking and traceability in manufacturing and logistics.
- Robot Guidance: Computer vision guides robots to perform precise tasks like picking and placing objects, welding, and painting in manufacturing processes.
- Process Monitoring and Control: Computer vision continuously monitors manufacturing processes to detect deviations and ensure consistency and quality.
- Predictive Maintenance: By analyzing visual data from sensors and cameras, computer vision can predict maintenance needs for machinery, reducing unplanned downtime.
- Safety and Compliance: Computer vision systems enhance safety by monitoring and enforcing compliance with safety protocols, including identifying unsafe conditions or behaviors.
- Defect Tracking and Root Cause Analysis: Computer vision systems can track defects back to their source in the manufacturing process, facilitating root cause analysis and process improvement.

- Traceability and Serialization: Computer vision helps manufacturers track the history and location of products, ensuring transparency in the supply chain and aiding in recalls.
- Customization and Personalization: Computer vision is used to recognize customer-specific features or preferences, enabling the mass customization of products.
- Inventory Management: Computer vision assists in tracking and managing inventory levels, reducing the risk of stockouts or overstocking.
- Visual Documentation and Record Keeping: It Capture visual data for record-keeping, documentation, and compliance purposes in manufacturing processes.
- Augmented Reality (AR): It Provide workers with AR assistance, displaying important information, such as assembly instructions or safety guidelines, directly within their field of view.
- Waste Reduction and Sustainability: It Reducing waste by minimizing defects and optimizing resource usage through data-driven insights from computer vision.

Hence, Computer vision in manufacturing is an ever-evolving field that provides substantial benefits, including improved product quality, reduced operational costs, enhanced safety, and increased production efficiency. As technology continues to advance, its applications in manufacturing will likely expand further, contributing to the growth and competitiveness of the industry.

22.3 Industrial Robotics in Smart Manufacturing

22.3.1 Industrial Robots: Introduction, Types, Benefits and Key Applications

Industrial robots are autonomous or semi-autonomous machines designed to perform tasks within industrial settings. They are versatile, programmable, and capable of carrying out a wide range of tasks with precision and consistency. Industrial robots have revolutionized manufacturing and automation across various industries [12]. Here's an introduction to industrial robots, the types, benefits, and key applications (refer Table 22.1):

Types of Industrial Robots:

- Articulated Robots: These robots have rotary joints and resemble a human arm with multiple segments. They are highly flexible and suitable for tasks that require reaching into confined spaces or complex movements.
- SCARA Robots (Selective Compliance Assembly Robot Arm): SCARA robots are known for their fast and precise horizontal movements. They are often used in assembly and pick-and-place applications.
- Delta Robots: Delta robots have a unique design with three arms connected to a common base. They are known for their high-speed, precision, and agility. They excel in tasks such as packaging and sorting.
- Cartesian Robots: Also known as gantry robots, these have a Cartesian coordinate system, making them suitable for applications that require precise linear movements, such as CNC machines and 3D printing.

- Cylindrical Robots: These robots have a single rotary joint that allows them to move in a circular path. They are useful for applications like material handling and sealing.

Table 22.1 Industrial robots: introduction, types, benefits and key applications.

Industrial Robots introduction, types, Benefits and Key Applications		
	Types of industrial robots	Articulated robots
		SCARA Robots (Selective Compliance Assembly Robot Arm):
		Delta robots
		Cartesian robots
		Cylindrical robots
		Collaborative robots (Cobots)
	Benefits of industrial robots	Increased productivity
		Precision and accuracy
		Flexibility
		Safety
		Cost savings
		Cost savings
		24/7 Operation
	Key applications of industrial robots	Manufacturing and assembly
		Material handling
		Packaging and palletizing
		Welding and cutting
		Inspection and quality control
		Medical and healthcare
		Food and beverage
		Pharmaceuticals
		Agriculture
		Logistics and e-commerce

- Collaborative Robots (Cobots): Cobots are designed to work alongside humans. They are equipped with safety features to ensure safe interaction with human workers. Cobots are increasingly used in various industries, including manufacturing.

Benefits of Industrial Robots:

- Increased Productivity: Robots can work continuously and without breaks, leading to increased productivity and shorter production cycles.
- Precision and Accuracy: Robots can perform tasks with high precision, reducing errors and ensuring consistent quality.
- Flexibility: Most industrial robots are programmable and can be quickly reconfigured to handle different tasks or products, making production lines more versatile.
- Safety: Robots can handle hazardous tasks, reducing the risk to human workers in dangerous environments.
- Cost Savings: While there is an initial investment in purchasing and programming robots, they can reduce long-term labor costs and improve operational efficiency.
- Consistency: Robots perform tasks consistently without variations due to fatigue or human error.
- 24/7 Operation: Robots can work around the clock, increasing the overall utilization of equipment and reducing downtime.

Key Applications of Industrial Robots:

- Manufacturing and Assembly: Robots are widely used in automotive manufacturing, electronics assembly, and other industries to perform tasks like welding, painting, and assembling components.
- Material Handling: Robots are used for the automated movement of materials and goods within warehouses, factories, and distribution centers.
- Packaging and Palletizing: Robots handle packaging tasks such as filling, sealing, labeling, and palletizing products.
- Welding and Cutting: Robots are used in welding and cutting processes, ensuring precision and consistency in the quality of welds and cuts.
- Inspection and Quality Control: Robots equipped with vision systems can inspect products for defects, ensuring high-quality production.
- Medical and Healthcare: In healthcare, robots assist in surgeries, automate medication dispensing, and provide rehabilitation support.
- Food and Beverage: Robots are used for food processing, such as sorting, slicing, and packaging.
- Pharmaceuticals: In pharmaceutical manufacturing, robots assist in filling, labeling, and packaging medications.
- Agriculture: Robots are employed in tasks like fruit picking, planting, and weeding in agriculture.

- Logistics and E-commerce: In logistics, robots assist in order picking, sorting, and shipping in warehouses and distribution centers.

Industrial robots continue to evolve and find new applications across diverse industries, providing solutions that enhance efficiency, quality, and safety in manufacturing and automation.

22.3.2 Robotic Automation and Collaborative Robots (Cobots) in Manufacturing

Robotic automation, including Collaborative Robots (Cobots), has revolutionized manufacturing by improving productivity, flexibility, and safety [13]. Here's an overview of how these technologies is used in manufacturing:

Robotic Automation in Manufacturing:

- Robotic automation in manufacturing involves the use of industrial robots to perform tasks traditionally carried out by human workers. These robots are programmable, versatile, and capable of executing a wide range of functions. Here's how they are utilized:
- Material Handling: Robots can move raw materials, components, and finished products within a manufacturing facility. This is particularly useful in automating logistics and reducing manual labor.
- Assembly: Robots are employed in assembly tasks, ensuring consistent precision and speed in assembling components or products, such as automotive assembly lines.
- Welding and Cutting: In metalworking industries, robots perform welding and cutting tasks with high accuracy, quality, and speed.
- Painting and Finishing: Robots are used for painting and finishing products like cars, appliances, and industrial equipment, ensuring even coats and reducing waste.
- Packaging and Palletizing: Robots package products, sort them, and stack them on pallets, streamlining the final stages of production.
- Quality Control and Inspection: Robots equipped with vision systems perform real-time quality control and inspect products for defects, ensuring high product quality.
- Material Removal: Robots can perform tasks like trimming, polishing, or deburring of materials in industries like aerospace and manufacturing.
- Machine Tending: Robots load and unload machines, such as CNC machines, 3D printers, and injection molders, increasing efficiency and reducing downtime.

Collaborative Robots (Cobots):

Collaborative robots, or Cobots, are a specific category of robots designed to work alongside human operators in a shared workspace [28, 29]. They are equipped with safety features to ensure safe interaction with humans. Cobots have unique advantages in manufacturing:

- Safety: Cobots are designed to stop or slow down when they come into contact with a human, minimizing the risk of injury. They are often equipped with force-sensing technology to detect any resistance or pressure.
- Flexibility: Cobots are easily reprogrammable and adaptable to various tasks. This flexibility allows them to quickly switch between different operations, making production lines more versatile.
- Ease of Programming: Many Cobots are designed to be user-friendly, with interfaces that enable non-experts to program them for specific tasks.
- Small Footprint: Cobots are often compact and lightweight, making them suitable for environments with limited space.
- Ergonomics: Cobots can handle repetitive and physically demanding tasks, reducing the risk of repetitive strain injuries for human workers.

Benefits of Robotic Automation and Cobots in Manufacturing:

- Improved Efficiency: Robots work consistently and without breaks, leading to increased productivity and reduced production cycle times.
- Quality Assurance: Automation ensures that tasks are performed with precision and consistency, reducing errors and defects in products.
- Cost Reduction: While there is an initial investment, robotic automation can lead to long-term labor cost savings and improved operational efficiency.
- Flexibility: Cobots can be easily reconfigured, making them suitable for a wide range of tasks and adaptable to changing production needs.
- Safety: Cobots enhance workplace safety by working alongside human employees without posing a significant risk.
- Sustainability: Automation can reduce waste and improve resource efficiency, contributing to environmental sustainability.
- Competitiveness: Manufacturers that embrace robotic automation and Cobots can stay competitive in a rapidly evolving marketplace.

The use of robotic automation and Cobots in manufacturing continues to grow, transforming the industry by optimizing production processes and improving working conditions for employees. These technologies provide a promising future for manufacturing and automation across various sectors.

22.4 Convergence of Computer Vision and Robotics in Today's Smart Era: Applications and Opportunities

The convergence of Computer Vision and Robotics in today's smart era provides a multitude of applications and opportunities across various industries. This synergy empowers machines to perceive, interpret, and interact with the physical world more intelligently. Here are some key applications and opportunities in this dynamic field:

a. Industrial Automation and Manufacturing: Computer vision integrated with robotics enables real-time inspection and quality control, reducing defects and ensuring consistent product quality. Also, Robots equipped with computer vision can locate, pick, and place objects with precision, streamlining logistics and manufacturing processes.

b. Healthcare and Medical Robotics: Computer vision enhances the precision and safety of minimally invasive surgeries by providing real-time imaging and guidance. Also, Vision systems aid in patient monitoring and assist in rehabilitation exercises for individuals with mobility issues.
c. Agriculture: Robots equipped with computer vision can identify and manage crops, monitor plant health, and automate tasks like weeding and harvesting.
d. Autonomous Vehicles: Combining computer vision with robotics allows autonomous vehicles to perceive and navigate the environment safely, contributing to the advancement of the automotive industry.
e. Logistics and Warehousing: Robots with computer vision guide themselves through warehouses, locate products, and pack orders for efficient e-commerce fulfillment.
f. Retail and Customer Service: Self-checkout kiosks and robots can streamline the shopping experience by recognizing products and processing payments. Also, Robots equipped with computer vision can assist customers by answering questions, providing directions, or guiding them to specific products.
g. Surveillance and Security: Surveillance robots with computer vision capabilities enhance security by monitoring and identifying security breaches or suspicious activities in real-time.
h. Construction and Infrastructure: Robots with computer vision can assist in construction tasks, such as bricklaying and site inspections, improving safety and efficiency.
i. Environmental Monitoring: Drones and robots equipped with computer vision can be used for monitoring wildlife, mapping ecosystems, and protecting the environment.
j. Smart Cities: Computer vision systems in traffic lights and cameras can optimize traffic flow, reduce congestion, and enhance road safety. Also, Robots equipped with computer vision can assist in public services such as waste management, street cleaning, and urban planning.

The convergence of computer vision and robotics provides a transformative potential that touches nearly every aspect of modern life. As technology continues to evolve, it opens new doors to innovative solutions, greater efficiency, and improved safety across a wide range of industries and applications.

22.5 Emerging Technologies for Computer Vision and Robotics in Smart Manufacturing

Emerging technologies are driving significant advancements in computer vision and robotics in the context of smart manufacturing (Industry 4.0). These technologies are transforming manufacturing processes by enhancing automation, precision, and efficiency [14]. Here are some of the key emerging technologies in this field (refer Figure 22.3):

A. Artificial Intelligence and Machine Learning:
   - Deep Learning: Deep neural networks are improving the accuracy and performance of computer vision algorithms, making it possible to handle complex tasks like object recognition, image segmentation, and anomaly detection.
   - Reinforcement Learning: AI-driven robots and systems can learn through trial and error, adapting to changing manufacturing environments and optimizing their actions over time.

B. 3D Vision and LiDAR:
   - 3D Vision Sensors: These sensors are being used to create detailed 3D models of objects and environments, enabling robots to work in unstructured settings with greater precision.
   - LiDAR (Light Detection and Ranging): LiDAR technology provides high-resolution 3D scanning, aiding in obstacle avoidance, navigation, and improved object recognition.

C. Augmented Reality (AR) and Virtual Reality (VR):
   - AR for Training: AR is used to train operators and technicians by overlaying digital instructions and information onto the real-world manufacturing environment.
   - VR for Simulation: Virtual reality enables simulation of manufacturing processes, helping in the design and testing of production lines and systems.

D. Edge Computing:
   - Real-time Processing: Edge computing enables real-time analysis of computer vision data directly at the point of capture, reducing latency and enabling faster decision-making for robots and automated systems.

E. 5G Connectivity:
   - Low Latency: 5G networks provide ultra-low latency, which is important for real-time communication between robots and central systems, making remote control and monitoring more efficient.

F. Human-Robot Collaboration:
   - Cobots: Collaborative robots (cobots) are becoming more sophisticated, with advanced safety features and improved human-robot interaction, allowing closer collaboration in manufacturing processes.
   - Wearable Devices: Workers can use wearable devices with AR and IoT capabilities to interact with robots and systems, increasing efficiency and safety.

G. Digital Twins:
   - Virtual Replicas: Digital twins create virtual replicas of physical assets, products, or systems, allowing manufacturers to simulate and optimize processes before implementing them in the real world.

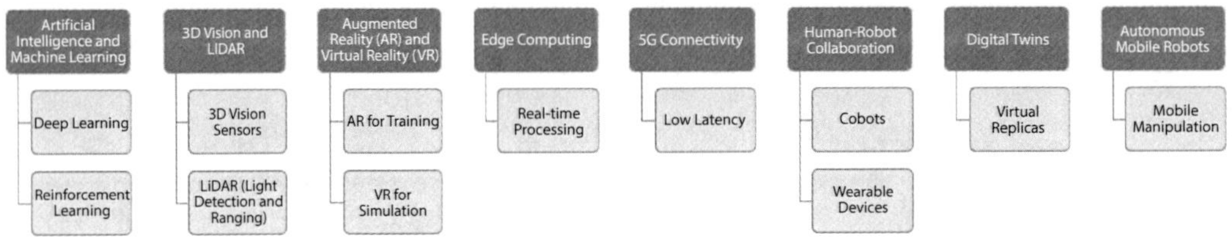

Figure 22.3 Emerging technologies for computer vision and robotics.

H. Autonomous Mobile Robots:
   - Mobile Manipulation: Autonomous mobile robots are equipped with manipulators for pick-and-place tasks, navigation capabilities, and obstacle avoidance systems.

I. IoT and Sensor Integration:
   - Sensor Networks: IoT and sensor data provide real-time information about equipment and environmental conditions, enabling predictive maintenance and process optimization.

J. Cloud Robotics:
   - Cloud-Based AI: Cloud robotics allows robots to use cloud-based AI for complex processing tasks, making them more capable and flexible.

K. Additive Manufacturing (3D Printing):
   - Custom Parts: 3D printing is used for manufacturing customized and complex components, with robots handling the printing process.

These emerging technologies are driving smart manufacturing to new heights by combining the power of computer vision and robotics. The result is increased efficiency, improved product quality, and enhanced adaptability, making smart manufacturing a reality for many industries.

22.6 Applications of Computer Vision and Robotics in Smart Manufacturing

The integration of computer vision and robotics in smart manufacturing is transforming industrial processes across various sectors [15]. Here are some applications where computer vision and robotics play an important role in smart manufacturing:

- Quality Control and Inspection: Computer vision systems are used for real-time quality control and inspection of products. They can identify defects, inconsistencies, and anomalies in products, reducing the likelihood of sub-standard items reaching consumers.
- Pick-and-Place: Robots equipped with computer vision can precisely locate and pick objects, components, or parts from a cluttered environment and place them in the desired position, streamlining assembly and packaging processes.
- Assembly and Manufacturing: Computer vision-guided robots assist in assembling components or products, ensuring precise positioning and

orientation. They can handle tasks that require a high degree of accuracy and consistency.

- Welding and Material Processing: Robots with computer vision capabilities are used in welding, cutting, and other material processing tasks, ensuring high-quality and efficient operations.
- CNC Machine Tending: Robots with computer vision can load and unload materials into CNC (Computer Numerical Control) machines, increasing productivity and reducing downtime.
- Packaging and Palletizing: Robots equipped with computer vision automate packaging and palletizing tasks, including recognizing product dimensions, configuring packaging materials, and stacking products efficiently.
- Collaborative Robots (Cobots): Cobots work alongside human workers, assisting with tasks such as assembly, testing, and materials handling. Their computer vision capabilities allow them to adapt to the dynamic manufacturing environment.
- Inventory Management: Computer vision systems are used for tracking and managing inventory in real time, optimizing stock levels, and preventing stockouts or overstocking.
- Autonomous Vehicles and AGVs: Autonomous vehicles and Automated Guided Vehicles (AGVs) in manufacturing facilities use computer vision for navigation and collision avoidance, facilitating material transport and logistics.
- Maintenance and Predictive Analytics: Computer vision and robotics are used for predictive maintenance, inspecting machinery and identifying signs of wear and tear, preventing costly breakdowns.
- Material Handling and Logistics: Robots and automated systems equipped with computer vision manage the movement and sorting of materials within warehouses, distribution centers, and logistics operations.
- Workforce Safety: Computer vision is used to enhance worker safety by detecting unsafe conditions, monitoring compliance with safety protocols, and identifying potential hazards.
- Human-Machine Interfaces: Augmented reality (AR) and virtual reality (VR) technologies with computer vision capabilities provide workers with real-time data, instructions, and visual aids to improve manufacturing processes and worker efficiency.
- Data Analytics and Process Optimization: Computer vision data is analyzed to identify patterns, improve production processes, and make data-driven decisions for enhanced manufacturing efficiency.
- Customization and Personalization: Computer vision systems enable the mass customization of products by recognizing customer-specific features or preferences and adjusting manufacturing processes accordingly.

These applications of computer vision and robotics in smart manufacturing provide increased efficiency, quality control, and adaptability in response to rapidly changing market demands, making manufacturing processes more streamlined, cost-effective, and agile.

22.7 Security and Data Privacy Towards Computer Vision and Robotics-Based Smart Manufacturing

Security and data privacy are paramount considerations in the context of computer vision and robotics-based smart manufacturing. The integration of these technologies brings many benefits, but it also introduces risks related to data protection, cyber threats, and privacy [16, 17]. Here's how security and data privacy are addressed in smart manufacturing (refer Figure 22.4):

- Data Encryption: Implement strong encryption techniques for data transmission, ensuring that sensitive data, such as images, videos, and operational data, are protected during transfer between devices and networks.
- Access Control: Restrict access to computer vision and robotics systems to authorized personnel only. Implement role-based access control (RBAC) to manage who can view, modify, or interact with the systems.
- Network Security: Secure the network infrastructure to prevent unauthorized access to data and control systems. Implement firewalls, intrusion detection systems, and regular security audits.
- Privacy by Design: Adopt a privacy-by-design approach when developing and deploying computer vision and robotics systems. This involves considering privacy and data protection from the outset of the design process.
- Data Minimization: Collect and store only the data necessary for the intended purpose. Avoid excessive data collection to minimize the risk of data breaches and unauthorized access.
- Secure Data Storage: Ensure that data storage systems, including cloud solutions, adhere to security best practices and data encryption to safeguard sensitive information.
- Regular Software Updates: Keep all software components, including operating systems, firmware, and applications, up-to-date with the latest security patches and updates to address vulnerabilities.

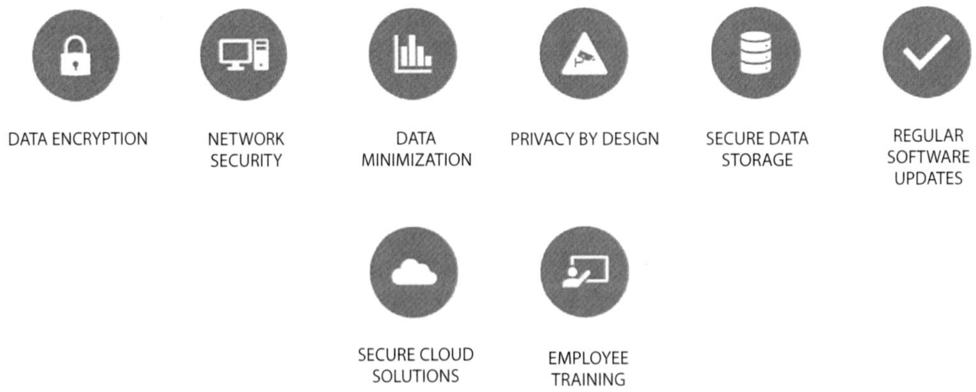

Figure 22.4 Security and data privacy towards computer vision and robotics based smart manufacturing.

- Secure Cloud Solutions: If using cloud services, choose providers that have robust security measures and comply with industry-standard security certifications.
- Data Ownership and Agreements: Clearly define data ownership and access rights in agreements with service providers and partners to ensure data protection and accountability.
- Data Retention Policies: Establish data retention and deletion policies to ensure that data is not stored longer than necessary, reducing the risk of unauthorized access.
- Regular Audits and Assessments: Conduct security audits, vulnerability assessments, and penetration testing to identify and address weaknesses in the system's security.
- Incident Response Plan: Develop a comprehensive incident response plan to address and mitigate security breaches, ensuring that personnel are well-prepared to respond to threats promptly.
- Employee Training: Train and raise awareness among employees about the importance of data privacy and security practices. Human error is a common cause of security breaches.
- Regulatory Compliance: Stay informed about relevant data protection regulations, such as GDPR, CCPA, and industry-specific standards. Ensure compliance with these regulations to avoid legal consequences.
- Privacy Impact Assessments (PIAs): Perform privacy impact assessments to identify and mitigate potential privacy risks associated with the use of computer vision and robotics in manufacturing.
- Data Anonymization: Anonymize or pseudonymize data to protect the privacy of individuals captured in computer vision systems while still allowing for meaningful analysis.

Hence, Balancing the benefits of computer vision and robotics in smart manufacturing with security and data privacy is essential [27, 30–33]. Implementing these measures helps ensure that the data generated and processed in manufacturing facilities is protected from unauthorized access and misuse, thereby building trust with employees, customers, and regulatory authorities.

22.8 Case Studies in Computer Vision and Robotics in Manufacturing

22.8.1 Computer Vision for Quality Control in Manufacturing

Computer vision plays an important role in quality control in manufacturing by automating the inspection and assessment of products, components, and materials. It enables the identification of defects, irregularities, and deviations from established quality standards, ensuring that only high-quality products are released to the market [18]. Here's how computer vision is used for quality control in manufacturing:

- Defect Detection: Computer vision systems are trained to identify defects in products, such as cracks, scratches, dents, discolorations, or irregular shapes. These systems can detect both visible and non-visible defects.
- Visual Inspection: Cameras and sensors capture images or video footage of products during various stages of production. Computer vision algorithms analyze these visuals for defects and deviations.
- Dimensional Analysis: Computer vision can measure the dimensions of products and components with high precision. This is essential for ensuring that items meet the required specifications and tolerances.
- Surface Inspection: Surface quality is assessed by examining the texture, finish, and appearance of products. Computer vision systems can detect imperfections in surfaces, such as bumps or rough spots.
- Color Inspection: Color-based inspection ensures that products have the correct color or shade. This is important in industries like automotive, textiles, and food production.
- Pattern Recognition: Computer vision algorithms recognize patterns and markings on products, such as labels, barcodes, or logos. This is used to verify branding and labeling accuracy.
- Anomaly Detection: Computer vision systems can identify anomalies or deviations from the expected shape or structure of a product, component, or material.
- Optical Character Recognition (OCR): OCR technology reads and verifies text, characters, and serial numbers on products to ensure they are correctly printed and legible.
- Packaging Inspection: Computer vision is used to assess the integrity of packaging, such as the presence of seals, labels, and packaging material defects.
- Defect Classification: Computer vision systems can classify defects based on predefined categories, allowing manufacturers to address specific issues efficiently.
- Real-time Feedback: Computer vision provides real-time feedback, allowing immediate corrective actions to be taken if defects or issues are detected.
- Data Logging and Reporting: The results of quality control inspections are logged and reported for quality assurance purposes, enabling manufacturers to track defects, trends, and process improvements.
- Integration with Automation: Computer vision systems can be seamlessly integrated with robots and automated systems to remove or correct defective items in real time, ensuring a streamlined production process.
- Continuous Monitoring: Computer vision can monitor the manufacturing process continuously, providing a continuous stream of data to identify and address quality issues.

In summary, Quality control through computer vision in manufacturing not only improves the quality and consistency of products but also reduces operational costs by automating inspection tasks. It also enhances competitiveness by ensuring that only high-quality items reach the market, improving customer satisfaction and trust.

22.8.2 Robotics in Automotive Manufacturing

Robotics plays a pivotal role in automotive manufacturing, transforming the industry with increased efficiency, precision, and flexibility. From assembly to quality control, robots are widely used to streamline production processes [19, 20]. Here's an overview of how robotics is employed in automotive manufacturing:

- Assembly Line Automation: Robots are extensively used in assembling vehicles. They can perform tasks such as welding, fastening bolts and screws, installing components like engines and transmissions, and applying sealant with high precision and speed.
- Welding: Robotic welding arms execute precise and consistent welds, ensuring the structural integrity of the vehicle. They are used for spot welding, arc welding, and laser welding in various automotive components.
- Painting and Finishing: Robots are employed for painting vehicle bodies, applying paint with precision and consistency. They also handle finishing tasks, such as sealing joints and applying clear coat.
- Material Handling and Logistics: Automated Guided Vehicles (AGVs) and mobile robots transport materials, parts, and components to various points on the assembly line. They can also transfer the assembled vehicles to the next stage of production.
- Inspection and Quality Control: Vision systems and robots perform quality control checks by inspecting components and assemblies for defects, ensuring that only high-quality vehicles move through the production process.
- Collaborative Robots (Cobots): Cobots work alongside human operators in tasks that require dexterity or flexibility, enhancing worker safety and efficiency. They are often used in assembling smaller parts or conducting inspections.
- Material Cutting and Machining: Robots are used in precision machining tasks, such as cutting and shaping metal parts and components, ensuring a high degree of accuracy and consistency.
- Testing and Quality Assurance: Robots conduct a variety of tests on vehicles, including crash tests, emission tests, and noise/vibration tests, ensuring that vehicles meet safety and quality standards.
- Screw and Nut Fastening: Robots with precision control systems fasten nuts and bolts in a uniform and controlled manner, contributing to the structural integrity of the vehicle.
- Engine and Transmission Manufacturing: Robots are utilized for the assembly of engines and transmissions, ensuring that important components are precisely aligned and secured.
- Electrical and Electronics Assembly: Robots handle the installation of electrical and electronic components, such as wiring harnesses, sensors, and control units, ensuring reliable connections and functionality.
- Maintenance and Predictive Maintenance: Robots equipped with sensors and cameras can perform routine maintenance tasks and monitor the health of machinery, predicting maintenance needs to reduce downtime.

- Customization: Robots can adapt to different vehicle configurations, allowing for greater customization in the manufacturing process, accommodating various features and options.
- End-of-Line Testing: At the end of the assembly line, robots and automated systems perform final checks and inspections to verify that vehicles are fully functional and meet quality standards before they are shipped to customers.

Hence, Robotics in automotive manufacturing increases efficiency, reduces labor costs, improves quality control, and enhances worker safety. The industry continues to innovate and adapt to new technologies, including the use of artificial intelligence and data analytics to optimize production and quality assurance processes.

22.8.3 Collaborative Robots in Electronics Assembly in Manufacturing

Collaborative robots, often referred to as cobots, have gained significant traction in the electronics assembly process within manufacturing. Their ability to work alongside human operators while providing high levels of precision, consistency, and flexibility makes them particularly well-suited for this industry [21, 22]. Here's how cobots are utilized in electronics assembly (refer Figure 22.5):

- Printed Circuit Board (PCB) Assembly: Cobots can assist with the placement of electronic components, such as resistors, capacitors, integrated circuits, and connectors, onto PCBs with remarkable precision. They can handle the delicate task of picking and placing small components.
- Soldering: Cobots equipped with soldering tips or soldering irons can perform soldering tasks, ensuring precise and reliable solder joints. They can manage reflow soldering and wave soldering processes as well.
- Inspection and Quality Control: Cobots equipped with computer vision systems and sensors can perform inspections of components and PCBs, checking for defects, soldering issues, and ensuring that components are correctly placed.
- Dispensing: Cobots are employed to apply adhesives, conformal coatings, or other materials to PCBs, connectors, and components with accuracy and consistency.
- Screw Tightening and Fastening: Cobots can handle the delicate task of tightening screws and fastening components onto PCBs and assemblies. They ensure uniform and precise tightening, which is essential for electronic products.

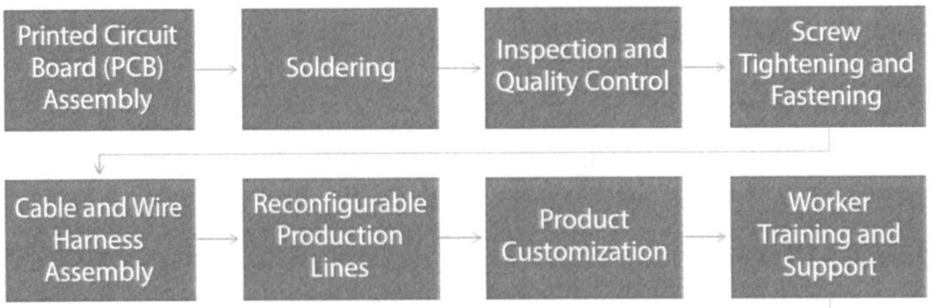

Figure 22.5 Collaborative robots in electronics assembly in manufacturing.

- Cable and Wire Harness Assembly: Cobots assist in the assembly of cable harnesses, ensuring that wires are correctly routed and connected, and that connectors are properly attached.
- Component Testing and Calibration: Cobots can be used for component testing and calibration, verifying the functionality and accuracy of components and instruments.
- Reconfigurable Production Lines: Cobots are flexible and easy to reprogram, making it simple to switch between different products and configurations on the assembly line.
- End-of-Line Testing: Cobots perform final testing and inspections of electronic products to ensure they meet quality standards before they are packaged and shipped.
- Collaborative Workstations: Cobots work alongside human operators in shared workstations, providing assistance and handling repetitive or physically demanding tasks, reducing worker fatigue and risk of injury.
- Product Customization: Cobots can adapt to different product configurations, facilitating customization in electronics assembly, accommodating various features and options.
- Worker Training and Support: Cobots can be used to train new employees by demonstrating assembly tasks and providing real-time guidance. They also enhance worker productivity by assisting in complex tasks.
- Compact and Space-Efficient: Cobots are often compact and do not require large amounts of space, making them suitable for electronics assembly environments with limited floor space.
- Safety Features: Cobots are equipped with safety features such as force sensing, collision detection, and emergency stop buttons to ensure a safe working environment in close proximity to human operators.

Note that Cobots in electronics assembly improve the efficiency and quality of the manufacturing process while also contributing to a safer and more collaborative working environment. They allow manufacturers to achieve high precision and consistency in the assembly of intricate and delicate electronic components.

22.9 Future Opportunities Towards Computer Vision and Robotics-Based Smart Manufacturing

The future of computer vision and robotics in smart manufacturing is filled with exciting opportunities as technology continues to advance [23, 24]. Here are some future opportunities and trends in this field:

- AI Integration: Increasing integration of artificial intelligence (AI) into computer vision and robotics systems, enabling machines to learn, adapt, and make real-time decisions autonomously.

- Edge Computing: The growth of edge computing, allowing data processing to occur closer to the source, reducing latency and enabling faster decision-making in real-time manufacturing environments.
- 5G Connectivity: Widespread adoption of 5G networks, enabling high-speed, low-latency communication between devices and central systems, facilitating remote monitoring and control.
- Autonomous Robots: Advancements in autonomous mobile robots and drones for materials handling, logistics, and transportation within smart manufacturing facilities.
- Digital Twins: Expansion of digital twins, creating virtual replicas of physical assets, allowing manufacturers to simulate and optimize processes for greater efficiency and cost reduction.
- Human-Machine Collaboration: Enhanced collaboration between humans and robots through improved safety features, ergonomic design, and natural language interfaces, leading to more interactive and adaptable production processes.
- Sustainability and Green Manufacturing: Integration of computer vision and robotics to improve resource efficiency, reduce waste, and enhance environmental sustainability in manufacturing.
- Customization and Personalization: Increased ability to customize products at scale, driven by AI and robotics, leading to highly personalized and tailored manufacturing processes.
- Cyber-Physical Systems: The growth of cyber-physical systems that seamlessly connect physical and digital components, creating intelligent, interconnected manufacturing ecosystems.
- Data-Driven Insights: Utilization of big data analytics and machine learning to gain deeper insights into manufacturing processes, helping optimize efficiency and predict maintenance needs.
- Enhanced Vision Systems: Development of advanced vision systems with improved accuracy, adaptability to variable lighting conditions, and the ability to detect a broader range of objects and materials.
- Augmented Reality (AR) and Virtual Reality (VR): Wider use of AR and VR for training, simulation, and real-time assistance, improving worker efficiency and reducing errors.
- Biometrics and Authentication: The application of biometric technologies to enhance security and access control in smart manufacturing facilities, ensuring only authorized personnel can access sensitive areas.
- Interconnected Supply Chains: Smart manufacturing will increasingly connect suppliers, manufacturers, and customers through IoT devices and data sharing, improving visibility and collaboration across the supply chain.
- 3D Printing Advancements: Advances in 3D printing technologies, including 4D printing, enabling rapid prototyping and production of complex, customized components.
- Regulatory Compliance: The development of standards and regulations specific to smart manufacturing, ensuring the security and privacy of data and the safe operation of robotics in manufacturing environments.

Hence, these future opportunities are poised to revolutionize the manufacturing industry by making it more efficient, sustainable, and adaptable to rapidly changing market demands. Smart manufacturing, driven by computer vision and robotics, will continue to play a pivotal role in the industry's ongoing transformation.

22.10 Sustainability and Environmental Impact Towards Computer Vision and Robotics-Based Smart Manufacturing

Computer vision and robotics-based smart manufacturing can significantly contribute to sustainability and reduce the environmental impact of industrial processes [25]. Here are ways in which these technologies can promote sustainability in manufacturing (refer Table 22.2):

Table 22.2 Sustainability and environmental impact towards computer vision and robotics-based smart manufacturing.

Sustainability and environmental impact towards computer vision and robotics-based smart manufacturing		
	Resource efficiency	Material optimization
		Energy efficiency
	Waste reduction	Quality control
		Recycling
	Predictive maintenance	Robotics and IoT sensors
	Smart inventory management	Computer vision and robotics
	Energy conservation	Automation through robotics
	Sustainable materials	Computer vision systems
	Reduced emissions	Smart manufacturing
	Water and resource management	Robotics and computer vision
	Remote monitoring and control	Manufacturing processes through robotics and computer vision
	Circular economy practices	Remanufacturing, refurbishment, and recycling
	Sustainable packaging	Minimize waste and environmental impact
	Compliance with environmental regulations	Environmental regulations, minimizing the risk of violations and associated penalties

- Resource Efficiency: Computer vision systems can assist in the precise cutting and use of materials, reducing waste and optimizing material consumption. And Robotics can be programmed to minimize energy consumption during production processes, and sensors can help identify and reduce energy waste in manufacturing facilities.
- Waste Reduction: Computer vision can identify defects and anomalies, reducing the production of faulty products and the subsequent disposal of defective items. And Robots can be employed to sort and process recyclable materials, contributing to the reduction of waste and the promotion of recycling practices.
- Predictive Maintenance: Robotics and IoT sensors can be used to monitor the health of machinery, enabling predictive maintenance to address issues before they lead to costly breakdowns. This reduces the need for resource-intensive repairs.
- Smart Inventory Management: Computer vision and robotics can optimize inventory management, reducing excess stock and minimizing the environmental impact of overproduction and storage.
- Energy Conservation: Automation through robotics can lead to more consistent production processes, reducing the energy consumption associated with stopping and starting equipment.
- Sustainable Materials: Computer vision systems can be used to identify and verify the use of sustainable and eco-friendly materials, ensuring compliance with sustainability goals.
- Reduced Emissions: Smart manufacturing can lead to reduced transportation emissions, as optimized processes and automation can minimize the need for transportation between production facilities.
- Water and Resource Management: Robotics and computer vision can be used in water and resource management, optimizing resource use in industries that rely on water, chemicals, and other inputs.
- Remote Monitoring and Control: Remote monitoring and control of manufacturing processes through robotics and computer vision can help reduce the need for on-site personnel, which can lead to fewer commuting-related emissions and energy use.
- Circular Economy Practices: Robotics and computer vision can facilitate the implementation of circular economy practices, such as remanufacturing, refurbishment, and recycling, extending the life of products and reducing waste.
- Sustainable Packaging: Computer vision can be employed to ensure that sustainable packaging materials are used and that packaging is designed to minimize waste and environmental impact.

The integration of computer vision and robotics in smart manufacturing contributes to more sustainable and environmentally responsible industrial processes. These technologies are instrumental in optimizing resource use, reducing waste, and mitigating the environmental impact of manufacturing, ultimately contributing to a greener and more sustainable future for the industry.

22.10.1 Overcoming Implementation Challenges Towards Computer Vision and Robotics in Smart Manufacturing

The implementation of computer vision and robotics in smart manufacturing can be highly beneficial, but it also comes with various challenges. Overcoming these challenges is essential to realize the full potential of these technologies [26]. Here are some strategies to address common implementation challenges:

A. High Initial Costs:
- Cost-Benefit Analysis: We perform a thorough cost-benefit analysis to justify the initial investment. Consider long-term savings in labor, increased efficiency, and improved quality.
- Modular Deployment: We implement the technology incrementally, starting with key areas to minimize upfront costs and learn from each phase before scaling up.

B. Integration with Legacy Systems:
- APIs and Middleware: We Use APIs and middleware to bridge the gap between legacy systems and new technology, enabling data exchange and coordination.
- Upgrade Legacy Systems: When feasible, consider upgrading or retrofitting legacy systems to support integration with computer vision and robotics.

C. Workforce Reskilling and Training:
- Training Programs: We invest in training and upskilling programs for existing employees to adapt to the new technology and ensure smooth implementation.
- Collaboration with Educational Institutions: We collaborate with educational institutions and technical schools to create a pipeline of talent with the necessary skills.

D. Data Security and Privacy:
- Data Encryption: We implement strong data encryption to protect sensitive information and ensure that data is transmitted securely.
- Data Governance: We establish clear data governance policies and practices to manage and protect data effectively.

E. Maintenance and Support:
- Service Contracts: We invest in service contracts and maintenance agreements with technology vendors to ensure prompt support and repairs.
- In-House Expertise: We develop in-house expertise to handle routine maintenance and troubleshooting.

F. Scalability:
- Scalable Architecture: We Design a scalable architecture that can easily accommodate future growth and changes in manufacturing needs.
- Flexible Robotics: We Choose robotics systems that are versatile and can be easily reconfigured or expanded as necessary.

G. Technical Challenges:
- Pilot Programs: We Begin with pilot programs to identify and address technical challenges before full-scale implementation.

- Collaboration with Technology Providers: We Work closely with technology providers to troubleshoot and fine-tune the technology for your specific needs.

H. Change Management:
- Change Management Plan: We develop a comprehensive change management plan to prepare employees for the transition and address any resistance.
- Communication: We make Transparent and regular communication with employees about the reasons for implementing the technology and the expected benefits.

I. Environmental Impact:
- Sustainability Goals: We ensure that the implementation aligns with your organization's sustainability goals and actively seeks to minimize the environmental impact.
- Life Cycle Analysis: We conduct a life cycle analysis to understand the environmental impact of the technology and find ways to reduce it.

J. Performance Metrics and KPIs:
- Key Performance Indicators (KPIs): We Define clear KPIs to measure the impact of the technology on productivity, quality, and efficiency.
- Continuous Improvement: We do regularly analyze KPIs and make continuous improvements to maximize the benefits of computer vision and robotics.

Hence, by addressing these challenges systematically and proactively, manufacturers can unlock the potential of computer vision and robotics in smart manufacturing while minimizing disruption and maximizing the benefits to their operations

22.11 Conclusion

In summary, the future of computer vision and industrial robotics in smart manufacturing holds immense promise and potential for revolutionizing the manufacturing industry. These technologies are at the forefront of the industry 4.0 transformation, reshaping how products are designed, produced, and delivered. Here are the key takeaways:

- Enhanced Efficiency and Quality: Computer vision and industrial robotics improve the efficiency and quality of manufacturing processes. They enable precise, repeatable tasks, reducing errors and defects while increasing productivity.
- Flexibility and Adaptability: Smart manufacturing systems with computer vision and robotics are inherently flexible and adaptable. They can reconfigure quickly to accommodate changing production demands and provide a level of customization that was previously unattainable.
- Data-Driven Decision-Making: These technologies provide a wealth of real-time data, which, when analyzed, empowers manufacturers with insights to make data-driven decisions, optimize processes, and reduce downtime.
- Human-Machine Collaboration: Collaborative robots (cobots) enable close interaction between humans and machines, creating safer, more productive

work environments. This collaborative approach is a cornerstone of future manufacturing.

- Sustainability: Computer vision and robotics contribute to sustainability by optimizing resource usage, reducing waste, and minimizing environmental impact. This aligns with global efforts to create more eco-friendly manufacturing processes.
- Customization and Personalization: Smart manufacturing allows for a higher degree of customization and personalization, meeting the increasing demands of consumers for unique products.
- Overcoming Challenges: The implementation of these technologies comes with challenges, including cost, integration, workforce adaptation, and data privacy. However, these challenges are surmountable with careful planning and investment.
- Continuous Innovation: The rapid pace of technological innovation continues to drive improvements in computer vision and robotics. Emerging technologies such as AI, 5G, edge computing, and digital twins will further enhance smart manufacturing capabilities.

Note that in the coming years, smart manufacturing will continue to evolve, providing companies with a competitive edge by delivering higher-quality products, reducing operational costs, and increasing flexibility in a rapidly changing market. The future of computer vision and industrial robotics in smart manufacturing is one of transformation, empowerment, and efficiency, heralding a new era of production for industries worldwide.

References

1. Arents, J. and Greitans, M., Smart industrial robot control trends, challenges and opportunities within manufacturing. *Appl. Sci.*, 12, 2, 937, 2022.
2. Zhou, L., Zhang, L., Konz, N., Computer vision techniques in manufacturing. *IEEE Trans. Syst. Man Cybern.: Syst.*, 53, 1, 105–117, 2022.
3. Sauter, T., The continuing evolution of integration in manufacturing automation. *IEEE Ind. Electron. Mag.*, 1, 1, 10–19, 2007.
4. Esmaeilian, B., Behdad, S., Wang, B., The evolution and future of manufacturing: A review. *J. Manuf. Syst.*, 39, 79–1005, 2016.
5. Zhou, L., Zhang, L., Konz, N., Computer vision techniques in manufacturing. *IEEE Trans. Syst. Man Cybern.: Syst.*, 53, 1, 105–117, 2022.
6. My, C.A., The role of big data analytics and AI in smart manufacturing: An overview, in: *Research in Intelligent and Computing in Engineering: Select Proceedings of RICE 2020*, pp. 911–921, 2021.
7. Peters, T.J., Rosen, D.W., Shapiro, V., A topological model of limitations in design for manufacturing. *Res. Eng. Des.*, 6, 223–233, 1994.
8. Alifui-Segbaya, F., Ituarte, I.F., Hasanov, S., Gupta, A., Fidan, I., Opportunities and Limitations of Additive Manufacturing, in: *Springer Handbook of Additive Manufacturing*, pp. 125–143, Cham: Springer International Publishing, 2023.
9. Jones, M.D., Hutcheson, S., Camba, J.D., Past, present, and future barriers to digital transformation in manufacturing: A review. *J. Manuf. Syst.*, 60, 936–9485, 2021.

10. Voulodimos, A., Doulamis, N., Doulamis, A., Protopapadakis, E., Deep learning for computer vision: A brief review. *Comput. Intell. Neurosci.*, 2018), 2018.
11. Zhou, L., Zhang, L., Konz, N., Computer vision techniques in manufacturing. *IEEE Trans. Syst. Man Cybern.: Syst.*, 53, 1, 105–117, 2022.
12. Hägele, M., Nilsson, K., Pires, J.N., Bischoff, R., Industrial robotics, in: *Springer handbook of robotics*, pp. 1385–1422, 2016.
13. Djuric, A.M., Urbanic, R.J., Rickli, J.L., A framework for collaborative robot (CoBot) integration in advanced manufacturing systems. *SAE Int. J. Mater. Manuf.*, 9, 2, 457–464, 2016.
14. Arents, J. and Greitans, M., Smart industrial robot control trends, challenges and opportunities within manufacturing. *Appl. Sci.*, 12, 2, 937, 2022.
15. Zhou, L., Zhang, L., Konz, N., Computer vision techniques in manufacturing. *IEEE Trans. Syst. Man Cybern.: Syst.*, 53, 1, 105–117, 2022.
16. Tuptuk, N. and Hailes, S., Security of smart manufacturing systems. *J. Manuf. Syst.*, 47, 93–1065, 2018.
17. Xia, K., Saidy, C., Kirkpatrick, M., Anumbe, N., Sheth, A., Harik, R., Towards semantic integration of machine vision systems to aid manufacturing event understanding. *Sensors*, 21, 13, 4276, 2021.
18. Villalba-Diez, J., Schmidt, D., Gevers, R., Ordieres-Meré, J., Buchwitz, M., Wellbrock, W., Deep learning for industrial computer vision quality control in the printing industry 4.0. *Sensors*, 19, 18, 3987, 2019.
19. Shi, J., Jimmerson, G., Pearson, T., Menassa, R., Levels of human and robot collaboration for automotive manufacturing, in: *Proceedings of the Workshop on Performance Metrics for Intelligent Systems*, vol. 95-100, 2012.
20. Li, M., Milojević, A., Handroos, H., Robotics in manufacturing—The past and the present, in: *Technical, Economic and Societal Effects of Manufacturing 4.0: Automation, Adaption and Manufacturing in Finland and Beyond*, pp. 85–95, 2020.
21. Lima, F., De Carvalho, C.N., Acardi, M.B.S., Dos Santos, E.G., De Miranda, G.B., Maia, R.F., Massote, A.A., Digital manufacturing tools in the simulation of collaborative robots: Towards industry 4.0. *Braz. J. Oper. Prod. Manage.*, 16, 2, 261–280, 2019.
22. Calvo, R. and Gil, P., Evaluation of collaborative robot sustainable integration in manufacturing assembly by using process time savings. *Materials*, 15, 2, 611, 2022.
23. Phuyal, S., Bista, D., Bista, R., Challenges, opportunities and future directions of smart manufacturing: a state of art review. *Sustain. Futures*, 2, 1000235, 2020.
24. Arents, J. and Greitans, M., Smart industrial robot control trends, challenges and opportunities within manufacturing. *Appl. Sci.*, 12, 2, 937, 2022.
25. Wuest, T., Romero, D., Khan, M.A., Mittal, S., The triple bottom line of smart manufacturing technologies: An economic, environmental, and social perspective, in: *Handbook of Smart Technologies: An Economic and Social Perspective*, pp. 310–330, Routledge, London, 2022, https://www. taylorfrancis. com/chapters/edit/10.4324/9780429351921-20.
26. Thoben, K.-D., Wiesner, S., Wuest, T., Industrie 4.0" and smart manufacturing-a review of research issues and application examples. *Int. J. Autom. Technol.*, 11, 1, 4–16, 2017.
27. Tyagi, A.K. and Tiwari, S., The Future of Artificial Intelligence in Blockchain Applications, in: *Machine Learning Algorithms Using Scikit and TensorFlow Environments*, IGI Global, 2024, doi: 10.4018/978-1-6684-8531-6.ch018.
28. Nair, M.M. and Tyagi, A.K., Chapter 11 - AI, IoT, blockchain, and cloud computing: The necessity of the future, in: *Distributed Computing to Blockchain*, R. Pandey, S. Goundar, S. Fatima (Eds.), pp. 189–206, Academic Press, 2023, ISBN 9780323961462, https://doi.org/10.1016/B978-0-323-96146-2.00001-2.

29. Singh, R., Tyagi, A.K., Arumugam, S.K., Imagining the Sustainable Future With Industry 6.0: A Smarter Pathway for Modern Society and Manufacturing Industries, in: *Machine Learning Algorithms Using Scikit and TensorFlow Environments*, IGI Global, 2024, doi: 10.4018/978-1-6684-8531-6.ch016.
30. Tyagi, A.K. and Richa, Digital Twin Technology: Opportunities and Challenges for Smart Era's Applications, in: *Proceedings of the 2023 Fifteenth International Conference on Contemporary Computing (IC3-2023)*, Association for Computing Machinery, New York, NY, USA, pp. 328–336, 2023, https://doi.org/10.1145/3607947.3608015.
31. Tyagi, A.K. and Richa, Smart Manufacturing Using Internet of Things, Artificial Intelligence, and Digital Twin Technology, in: *Global Perspectives on Robotics and Autonomous Systems: Development and Applications*, 2023.
32. Tyagi, A.K., Lakshmi, P.R., Mishra, A.K., Balamurugan, G., Industry 5.0: Potentials, Issues, Opportunities, and Challenges for Society 5.0, in: *Privacy Preservation of Genomic and Medical Data*, vol. 17, pp. 409–32, 2023 Oct.
33. Tyagi, A.K., Mishra, A.K., Vedavathi, N., Kakulapati, V., Sajidha, S.A., Chapter 21 Futuristic Technologies for Smart Manufacturing, in: *Research Statement and Vision for the Future*, Tyagi, A.K. (Ed.), 2023, https://doi.org/10.1002/9781394213948.ch21.

23

The Future of Manufacturing with AI and Data Analytics

Neel Shah[1]*, Sneh Shah[1], Janvi Bhanushali[1], Nirav Bhatt[1], Nikita Bhatt[2] and Hiren Mewada[3]

[1]*Department of Artificial Intelligence and Machine Learning, Chandubhai S. Patel Institute of Technology, Charotar University of Science and Technology, Changa, India*
[2]*Department of Computer Engineering, Chandubhai S. Patel Institute of Technology, Charotar University of Science and Technology, Changa, India*
[3]*Prince Mohammad Bin Fahd University, Kingdom of Saudi Arabia*

Abstract
This chapter explores the potential of applying AI and data analytics to transform manufacturing. It provides an overview of new research trends in smart manufacturing, including the use of IoT, big data, and advanced AI technologies like machine learning and digital twins. The conceptual background of relevant AI approaches is discussed, including deep learning, reinforcement learning, unsupervised learning, and state-of-the-art models. A key focus is examining the role of AI in predictive maintenance through data-driven techniques for remaining useful life estimation, anomaly detection, prognostics, and optimizing maintenance strategies. Challenges and limitations such as noisy data, imbalanced datasets, and high computational requirements are addressed. The opportunities enabled by AI in manufacturing are highlighted, spanning synthetic data generation, real-time prediction, and enhancing asset utilization. The chapter concludes that transformative gains in productivity, sustainability, and resilience will arise from thoughtfully leveraging AI and data to inform decision-making in industrial settings. Adoption remains in the early stages, and realizing the full potential will require interdisciplinary collaboration and purposeful innovation.

Keywords: Artificial intelligence, machine learning, modern manufacturing, predictive analysis, IoT, IIoT, synthetic data generation

23.1 Introduction

Manufacturing is undergoing a digital transformation. Cutting-edge technologies such as artificial intelligence (AI), machine learning, data analytics, and digital twins are enabling manufacturers to enhance productivity, reduce downtime, cut costs, and improve overall equipment effectiveness. The transformation driven by technology toward intelligent and interconnected manufacturing is commonly denoted as Smart manufacturing (Industry

**Corresponding author*: neeldevenshah@gmail.com

Amit Kumar Tyagi, Shrikant Tiwari, Senthil Kumar Arumugam and Avinash Kumar Sharma (eds.) Artificial Intelligence-Enabled Digital Twin for Smart Manufacturing, (541–564) © 2024 Scrivener Publishing LLC

4.0). At its core, it utilizes data and analytics to gain insights that facilitate predictive maintenance, better decision-making, and innovation.

This chapter explores how AI, big data analytics, prognostics, digital twin technologies [1], and other Industry 4.0 innovations are modernizing the manufacturing sector. Specifically, it focuses on how these technologies allow for predictive maintenance and overall equipment health monitoring to minimize unplanned downtime and boost productivity. The use of simulations, digital thread, and digital twin models for virtual design, prototyping, and process optimization is also discussed [2, 3].

By employing data-driven analytical methods, it becomes feasible to discover crucial findings for strategic decision-making, offering benefits such as decreased maintenance costs, reduced machine faults, minimized spare parts inventory, and heightened production. Maintenance assumes a pivotal role in the industrial sector, as its expenditures may constitute a substantial portion of a company's production costs [4]. Efficient maintenance strategies mitigate unexpected production halts, lower costs, and potentially extend the operational lifespan of industrial machinery. Due to these considerations, maintenance approaches have transformed as a result of the dedication and efforts of researchers, engineers, technicians, and experts.

Predictive maintenance is one of the most impactful applications of AI and data analytics in smart manufacturing. Predictive maintenance solutions can detect anomalies in vibration, temperature, or other sensor data that may indicate wear, cracks, contamination, or misalignments [5]. Advanced machine learning algorithms compare real-time operating conditions to baseline models to flag deviations from normal behavior.

In summary, this chapter explores the ongoing revolution in manufacturing driven by AI, machine learning, prognostics, digital twins, and advanced analytics. The introductory section presents the background and motivation behind this trend towards data-driven smart manufacturing. The next sections provide an overview of maintenance strategies, research directions, and key technologies powering Industry 4.0 innovations. Details on the conception and implementation of predictive maintenance using AI along with its limitations and challenges follow. The chapter culminates in a discussion of the promising future opportunities enabled through these advanced technologies and their ongoing adoption to optimize manufacturing operations, equipment health, quality assurance, and global competitiveness. In totality, it underlines the transformative nature of data-driven, IIoT, and AI-based smarter manufacturing taking shape through Industry 4.0 developments in factories across the industrial spectrum.

23.2 Different Types of Maintenance Strategies

Maintenance is a critical aspect of industrial equipment management impacting overall availability, performance, and operational expenditure. While traditional maintenance was reactive and breakdown-focused, modern smart manufacturing facilities are leveraging connectivity, analytics, and automation capabilities to adopt more proactive, optimized, and intelligent maintenance approaches. Figure 23.1 shows the abstract benefits and the difference between the maintenance methods and Figure 23.2 shows the relationship between maintenance frequency and cost, and how prescriptive maintenance can help to optimize both. The four overarching equipment maintenance methodologies are discussed [6–8]:

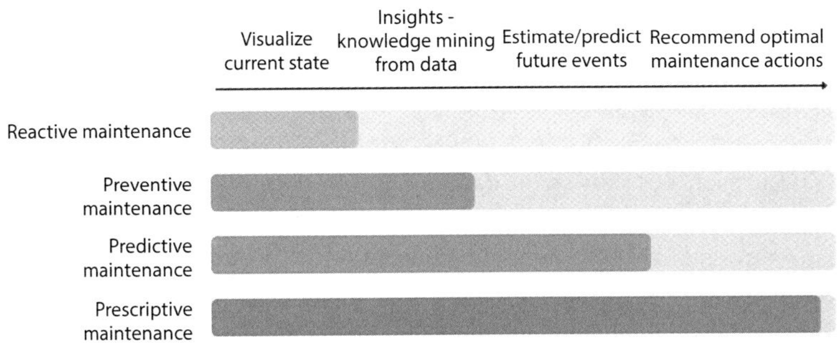

Figure 23.1 Schematic representation for different type of maintenance with potential benefits.

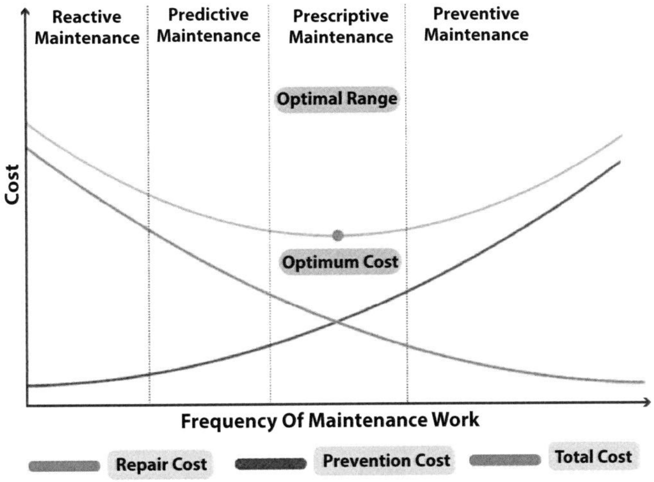

Figure 23.2 Schematic representation of the frequency of the work vs. cost for different types of maintenance.

- Reactive Maintenance
  Reactive maintenance also known as corrective or run-to-failure maintenance, is a conventional maintenance tactic that involves operating equipment until failure and then repairing or replacing it reactively. No actions are taken to detect or prevent impending failures. Reactive maintenance allows for maximizing equipment lifetime usage without investment in predictive capabilities. However, process downtimes and costs of secondary equipment or product damage from unexpected breakdowns remain high [6, 7].
- Preventive Maintenance
  Preventive maintenance or time-based maintenance relies on scheduled overhaul, parts replacement, or lubrication based on historic failure patterns rather than the actual condition. While this improves upon purely reactive approaches, activities are not necessarily aligned with the needs of an asset. Performing maintenance too frequently drives inefficiency through excess downtime disruption. Infrequent maintenance can still risk unexpected failures. Real-time condition visibility remains low [6, 7].

- Predictive Maintenance
  It leverages the Internet of Things sensors, equipment monitoring, diagnostics software, and analytics to identify failure patterns or degradation rates of in-service equipment. By recognizing when equipment needs maintenance before failure, issues can be addressed proactively to minimize downtime. Machine learning algorithms comparing real-time data to baseline models of equipment health play a key role here. While implementation requires upfront investment in monitoring infrastructure, efficiencies through failure avoidance provide ROI in months [6, 7].
- Prescriptive Maintenance
  This AI-enabled maintenance strategy provides specific decision support recommendations for when and how to best maintain assets. By combining predictive capabilities with optimization algorithms, prescriptive maintenance improves upon predictive outputs to calculate and suggest the most cost-efficient maintenance actions dynamically. This shifts maintenance from preventive schedules or reactionary practices to data-based, optimized decisions balancing uptime, spending, and lifespan [7, 8].
  While both predictive and prescriptive maintenance strategies rely on data and analytics to improve upon conventional maintenance approaches in manufacturing, there is a key difference between the two methodologies that the predictive maintenance focuses just on the projection of impending failures based on digital models and diagnostic analytics while prescriptive maintenance supplements reliability estimations with actionable and scenario-specific decision support for maintenance teams [8]. The evolution from predictive to prescriptive strategies marks a shift towards AI optimizing the true end goal - maximizing manufacturing productivity and performance through minimized downtime and maintenance costs.

23.3 New Research Trends in Manufacturing

AI represents the next wave of smart technology advancing manufacturing through increasingly connected, transparent, and autonomous operations. From computer vision for flawless quality control to machine learning algorithms optimizing complex supply chains, AI and data analytics are enabling a new era – the cyber-physical production system. Internet of Things (IoT) sensors linked across plant infrastructure and equipment feed enterprise data lakes. Advanced analytics extract embedded insights to drive predictive capabilities, while machine learning automation handles intricate manufacturing tasks [9, 10]. Together, AI and Big Data fuel the self-correcting smart factory. They grant manufacturers the flexibility to cost-effectively address demand fluctuations, customization appetite, and sustainability imperatives [11]. Figure 23.3 highlights the potential of automated maintenance to optimize equipment health using sensor data and AI methods to proactively predict and prevent failures, reducing downtime and maintenance costs.

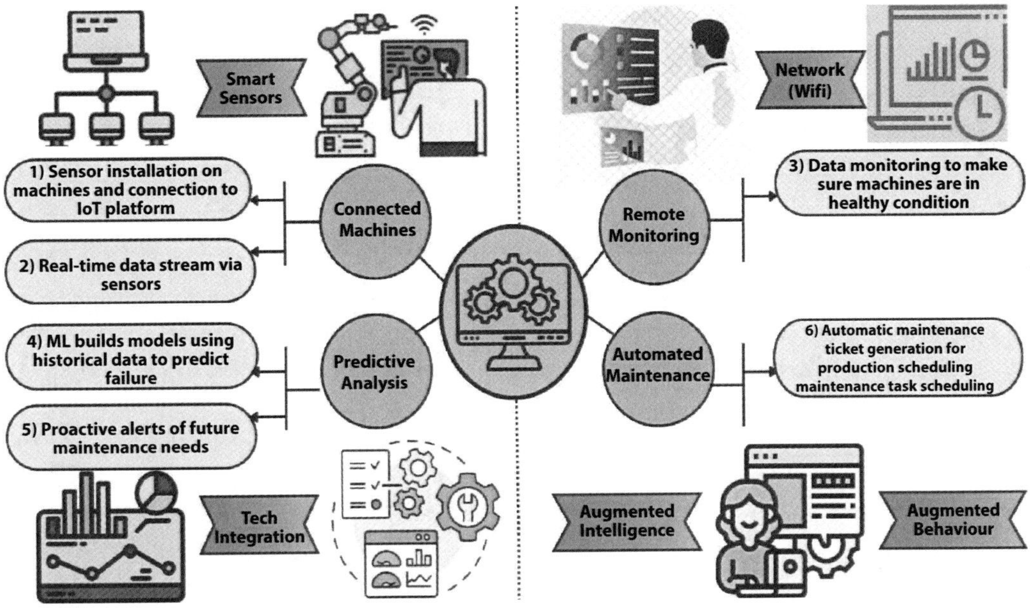

Figure 23.3 Schematic representation of an automated maintenance process using IIoT, big data, and AI to predict and prevent equipment failures.

23.3.1 Usage of IoT in Modern Manufacturing

The proliferation of low-cost sensors, connectivity, and cloud analytics has fueled the expansion of the industrial Internet of Things (IIoT) across factory floors. Networked sensors embedded throughout manufacturing infrastructure, equipment, and products generate immense volumes of previously inaccessible data. These IIoT devices measure, collect, and transmit operational statistics, machine health indicators, product locations, temperatures, vibrations, and more [9].

The dawn of IIoT marks the beginning of a manufacturing industry transformation - from dated production limited by data scarcity to analytics-optimized, intelligent, flexible manufacturing guided by abundant, meaningful insights extracted from IIoT sensor flows. When contextualized using advanced analytics, this high-fidelity visibility promises to propel manufacturing toward new horizons of quality, reliability, and efficiency [11].

23.3.2 Usage of Big Data in Modern Manufacturing

The manufacturing industry is undergoing a silent revolution fueled by an unlikely source: big data. This vast ocean of information, encompassing everything from sensor readings to customer feedback, is transforming the way products are designed, built, and delivered [12].

Modern manufacturing generates a cacophony of data - sensor readings, machine logs, production metrics, customer feedback, and even environmental data. This unstructured data, traditionally ignored, holds hidden gems. Big data tools like Hadoop and Spark tame this chaos, enabling analysis and extraction of valuable insights.

Traditional manufacturing relied on historical data, often lagging behind reality. Big data platforms like Apache Kafka and RabbitMQ enable real-time data streaming, feeding insights directly into production processes. Imagine machines adjusting parameters based on live sensor readings, or predictive maintenance triggered by real-time anomaly detection [12, 13].

23.3.3 Development of AI Technologies in Manufacturing

The field of AI has experienced rapid transformations over the past decade. Advances in computational power, the availability of large datasets, and innovations in machine learning algorithms have led to AI solutions that rival or exceed human capabilities in a range of focused tasks.

Within manufacturing, AI leverages the proliferation of industrial internet-of-things (IIoT) sensors as well as industry data sets to deliver next-generation capabilities. Two prime use cases are applying computer vision for automating quality control assessment and leveraging deep learning algorithms to accurately forecast equipment failures and recommend predictive maintenance actions. AI can be divided into machine learning, deep learning, and reinforcement learning [6]. There are also many diverse technologies in AI that are used in the modern manufacturing industry, some of which are shown in Figure 23.4. The AI is further discussed in detail in section 23.4.

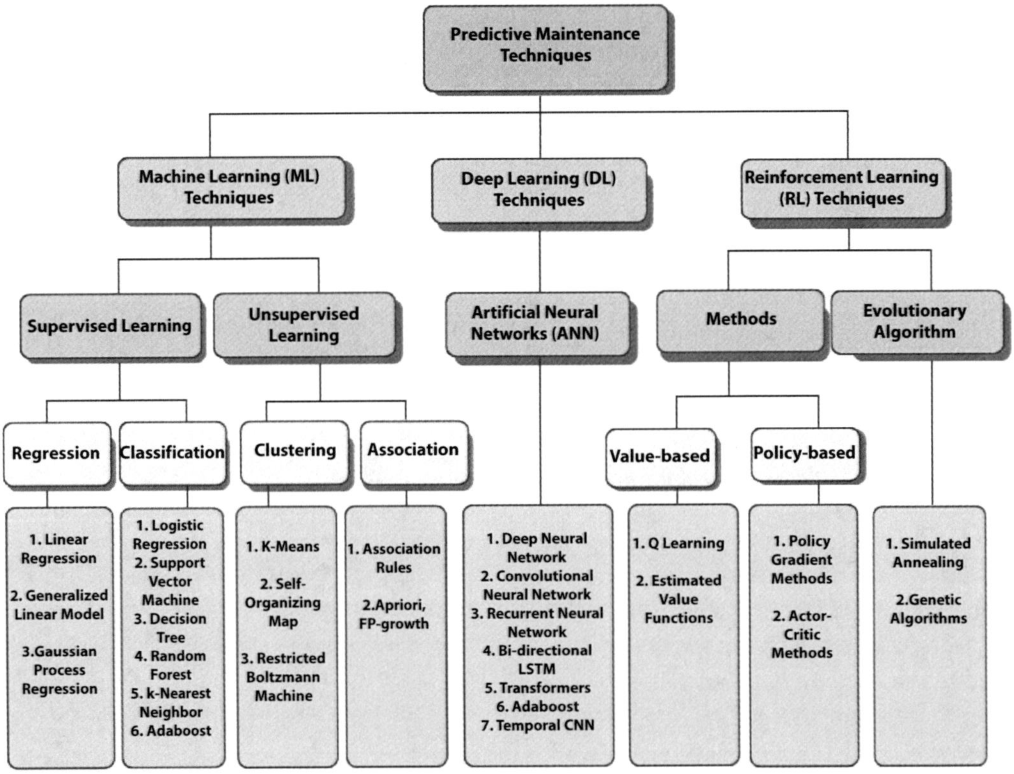

Figure 23.4 Schematic representation showing the hierarchy of different AI methods used in different literature for predictive maintenance for the manufacturing industry.

23.3.4 Digital Twin

Digital twins are emerging as a pivotal technology for enabling predictive maintenance and driving operational excellence. These virtual models mirror physical manufacturing assets, replicating behaviors based on real-time data from their counterparts on the factory floor. Digital twins leverage tens of thousands of IoT sensors plus computer vision, vibration monitoring, and other inputs to simulate component conditions and overall machine health. Advanced analytics extract insights from these ever-updating virtual constructs to identify signs of future failures and degradation far earlier than reactive approaches. Digital Twin is further discussed in section 23.5.

23.4 Conception of Different AI Technologies

23.4.1 Artificial Intelligence and Machine Learning

AI pertains to computer systems capable of executing functions that usually necessitate human intelligence. AI is revolutionizing products, services, and scientific discovery through technologies that learn and improve based on data and experience [14]. In recent years, a subfield within AI that has witnessed substantial innovation is machine learning. Machine learning involves algorithms and statistical models capable of learning patterns from data, enabling them to make predictions or decisions without explicit programming for those tasks. These algorithms iteratively learn from data, discover insights, and enhance their analytical capabilities over time [15].

Machine learning has enabled solutions that were once deemed unattainable across various sectors, such as manufacturing, healthcare, and finance. Within the field of machine learning deep learning, and reinforcement learning represent more advanced techniques that can overcome some of the limitations of more basic machine learning approaches [6, 15], The schematic representation of the interconnections between AI, machine learning, and deep learning is shown in Figure 23.5.

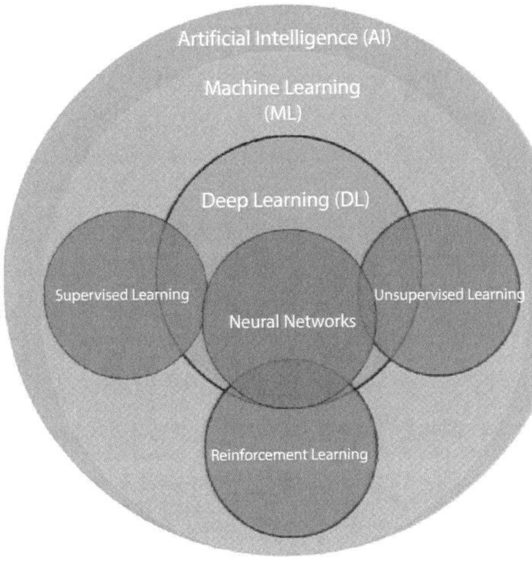

Figure 23.5 Schematic representation showing a brief overview of the different branches of AI and the relation between sub-branches.

23.4.2 Deep Learning

Deep learning has emerged as a revolutionary method for autonomously acquiring intricate patterns from extensive datasets. Drawing inspiration from the neural networks in the human brain, deep learning models employ multiple layers of processing to gradually extract more advanced features from raw input data. This hierarchical learning strategy distinguishes deep learning from earlier machine learning approaches that depended on human-engineered feature extraction [15, 16].

Convolutional neural networks represent the most widespread deep learning architecture adopted for manufacturing use cases. With a structure optimized for processing pixel imagery input, convolutional neural networks can effectively analyze scans and photographs to identify production defects and maintenance issues without any need for human assistance [15, 6]. Companies are already realizing millions of dollars in savings from convolutional neural network-enabled scrap reduction and quality improvements.

While showing great promise, deep learning poses multiple challenges including data preprocessing requirements, intense computational demands for training complex models, and interpretability issues. Some of the widely used models like the convolution layers and the transformer architectures are explained briefly in the following sections.

23.4.2.1 Convolution Layers (CNNs)

Convolutional neural networks (CNNs) are deep learning models known for using shared weights and being effective at representing local patterns. They are skilled at extracting local features from input data and incrementally combining them through layers to generate high-level features. A typical CNN structure contains an input layer, convolutional layer, pooling layer, and fully connected layer. As illustrated in Figure 23.6, a conventional CNN structure usually includes a fully connected layer, convolutional layer and pooling layer.

The input layer can be represented in a two-dimensional format, such as a time-frequency spectrum, or a one-dimensional format, such as time series data. For instance, the input data can be denoted as $X \in R\ A \times B$, where A and B are the dimensions of the input data [6, 17]. In the convolutional layer, the convolution kernel (filter) convolves the input data from

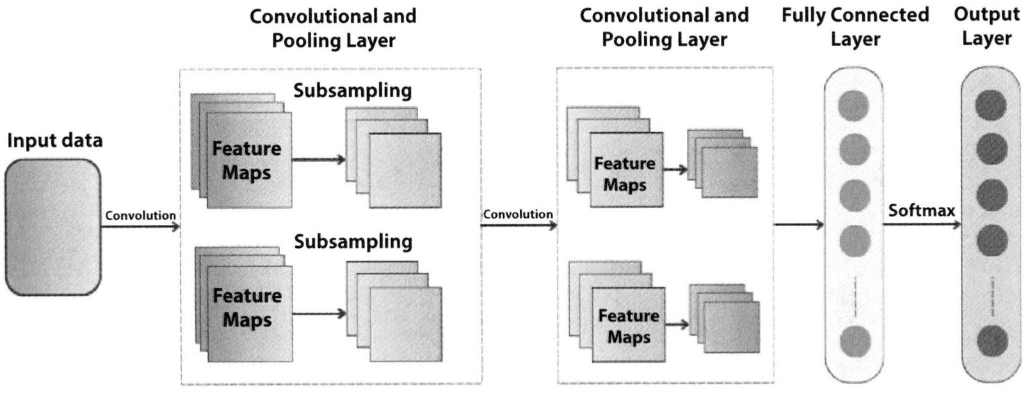

Figure 23.6 Schematic representation showing brief working of the convolutional neural network (CNN).

the previous layer using a set of weights, producing a feature output commonly referred to as a feature map. The output of the convolutional layer can be computed as

The input layer can represent data in 2D like a time-frequency spectrum or 1D like a time series. For example, the input data X could be a matrix with dimensions A x B. In the convolutional layer, the convolution kernel slides over the input, applying a set of weights to produce feature maps. The convolutional layer output is calculated using a convolution operation indicated by the * symbol, where cn is the number of filters. Wcn represents the weight matrix for filter cn, bcn is the bias, and f is an activation function like ReLU.

$$Y_{cn} = f(X * W_{cn} + b_{cn})$$

And the Pooling layers subsample the data to reduce parameters and prevent overfitting while retaining key information. Max pooling takes the maximum value from each feature map.

$$P_{cn} = max_{S^{MXN}} f(Y_{cn})$$

Here, S is an M x N scale matrix used in pooling. M and N denote the dimensions of S. Following various combinations of convolutional and pooling layers, multiple fully-connected layers are employed. These layers are responsible for converting the filter matrix into either a column or a row format. Ultimately, a classification or regression layer can be appended to accomplish specific objectives.

23.4.2.2 Transformer Architecture

While deep learning methods have demonstrated effectiveness in addressing the issue, an ongoing research challenge involves developing Predictive Maintenance (PdM) techniques that are not only computationally efficient but, more crucially, suitable for real-world IoT scenarios. In such scenarios, the methods need to be executable directly on the limited hardware of devices. With their attention mechanism and parallelizability in transformer blocks, transformers are a neural architecture that bridges the gap between compute efficiency and representational power – unlocking deep learning scalability for manufacturing [18].

Transformers are a novel neural network architecture first proposed in the literature by [19] that have become ubiquitous across natural language processing tasks and are now being adapted for image, video, and signal processing use cases. Transformers introduced the transformer block - a component built solely using attention mechanisms without convolution or recurrence. Attention enables models to focus processing only on relevant parts of a large input vector, significantly improving computational efficiency. The architecture of the transformer is illustrated in Figure 23.7.

The conventional transformer comprises both an encoder and decoder stack. The encoder transforms an input sequence into a continuous vector representation, and the decoder produces the elements of the target sequence incrementally based on the encoder outputs. Each utilizes transformer blocks applying multi-headed attention - running multiple parallel

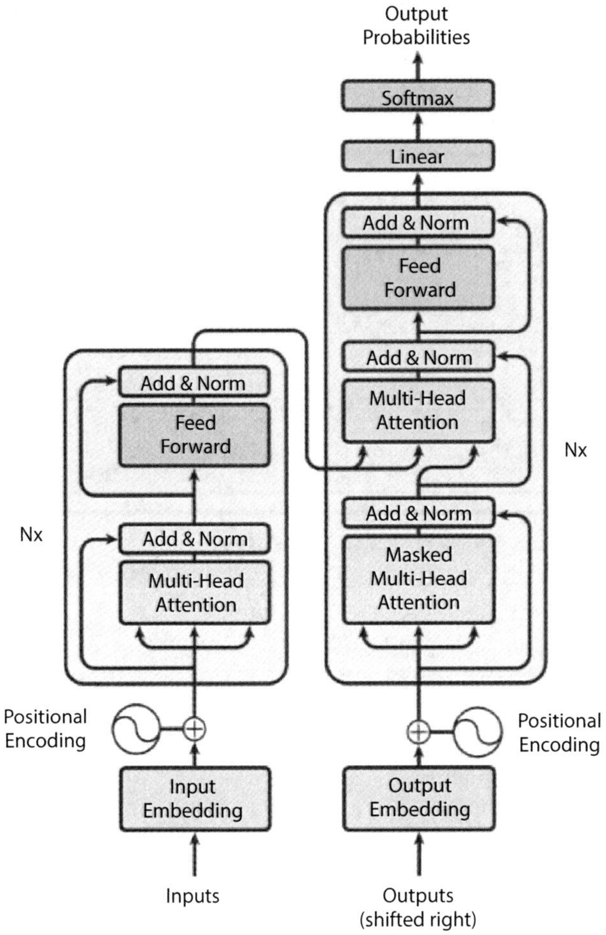

Figure 23.7 Schematic representation showing transformer architecture introduced in [24].

attention layers to connect distant input signals [18]. Transformers also employ residual connections adding the block input to its output as well as layer normalization stabilizing the internal layers' activations.

For the research community in manufacturing, transformers offer promising possibilities. Early manufacturing investigations successfully show accuracy improvements in fault diagnosis classification and retrieval based on technical service bulletins. Transformers have also demonstrated an aptitude for handling sensor measurements over time for predictive maintenance forecasting.

23.4.3 Reinforcement Learning

Reinforcement learning algorithms can make decisions by interacting with an environment. This means an AI applied to predictive manufacturing maintenance can learn directly from the observations and outcomes experienced in the actual production facility. It can experiment with different maintenance policies to minimize total downtime over the long run [7].

By formalizing the maintenance planning challenge as a reinforcement learning problem, the objective becomes optimizing a reward function. This reward function can incorporate factors like maximizing uptime, minimizing technician time and costs, or planning around fluctuations in product demand [7]. The AI learns which conditions should trigger a maintenance action through its past experiences and incremental improvements.

Here is one way/example for which reinforcement learning can be utilized for optimizing predictive manufacturing maintenance, and is described in brief and also shown in the pictorial form in Figure 23.8.

Environment – The environment consists of the manufacturing system the agent must learn to operate in. This includes the machines, production schedule, sensors collecting data on equipment condition, maintenance technicians, repair parts inventories, etc. The environment evolves dynamically over time, presenting new situations to the learning agent.

States – The state portrays the current state of the environment as observed by the learning agent. The state encodes salient information useful for maintenance decisions - e.g. machine temperature or vibration thresholds crossed, time since service on components, availability of technicians, downtime costs, etc. The state evolves based on the environment and actions taken.

Actions – Possible actions could include scheduling/canceling maintenance work orders, assigning limited technician resources, ordering replacement parts, changing operating conditions, making no changes continuing monitoring, etc. The policy will determine what action the agent takes in each state encountered.

Policy – The policy defines the agent's strategy for which action to take given the current state. Improving this predictive maintenance policy to maximize long-run reward is the objective of reinforcement learning.

Rewards – A reward function assigns a numeric value representing the desirability of outcomes to provide feedback to the agent. Positive rewards could come from minimizing machine downtime and repair costs. Negative rewards may result from excessive preventive maintenance or unplanned breakdowns. The updates guide the agent toward an optimal policy.

Reinforcement learning (RL) addresses the limitations of traditional control and planning methods for complex systems like PdM [7]. RL builds optimal solutions without

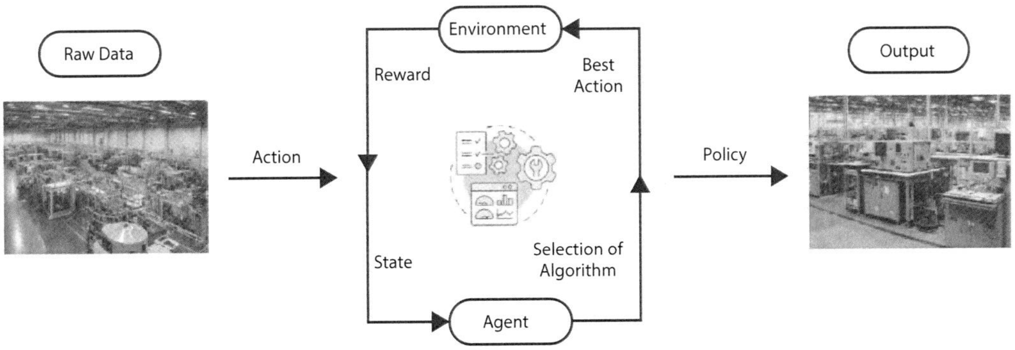

Figure 23.8 Illustration of an approach where reinforcement learning (RL) can be utilized for predictive maintenance this.

needing a model of the system, making it suitable for non-linear processes characteristic of PdM [7]. RL is better suited for complex systems than analytical methods because it learns from interaction and feedback, not labels or models [7]. The RL "feedback loop" involves the agent taking actions, receiving rewards, and adapting its policy, effectively learning through trial and error [7]. The PdM "agent" is the planner, and the "environment" includes sensors, operators, and external data, emphasizing the real-world context of RL for predictive maintenance [7].

Some of the key advantages of using reinforcement learning are:

1. Adaptability: Where traditional approaches might fail, RL agents can learn to navigate complicated, dynamic settings.
2. Scalability: RL algorithms are applicable to a broad range of activities and domains, irrespective of the problem's complexity or magnitude.
3. Sample efficiency: Compared to supervised learning techniques, RL agents can learn from a minimal quantity of data.

23.4.4 Unsupervised Learning

While much AI focus goes on supervised techniques like deep learning, unsupervised methods hold significant transformative potential for discovering hidden insights in manufacturing data [18]. Unsupervised learning aims to model the underlying structure of data without labeled examples guiding model training. Key unsupervised approaches include cluster analysis for grouping similar data points and dimensionality reduction for visualizing complex data spaces [15].

Combined with anomaly detection methods, reducing manufacturing data also shows promise for predictive maintenance to accurately pinpoint emerging equipment faults without tightly supervised failure data. Emerging generative modeling techniques such as variational autoencoders, generative adversarial networks, and restricted Boltzmann machines are at the forefront of unsupervised industrial AI [18, 15]. By learning to generate synthetic manufacturing data streams, models can address monitoring limitations like the lack of low probability failure mode training data while preserving production line privacy and IP. As algorithms mature, the enhanced simulation will provide decision support for production planning and optimization scenarios difficult to recreate offline.

A recognized challenge in supervised predictive maintenance is having access to only a small amount of labeled data. A previous study investigated using unsupervised pre-training to improve remaining useful life predictions in a semi-supervised framework. Specifically, they employed restricted Boltzmann machines (RBMs) for unsupervised feature extraction from unlabeled inputs as the initial pre-training phase. This enabled automatic discovery of abstract features and set network weights at a near-optimal starting point. The model was then fine-tuned through supervised training end-to-end with the extracted features as input.

23.4.5 Transfer Learning

Transfer learning is often used when there are insufficient labeled examples, especially of faulty behavior, for the target industrial systems. Optimal deep learning predictive maintenance models need substantial data encompassing both normal running and failure instances. However, faults are inherently rare in production environments since prolonged faulty operation risks severe breakdowns and is costly. Moreover, the gradual deterioration ultimately causing failure further limits available defect data. Failures are uncommon in industrial settings, making labeled data scarce to train predictive models. Transfer learning provides a way to overcome this by leveraging labeled data from a separate source system to initialize the target model before final tuning on the limited target data [6].

To address this challenge, one approach involves transfer learning. In transfer learning the tasks share underlying principles, the transferred knowledge can enhance the performance of the target domain, for instance, by reducing the required number of samples to achieve near-optimal performance. The concept of transfer learning can be visualized by referring to Figure 23.9.

Therefore, utilizing labeled data from the source domain along with unlabeled data from the target domain enables domain adaptation algorithms to mitigate the distribution mismatch between the two domains. Manufacturers benefit from employing a pre-trained deep learning model, eliminating the necessity to construct extensive labeled datasets for training complex models from the ground up. Instead, by using a modest manufacturing-specific dataset, it is possible to train a few additional neural network layers to adjust the imported feature maps and knowledge for the target tasks of predictive maintenance, quality assurance, or forecasting [6]. With the advancement of computing capabilities, transfer learning has played a key role in fostering growth in industrial AI applications by reducing data-related hurdles.

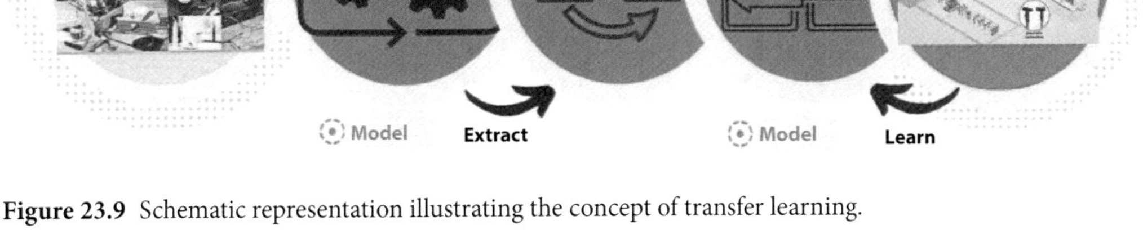

Figure 23.9 Schematic representation illustrating the concept of transfer learning.

Table 23.1 State-of-the-art AI models and approaches utilized in the manufacturing industry.

References	Authors	Year	Brief description
[20]	Jin *et al.*	2023	The article suggests an optimized IoT Big Data ecosystem designed for predictive maintenance in smart manufacturing, integrating edge computing and a deep learning technique based on autoencoders.
[21]	Murugiah *et al.*	2023	This paper presents a new predictive manufacturing system for monitoring industrial machines in the Industry 4.0 era. The approach uses a Multi-Scale Dilation Attention Convolutional Neural Network (MSDA-CNN) to extract descriptive features. These features are weighted and input into an Optimized Hybrid Fault Detection (OHFD) procedure carried out jointly by a Deep Neural Network (DNN) and a Deep Belief Network (DBN).
[22]	Netisopakul *et al.*	2022	In this paper, AI technology was utilized in the manufacturing sector, specifically to predict the temperature and insulation values of motors from the CNC machine. The authors discovered that the Bi-LSTM model produced the lowest RMSE and MAE values, making it the preferred choice.
[23]	Suawa *et al.*	2023	In this study, a technique involving noisy training was employed to enhance the resilience and precision of convolutional deep learning models in monitoring industrial equipment. Despite introducing noise during the test phase, the approach achieved an accuracy level of 95%.
[24]	Shoorkand *et al.*	2023	The article suggests a consolidated framework for production planning and predictive maintenance through the application of deep learning. It employs a long short-term memory model to precisely forecast the machine's health status, aiding in the identification of suitable preventive maintenance measures.
[25]	Chen *et al.*	2023	This paper puts forward a predictive maintenance approach applying an ensemble model that integrates two deep learning techniques - deep autoencoders (DAEs) and long short-term memory networks (LSTMs). This combined model is used to forecast system failures.

23.4.6 State-of-the-Art Models

Some of the state-of-the-art models and approaches used for predictive maintenance are described in Table 23.1.

23.5 Digital Twins

A digital twin is a virtual representation that acts as a real-time artificial intelligence counterpart to a physical object or process. Unlike other tools, digital twins emphasize analyzing maintenance logs and presenting analytics on smart dashboards rather than just processing sensor data. A hybrid system paradigm has been proposed that supplements physics-based modeling with data-driven analytics. This allows leveraging real-time sensor feeds from industrial equipment into a digital twin representation that continuously adapts to operational changes, enhancing autonomy. Digital twins move beyond sensor data processing to incorporate textual data analysis and dashboard tools for providing intelligent analytics. Their ability to synergize first-principles and data-driven modeling in a responsive virtual representation of physical assets can enhance autonomous adaptation.

Digital twins can enable hybrid predictive maintenance in cyber-physical systems. A digital twin is comprised of a physics-based model of physical equipment along with real-time sensor feeds and historical operational records. This creates a detailed digital representation that mimics the physical asset's conditions and evolution over time based on physical principles. Additionally, digital twins can provide reliable, contextualized data via intelligent analysis of sensor measurements as well as mining of operational logs. Their multi-domain modeling facilitates high fidelity representations while model consistency maintenance ensures dynamic updating with the latest observations.

Utilizing highly precise digital twin models of industrial machines enables manufacturers to shift to more proactive, predictive maintenance strategies. By mirroring the real-time status and performance of equipment based on sensor data from its physical counterpart, the digital twin provides the necessary visibility.

As problems are spotted through predictive analytics of digital twin data, manufacturers can act ahead of time to schedule maintenance at the optimal point. Technicians are equipped with insights on the likely root cause, location, and other specifics to streamline addressing the issue thanks to the intelligence within the digital twin [26]. The result is less downtime and greater productivity due to the prevention of equipment failures before they escalate or cause line stoppages. Table 23.2 contains some of the approaches that have implemented the digital twins for predictive maintenance in the manufacturing industry.

23.6 Role of Artificial Intelligence in Predictive Maintenance

AI is advancing predictive maintenance in manufacturing in five key areas: 1) Processing diverse data from sensors and imagery related to equipment health; 2) Using generative techniques like GANs to expand limited real-world training data; 3) Applying prognostic health models to estimate component life; 4) Detecting anomalies in sensor streams and vision outputs that may indicate emerging issues; 5) Optimizing maintenance strategies

Table 23.2 State-of-the-art digital twins utilized in the manufacturing industry.

References	Authors	Year	Brief description
[27]	Luo *et al.*	2020	This study puts forth a combined approach for predictive maintenance of computer numerical control (CNC) machine tools, utilizing both a Digital Twin model and data. It includes a case study concentrating on forecasting the remaining useful life of cutting tools.
[28]	Siddiqi *et al.*	2023	This article presents a predictive maintenance algorithm to detect anomalies in automation systems, aiming to prevent asset failures. The approach employs a Digital Twin powered by artificial intelligence to spot abnormalities early, thereby reducing the risk of critical equipment breakdowns and associated consequences.
[26]	Mourtzis *et al.*	2023	The paper introduces an approach for optimizing the reliability of robotic cells through the integration of digital twin technology and predictive maintenance. It discusses leveraging a digital twin for simulation and near-real-time monitoring of the robot, combined with predictive maintenance strategies to identify and categorize component malfunctions.

by dynamically adjusting schedules and allocating resources. Together, these AI capabilities allow for increased uptime and optimized decisions through sophisticated analysis of equipment data and insights. All of these areas are discussed in brief in the following sub-sections.

23.6.1 Types of Data Used for Predictive Maintenance

Emphasizing the multifaceted nature of manufacturing data, the section explores four key data types: time series, natural language, knowledge graph, and image-based data [29]. Time series data provides a chronological sequence of events, offering insights into machinery performance over time. Natural language data facilitates the integration of textual information, enabling the analysis of maintenance reports and documentation. Knowledge graphs contribute to a holistic understanding of interconnected data points, fostering predictive insights. Additionally, image-based data is pivotal for visual analysis, allowing for the identification of equipment anomalies and defects. Furthermore, Table 23.3 contains an example of how one can utilize the specific data type for some cases with reference to the literature that has used them.

Table 23.3 Table illustrating different data types utilized for the predictive maintenance approach in manufacturing industry.

Data type	Sample usage of data type	References in which data types used
Time Series	Sensor data (temperature, pressure, vibration, etc.) recorded over time from equipment and machines. This can show patterns and trends that indicate potential issues.	[30, 31]
Natural language	Operator logs and technician notes documenting observed issues with machines or quality problems. Natural Language Processing can extract insights.	[32]
Knowledge Graph	Structured data on machine and component relationships, configurations, and hierarchies on the factory floor. Allows tracking issues propagating between connected equipment.	[1, 33, 34]
Based	Camera feeds monitor production lines, equipment, and products and can supply images and video to AI systems to detect defects, abnormalities, etc.	[5, 35]

23.6.2 Synthetic Data Generation

With the wider adoption of sensors and industrial internet-of-things (IIoT) devices, manufacturers have greater access than ever before to detailed operations data. However, developing accurate machine learning models for predictive maintenance requires abundant quality data covering all potential failure modes – which rarely exists in the necessary volumes [36]. Synthetic data generation offers a solution for producing large supplemental datasets that can augment real-world histories and improve reliability predictions.

23.6.2.1 Generative Adversarial Net (GAN)

One method for creating synthetic data uses Generative Adversarial Networks (GANs), first introduced in [37]. A GAN has two components - a generator (G) and a discriminator (D). As illustrated in Figure 23.10, the generator G takes random input from a latent space and generates fake samples, like time series sequences. It provides these fabricated examples to the discriminator D. The job of D is to differentiate between the fake samples created by G and real samples from the original dataset. GANs work adversely, with G trying to better fool D by synthesizing increasingly realistic data, while D tries to improve at detecting the fakes.

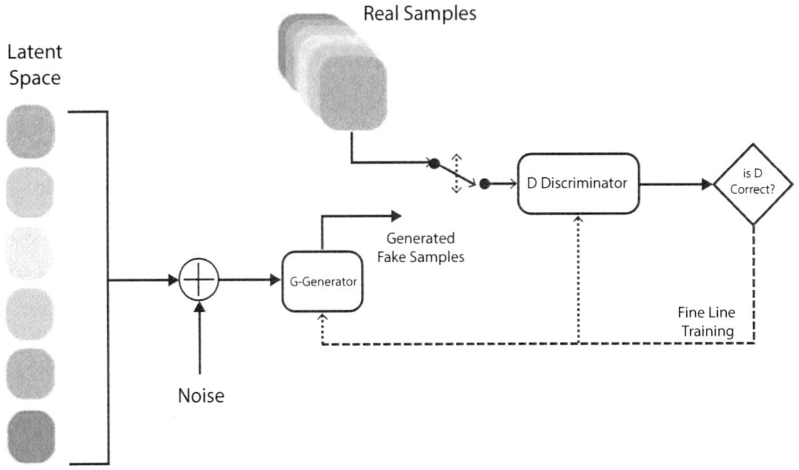

Figure 23.10 Schematic representation illustrating architecture of the generative adversarial nets (GANs).

This back-and-forth competition forces both models to become progressively enhanced - G at fabricating plausible synthetic data, and D at distinguishing real from fake data.

Originally, GANs were applied in predictive maintenance as a data augmentation technique to handle class imbalance. Research has demonstrated that GANs can effectively generate oversampled data when the imbalance ratio between classes is low [38]. However, when the number of samples from the minority class is very limited, the GAN generator can struggle to produce meaningful synthetic data reflecting true characteristics. To address this limitation for highly skewed datasets, a hybrid oversampling approach combining adaptive synthetic sampling (ADASYN) with GANs was proposed. This leverages GAN-produced data in conjunction with strategically created synthetic samples based on adaptive algorithms.

23.6.2.2 Diffusion Models

While GANs have shown promise for generating synthetic data useful for training AI models, diffusion models are emerging as an alternative generation approach with some distinct advantages [39]. Diffusion models work by adding noise to real data over repeated iterations and learning these noise distributions to allow "denoising" back to realistic outputs. This offers greater control, interpretability, and the ability for conditional steering not found in GANs [40].

Diffusion-based augmentation with tight conditioning guardrails results in high-quality synthetic images, signals, and data that maintain the integrity and distributions of real documented examples from the factory floor [40]. This leads to safer and more effective adoption of AI for critical predictive maintenance tasks. As algorithms evolve, diffusion models are expected to become an essential data synthesis method within smart factories - significantly multiplying scarce real-world examples into abundant quality data for powering advanced predictive maintenance driven by AI.

23.6.3 Prognostic and Health Management (PHM)

PHM refers to techniques that enable predicting and preventing unexpected equipment failures through proactive condition monitoring and diagnostics. In manufacturing, PHM uses data from sensors on industrial machines as well as histories of operation and failure to ascertain equipment health and anticipate maintenance needs before breakdowns occur. This avoids costly downtime and prevents secondary damage from failures [6, 41].

For example vibration sensors on motors or pumps connected to cloud analytics that identify increasing vibration thresholds indicating wearing bearings or introduction of friction - allowing for proactive repair or part replacement before failure.

23.6.3.1 Remaining Useful Time (RUL)

RUL signifies the duration during which a system or component is expected to function before reaching a critical degradation level that requires maintenance or replacement. In manufacturing predictive maintenance, AI algorithms analyze all available data on usage, wear, and tear indicators, operating conditions, etc. to dynamically estimate component RUL at any given point [6, 29].

23.6.3.2 End of Life (EOL)

EOL refers to when an equipment part or system reaches the end of its effective operational life based on its design parameters or usable lifespan considering found damage, fatigue and applied stress over time. In manufacturing, predictive analytics tracking degradation can forecast EOL and cue replacement ordering, migrations, etc. while the system is still functioning without issue. This buffers against end of runtime surprises [6, 29].

23.6.4 Anomaly Detection

Anomaly detection refers to finding data points that diverge markedly from expected patterns. Anomalies can stem from issues in the data acquisition pipeline, like sensor malfunctions, low power, or transmission errors [41, 42]. They may also reflect real events in the industrial process, such as equipment failure or line changes. While the latter anomalies provide useful analytical signal, the former sensor-induced anomalies constitute noise that can impede analysis [43]. However, the distinction between anomalous events and noise depends on the data properties. As such, what is considered noise versus useful outlier information requires evaluation based on the specific analytical context and data type.

A key capability provided by AI is automated anomaly detection on the manufacturing floor [44–46]. Rather than relying solely on human monitoring and vigilance to spot equipment issues or product defects, anomaly detection algorithms leverage sensor streams, imaging, and other data to flag abnormalities. Computer vision techniques enable scrutinizing video feeds of production lines for shape, color, or finish flaws invisible to the human eye. The vision systems automatically recognize and report aberrations as they occur to speed remediation. AI introspection spots inconsistencies in quality control test results indicating potential instrumentation drifts.

23.6.5 Maintenance Strategy Optimization

The integration of AI in maintenance strategy optimization introduces real-time monitoring capabilities that continuously assess the health and performance of manufacturing assets. Sensors and IoT devices feed a constant stream of data into AI algorithms, allowing for the detection of anomalies and subtle patterns indicative of potential issues. In response to these insights, AI systems can dynamically adjust maintenance schedules, allocating resources where they are most urgently required. This adaptive approach not only maximizes operational uptime but also enhances overall system reliability [47].

AI-driven maintenance strategy optimization goes beyond mere prediction by incorporating advanced decision-making capabilities. AI models analyze vast datasets to identify optimal maintenance strategies based on factors such as equipment condition, historical performance, and production demands. This data-driven decision-making process ensures that resources are allocated efficiently, reducing unnecessary maintenance costs while maintaining the integrity of critical assets. As a result, manufacturers can achieve a delicate balance between minimizing expenses and maximizing the reliability of their production infrastructure [6, 36].

23.7 Limitations and Challenges

This section examines the limitations and challenges by presenting a balanced view, this section contributes to a comprehensive understanding of the multifaceted implications of integrating AI into manufacturing, highlighting the need for strategic considerations and careful implementation to navigate both the advantages and obstacles in this transformative journey.

- Synthetic Dataset: Creating realistic synthetic datasets for training AI models in manufacturing is challenging, requiring accurate representation of real-world processes.
- Complex Systems Prediction: Shifting from predicting individual components to behaviors in complex manufacturing systems demands AI models to adapt and account for multifaceted interactions, demanding advanced modeling techniques and increased computational resources.
- High-End Computation: AI implementation in manufacturing, especially for large-scale data analytics, requires significant computational power, adding complexity and cost.
- Non-Real-Time Prediction: Achieving real-time prediction in manufacturing is hindered by complexities and rapid dynamics, impacting immediate responsiveness.
- Data Heterogeneity Issues: Effective AI model training depends on diverse datasets; challenges arise when data lacks diversity or is unbalanced, potentially leading to biased models.
- Noisy Data in Manufacturing: Noisy data, with inaccuracies or outliers, is common in manufacturing environments, posing a challenge for AI algorithms and requiring robust preprocessing techniques.

23.8 Opportunities and Future Scope

As AI technologies advance, manufacturing processes can gain efficiency, predictive maintenance, and optimized resource allocation. The synergy of AI and data analytics has the potential to revolutionize production workflows, enabling real-time decision-making and adaptive manufacturing systems. This section emphasizes the transformative impact of AI on manufacturing, exploring myriad possibilities for innovation and growth in a new era driven by intelligent technologies.

- Collaborative Dataset Efforts: Enhancing the effectiveness of PdM relies on extensive datasets, but data collection can be challenging. Collaborative efforts within the PdM community to gather and share large-scale datasets are crucial for overcoming this obstacle.
- Optimization of Maintenance Strategies: While current research focuses on fault diagnosis and prognosis, there is a need to optimize maintenance strategies. Efficient planning of maintenance activities using AI can lead to automation, cost reduction, and minimized downtime.
- Addressing Class Imbalance: In production systems, class imbalance due to infrequent failure events poses a challenge. Addressing this issue in data collection is essential for robust AI-based PdM models.
- Digital Twin Integration: Utilizing constantly refreshed digital twins provides valuable run-to-failure data for crucial components, enhancing fault detection and prediction capabilities in PdM.
- Reinforcement Learning for Proactive Maintenance: Applying reinforcement learning for proactive and cost-effective PdM in manufacturing allows systems to learn optimal maintenance strategies through experiential learning, adapting in real-time to changing equipment health and operating contexts.
- Multi-component System Challenges: With the increasing complexity of manufacturing systems involving numerous components, existing AI approaches often focus on individual components. Developing proficient AI-based PdM algorithms for multi-component systems remains a significant unresolved concern, presenting opportunities for future research.

23.9 Concluding Remarks

This chapter has delved into the immense possibilities that AI presents for revolutionizing the manufacturing sector. Through applications such as predictive maintenance, process enhancement, and quality assurance, AI has the capability to significantly reshape the industry and its operations. When thoughtfully implemented, these technologies can provide invaluable insights from data to enhance decision-making, asset utilization, and sustainability. However, successfully leveraging AI in manufacturing will require interdisciplinary collaboration. Domain experts must partner with data scientists and ML engineers to properly frame business challenges, curate quality datasets, and iteratively refine AI solutions. While AI adoption in manufacturing remains in the early stages, rapid advances in enabling technologies are unfolding. By combining strengths in applications, algorithms, and infrastructure, manufacturers can traverse the AI maturity curve in manageable steps.

References

1. Hansen, E.B. and Bøgh, S., Artificial intelligence and internet of things in small and medium-sized enterprises: A survey. *J. Manuf. Syst.*, 58, 362–372, 2021.
2. Singh, R.R., Bhatti, G., Kalel, D., Vairavasundaram, I., Alsaif, F., Building a digital twin powered intelligent predictive maintenance system for industrial AC machines. *Machines*, 11, 8, 796, 2023.
3. Zhang, W., Yang, D., Wang, H., Data-driven methods for predictive maintenance of industrial equipment: A survey. *IEEE Syst. J.*, 13, 3, 2213–2227, 2019.
4. Bevilacqua, M. and Braglia, M., The analytic hierarchy process applied to maintenance strategy selection. *Reliab. Eng. Syst. Saf.*, 70, 1, 71–83, 2000.
5. Shin, W., Han, J., Rhee, W., AI-assistance for predictive maintenance of renewable energy systems. *Energy*, 221, 119775, 2021.
6. Ran, Y., Zhou, X., Lin, P., Wen, Y., Deng, R., A survey of predictive maintenance: Systems, purposes and approaches, *arXiv preprint arXiv:1912.07383*, 2019.
7. Siraskar, R., Kumar, S., Patil, S., Bongale, A., Kotecha, K., Reinforcement learning for predictive maintenance: a systematic technical review. *Artif. Intell. Rev.*, 1-63, 2023.
8. Esa, M.A.M. and Muhammad, M., Adoption of prescriptive analytics for naval vessels risk-based maintenance: A conceptual framework. *Ocean Eng.*, 278, 114409, 2023.
9. Bi, Z., Jin, Y., Maropoulos, P., Zhang, W.J., Wang, L., Internet of things (IoT) and big data analytics (BDA) for digital manufacturing (DM). *Int. J. Prod. Res.*, 61, 12, 4004–4021, 2023.
10. Khan, S., II, Kaur, C., Al Ansari, M.S., Muda, I., Borda, R.F.C., Bala, B.K., Implementation of cloud based IoT technology in manufacturing industry for smart control of manufacturing process. *Int. J. Interact. Des. Manuf.*, 1–13, 2023.
11. Rath, K.C., Khang, A., Roy, D., The Role of Internet of Things (IoT) Technology in Industry 4.0 Economy, in: *Advanced IoT Technologies and Applications in the Industry 4.0 Digital Economy*, pp. 1–28, CRC Press, 2024.
12. Juma, M., Alattar, F., Touqan, B., Securing Big Data Integrity for Industrial IoT in Smart Manufacturing Based on the Trusted Consortium Blockchain (TCB). *IoT*, 4, 1, 27–55, 2023.
13. Schmitt, M., Big Data Analytics in the Metaverse: Business Value Creation with Artificial Intelligence and Data-Driven Decision Making, 2023, Available at SSRN 4385347.
14. Çınar, Z.M., Abdussalam Nuhu, A., Zeeshan, Q., Korhan, O., Asmael, M., Safaei, B., Machine learning in predictive maintenance towards sustainable smart manufacturing in industry 4.0. *Sustainability*, 12, 19, 8211, 2020.
15. Ar ena, F., Collotta, M., Luca, L., Ruggieri, M., Termine, F.G., Predictive maintenance in the automotive sector: A literature review. *Math. Comput. Appl.*, 27, 1, 2, 2021.
16. Mohamed Almazrouei, S., Dweiri, F., Aydin, R., Alnaqbi, A., A review on the advancements and challenges of artificial intelligence based models for predictive maintenance of water injection pumps in the oil and gas industry. *SN Appl. Sci.*, 5, 12, 391, 2023.
17. Li, Z., Liu, F., Yang, W., Peng, S., Zhou, J., A survey of convolutional neural networks: analysis, applications, and prospects. *IEEE Trans. Neural Netw. Learn. Syst.*, 2021.

18. De Luca, R., Ferraro, A., Galli, A., Gallo, M., Moscato, V., Sperli, G., A deep attention based approach for predictive maintenance applications in IoT scenarios. *J. Manuf. Technol. Manage.*, 34, 4, 535–556, 2023.
19. Vaswani, A., Shazeer, N., Parmar, N., Uszkoreit, J., Jones, L., Gomez, A.N., Polosukhin, I., Attention is all you need. *Adv. Neural Inf. Process. Syst.*, 30, 2017.
20. Yu, W., Liu, Y., Dillon, T., Rahayu, W., Edge computing-assisted IoT framework with an autoencoder for fault detection in manufacturing predictive maintenance. *IEEE Trans. Ind. Inf.*, 19, 4, 5701–5710, 2022.
21. Murugiah, P., Muthuramalingam, A., Anandamurugan, S., A design of predictive manufacturing system in IoT-assisted Industry 4.0 using heuristic-derived deep learning. *Int. J. Commun. Syst.*, 36, 5, e5432, 2023.
22. Netisopakul, P. and Phumee, N., AI-Enhanced Predictive Maintenance in Manufacturing Processes, in: *2022 22nd International Conference on Control, Automation and Systems (ICCAS)*, 2022, November, IEEE, pp. 1107–1112.
23. Suawa, P., Halbinger, A., Jongmanns, M., Reichenbach, M., Noise-robust Machine Learning Models for Predictive Maintenance Applications. *IEEE Sens. J.*, 2023.
24. Dehghan Shoorkand, H., Nourelfath, M., Hajji, A., A deep learning approach for integrated production planning and predictive maintenance. *Int. J. Prod. Res.*, 1-20, 2023.
25. A Predictive Maintenance Strategy Using Deep Learning Quantile Regression and Kernel Density Estimation for Failure Prediction.
26. Mourtzis, D., Tsoubou, S., Angelopoulos, J., Robotic Cell Reliability Optimization Based on Digital Twin and Predictive Maintenance. *Electronics*, 12, 9, 1999, 2023.
27. Luo, W., Hu, T., Ye, Y., Zhang, C., Wei, Y., A hybrid predictive maintenance approach for CNC machine tool driven by Digital Twin. *Rob. Comput. Integr. Manuf.*, 65, 101974, 2020.
28. Siddiqui, M., Kahandawa, G., Hewawasam, H.S., Artificial Intelligence Enabled Digital Twin For Predictive Maintenance in Industrial Automation System: A Novel Framework and Case Study, in: *2023 IEEE International Conference on Mechatronics (ICM)*, 2023, March, IEEE, pp. 1–6.
29. Stanton, I., Munir, K., Ikram, A., El-Bakry, M., Predictive maintenance analytics and implementation for aircraft: Challenges and opportunities. *Syst. Eng.*, 26, 2, 216–237, 2023.
30. Liu, X., Liu, L., Liu, D., Wang, L., Guo, Q., Peng, X., A hybrid method of remaining useful life prediction for aircraft auxiliary power unit. *IEEE Sens. J.*, 20, 14, 7848–7858, 2020.
31. Liao, L., Jin, W., Pavel, R., Enhanced restricted Boltzmann machine with prognosability regularization for prognostics and health assessment. *IEEE Trans. Ind. Electron.*, 63, 11, 7076–7083, 2016.
32. Dangut, M.D., Skaf, Z., Jennions, I.K., An integrated machine learning model for aircraft components rare failure prognostics with log-based dataset. *ISA Trans.*, 113, 127–139, 2021.
33. Zhang, G., Cao, X., Zhang, M., A Knowledge Graph System for the Maintenance of Coal Mine Equipment. *Math. Probl. Eng.*, 2021, 1–13, 2021.
34. Doğru, A., Bouarfa, S., Arizar, R., Aydoğan, R., Using convolutional neural networks to automate aircraft maintenance visual inspection. *Aerospace*, 7, 12, 171, 2020.
35. Kiangala, K.S. and Wang, Z., An effective predictive maintenance framework for conveyor motors using dual time-series imaging and convolutional neural network in an industry 4.0 environment. *IEEE Access*, 8, 121033–121049, 2020.
36. Klein, P. and Bergmann, R., Generation of Complex Data for AI-based Predictive Maintenance Research with a Physical Factory Model, in: *ICINCO*, 2019, July, vol. 1, pp. 40–50.
37. Goodfellow, I., Pouget-Abadie, J., Mirza, M., Xu, B., Warde-Farley, D., Ozair, S., Bengio, Y., Generative adversarial nets. *Adv. Neural Inf. Process. Syst.*, 27, 2014.

38. Lee, Y.O., Jo, J., Hwang, J., Application of deep neural network and generative adversarial network to industrial maintenance: A case study of induction motor fault detection, in: *2017 IEEE international conference on big data (big data)*, 2017, December, IEEE, pp. 3248–3253.
39. Stypułkowski, M., Vougioukas, K., He, S., Zięba, M., Petridis, S., Pantic, M., Diffused heads: Diffusion models beat gans on talking-face generation, in: *Proceedings of the IEEE/CVF Winter Conference on Applications of Computer Vision*, pp. 5091–5100, 2024.
40. Wang, H., Zhang, W., Han, C., Fu, Z., Song, L., A bearing fault diagnosis method with an improved residual Unet diffusion model under extreme data imbalance. *Meas. Sci. Technol.*, 35, 4, 046113, 2024.
41. Nunes, P., Santos, J., Rocha, E., Challenges in predictive maintenance–A review. *CIRP J. Manuf. Sci. Technol.*, 40, 53–67, 2023.
42. Erhan, L., Ndubuaku, M., Di Mauro, M., Song, W., Chen, M., Fortino, G., Liotta, A., Smart anomaly detection in sensor systems: A multi-perspective review. *Inf. Fusion*, 67, 64–79, 2021.
43. Keogh, E., Lin, J., Fu, A., Hot sax: Efficiently finding the most unusual time series subsequence, in: *Fifth IEEE International Conference on Data Mining (ICDM'05)*, 2005, November, IEEE, p. 8.
44. Ripley, B.D., *Pattern recognition and neural networks*, Cambridge university press, 2007.
45. Bezdek, J.C., Ehrlich, R., Full, W., FCM: The fuzzy c-means clustering algorithm. *Comput. Geosci.*, 10, 2-3, 191–203, 1984.
46. Labrín, C. and Urdinez, F., Principal component analysis, in: *R for political data science*, pp. 375–393, Chapman and Hall/CRC, 2020.
47. Hesabi, H., Nourelfath, M., Hajji, A., A deep learning predictive model for selective maintenance optimization. *Reliab. Eng. Syst. Saf.*, 219, 108191, 2022.

24

Artificial Intelligence Techniques in Predictive Maintenance, Their Applications, Challenges, and Prospects

Akriti Rai, Jyotika shastri and Hina Bansal*

Centre for Computational Biology and Bioinformatics, Amity Institute of Biotechnology, Amity University, Noida, Uttar Pradesh, India

Abstract
Predictive maintenance refers to a preventative care technique which finds flaws or issues before they lead to equipment failures. Using data gathered from sensors on equipment, predictive maintenance builds models that anticipate when the component of equipment is likely to break. Preventive measures can be implemented to avert an issue after a failure is anticipated. By preventing unanticipated failures and the expenses that come with them, predictive maintenance can assist businesses in becoming more cost-effective. There are several predictive maintenance techniques that can be used, depending on the available type of data. Some common techniques include time-series analysis, probability forecasting and artificial intelligence that is linked to the internet of things (IoT). Time-series analysis looks at how past values of a variable change over time. It can be used to predict future values of the variable. Probability forecasting uses historical data to calculate the probability that a particular event will occur. This can be used to predict when a particular component is likely to fail. Artificial intelligence can also be used to create models that learn how equipment behaves and how it is likely to fail. The present chapter gives a detail insight into the above predictive maintenance techniques, their applications and challenges.

***Keywords*:** Predictive maintenance, sensor data, preventive measures, cost-effective, time-series analysis, probability forecasting, artificial intelligence, internet of things (IoT)

24.1 Introduction

Artificial intelligence (AI) tools can examine the state of an equipment's operation and search for signs that it might break down in the future, even if there have not been any obvious problems with it [1]. Statistical analysis is a subset of data analytics that primarily employs modelling, information extraction, robotics, and machine learning techniques. It is used to analyze current and historical data to estimate future events. By examining past and present healthcare data, practitioners can utilize predictive analytics to uncover opportunities to

**Corresponding author*: hinabansal@gmail.com

Amit Kumar Tyagi, Shrikant Tiwari, Senthil Kumar Arumugam and Avinash Kumar Sharma (eds.) Artificial Intelligence-Enabled Digital Twin for Smart Manufacturing, (565–580)

make superior and more effective medicinal and operational decisions, foresee trends, and even restrict the spread of infections [2]. The healthcare sector generates immense quantities of data, but it struggles to translate that data as information that can be utilized to improve patient outcomes [3]. The purpose of analyzing data in the medical field is that it can be used for all aspects of administration of operations and providing care to patients. It is used to seek ways to reduce treatment costs, predict disease outbreaks, improve patient care, and numerous other purposes. From a commercial perspective, analytics can assist healthcare organizations in streamlining internal processes, optimizing resource allocation, and enhancing the efficiency and coordination of care teams [3].

Artificial intelligence (AI) systems can identify minor changes in machine performance that might indicate a requirement for maintenance by comparing the machine's current performance to baseline data [4]. Predictive maintenance is an area where artificial intelligence is used to bring vitality to the formerly stagnant field of equipment maintenance currently [5]. Two essential elements for the proper operation of AI systems are machine learning (ML) techniques and dependable data access.

Artificial intelligence (AI) systems need ongoing access to both past and current data to assess their efficiency and satisfy the requirement for maintenance. This contains details about the functionality of a particular machine, how it functions as part of a wider manufacturing line, and any information regarding abnormalities. In the meanwhile, algorithms are crucial to machine learning and predictive maintenance [2]. These algorithms, which define the steps of ingesting, analyzing, and connecting data sets, are collections of predetermined guidelines and directives. The algorithms can learn over the course of time as they start to identify patterns in the data, as suggested by their name. To distinguish the information from the noise, machine learning methods are also essential. All other data points are noise, whereas signals provide pertinent information.

24.2 Techniques of Predictive Maintenance

24.2.1 Time Series Analysis

Time series analysis is the powerful tool that can be used for the identify the pattern and the terns and the performance and health of data of vehicles [6]. By analyzing the data series, it's possible to detect the change in vehicle performance and help us to predict when they are likely to fail. Additionally, statistical analysis of time series can be used to pinpoint a failure's primary cause, which enables businesses to create maintenance plans that are more successful. Time series analysis for maintenance prediction can employ a few methods, including statistical process control, spectral analysis, and trend analysis. While spectrum analysis is used to find patterns in material that repeat over time, trends analysis is used to find persistent trends in the data. The technique of statistical process control can be employed to find patterns in the data that are not typical. In addition, time series data analysis and forecasts about the vehicle's remaining useful life can be made using methods like models of forecasting, Raman restrictions, and Hidden Markov Models. Using machine learning algorithms is one of the most widely utilized methods in analyzing time series for future maintenance. By using past data, such techniques can be trained to anticipate future performance and spot any problems.

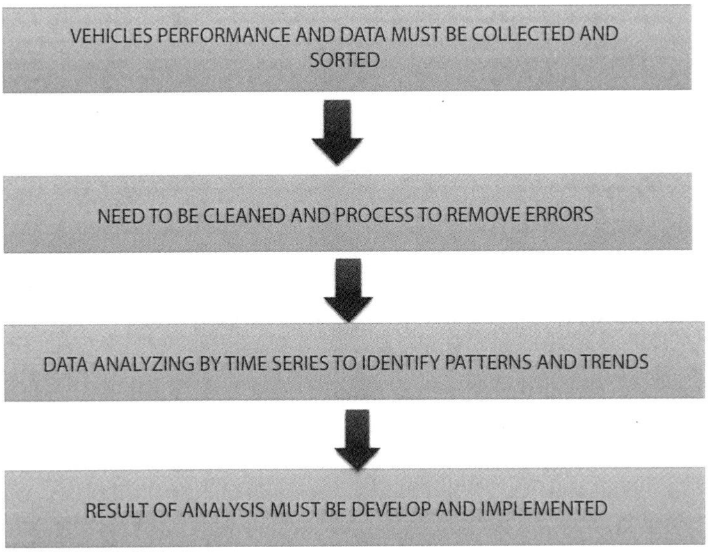

Figure 24.1 Implementation of time screening.

Algorithms like Random Forest, a Support Vector Machine, Artificial Neural Network, and Long Short-Term Memory are frequently utilized. Numerous data sources, including log data, system-wide data, and sensor data, can be used to train these algorithms [2]. Utilizing a variety of data sources helps improve the accuracy of the predictions. The essential features needed to carry out predictive analytics are provided by a time series database. Constructed especially to manage time-ordered data points, it facilitates the rapid storage, retrieval, and processing of time-stamped data in large quantities. Time series databases are especially well-suited for applications like discovering anomalies and predictive maintenance because of these features. All kinds of time series data, such as metrics, events, logs, and traces, can be stored using the time series storage and platform InfluxDB 3.0. Figure 24.1 represents various steps involved in the process of time series analysis.

24.2.2 Probability Forecasting

The likeness of an event, or its likelihood of occurring, is known as probability. On the other hand, forecasting is the process of predicting the occurrence of an event based on similar past events. An entire range of potential future possibilities are assigned a set of probabilities in a probabilistic forecast, as opposed to focusing only on one specific possibility. A supply chain analysis's two most important variables—demand and lead time—can be best addressed with probabilistic forecasting [7]. Since a probabilistic approach is predicated on the likelihood of the outcomes, it is well suited to the economic prioritization of an alternative based on its predicted but uncertain returns. Numerous statistical models are available for use in creating probabilistic forecasts. While some have different structural approaches, some are structurally comparable to their deterministic counterparts. In contrast, a probabilistic forecast requires the measurement of multiple parameters to be considered accurate to a deterministic data-based one [8]. It is necessary to build specialized techniques that are different from what is used to analyze deterministic forecasts to fully utilize probabilistic forecasts.

24.2.2.1 Steps in Probability Forecasting

1. Data collection - The goal is to project outcomes and then collect pertinent data for forecast training and probability development.
2. Data processing - In order to make data more reliable and ready for analysis, it must be cleaned up by removing outliers, missing numbers, and plagiarism.
3. Selection of model - Choose the algorithm that, considering the available data and anticipated accuracy, best fits the issue description. According to data valuation, the model might also be chosen based on a pattern that proved effective in a previous investigation.
4. Model fitting - To estimate the model's criteria, train it on the data. Data can be used to test it and see how the model performs in real-world scenarios. Take Dry Run, for instance.
5. Evaluation of model - To assess the model's performance, a comparison between the model's predictions and the actual outcomes is necessary. It also describes the long-term viability of the probability model.
6. Model improvement - enhancing a model iteratively or repeatedly by acquiring more data, changing its parameters, or utilizing an alternative method.
7. Prediction- By regularly feeding the latest information into the algorithms to stay up to speed with current occurrences, the trained model becomes capable of making predictions.
8. Post -processing -The forecasts ought to be provided in a way that makes sense by providing thorough explanations. When doing probability forecasting, there is a lot of knowledge to consider. To obtain the required level of accuracy, it is common practice to repeat steps 4–7 several times.

24.2.3 AI Technologies Used in Predictive Maintenance

Predictive maintenance is being revolutionized by AI technologies, which allow businesses to manage their assets more proactively and minimize downtime. Figure 24.2 shows a few most important AI tools for predictive maintenance.

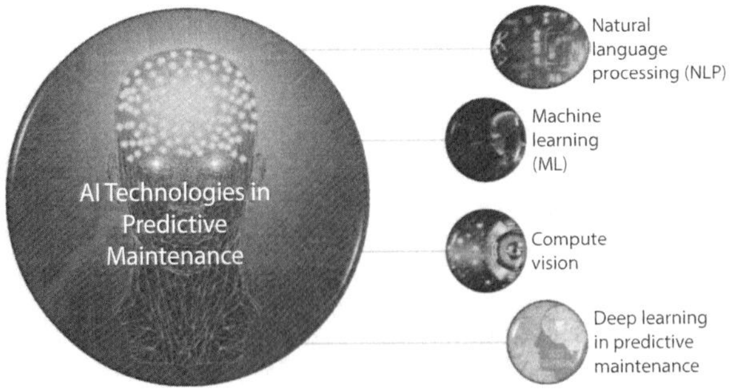

Figure 24.2 AI technologies in predictive maintenance.

24.2.3.1 Natural Language Processing (NLP)

Natural language processing is a newly developed technology in the realm of predictive maintenance that provides companies with additional capabilities for text data management and more effective maintenance schedules [9]. Natural language processing, also known as NLP, has made it possible for machines to comprehend and speak human language. This is particularly useful for extracting data from academic notes, monthly states and equipment manuals. A Key application of NLP which helps in the extraction of information. NLP algorithm can access large volumes of unstructured data such as reports and data, to identify the important information about the equipment issues, repairs and performance.

Sentiment analysis is another application of NLP [10]. It may interpret the attitudes stated regarding the state of the equipment and maintenance tasks by analyzing the language used in technician notes and maintenance reports. For instance, NLP could identify annoyance in a technician's report regarding persistent machine problems. Decision-makers will discover this knowledge useful in prioritizing maintenance work according to the sentiment conveyed [11]. NLP also aids in the creation of extensive knowledge bases. NLP systems can build databases that hold a multitude of equipment-related information by structuring textual input. Databases such as this are useful for maintenance teams, allowing simple access to previous maintenance records, troubleshooting tips, and best practices. Moreover, NLP improves maintenance staff communication.

With natural language processing (NLP) capabilities, robots and virtual assistants may respond instantly to questions from experts and maintenance personnel. Making better decisions is made possible by the constant availability of vital information, which is ensured by this real-time support [12].

24.2.3.2 Machine Learning

Machine learning, including random forest models like XGBoost and LightGBM, K Clustering, Decision Trees, vector methods, unplanned forests, and long short term memories (LSTM), plays a significant role in predictive maintenance [6]. The algorithms that use machine learning analyze historical data, such as sensor measurement results and equipment performance logs, in-depth to identify patterns and relationships. By learning from past data, machine learning models predict when care is probably not needed, enabling businesses to address equipment issues before they result in breakdowns [13]. Machine learning, or ML, is crucial for transforming reactive maintenance approaches into proactive ones that reduce downtime, save costs, and boost operational efficiency because it is based on data-driven insights.

24.2.3.3 Compute Vision

Compute vision whereby helps us to provide the ability to visually inspect equipment and identify potential issues. This process includes imaging processing techniques and the deep learning techniques which help us to analyze and visualize images [14]. Visual inspection is one of the most significant applications of visual learning and machine learning in predictive maintenance. Sensor assemblies are placed precisely on equipment to capture image information in real time. Computer vision algorithms examine this data to look for anomalies, defects, or signs of mechanical breakdown. For example, it can identify a worn-out conveyor belt, a loose bolt, or a damaged valve [15]. Early detection allows maintenance staff to take corrective action before a breakdown occurs, hence reducing the price of repair and downtime. Computer vision also offers the benefit of remote device monitoring. By installing monitoring and CCTV devices on essential assets, maintenance personnel may do remote equipment inspections, even in hazardous or challenging-to-reach locations. This continuous observation provides valuable information on the state of the machinery and allows for timely maintenance actions [15]. Predictive modelling can also be supported by computer vision since one of the information streams is visual data. When visual data is combined with other types of information, such as sensor readings and historical maintenance records, the accuracy of maintenance forecasting software is improved [4]. For example, it can help anticipate when parts, like valves or gears, are likely to break based on how they seem. Furthermore, computer vision technology facilitates the creation of digital twins for equipment. A digital twin is an electronic duplicate of an object of value, and computer vision is essential to creating and maintaining these electronic representations [16]. Organizations can make preventive maintenance decisions by gaining a thorough understanding of the performance and condition of their equipment through the continual updating of digital twins providing real-time visual data [1].

24.2.3.4 Deep Learning

One of deep learning's greatest strengths is its extraordinary capacity to absorb high dimensional, complicated data and identify subtle connections and patterns within it. Because of this special quality, deep learning is especially useful for predicting equipment breakdowns and improving maintenance plans. Its main application lies in anomaly detection [2].

To find abnormalities in sensor data, deep learning models such as autonomous encoders, recurrent neural networks, also called are used. These models are skilled at identifying even the slightest departures from the typical operation of machinery, alerting users to possible problems before they develop into expensive malfunctions [1]. Time-series analysis is another crucial application of deep computing in predictive maintenance. Time-series data is particularly well-suited for neural network models (RNNs), including their specialized forms such as LSTM (long-short-term memory) networks. They are excellent at identifying patterns and dependencies in data, which makes it possible to estimate maintenance needs accurately and aids in more effective planning of maintenance tasks for businesses [17]. Image analysis becomes essential when visual data is available, such as pictures of equipment components. In such cases, Convolutional Neural Network (CNNs) are used. Renowned for their abilities in image identification, CNNs can identify wear, damage, or faults on equipment surfaces, giving maintenance crews important information [18]. Developing large forecasting models that consider many variables and their intricate interconnections is another area in which deep learning is crucial.

These models enable organizations to make well-informed decisions about maintenance tasks by forecasting the equipment's remaining usable life [19]. The versatility of deep computing is one of its notable features. Models can continuously learn from newly available data. Because of their flexibility, predictive maintenance models are guaranteed to be precise and functional even when equipment circumstances vary over time.

24.2.3.5 Intelligent Teaching

Intelligent educational systems (ITSs) are becoming increasingly potent instruments in the field of education by utilizing artificial intelligence's (AI) capabilities to deliver individualized instruction. With the ability to provide individualized instruction and assistance without the need for a teacher, these technologies possess the potential to completely transform the way that students learn [19]. Visualization helps us to make the study more interesting and motivate the learners like in game-based learning. Virtual reality provides stimulated learning that can challenge real life. Affection computing, which analyses human emotions and feelings using sensors and affective algorithms, can improve human–computer interaction [20]. To give students timely emotional feedback, ITS was updated to include a recognition of students' emotional status. This system provides interaction as they promote the AI tutor that can collaborate with the learners and help us to evaginate them turn by turn and help us to adapt the surrounding environments [21]. The Figure 24.3 supplied represents a conceptual diagram of an intelligent tuition system's (ITS) architecture. On top of providing personal feedback, AI tools help in providing the learning medium. While in most cases they only provide notes and videos but in cases of AI they help to provide tutoring system with the text, multimedia, stimulation and even games also.

Unlike the human teacher the AI consider their emotion and the style of their learning will be taken into consideration also. AI tutors help us to track the progress of students and help us to give feedback so that they can be able to know about the mistake they are facing. Usually, human's teacher only teaches us one method for solving the problem but in AI, they help us to provide more than one method to solve the problems [22].

The intelligent of teaching in the application of AI can be integrated into the four-interacting process:

I. Knowledge Base
This houses which help us to give the subject specific information.
II. Student Model
This specifies the student the current degree of understanding.
III. Domain Model
The educational tactics that are adapted during the learning stages.

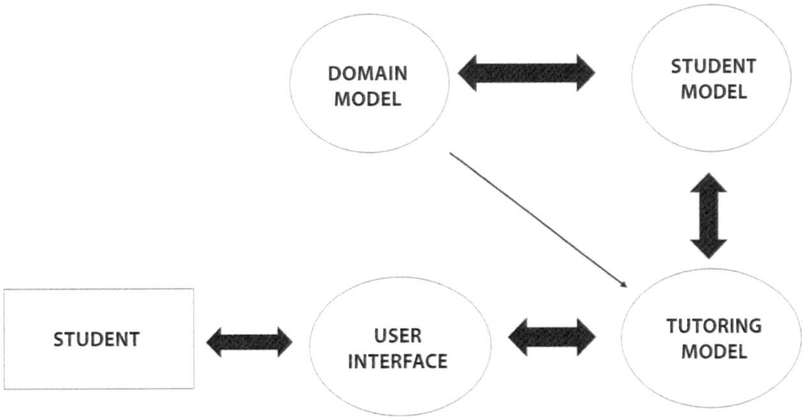

Figure 24.3 Intelligence in AI.

IV. User Interface
This interfaces which help to provide the better connection between the user and the student and help us to explore your interest.

ITSs can help with today's educational problems, most notably the requirement for individualized instruction in a system that frequently depends on standardized assessment. Effective teaching techniques are promoted by AI-driven systems such as ITS, which adapt to individual learning preferences and aptitudes. With countries moving more and more towards customized education, ITSs will be essential in transforming digital learning environments and classrooms. Intelligent tutoring systems represent a groundbreaking fusion of AI and education [23]. By offering personalized instruction and guidance, ITSs pave the way for more effective learning experiences. The journey of ITSs from conceptualization to implementation demonstrates their potential to transform education, making it more adaptive and individual-focused. As the educational landscape evolves, ITS stands as a beacon of innovation, ushering in a new era of tailored learning [24].

Domain Model: The ITS is programmed with a library (domain model) containing data on the subject area that the learner is studying. This domain model may contain factual knowledge, regulations, procedures, or other sorts of information that this ITS can use to respond to student inquiries and provide feedback.
Student Model: The ITS also produces a student model to monitor the student's development and comprehension of the topic.expand_more This student model could be dependent on the student's responses to questions, progress on jobs, or other types of engagement with the ITS.
Tutoring Model: The ITS employs a pedagogic model to determine what actions to conduct in response towards the student's behaviour. This tutoring model may employ a range of methodologies, including governed by rules reasoning, decision trees, and learning algorithms, to choose the best instructional plan for the learner.
User Interface: The ITS provides a user input that enables students to engage with the system. This user interface may comprise a number of elements, such as text fields, prices, or graphical displays, which allow the student to inquire, answer, and complete tasks.
The tutoring model is essential to an ITS's intelligence because it enable the system to tailor its instruction to each individual learner. Using the information from the student framework the tuition model can identify instructional tactics that are probably going to be beneficial for that specific student. This may involve:
Suggestions or superstructure: The ITS may give students with suggestions or scaffolding to assist them in completing a task or solving a problem.
Adjusting the learning level: The instructional technology system can change the level of challenge of the material in question to match the pupil's present stage of comprehension.

Choosing various scenarios or justifications: The ITS can tailor its instances or explanations to the pupil's teaching style.

24.2.4 Implementation of Predictive Maintenance in AI

24.2.4.1 Manufacturing

AI predictive maintenance works wonders for a variety of equipment throughout discrete and process production, such as robotic arms, milling machines, hydraulic printing presses, and conveyor systems. Think about the milling machines that create precise components out of raw materials. By outfitting these devices with vibration gauges and cross-referencing sensor measurements with past failure data, it will be possible to identify spindle wear indicators and misalignment [19]. Additionally, by monitoring the temperature, you could notice overheating sign that indicate an impending malfunction. There is even more to the potential of predictive maintenance based on machine learning [25]. It is possible to keep an eye out for leaks and obstructions in hydraulic presses that indicate a potential problem. Furthermore, by continuously monitoring temperature, predictive maintenance solutions can identify overheating, which is a warning indication of potential problems in the hydraulic system [7]. Additionally, artificial intelligence (AI) predictive maintenance may be able to precisely anticipate when components could need maintenance due to their cycle counting capability. This would allow for prompt maintenance interventions and the avoidance of possible breakdowns. Future maintenance tools may track belt tension on conveyor belts to stop premature wear.

Additionally, by keeping an eye on motor pressure, it may guarantee effectiveness and spot any warning signs that can result in unplanned downtime [8]. One of the biggest producers of pulp and solid wood, Mercer Celgar, is one of the manufacturers employing predictive maintenance. A crucial component of the business's manufacturing strategy is now predictive maintenance. Mercer Celgar was able to monitor various production lines and acquire individual reports for every tool type within a single platform after implementing the based on artificial intelligence predictive maintenance platform. This gave them the ability to monitor and assess performance and gave their maintenance crews early warning signs of possible malfunctions and breakdowns [26].

24.2.4.2 Health Care

The healthcare sector is undergoing a transformation because of machine learning predictive maintenance, which makes sure that vital medical equipment keeps working properly. Imagine a hospital with an MRI machine. Using sensor data, such as pressure, temperature, and electromagnetic field intensity, predictive maintenance computers can identify patterns, anomalies, and deviations that may point to possible problems. To minimize disruptions to patient care, hospitals may subsequently proactively reschedule maintenance tasks during times of low demand [9]. This method improves patient outcomes, decreases equipment downtime, and streamlines operations. A healthcare institution that has used this strategy is Kashiwa Health Assessment Clinic.

To stop MRI equipment breakdowns, they have teamed up with Hitachi to put in place a predictive maintenance system. After examining the last three years of data from sensors from 100 MRI devices, the technology partner developed a method to look into the trends that result in device failures [27].

24.2.4.3 Oil and Gas

Predictive maintenance based on machine learning is very beneficial to the oil and gas industry, which is recognized for having sophisticated and costly equipment. Envision an offshore digging rig equipped with many turbines, compressors, and pumps. Predictive maintenance algorithms could identify irregularities and anticipate future breakdowns by means of constant parameter monitoring, including temperatures, pressure, and flow rate [28]. Businesses may maximize equipment performance, prevent expensive unscheduled shutdowns, and guarantee worker safety by adopting a proactive strategy. Aramco, a major force in the manufacturing sector, has employed predictive maintenance to stop excess water from wells containing gas and oil from causing rust, scaling, and other types of damage to equipment and pipelines. The organization was able to obtain real-time insight into the water flow in the wells after implementing an AI-powered flow meter [10]. This meter predicts pipe flow using an AI-powered virtual model, which it then compares to the measured data. The meter can now automatically adapt to different operating circumstances thanks to this procedure, which significantly reduces the need for recalibration.

24.2.4.4 Automotive

Predictive maintenance based on machine learning is frequently utilized in the automotive sector to lower warranty costs and improve dependability. Automakers can identify possible problems before they become serious malfunctions, for example, by keeping an eye on sensor data from field-based vehicles [7]. Think of a group of delivery vans, predictive maintenance allows firms to prolong vehicle lifespans, optimize maintenance schedules, and prevent breakdowns by continuously monitoring fuel economy, tire condition, and engine performance. Predictive maintenance is being used by a few automakers to foresee maintenance needs, keep customers happy, and entice them to use the official network of the car manufacturer for maintenance services [26]. Opel is one such business; it has fitted its cars with a predicted maintenance system that uses past data analysis to recommend the best possible inspection intervals. The system's primary goal is to increase customer retention, therefore in order to deter consumers from using outside repair services, it warns customers when maintenance is necessary and schedules appointments with Opel's official network [9].

24.2.5 Challenges

24.2.5.1 Data Security

The data that powers these intelligent systems is more valuable than ever as artificial intelligence permeates every aspect of our daily life. But as its value rises, so do the associated risks. In the current digital era, protecting sensitive data has grown more crucial because AI systems have access to enormous volumes of it for tasks like corporate analytics and personalized suggestions. Data security is one of the biggest issues facing society today, and its effects go far past the IT department to include a wider range of interests. In AI systems, data security encompasses more than just information protection; it also involves upholding privacy rights, fostering trust, and guaranteeing the accuracy of AI decision-making procedures. Everyone who deals with data in any way is accountable, not just database administrators and network engineers [29, 30].

Every interaction with data, whether it be for producing, managing, or accessing it, could be a weakness in an organization's security protocol. Artificial intelligence is growing at a rate that cannot be ignored, with breakthroughs occurring at an unprecedented rate. Security of data in AI systems is becoming more and more crucial as AI keeps rewriting many economic sectors. Due to the significant volume of sensitive data that large organizations and their networks handled, data security was historically primarily an issue for them [24]. But things have changed as AI programmers have become more popular. Because AI, and particularly generative AI, depends so much on data for both training and decision-making, there is a chance that it will be compromised. Since many AI programmers assume that pre-existing safety procedures are sufficient, they have failed to recognize the need for data integrity [31]. This strategy, meanwhile, ignores the possibility of deliberate, malevolent attacks over AI systems. The following three strong arguments demonstrate why the safety of information in AI systems is so important:

1. Thread of Model Poising
   Model poising is the modern concern in terms of AI. This can lead to mistreating data into AI training sets. In the early stages of AI inaccurate data can lead to misinterpretation. But if AI develops and gets more complex, these mistakes could be used for more nefarious ends, which would negatively affect companies that deal with things like fraud detection security debugging. It is even possible to employ model poisoning as a diversion, spending resources while actual risks go unchecked [32]. Thus, complete data security is necessary to shield companies from these kinds of damaging cyberattacks.
2. Data Privacy in Paramount
   As consumers are increasingly aware about the data privacy rights business need to prioritize their security also. Businesses must priorities their data security procedures as customers' awareness of their rights to data privacy grows. Businesses need to make sure that their AI models adhere to privacy regulations and show openness about how they utilize data. Nevertheless, not all businesses now make their data usage practices very apparent. Ensuring regulatory compliance and fostering consumer trust can be achieved through streamlining privacy rules and effectively communicating data usage plans. To keep sensitive information out of the wrong hands, data security is essential.
3. Insider Threatt's
   As AI develops, there is a greater chance that workers who are replaced by technology will become resentful, which could result in internal threats. These internal problems are beyond the scope of traditional cybersecurity procedures, which concentrate mostly on external threats. These dangers can be reduced by implementing agile security measures like time-limited access controls and Zero Trust policies [33]. Additionally, a clear communication strategy and a well-thought-out AI adoption roadmap can comfort colleagues and

present chances for role-transitioning or upskilling. It is critical to present AI as a productivity-boosting tool rather than a danger to employment stability.

24.2.6 Devices

24.2.6.1 EMR's (Electronic Medical Records)

Many efforts are currently underway to use EMRs in the development of AI systems. AI technology is being used by EMR businesses in the United States.

24.2.6.1.1 Medical Image Analysis

Most medical professions that employ images, including pathologists, the field of dermatology cardiology, gastrointestinal, and ophthalmology, use machine learning algorithms for medical image analysis in addition to radiology [15]. To diagnose or categorize the severity of the disease, machine learning algorithms specifically use information from computed tomographic (Connecticut), positron emission tomography (magnetic resonance ultrasound imaging, pathology image, and endoscope.465 microscopic polyps were examined using a machine learning method during a live colonoscopy; the findings revealed with 94 per cent correct diagnosis & a 95percent negative predictive value [34]. An outstanding record in visual pattern analysis is maintained by the convolution neural networks method, one of the numerous deep learning algorithms that has been shown to be useful in determining the meaning of medical images with intricate patterns. Siemens Healthcare developed the artificially intelligent AI-Rad Multimodal Chest CT software to help physicians diagnose chest CT cases. Furthermore, using artificial intelligence, GE Healthcare is developing medical image processing technology [1]. A platform for the development and application of AI called Discovery was also created by Philips Healthcare [31]. With the FDA certification of Artery s' software for the Heart, Liver, and Lung, the business was able to develop the medical image processing Cloud AI platform. A number of businesses, such Vertex Medical Vison and Aidoc, are attempting to market AI-based medical image processing in addition to the cases previously mentioned software for analysis [20]. A number of Korean businesses, such Examination and Deepnoid, are in the process of receiving licenses from the Department of Safety of Food and Drugs to start marketing powered by artificial intelligence medical image analysis solutions [18].

24.2.6.2 Smart Devices, Signals, and In Vitro Diagnosis

Giants in the technology industry such as IBM but Google, Apple, Motorola, and Samsung are vying with each other to create and market items and offerings that can improve user health by utilizing trackers and the Network of Things (also known as IoT) to collect health data from everyday life. In 2017, Apple upgraded its smartwatch with an FDA-certified deep learning technology that may identify atrial fibrillation [17]. It determines the user's typical heart rates both during rest and activity through the application of accelerometer and photoplethysmography sensors.

The gadget alerts the user if there is a big difference in the projected numbers. Additionally, the deep learning approach evaluates electrocardiograms (ECGs) with amazing accuracy. In a recent study, a deep learning system was used to evaluate 91,232 single-lead ECGs. The results demonstrated exceptional diagnostic performance comparable to that of cardiologists [2]. Patients with chronic renal illness and hyperkalemic blood levels would benefit most from the application that applies the ECG patter analysis method to smart devices [3]. Non-invasive glucose meters are among the cutting-edge medical equipment that multinational corporations are now manufacturing. In addition to detecting and identifying infections based on the collected real-time data, these bio-signal sensors can be used for disease administration and therapy in a variety of hospital situations, including the critical care center, operation space emergency department, and recovery room [19]. IBM worked with Medtronic to create Sugar. One well-known instance of this is.IQ, which combined IBM's AI Watson with Medtronic's continuous glucose, sweat monitoring. It has been demonstrated that IQ directly helps diabetics by lowering the number of symptoms associated with both hyperglycemia and hypoglycemia, aids in readiness for unanticipated and deadly circumstances by employing wearable biosensors to continuously check the patient's condition and by observing subtle changes in the surroundings [20].

24.3 Conclusion

It is anticipated that AI-powered medical technology would completely transform both acute and long-term healthcare. In present circumstances, AI-based hospitals have demonstrated impressive performance in precisely recognizing and classifying patient illnesses as well as forecasting the course of diseases utilizing the gathered medical data. It is therefore anticipated that these advances in technology will assist medical professionals to make treatment decisions and enhance the outcomes of such decisions. However, there are currently a variety of privacy, safety, liability, and dependability issues with AI-based health care technology. Before AI technologies are used more extensively in the healthcare sector, they should not only develop technologically but also increase public awareness, create consistent standards, and make continuous growth.

References

1. Pundir, A.K., Ganapathy, L., Maheshwari, P., Thakur, S., Interpretive Structural Modelling to assess the enablers of blockchain technology in supply chain, in: *2020 11th IEEE Annual Information Technology, Electronics and Mobile Communication Conference (IEMCON)*, IEEE., pp. 0223–0229, 2020b.
2. Pundir, A.K., Ganapathy, L., Maheshwari, P., Kumar, M.N., Machine learning for revenue forecasting: A case study in retail business, in: *2020 11th IEEE Annual Information Technology, Electronics and Mobile Communication Conference (IEMCON)*, pp. 0201–0207, 2020a.
3. Maheshwari, P., Kamble, S., Pundir, A., Belhadi, A., Ndubisi, N.O., Tiwari, S., Internet of things for perishable inventory management systems: an application and managerial insights for micro, small and medium enterprises. *Ann. Oper. Res.*, 1-29, 2021.
4. Maheshwari, P., Kamble, S., Mani, V., Belhadi, A., Pundir, A., Digital twin implementation for performance improvement in process industriesA case study of food processing company. *Int. J. Prod. Res.*, 2022.
4. Mohanty, A. and Ranjana, P., Usage of predictive research on further business. *Int. J. Innov. Technol. Exploring Eng.*, 8, 11, 3464–3466, 2019.

5. Gautam, M., Ramu, V.S., Sinha, S., Reddy, P.K., Kondur, M., Kumar, S.S., Augmenting the existing CBPM maintenance philosophy currently in use in the marine sector with intelligent Predictive—CBPM philosophy, 2021.
6. Hinz, M., Brueggemann, D., Bracke, S., On the application of machine learning techniques in condition monitoring systems of complex machines, in: Paper presented at the *PSAM 2018 - Probabilistic Safety Assessment and Management*, 2018.
7. Bianchi, F., Masillo, F., Castellini, A., Farinelli, A., XM_HeatForecast: Heating load forecasting in smart district heating networks, 2020.
8. Bulla, C. and Birje, M.N., Improved data-driven root cause analysis in fog computing environment. *J. Reliab. Intell. Environ.*, 2021.
9. Cindy Tan, H.F., Lok Woo, W., Sharma, A., Logenthiran, T., Kumar, D.S., Study of smart condition monitoring using deep neural networks with dropouts and cross-validation, in: Paper presented at the *2019 IEEE PES Innovative Smart Grid Technologies Asia, ISGT*, pp. 3965–3969, 20192019.
10. Miner, G., Miner, L., Goldstein, M. *et al.*, Practical Predictive Analytics and Decisioning Systems for Medicine, in: *Informatics Accuracy and Cost-Effectiveness for Healthcare Administration and Delivery Including Medical Research*, 1st edition, Academic Press, Inc., Cambridge, MA, USA, 2016.
11. Hanshuo, M., Xiaodong, Z., Xuan, T., Fei, Q., Research on fault prediction method of electronic equipment based on improved SVR algorithm. Paper presented at the *Proceedings - 2020 Chinese Automation Congress, CAC*, vol. 2020, pp. 3092–3096, 2020.
12. Hermsen, M., Volk, V., Bräsen, J.H., Geijs, D.J., Gwinner, W., Kers, J., van der Laak, J.A.W.M., Quantitative assessment of inflammatory infiltrates in kidney transplant biopsies using multiplex tyramide signal amplification and deep learning. *Lab. Invest.*, 101, 8, 970–982, 2021.
13. Joshi, A.V., *Machine learning and artificial intelligence*, Cham: Springer, 2020.
14. Li, D., Piao, C., Wang, D., Zhao, D., Reliability prediction of CNC machine tools based on GM(1,1)- SVM combined model. Paper presented at the *Proceedings - 2020 5th International Conference on Mechanical, Control and Computer Engineering, ICMCCE*, vol. 2020, pp. 35–40, 2020.
15. Lee, W.J., Wu, H., Yun, H., Kim, H., Jun, M.B.G., Sutherland, J.W., Predictive Maintenance of Machine Tool Systems Using Artificial Intelligence Techniques Applied to Machine Condition Data. *Proc. CIRP*, *80*, 506–511, 2019, doi: 10.1016/j.procir.2018.12.019.
16. Ognjanovic, I., Artificial intelligence in healthcare. 274, 2020. Patel, H.K., Shah, D., Raghuwanshi, A., Real time machine health monitoring and vibrational analysis using FFT approach. *Int. J. Eng. Technol.*, 8, 5, 1833–1836, 2019.
17. Partridge, D., *Artificial intelligence and software engineering: Understanding the promise of the future*, 2013. Rogers, R., *How video games impact players: The pitfalls and benefits of a gaming society*, Lexington Books, 2016.
18. Rogers, R., *How video games impact players: The pitfalls and benefits of a gaming society*, Lexington Books, 2016.
19. Shah, D., Let's talk: Iiot enabled testbeds and big data analytics applications for smart manufacturing, in: Paper presented at the *Meet the Industry Candidates Poster Session 2019*, Held at the AIChE Spring Meeting and 15th Global Congress on Process Safety, pp. 22–23, 2019.
20. Silva, C., da Silva, M.F., Rodrigues, A., Silva, J., Santos Costa, V., Jorge, A., Dutra, I., *Predictive maintenance for sensor enhancement in industry 4.0*, 2021, doi: 10.1007/978-981-16-1685-3_33.
21. Chatterjee, S., Chaudhuri, R., Shah, M., Maheshwari, P., Big data driven innovation for sustaining SME supply chain operation in post COVID-19 scenario: Moderating role of SME technology leadership. *Comput. Ind. Eng.*, 168, 108058, 2022.

22. Shankar, P.R., Artificial Intelligence in Health Professions Education. *Arch. Med. Health Sci.*, 10, 2, 256–261, Jul–Dec 2022, doi: 10.4103/amhs.amhs_234_22 Virvou M.
26. Tsihrintzis, G.A. and Tsoukalas, L.H., Advances in Artificial Intelligence-based Technologies, in: *Selected Papers in Honour of Professor*, vol. 1.
23. Aji, A.S., Sashiomarda, J.A., Handoko, D., Predictive maintenance magnetic sensor using random forest method. Paper presented at *J. Phys. Conf. Ser.*, 1528, 1, 2020.
24. Andrei, A.G., Balasa, R., Costea, M.L., Semenescu, A., Building a blockchain for aviation maintenance records. Paper presented at *J. Phys. Conf. Ser.*, 1781, 1, 2021.
25. Au-Yong, C.P., Ali, A., Ahmad, F., *Predictive Maintenance: Monitoring Tools and Equipment*, vol. 49, pp. 6–11, The Malaysian Surveyor, 2014.
26. Cetiner, M. and Sahingoz, O.K., A comparative analysis for machine learning based software defect prediction systems. Paper presented at the *2020 11th International Conference on Computing, Communication and Networking Technologies, ICCCNT*, 2020.
27. Fablillah, F., Machine learning tool of pattern recognition in predicting wells safety critical element failure. Paper presented at the *International Petroleum Technology Conference 2020, IPTC*, 2020, doi: 10.2523/iptc-20089- abstract.
28. Fernández-García, M., Larrey-Ruiz, J., Ros-Ros, A., Figueiras-Vidal, A.R., Sancho-Gómez, J., Machine-health application based on machine learning techniques for prediction of valve wear in a manufacturing plant, 2019.
29. Abishekraj, N., Jeeva Prashanna, G.R., Suriyaa, M.S., Barathraj, T., Mohanraj, D., Condition based monitoring for fault detection in windmill gear box using artificial neural network. Paper presented at the *IOP Conference Series: Materials Science and Engineering*, vol. 912, 2020.
30. Lee, W.J., Wu, H., Yun, H., Kim, H., Jun, M.B.G., Sutherland, J.W., Predictive Maintenance of Machine Tool Systems Using Artificial Intelligence Techniques Applied to Machine Condition Data. *Proc. CIRP*, *80*, 506–511, 2019, doi: 10.1016/j.procir.2018.12.019.
31. Lee, W.J., Wu, H., Yun, H., Kim, H., Jun, M.B.G., Sutheralnd, J.W., Predictive Maintenance of Machine Tool Systems Using Artificial Intelligence Techniques Applied to Machine Condition Data. *Proc. CIRP*, *80*, 506–511, 2019, doi: 10.1016/j.procir.2018.12.01910.1016/j.procir.2018.12.019.
32. Virvou, M., Tsihrintzis, G.A., Tsoukalas, L.H., dvances in Artificial Intelligence-based Technologies, in: *Selected Papers in Honour of Professor*, vol. 1.
33. Hassanien, A.E., Advances in Intelligent Systems and Computing 723. *The International Conference on Advanced Machine Learning Technologies and Applications (AMLTA2018)*, 2020.
34. Aji, A.S., Sashiomarda, J.A., Handoko, D., Predictive maintenance magnetic sensor using random forest method. Paper presented at *J. Phys. Conf. Ser.*, 1528, 1, 2020.

Index

Printed in the USA/Agawam, MA
September 21, 2026

910661.178